Endogenous Peptides
and
Learning and Memory Processes

BEHAVIORAL BIOLOGY

AN INTERNATIONAL SERIES

Series editors

James L. McGaugh

Department of Psychobiology
University of California
Irvine, California

John C. Fentress

Department of Psychology
Dalhousie University
Halifax, Canada

Joseph P. Hegmann

Department of Zoology
The University of Iowa
Iowa City, Iowa

Endogenous Peptides and
Learning and Memory Processes

EDITED BY

Joe L. Martinez, Jr.
Robert A. Jensen
Rita B. Messing

Psychobiology Department
University of California, Irvine

Henk Rigter

Scientific Development Group
CNS Pharmacology Department
Organon, Oss, The Netherlands

James L. McGaugh

Psychobiology Department
University of California, Irvine

ACADEMIC PRESS

A Subsidiary of Harcourt Brace Jovanovich, Publishers

New York   London   Toronto   Sydney   San Francisco   1981

ACADEMIC PRESS, INC.
111 Fifth Avenue, New York, New York 10003

United Kingdom Edition published by
ACADEMIC PRESS, INC. (LONDON) LTD.
24/28 Oval Road, London NW1 7DX

Library of Congress Cataloging in Publication Data
Main entry under title:

Endogenous peptides and learning and memory processes.

(Behavioral biology)

Includes bibliographies and index.

1. Learning--Physiological aspects. 2. Memory--
Physiological aspects. 3. Peptides. 4. Nerve proteins.
I. Martinez, J. L. (Joe L.) II. Series: Behavioral bio-
logy (New York, N. Y. : 1978)
QP408.E5   615' .78   81-12691
ISBN 0-12-474980-1   AACR2

PRINTED IN THE UNITED STATES OF AMERICA

81 82 83 84   9 8 7 6 5 4 3 2 1

Contents

CHAPTER 6
ACTH and the Reminder Phenomena 117

David C. Riccio and James T. Concannon

CHAPTER 7
ACTH$_{4-9}$ Analog (ORG 2766) and Memory Processes in Mice 143

Bernard Soumireu-Mourat, Jacques Micheau, and Cécile Franc

CHAPTER 8
Pituitary–Adrenocortical Effects on Learning and Memory in Social Situations 159

Alan I. Leshner, Dennis A. Merkle, and James F. Mixon

CHAPTER 13
Endogenous Opioids, Memory Modulation, and State Dependency

269

Iván Izquierdo, Marcos L. Perry, Renato D. Dias, Diogo O. Souza, Elaine Elisabetsky, María A. Carrasco, Otto A. Orsingher, and Carlos A. Netto

CHAPTER 14
Facilitation of Long-Term Memory by Brain Endorphins

291

James D. Belluzzi and Larry Stein

CHAPTER 15
Endorphin and Enkephalin Effects on Avoidance Conditioning: The Other Side of the Pituitary–Adrenal Axis

305

Joe L. Martinez, Jr., Henk Rigter, Robert A. Jensen, Rita B. Messing, Beatriz J. Vasquez, and James L. McGaugh

CHAPTER *22*
Changes in Brain Peptide Systems and Altered Learning and Memory Processes in Aged Animals   463

Robert A. Jensen, Rita B. Messing, Joe L. Martinez, Jr., Beatriz J. Vasquez, Vina R. Spiehler, and James L. McGaugh

CHAPTER *23*
A Two-Process Model of Opiate Tolerance   479

Raymond P. Kesner and Timothy B. Baker

PART *V*
Other Neuropeptides   519

CHAPTER *24*
Substance P and Its Effects on Learning and Memory   521

Joseph P. Huston and Ursula Stäubli

Contributors

Numbers in parentheses indicate the pages on which the authors' contributions begin.

Peter W. Achterberg* (7), Netherlands Institute for Brain Research, 1095 KJ Amsterdam, The Netherlands

William H. Bailey (371), The Rockefeller University, New York, New York 10021

Timothy B. Baker (479), Department of Psychology, University of Wisconsin, Madison, Wisconsin 53706

James D. Belluzzi (291), Department of Pharmacology, College of Medicine, University of California, Irvine, California 92717

Floyd E. Bloom (199, 249), A. V. Davis Center for Behavioral Neurobiology, The Salk Institute, San Diego, California 92138

Gerard J. Boer (7), Netherlands Institute for Brain Research, 1095 KJ Amsterdam, The Netherlands

Béla Bohus (59, 413), Rudolf Magnus Institute for Pharmacology, Medical Faculty, University of Utrecht, 3521 GD Utrecht, The Netherlands

* PRESENT ADDRESS: Cardiochemical Laboratory, Thorax Center, Erasmus University, Rotterdam, The Netherlands.

María A. Carrasco (269), Departamento de Bioquímica, Instituto de Biociências, Universidade Federal do Rio Grande do Sul (centro), 90.000 Pôrto Alegre, RS, Brasil

James T. Concannon* (117), Department of Psychology, Kent State University, Kent, Ohio 44242

David H. Coy (563), Veterans Administration Medical Center, Tulane University School of Medicine; and University of New Orleans, New Orleans, Louisiana 70146

Richard L. Delanoy (79), Department of Psychology, University of Virginia, Charlottesville, Virginia 22901

David de Wied (37, 59, 231, 397, 413), Rudolf Magnus Institute for Pharmacology, Medical Faculty, University of Utrecht, 3521 GD Utrecht, The Netherlands

Renato D. Dias (269), Departamento de Bioquímica, Instituto de Biociências, Universidade Federal do Rio Grande do Sul (centro), 90.000 Pôrto Alegre, RS, Brasil

Jan Dogterom (7), Netherlands Institute for Brain Research, 1095 KJ Amsterdam, The Netherlands

Elaine Elisabetsky (269), Departamento de Bioquímica, Instituto de Biociências, Universidade Federal do Rio Grande do Sul (centro), 90.000 Pôrto Alegre, RS, Brasil

Cécile Franc (143), Laboratoire de Psychophysiologie, Université de Bordeaux I, 33405 Talence Cedex, France

Anthony W. K. Gaillard (181), Institute for Perception TNO, 3769 ZG Soesterberg, The Netherlands

Michela Gallagher (445), Department of Psychology, The University of North Carolina at Chapel Hill, Chapel Hill, North Carolina 27514

Willem H. Gispen (37), Division of Molecular Neurobiology, Institute of Molecular Biology, Utrecht, The Netherlands

Paul E. Gold (79), Department of Psychology, University of Virginia, Charlottesville, Virginia 22901

Joseph P. Huston (521), Institute of Psychology, University of Düsseldorf, 4000 Düsseldorf, Federal Republic of Germany

Iván Izquierdo (269), Departamento de Bioquímica, Instituto de Biociências, Universidade Federal do Rio Grande do Sul (centro), 90.000 Pôrto Alegre, RS, Brasil

Robert A. Jensen† (305, 431, 463), Department of Psychobiology, School of Biological Sciences, University of California, Irvine, California 92717

Bruce S. Kapp (445), Department of Psychology, The University of Vermont, Burlington, Vermont 05405

---

* PRESENT ADDRESS: Department of Pharmacology, College of Medicine, Northeastern Ohio Universities, Rootstown, Ohio 44272.

† PRESENT ADDRESS: Department of Psychology, Southern Illinois University, Carbondale, Illinois 62901.

Abba J. Kastin (563), Veterans Administration Medical Center, Tulane University School of Medicine, New Orleans, Louisiana 70146

Raymond P. Kesner (479), Department of Psychology, University of Utah, Salt Lake City, Utah 84112

George F. Koob (249), A. V. Davis Center for Behavioral Neurobiology, The Salk Institute, San Diego, California 92138

Gábor L. Kovács (231), Institute of Pathophysiology, University Medical School, Szeged, Semmelweis u.1., H-6701, Hungary

Michel Le Moal (249), Laboratoire de Neurobiologie des Comportements, Université de Bordeaux II, 33076 Bordeaux Cedex, France

Alan I. Leshner (159), Department of Psychology, Bucknell University, Lewisburg, Pennsylvania 17837; and National Science Foundation, Washington, D.C. 20550

James T. Martin (99), Department of Natural Sciences and Mathematics, Richard Stockton State College, Pomona, New Jersey 08240

Joe L. Martinez, Jr. (1, 305, 431, 463), Department of Psychobiology, School of Biological Sciences, University of California, Irvine, California 92717

James L. McGaugh (1, 305, 431, 463), Department of Psychobiology, School of Biological Sciences, University of California, Irvine, California 92717

Jacqueline F. McGinty (199), A. V. Davis Center for Behavioral Neurobiology, The Salk Institute, San Diego, California 92138

Dennis A. Merkle* (159), Department of Psychology, Bucknell University, Lewisburg, Pennsylvania 17837

Rita B. Messing (305, 431, 463), Department of Psychobiology, School of Biological Sciences, University of California, Irvine, California 92717

Jacques Micheau (143), Laboratoire de Psychophysiologie, Université de Bordeaux I, 33405 Talence Cedex, France

James F. Mixon (159), Department of Psychology, Bucknell University, Lewisburg, Pennsylvania 17837

Carlos A. Netto (269), Departamento de Bioquímica, Instituto de Biociências, Universidade Federal do Rio Grande do Sul (centro), 90.000 Pôrto Alegre, RS, Brasil

Richard D. Olson (563), Department of Psychology, University of New Orleans, New Orleans, Louisiana 70122

Otto A. Orsingher (269), Departamento de Farmacología, Facultad de Ciências Químicas, Universidad de Córdoba, Argentina

Marcos L. Perry (269), Departamento de Bioquímica, Instituto de Biociências, Universidade Federal do Rio Grande do Sul (centro), 90.000 Pôrto Alegre, RS, Brasil

David C. Riccio (117), Department of Psychology, Kent State University, Kent, Ohio 44242

---

* PRESENT ADDRESS: Eastern Pennsylvania Psychiatric Institute, Philadelphia, Pennsylvania 19159.

Henk Rigter* (305), Scientific Development Group, CNS Pharmacology Department, Organon, Oss, The Netherlands

Aryeh Routtenberg (541), Cresap Neuroscience Laboratory, Departments of Psychology and Biological Sciences, Northwestern University, Evanston, Illinois 60201

Jacqueline Sagen (541), Department of Pharmacology, University of Illinois Medical Center, Chicago, Illinois 60680

Curt A. Sandman (563), Department of Psychiatry and Human Behavior, University of California, Irvine, Medical Center, Orange, California 92668; and Fairview Hospital Research, Costa Mesa, California 92626

Michael V. Sofroniew† (327), Department of Anatomy, Ludwig-Maximilians University, D-8000 Munich 2, Federal Republic of Germany

Diogo O. Souza (269), Departamento de Bioquímica, Instituto de Biociências, Universidade Federal do Rio Grande do Sul (centro), 90.000 Pôrto Alegre, RS, Brasil

Vina R. Spiehler (431, 463), Department of Psychobiology, School of Biological Sciences, University of California, Irvine, California 92717

Ursula Stäubli (521), Institute of Psychology, University of Düsseldorf, 4000 Düsseldorf, Federal Republic of Germany

Larry Stein (291), Department of Pharmacology, College of Medicine, University of California, Irvine, California 92717

Dick F. Swaab (7), Netherlands Institute for Brain Research, 1095 KJ Amsterdam, The Netherlands

Bernard Soumireu-Mourat (143), Université de Provence, Laboratoire de Neurobiologie des Comportements, Centre de Saint-Jérôme, 13397 Marseille Cedex 13, France

Frederik W. van Leeuwen (7), Netherlands Institute for Brain Research, 1095 KJ Amsterdam, The Netherlands

Jan M. van Ree (397), Rudolf Magnus Institute for Pharmacology, University of Utrecht, Medical Faculty, 3512 GD Utrecht, The Netherlands

Tjeerd B. van Wimersma Greidanus (413), Rudolf Magnus Institute for Pharmacology, University of Utrecht, 3512 GD Utrecht, The Netherlands

Beatriz J. Vasquez (305, 431, 463), Department of Psychobiology, School of Biological Sciences, University of California, Irvine, California 92717

Adolf Weindl (327), Department of Anatomy, Ludwig-Maximilians University, D-8000 Munich 2, Federal Republic of Germany

Jay M. Weiss (371), The Rockefeller University, New York, New York 10021

Albert Witter (37), Rudolf Magnus Institute for Pharmacology, Medical Faculty, University of Utrecht, 3512 GD Utrecht, The Netherlands

* PRESENT ADDRESS: Dutch Science Council, 2502 EM The Hague, The Netherlands.
† PRESENT ADDRESS: Department of Human Anatomy, University of Oxford, Oxford OX1 3QT, England.

Preface

Spectacular progress has been made in the past several years in understanding the role that neuropeptides play in regulating complex behaviors. A major focus of this research has been investigations of the role of peptides in learning and memory processes. Pioneering research in this field began almost twenty years ago with studies of the effects of the pituitary peptides adrenocorticotropin (ACTH) and vasopressin on learning and memory. With the discovery of the endogenous opioid peptides, the pace of research activity has accelerated dramatically.

With this in mind, the editors felt that a comprehensive statement about endogenous peptides and learning and memory was needed to provide a basis for integrating many diverse lines of research findings that have evolved. This book presents work from laboratories in North and South America and Europe dealing with the role of pituitary and central nervous system peptidergic systems in the modulation of learning and memory. Descriptive chapters head each of the sections of the book, providing general overviews of the brain distribution of ACTH, melanocyte-stimulating hormone (MSH), vasopressin, oxytocin, and the enkephalins and endorphins. Specific experimental findings concerning the role of these peptides in attention, conditioning, memory, amnesia, and opiate tolerance are also presented. The book is organized into five major sections that reflect both historical strategies used in the study of peptides and learning and memory and the biological relatedness of the peptides being considered: Part I, ACTH and MSH; Part II, Endorphins; Part III, Vasopressin and Oxytocin; Part IV, Opiates; and Part V, Other Neuropeptides. Specific theoretical issues that are addressed include questions of (1) the specificity of peptidergic effects on information processing, (2) the generality of hormone effects on emotional and motivational systems that affect learning and memory, and (3) the relationship between drug tolerance development and conditioning and habituation.

This book will provide both specialists and nonspecialists, including teachers, clinicians, and researchers in the fields of behavioral pharmacology, neuropharmacology, psychopharmacology, and experimental psychology, with an up-to-date survey in an area of neuroscience research that has captured the creative energies of an increasingly large number of scientists. Finally, we hope that this book will help lay the foundation for further discoveries related to the function of peptide systems and behavior.

Joe L. Martinez, Jr.
Robert A. Jensen
Rita B. Messing
Henk Rigter
James L. McGaugh

INTRODUCTION

Learning Modulatory Hormones: An Introduction to Endogenous Peptides and Learning and Memory Processes

James L. McGaugh and Joe L. Martinez, Jr.

Extensive evidence that learning and memory are influenced by endocrine hormones has been provided by research of the last 25 years. In laboratory animals, learning is impaired by removal of some of the sources of hormones, that is, by hypophysectomy or adrenalectomy, and learning and memory can be either enhanced or impaired by administration of a variety of hormones, including catecholamines, and hormones of the pituitary (see Part I), as well as other peptide hormones (see Parts II, III, and V). Thus, given the tendency to christen hormones according to their known effects, it is quite appropriate to call some hormones "learning modulatory hormones" or LMHs.

It is probably not surprising, either now or in retrospect, that learning is impaired by removal of the pituitary or the adrenal glands, and that this impairment is attenuated by administering appropriate hormones. Such effects would be expected simply because the hormones of the pituitary and adrenal glands influence many physiological functions. It is somewhat more interesting to know that a hormone such as adrenocorticotropin (ACTH) affects learning processes even in the absence of the adrenal gland, and that learning is also influenced by peptide analogues of ACTH that do not stimulate the adrenal cortex (see Chapters 2, 3, 5, 6, 7, 8, and 25). Such findings indicate that the effects of peptide hormones on learning and memory may be independent of their classical actions on target organs.

The recent discovery of the endogenous opioid peptides, the endorphins and enkephalins, dramatically increased interest in peptide influences on behavior (see Part II). The finding that these peptides bind to opiate receptors suggested that endorphins and enkephalins may be involved in the regulation of pain.

1

ENDOGENOUS PEPTIDES
AND LEARNING AND MEMORY PROCESSES

However, given their widespread distribution in the brain as well as in other organs, it seems likely that these peptides may have many functions unrelated to pain. In fact, opioid systems appear not to be limited to the regulation of pain. The research findings discussed in this book provide extensive evidence that learning and retention are influenced by opioid peptides as well as by opiate drugs such as morphine and naloxone (see Parts II and IV).

It is now known that there are many peptide hormones in the brain. While important details concerning the source and actions of the hormones are presently lacking, the evidence of their presence in the brain as well as the evidence that they affect brain activity and behavior has resulted in the peptides being termed neuropeptides. Some of the neuropeptides, such as ACTH, already have functional names. Of course, the labeling of hormones in terms of their first identified effect, as in the case of ACTH, or by presumed function, as in the case of endorphin, is quite an accidental or arbitrary matter. Any particular peptide may have, and most peptides no doubt do have, a variety of effects. The function of a peptide is not determined by its structure or even its receptor, but rather by the endocrine or neural system activated by the hormone. In this sense, asking the question, "What is the function of ACTH or endorphin?" is like asking the question, "What is the function of acetylcholine or norepinephrine?"

The research findings discussed in this book provide strong evidence that the processes underlying learning and memory are influenced by endogenous peptide hormones. Peptides shown to influence aspects of learning and retention include ACTH and melanocyte-stimulating hormone (MSH), endorphins and enkephalins, vasopressin, oxytocin (and analogues of these peptides), angiotensin, and substance P. A good case is made for the existence of LMHs.

Research on this problem has, however, only begun to address some of the important issues. What is the site of action of the peptides? Is the brain the "target organ"? While it seems reasonable to presume that the peptides affect learning by directly affecting the brain, many of the peptides are known to affect learning when administered peripherally. Because the peptides may or may not readily pass into the brain, it is not clear under what conditions peptides of peripheral origin reach the brain. If the effects of the peptides are not directly due to effects on the brain, then where are the peripheral receptors that constitute the target organs, and how, ultimately, do the peptides affect brain functioning underlying learning and memory?

Why is it that learning and memory are influenced by so many peptide hormones? One possibility is that each peptide might affect a different aspect of the learning process. There is, in fact, some evidence that the peptides differentially affect arousal, attention, acquisition, consolidation, and retrieval (see Chapters 3, 5, 6, 9, 11, 13, and 26). It seems likely that learning involves the action of many different neuronal systems. Thus, neuropeptide modulation of any of the neuronal systems might result in an effect on learning. Peptides might act in a variety of ways to produce a common influence on general regulatory neuronal or hormonal systems such as those involved in arousal or stress.

How are peptide hormones involved in the "normal" physiology of memory? It seems clear that learning and memory can be enhanced or impaired by peptide hormones or treatments that affect peptide hormones. But the important question is that of the role of endogenous peptide hormones in memory. Do the hormones have a critical, essential role in learning? If so, what is the process (or what are the processes) that require neuropeptide action? If not, do the hormones normally have a modulatory role in memory, and if so, what is the nature of the modulating influence?

On the basis of recent research progress it seems clear that further investigation of these problems will contribute significantly to an understanding of the role of endogenous peptide hormones in the physiology of memory.

PART I

ACTH and MSH

CHAPTER 1

The Distribution of MSH and ACTH in the Rat and Human Brain and Its Relation to Pituitary Stores

Dick F. Swaab, Peter W. Achterberg, Gerard J. Boer,
Jan Dogterom, and Frederik W. van Leeuwen

Substances with melanocyte-stimulating activity were found 50 years ago in the pars intermedia of the pituitary and in the periventricular regions of the human brain. The pituitary was, however, long considered to be the sole source of "intermedin," which was thought to be transported to the brain via the portal vessels or the cerebrospinal fluid. Although this view is still held by some, evidence is rapidly accumulating that melanocyte-stimulating hormone (MSH), adrenocorticotropic hormone (ACTH), and a family of related peptides are not only produced by the pituitary, but also by the nervous system. This family of peptides has been shown, at least in the pituitary, to be derived from a 30 kdalton precursor.

The adult rat pituitary contains about three times more α-MSH than ACTH; 97% of the total α-MSH content is found in the neurointermediate lobe and 95% of the total ACTH content is found in the pars distalis. Comparable assay data do not exist for the human pituitary. The rat brain has high levels of α-MSH and ACTH in the area of the arcuate nucleus of the hypothalamus, but these peptides are found, in addition, in cell bodies and fibers in a number of extrahypothalamic areas.

ENDOGENOUS PEPTIDES
AND LEARNING AND MEMORY PROCESSES

Immunocytochemical observations are consistent with the notion that opiomelanocortin peptides are present in neurons of the arcuate nucleus and in pathways running from this nucleus to the limbic system, midbrain, and hindbrain. However, extrahypothalamic cell bodies positive for anti-α-MSH or anti-ACTH have been described as well. Cross-reaction of the antibodies to as yet unidentified, related compounds is the most probable explanation for this discrepancy and illustrates that specificity is a crucial problem in immunocytochemistry.

α-MSH, ACTH, and other peptides of the opiomelanocortin family are present early in the development of fetal brain and pituitary of man and rat and might influence development and labor.

Evidence indicating that peptides related to opiomelanocortin are synthesized in the vertebrate nervous system has been obtained recently by means of bio- and radioimmunoassays, biochemistry, immunocytochemistry, hypophysectomy, mechanical and chemical brain lesions, and tissue culture. Other data indicate the presence of these peptides in neuronal tissue in lower animals. The importance of peptides of pituitary origin for brain peptide content and function remains to be determined.

I. HISTORICAL INTRODUCTION

A pituitary hormone was discovered 50 years ago that was capable of inducing changes in the skin color of some species, such as bridal staining in the fish and the darkening of frog skin. This hormone is also found in the mammalian pituitary gland. In the human pituitary, it is present in similar amounts in Blacks and Caucasians. Because the highest concentration of this hormone is in the pars intermedia of the pituitary, it was called "intermedin" (Zondek, 1935; Zondek & Krohn, 1932). The locus of intermedin in the pars intermedia was established in 1930 when Bayer found one light-colored frog among a shipment of 70 esculenta. The color of this animal remained light even after many days in the dark, and histological investigation revealed that its pars intermedia had been selectively destroyed by a parasite (see Zondek, 1935). The high amount of chromatophoric activity found by Zondek can, however, certainly not be attributed solely to α-melanocyte-stimulating hormone (α-MSH, see Section II), as this hormone represents only 2% of the bioassayable melanotropic activity in the human pituitary (Abe, Island, Liddle, Fleischer, & Nicholson, 1967).

In addition to the localization of intermedin in the pituitary, Zondek reported the presence of this hormone in the human fetal, neonatal, and adult brain. Although he found intermedin in the choroid plexus of the brain of eclamptic women, and in areas around the third ventricle, Zondek could not find any hormone in the cerebrospinal fluid itself. The earlier identification of inter-medin in this fluid by Trendelenburg and Ehrhardt (see Zondek, 1935) were therefore attributed by Zondek to aspecific effects due to their extraction procedure. Zondek concluded that intermedin was transported from the pituitary via the pituitary stalk to the vegetative centers in the periventricular brain area, where it exerted its actions. He concluded that transport of intermedin to the brain is very rapid because little intermedin was found in the pituitary stalk itself. Popa (1938) proposed that the transport of "xanthomélanophorine" from the pituitary to the brain takes place via both the portal vessels and the cerebrospinal fluid. Little further work was done until 1958 when Mialhe-Voloss reported the disappearance of melanotropic activity from the rat brain after hypophysectomy, and restoration of the activity after intermedin substitution via the carotid artery. She thus came to the conclusion that the pituitary was the sole source of intermedin. Later, however, Rudman and coworkers provided evidence by means of similar experiments that the melanotropic peptides in the brain were not derived from the pituitary but from the central nervous system (CNS) itself (Rudman, Del Rio, Hollins, Houser, Keeling, Sutin, Scott, Sears, & Rosenberg, 1973; Rudman, Scott, Del Rio, Houser, & Sheen, 1974). Meanwhile, Guillemin, Schally, Lipscomb, Andersen, and Long (1962) had found, in addition to melanotropic activity, adrenocorticotropic activity in the hypothalamus and proposed a diencephalic origin for such peptides. Despite the discovery in the 1930s of intermedin in the brain, little attention was paid to its possible central functions until the mid-1950s, when a number of central actions of MSH and ACTH were discovered, such as the effects of ACTH and α-MSH on shuttlebox avoidance behavior (Applezweig & Moeller, 1959; de Wied, 1964, 1965; Murphy & Miller, 1955), facilitation of monosynaptic spinal reflexes by β-MSH (Guillemin & Krivoy, 1960), effects of MSH and ACTH on grooming, scratching, and stretching (Ferrari, Gessa, & Vargiu, 1963), and alterations of the electroencephalogram by α- and β-MSH (Sandman, Denman, Miller, Knott, Schally, & Kastin, 1971). We became interested in the possibility that the brain produces peptides related to MSH when we found that antibodies raised against α-MSH and used for immunofluorescence microscopy also persistently stained homogenized brain tissue that was put around fetal pituitaries as support for the cryostat sections. The immunoreactivity of MSH was indeed found in nerve cells and fibers, suggesting a central origin of such compound(s) (Swaab, 1976; Swaab & Visser, 1977). Although the "classical" view that the pituitary is the only source for MSH, ACTH, and related peptides in the brain has again recently been advocated (Moldow & Yalow, 1978a), evidence is accumulating that a family of peptides related to MSH and ACTH is not only produced by the pituitary but also by the nervous system.

II. PRO-OPIOMELANOCORTIN, MSH, AND ACTH

During the last decades, it has become apparent that MSH and ACTH belong to a family of related peptides that are all derived from one large precursor.

The possible chemical relationship between MSH and ACTH was suggested as early as the 1950s by the observation that in patients with Addison's disease or Cushing's syndrome, the levels of both MSH and corticotropic activity were elevated. Two forms of MSH, α- and β-MSH, were quickly isolated. Structural elucidation revealed that their amino acid sequences of 13 and 18 amino acids, respectively, had a central core of 7 amino acids (ACTH/α-MSH$_{4-10}$) in common with each other and with ACTH. In addition, the entire sequence of α-MSH was found to be present within the ACTH molecule (cf. Harris, 1960). The sequence of β-MSH was later retrieved in β-lipotropic hormone (β-LPH) (Li, Barnafi, Chrétien, & Chung, 1965). The factor necessary for both melanocyte-expanding and corticosteroidogenic activity was found to be ACTH/α-MSH$_{6-9}$ (cf. Schwyzer & Eberle, 1977). These findings suggested the existence of a family of related peptides containing two classes: (1) ACTH and its cleavage products of α-MSH and corticotropin-like intermediate lobe peptide (CLIP); and (2) β-LPH and its fragments γ-LPH and β-MSH (Scott, Ratcliffe, Rees, Landon, Bennett, Lowry, & McMartin, 1973). It is not surprising, therefore, that ACTH and β-LPH were immediately regarded as precursor molecules. A remarkable difference from prohormones that were already known, such as insulin (Steiner, Rubenstein, Peterson, Kemmler, & Tager, 1973) and vasopressin (Sachs & Takabatake, 1964), was that the "prohormone," ACTH itself, also had a known biological function.

Interest in the peptides of the ACTH/LPH family was further heightened when Hughes, Smith, Kosterlitz, Fothergill, Morgan, and Morris (1975) reported the structure of a morphine-like pentapeptide in the brain, which they named *enkephalin*, and of which the Met-derivative turned out to be present in its entirety in the β-LPH molecule. The part of β-LPH remaining after splitting off γ-LPH appeared to have even more potent opiate-like properties and was named β-endorphin (Bradbury, Smyth, & Snell, 1976; Guillemin, Ling, & Burgus, 1976). β-Endorphin was thought to fragment successively into γ- and α-endorphin, which in turn were both thought to be precursors of enkephalin (Austen, Smyth, & Snell, 1977). Thus, β-LPH is a prohormone for melanocyte-expanding and opiate activities and can be named pro-opiomelanin.

Studies of high-molecular-weight forms of ACTH, first described by Yalow and Berson (1971) as "big ACTH," indicated the existence of a larger precursor protein, in which both immunoreactive ACTH and β-endorphin were present. Other observations pointing to a 30 kdalton precursor came from studies of a pituitary tumor cell line (Mains, Eipper, & Ling, 1977; Roberts & Herbert, 1977), of bovine pituitaries (Nakanishi, Inoue, Taii, & Numa, 1977), and of human pulmonary carcinoma cells (Bertagna, Nicholson, Sorenson, Pettengill, Mount, & Orth, 1978). Recently, the entire nucleotide sequence of a cloned DNA insert encoding the bovine corticotropin-β-lipotropin precursor has been determined (Nakanishi, Inoue, Kita, Nakamura, Chang, Cohen, & Numa,

1979), which was named *pro-opiomelanocortin* (Chrétien, Benjannet, Gossard, Gianoulakis, Crine, Lis, & Seidah, 1979). This sequencing study defined not only the precise locations of ACTH and β-LPH (Fig. 1) but also the amino acid composition of the remaining NH_2-terminal portion, which was known to exist but had never been determined (Mains et al., 1977; Roberts & Herbert, 1977) and in which the sequence -His-Phe-Arg-Trp- (ACTH/α-MSH$_{6-9}$ and β-MSH$_{9-12}$) appeared to be present for a third time. On the basis of the occurrence of two adjacent pairs of basic amino acids, which are apparent sites of chain splitting, Nakanishi et al. (1979) postulated this sequence to be part of a not yet identified decapeptide that they called γ-MSH (Fig. 1).

The rat intermediate lobe pituitary cells split off the 30 kdalton precursor first between the sequences of ACTH and β-LPH, whereby a 20 kdalton fragment (promelanocortin) and β-LPH are produced. From the 20 kdalton fragment, α-MSH, CLIP plus a 16 kdalton fragment are derived, and from β-LPH, γ-LPH plus β-endorphin are derived (Mains & Eipper, 1979). The pro-opiomelanocortin molecule of the mouse pituitary tumor cell line has been shown to fragment, after the same first cleavage into ACTH, β-LPH, and β-endorphin molecules (Mains et al., 1977). This seems to suggest the possibility that different cells (e.g., pars intermedia versus pars distalis of the pituitary or versus brain) or cells under different circumstances (e.g., in the course of development) might process the precursor in different ways. The large number of possible sites of fragmentation (i.e., between each of the pairs of basic amino acids of pro-opiomelanocortin; Fig. 1) may account for the high number of different ACTH- and MSH-like compounds that appear to exist (see Section III-B). This, and the great number of members of the pro-opiomelanocortin family, as revealed so far from pituitary sources (Fig. 1), indicates that problems of specificity will be encountered in radioimmunoassay and immunocytochemical studies.

III. ASSAY OF α-MSH AND ACTH IN THE PITUITARY AND THE BRAIN

As has already been suggested by Zondek (1935), peptides of the opiomelanocortin family are present both in the pituitary and in the brain.

A. Pituitary Content of α-MSH and ACTH

Reports by various authors on the rat pituitary content of α-MSH and ACTH are in reasonable agreement (Table 1), although assay data vary according to a number of factors, such as the strain of rat, the ages and sexes of the animals,

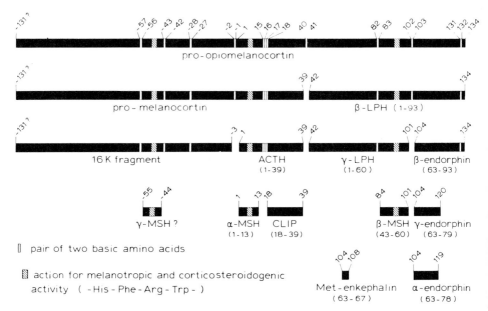

Fig. 1. The peptides of the pro-opiomelanocortin family. Bovine pro-opiomelanocortin is a protein of presumably 265 amino acids numbered according to Nakanishi et al. (1979). A pituitary extract splits the molecule at site 40/41, resulting in promelanocortin and pro-opiomelanin (β-LPH) parts. Thus far, no physiological function is known for either of these peptides, other than being prohormones, that may serve as a storage mechanism until further cleavage is needed. Promelanocortin can give rise to either ACTH and a 16 kdalton fragment (e.g., in the pars distalis of the pituitary) or, in addition, to α-MSH and CLIP as further cleavage products of ACTH (e.g., in the pars intermedia). The pro-opiomelanin molecule is physiologically cleaved at site 102/103, liberating γ-LPH and β-endorphin.

Successive cleavage of pro-opiomelanocortin probably occurs at sites of paired basic amino acids, whereby direct amidation of the terminal carboxyl group is concomitant (cf. Bradbury et al., 1976). It is assumed, therefore, that the pituitary β-MSH is liberated from the γ-LPH fragment and, consequently, both ACTH and γ-LPH might be called promelanins.

The recognition of a third ACTH$_{6-9}$ sequence (-51/-48) in the pro-opiomelanocortin as well as pairs of basic amino acids around this sequence have been regarded as possible evidence for the existence of more "family members" than have been recognized so far. The 16 kdalton fragment might split off γ-MSH (-55/-44; Nakanishi et al., 1979), but a fragment -55/-29 or -55/-3 is also possible.

The peptides γ- and α-endorphin and Met-enkephalin can be liberated *in vitro* from β-endorphin. However, because of the absence of paired basic amino acids, processing of β-endorphin is enzymatically different (the site 131/134 has not yet been mentioned as a site of cleavage). Moreover, in this scheme, bovine β-LPH contains 93 amino acids instead of the usually mentioned 91 (Li et al., 1965). Two additional amino acid residues (Ala and Glu-β-LPH$_{35-36}$) have been added on the basis of DNA-sequencing (Nakanishi et al., 1979).

TABLE 1

Summary of Data on the α-MSH and ACTH Content of the Rat Pituitary

	α-MSH(ng/mg)	Assay[a]	Reference	ACTH(ng/mg)	Assay[a]	Reference
Total pituitary	210	(BIO)	Shapiro et al., 1972	85	(RIA)	Usategui et al., 1976
	200[b]	(BIO)	Tilders, 1975	38	(RIA)	Moldow and Yalow, 1978a
	220	(RIA)	Usategui et al., 1976	100	(RIA)	Vaudry et al., 1978
	140	(RIA)	Oliver and Porter, 1978			
	610	(RIA)	Vaudry et al., 1978			
	80	(RIA)	Eskay et al., 1979a			
	260	(RIA)	Visser and Swaab (see Fig. 4)			
(Mean)	(250)			(74)		
Pars distalis	17	(BIO)	Shapiro et al., 1972	50	(BIO)	Scott et al., 1974
	4	(RIA)	Shapiro et al., 1972	95	(RIA)	Scott et al., 1974
	0,7[b]	(RIA)	Thody et al., 1975	110	(RIA)	Scott et al., 1974
	6[b]	(BIO)	Tilders, 1975	65	(BIO)	Moriarty and Moriarty, 1975
	18	(RIA)	Usategui et al., 1976	92	(RIA)	Usategui et al., 1976
	5	(RIA)	Oliver and Porter, 1978	105	(RIA)	Vaudry et al., 1978
	13	(RIA)	Vaudry et al., 1978	87	(RIA)	van Dijk, 1979
(Mean)	(9)			(86)		
Neurointermediate lobe	3200	(BIO)	Thody, 1974	18	(BIO)	Scott et al., 1974
	3560	(BIO)	Thody et al., 1975	63	(RIA)	Scott et al., 1974
	2770[b]	(RIA)	Thody et al., 1975	15	(BIO)	Moriarty and Moriarty, 1975
	3350[b]	(BIO)	Tilders, 1975	33	(RIA)	Usategui et al., 1976
	3490[b]	(BIO)	Kraicer et al., 1977	60	(RIA)	Vaudry et al., 1978
	1920	(BIO)	Vaudry et al., 1978			
	6220	(RIA)	Vaudry et al., 1978			
	1100	(RIA)	Oliver and Porter, 1978			
(Mean)	(3200)			(38)		

[a] Bioassay (BIO); radioimmunoassay (RIA).

[b] Wherever total content is given, ng/mg was calculated based on the assumption that the weight of the pars distalis was 9 mg and that of the neurointermediate lobe was 1 mg.

the time of sacrifice, the stage of the estrous cycle, the tissue extraction procedure, the type and specificity of the assay, and the standards used (e.g., Celis,
1975; Tilders, 1975; Tilders & Smelik, 1975).

The rat pituitary contains about three times as much α-MSH as ACTH on
a weight basis; 97% of the total α-MSH content is found in the neurointermediate lobe and 95% of the total pituitary ACTH content is in the pars
distalis. Some extremely low α-MSH concentrations (500 ng/mg by bioassay
and 59 ng/mg by radioimmunoassay, 0.04 ng/mg for the pars distalis, and 72
ng/mg for the intermediate lobe) were reported by Shapiro, Nicholson, Orth,
Mitchell, Island, and Liddle (1972) and Dubé, Lissitzky, Leclerc, and Pelletier
(1978), respectively. The anomalous results may be explained by the neutral
extraction medium used by these investigators (P. Janssens, unpublished observations). Some exceptionally high ACTH levels have been reported in the
neurointermediate lobe (890 ng/mg: Moriarty & Moriarty, 1975; 2190 ng/mg:
Scott, Lowry, Ratcliffe, Rees, & Landon, 1974; and 480 ng/mg: van Dijk,
1979). The apparent high levels may be due to the use by these authors of
antibodies directed to the C-terminal part of ACTH, which probably cross-react
with CLIP (Moriarty & Moriarty, 1975; van Dijk, 1979). This could explain
why these levels are in the same range as the α-MSH content of this structure
(cf. Table 1). No assay data exist for *human* pituitaries that allow an estimation
of the distribution of ACTH and α-MSH. On the basis of the bioassay data
of Morris, Russell, Landgrebe, and Mitchell (1956) and the individual weights
of the three parts of the human pituitary (Romeis, 1940; Table 1), it appears
that the distribution of melanotropic and corticotropic activity in the pars
distalis, pars intermedia, and posterior lobe is very similar (96%; 0.2%; 3.8%,
and 89.9%; 0.1%; 10%, respectively). However, because the α-MSH content
of the human pituitary was estimated to represent only 2% of the bioassayable
activity (Abe et al., 1967), Morris's melanotropic activity data do not reflect
the distribution of α-MSH in the human pituitary. The only information on
this subject is based upon immunocytochemistry (see Section V-B).

B. Brain Content of α-MSH and ACTH

Consistently high levels of α-MSH have been found in the arcuate nucleus
in the rat (Table 2). The same nucleus was also found to contain relatively
large amounts of ACTH (van Dijk, 1979). Assay data for ACTH are highly
variable, as can be seen by the more than hundredfold difference in hypothalamic ACTH content found in different studies. Table 2 also reveals that there
is not a complete overlap in distribution of α-MSH and ACTH. In this respect,
it is remarkable that the ventral hypothalamus seems to contain more radioim-

munoassayable ACTH than does the dorsal hypothalamus, whereas the reverse is found for α-MSH. The huge differences sometimes reported in the MSH/ ACTH contents of the various brain areas might be explained by the assay variables mentioned before (see also O'Donohue, Miller, Pendleton, & Jacobowitz, 1979c), particularly in view of the low amounts that are present in the brain, as compared to the pituitary, and the variability in the way the brain areas are dissected. An additional source of variability might be the different sensitivities of the various assays for the known or unknown members of the opiomelanocortin family. For instance, there are good reasons to assume that α-MSH itself is not present in the pineal gland but that, instead, an unknown related immunoreactive compound is synthesized in this structure (Pévet, Ebels, Swaab, Mud, & Arimura, 1980). The differences between the pineal α-MSH content (< 2 up to 380 pg/mg—see Table 2) might therefore be related to the differences in sensitivity for the unknown compound(s) of the various assays. Other compounds have been demonstrated by Loh and Gainer (1977) and Loh, Zucker, Verspaget, and van Wimersma Greidanus (1979), who found most of the bioassayable melanotropic activity in the rat brain and pituitary to be due to compounds that showed distinct cross-reactivity in a radioimmunoassay for α-MSH. These compounds appeared to be identical neither to α-MSH, $ACTH_{4-10, \ 1-10, \ 1-24}$, nor to $N-Ac-ACTH_{1-10}$, a conclusion fully consistent with our immunocytochemical data (see Section IV). High-performance liquid chromatography (HPLC) also indicates the existence of radioimmunoassayable peptides that are not identical to α-MSH (O'Donohue, Miller, & Jacobowitz, 1979a). Rudman, Chawla, and Hollins (1979) showed that an α-MSH-like peptide of bovine, rabbit, guinea pig, and rat pituitaries is N,O-diacetyl serine-α-MSH, and suggested that α-MSH may need to be acetylated before it is released from the pituitary. There may, in fact, be a host of converted peptides of the opiomelanocortin family in the brain and pituitary. One of the two peaks found by O'Donohue et al. (1979a) using HPLC might represent the compound recently described by Rudman et al. (1979). The presence of the melanotropic peptide "II F," as claimed earlier by Rudman et al. (1974), has been shown by means of HPLC to be due to a minor (1%) contamination of α-MSH in this fraction (Rudman et al., 1979). However, γ-MSH (Nakanishi et al., 1979) might be another candidate for an α-MSH-like peptide (see Fig. 1).

In the human brain, β-endorphin concentrations of about 100 pg/100 g protein were recently reported in the arcuate nucleus, median eminence, suprachiasmatic area, preoptic nucleus, supraoptic and paraventricular nuclei, and ventro- and dorsomedial nuclei. The highest concentrations were observed in the ventral hypothalamus (Wilkes, Watkins, Stewart, & Yen, 1980). Other quantitative data on the opiomelanocortin family in the human brain are not presently available.

TABLE 2

Summary of Data on the α-MSH and ACTH Content of the Rat Brain

	α-MSH(pg/mg)	Assay[a]	Reference	ACTH(pg/mg)	Assay[a]	Reference
Hypothalamus (total)	645	(RIA)	Vaudry et al., 1978	1500–4500[b]	(RIA)	Krieger et al., 1977a
	140	(RIA)	Oliver and Porter, 1978	800	(RIA)	Krieger et al., 1977b
	35	(RIA)	Dubé et al., 1978	1000[b]	(BIO)	Krieger et al., 1977a
	250	(RIA)	Eskay et al., 1979b	450	(BIO)	Krieger et al., 1977b
	80	(RIA)	Orwoll et al., 1979	14	(RIA)	Moldow and Yalow, 1978a
	180	(RIA)	Warberg et al., 1979	<8	(BIO)	Vaudry et al., 1978
				29	(RIA)	Orwoll et al., 1979
				54	(RIA)	van Dijk, 1979
Ventromedial nucleus	320[b]	(RIA)	O'Donohue et al., 1979a	40	(RIA)	Moldow and Yalow, 1978a
	290[b]	(RIA)	Eskay et al., 1979b	61[b]	(RIA)	van Dijk, 1979
Dorsomedial nucleus	895[b]	(RIA)	O'Donohue et al., 1979a	10	(RIA)	Moldow and Yalow, 1978a
	550[b]	(RIA)	Eskay et al., 1979b	36[b]	(RIA)	van Dijk, 1979
Arcuate nucleus	220	(RIA)	Eskay et al., 1979a	105[b]	(RIA)	van Dijk, 1979
	690[b]	(RIA)	Eskay et al., 1979b			
	870[b]	(RIA)	O'Donohue et al., 1979a			
Thalamus	25	(RIA)	Oliver and Porter, 1978	24[b]	(RIA)	Krieger et al., 1977a
	30	(RIA)	Vaudry et al., 1978	4	(RIA)	Moldow and Yalow, 1978a
	6–12[b]	(RIA)	Eskay et al., 1979a, b	<8	(RIA)	Vaudry et al., 1978
	42–719[b]	(RIA)	O'Donohue et al., 1979a	31[b]	(RIA)	van Dijk, 1979
Hippocampus	0.2–1.0	(RIA)	Eskay et al., 1979b, a	105[b]	(BIO)	Krieger et al., 1977b
	25[b]	(RIA)	O'Donohue et al., 1979a	120[b]	(RIA)	Krieger et al., 1977b
				120[b]	(RIA)	Krieger et al., 1977a
				3	(RIA)	Moldow and Yalow, 1978a
				<8	(RIA)	Vaudry et al., 1978
				15[b]	(RIA)	van Dijk, 1979
Septum	32	(RIA)	Eskay et al., 1979a	150[b]	(BIO)	Krieger et al., 1977b
	47	(RIA)	Eskay et al., 1979b	190[b]	(RIA)	Krieger et al., 1977b

Region						
Amygdala	12–87[b]	(RIA)	Eskay et al., 1979b	12[b]	(RIA)	van Dijk, 1979
	15–130[b]	(RIA)	O'Donohue et al., 1979a			
	90–355[b]	(RIA)	O'Donohue et al., 1979a			
Cerebellum	6	(RIA)	Eskay et al., 1979a	180[b]	(BIO)	Krieger et al., 1977b
	17–19	(RIA)	Eskay et al., 1979b	260[b]	(RIA)	Krieger et al., 1977b
	2	(RIA)	Oliver and Porter, 1978	3	(RIA)	Moldow and Yalow, 1978a
	4.1	(RIA)	Vaudry et al., 1978	20[b]	(BIO)	Krieger et al., 1977b
	0.2	(RIA)	Eskay et al., 1979b	40[b]	(RIA)	Krieger et al., 1977b
	8.5	(RIA)	Achterberg (unpubl.)	3.4[b]	(RIA)	Krieger et al., 1977a
Cortex (parietal or total)	1.5–3.2	(RIA)	Vaudry et al., 1978	40[b]	(BIO)	Krieger et al., 1977b
	3.6	(RIA)	Dubé et al., 1978	50[b]	(RIA)	Krieger et al., 1977b
	5.7	(RIA)	Oliver and Porter, 1978	46[b]	(RIA)	Krieger et al., 1977a
	0.07–1.4[b]	(RIA)	Eskay et al., 1979a,b	<1	(RIA)	Moldow and Yalow, 1978a
	1.8	(RIA)	Achterberg (unpubl.)	<4[b]	(RIA)	van Dijk, 1979
Brain stem/medulla	11	(RIA)	Oliver and Porter 1978	<10[b]	(BIO)	Krieger et al., 1977b
	3.6	(RIA)	Eskay et al., 1979a,b	<20[b]	(RIA)	Krieger et al., 1977b
	6.9	(RIA)	Achterberg (unpubl.)	3	(RIA)	Moldow and Yalow, 1978a
Spinal cord	0.6	(RIA)	Eskay et al., 1979b			
	2.4	(RIA)	Achterberg (unpubl.)			
Spinal ganglia	27	(RIA)	Achterberg (unpubl.)			
Pons	2.4	(RIA)	Eskay et al., 1979a,b	23[b]	(BIO)	Krieger et al., 1977b
				23[b]	(RIA)	Krieger et al., 1977b
				4	(RIA)	Moldow and Yalow, 1978a
Pineal	2.2	(RIA)	Dubé et al., 1978	<8	(BIO)	Vaudry et al., 1978
	160	(RIA)	Oliver and Porter, 1978			
	380	(RIA)	Vaudry et al., 1978			
	27	(RIA)	Eskay et al., 1979b			
	240	(RIA)	O'Donohue et al., 1979a			
	<2	(RIA)	Achterberg (unpubl.)			

[a] Radioimmunoassay (RIA); bioassay (BIO).

[b] Originally expressed as pg/mg protein and transferred to pg/mg wet tissue by multiplication by 0.1. The data of Krieger et al. (1977a,b) were obtained by the paradoxical binding phenomenon.

Although the exact nature of the compounds measured radioimmunochemically in the brain as α-MSH and ACTH is not known (see, e.g., the second α-MSH-like material containing peak in O'Donohue et al., 1979a), the literature is consistent in that these compounds do not disappear from the brain after hypophysectomy and are, thus, most probably synthesized (see Section VI) by the brain itself (Krieger, Liotta, & Brownstein, 1977a,b; Liotta, Hauser, Brownstein, & Krieger, 1978; O'Donohue, Holmquist, & Jacobowitz, 1979b; Oliver & Porter, 1978; Vaudry, Tonon, Delarue, Vaillant, & Kraicer, 1978; van Dijk, 1979).

IV. IMMUNOCYTOCHEMICAL LOCALIZATION OF α-MSH AND ACTH IN THE BRAIN

Immunocytochemical techniques, in principle, enable the precise localization of the opiomelanocortin peptides in the brain. However, due to the current limitations of these techniques, a number of questions still remain unanswered. In agreement with the assay data (Table 2), cell bodies immunoreactive with antisera raised against $ACTH_{1-24,\ 17-39}$ and $ACTH_{1-39}$ and α-MSH have all been found in the arcuate nucleus, the periarcuate area, and the ventromedial nuclei of the *rat* hypothalamus. The same cells can also be stained with anti-β-MSH, β-LPH, and α- and β-endorphin (Bloch, Bugnon, Fellmann, Lenys, & Gouget, 1979; Sofroniew, 1979; Watson & Akil, 1980; Watson & Barchas, 1979). These neurons are distinct from the monoamine-containing cells in the same region (Bugnon, Bloch, Lenys, Gouget, & Fellmann, 1979a). Peptide-containing hypothalamic cell bodies send projections positive for α-MSH, ACTH, and β-LPH (Fig. 2) to the medial and lateral hypothalamus, zona incerta, lateral septum, nucleus accumbens, periventricular thalamus, periaqueductal grey, locus coeruleus, nucleus tractus solitarius, reticular formation, stria terminalis, and medial amygdala (Bloch et al., 1979; Dubé et al., 1978; Jacobowitz & O'Donohue, 1978; O'Donohue et al., 1979a; Sofroniew, 1979; Watson & Barchas, 1979). In addition, ACTH immunoreactive fibers were observed in the spinal cord (Pelletier & Leclerc, 1979) and occasionally in the hippocampus (Larsson, 1978).

Jacobowitz and O'Donohue (1978) have shown that α-MSH and β-LPH do not always occur together. Positive β-LPH fibers were found in the locus coeruleus and substantia nigra zona compacta, but essentially no α-MSH-containing fibers were seen. However, it is not known whether the obvious overlap in the localization of the ACTH–MSH–LPH immunoreactivity is due to the presence of all of these compounds and their family members in the same cells and fibers, or to cross-reaction of the different antibodies with one or more related or unrelated compounds, because it is impossible at present

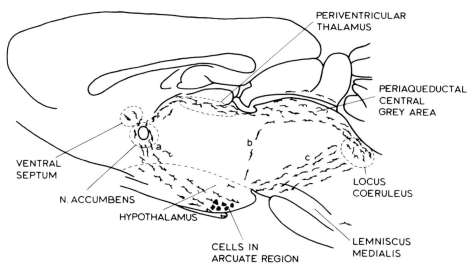

Fig. 2. Parasagittal schematic illustration of the arcuate cells and fibers that contain peptides of the opiomelanocortin family in the rat brain (after Watson & Barchas, 1979, with permission). (a) Cells are found in and near the arcuate nucleus with fibers running via the n. accumbens and periventricular thalamus to the periaqueductal central grey. (b) A caudo–dorsally directed pathway; and (c)·a centro–caudo–laterally running pathway, arising in the arcuate region and running dorsally via the nucleus hypothalamicus posterior to the nucleus cuneiformis where the fibers join the dorsal ones. (Data from O'Donohue et al., 1979a.)

to sufficiently elaborate specificity in immunocytochemical studies (see the following).

A striking coincidence exists between the distribution of opiomelanocortin-containing fibers and noradrenergic projections (Jacobowitz & O'Donohue, 1978), on the one hand, and the localization of exohypothalamic vasopressin and oxytocin pathways, on the other (Buijs, 1978; Buijs, Swaab, Dogterom, & van Leeuwen, 1978). There is indeed pharmacological evidence for a functional relationship between the brain levels of these peptides and catecholamine metabolism (Kovács, Bohus, & Versteeg, 1979; Versteeg, de Kloet, & de Wied, 1979). Exactly how these systems are related in the various brain areas is not clear.

Apart from the opiomelanocortin system arising from the arcuate nucleus, α-MSH and ACTH cell bodies and fibers have been demonstrated in a number of other brain sites. Anti-α-MSH-positive cell bodies that did not react with antibodies to $ACTH_{1-39, 1-24}$ and $ACTH_{4-10}$ were observed by Swaab and Fisser (1978) in the spinal cord motoneurons, the pineal gland, dorsal root ganglion cells, and neurons of the nucleus reticularis gigantocellularis (Fig. 3). Cerebellar basket cell fibers also stained positively. It is of interest that the cells in the

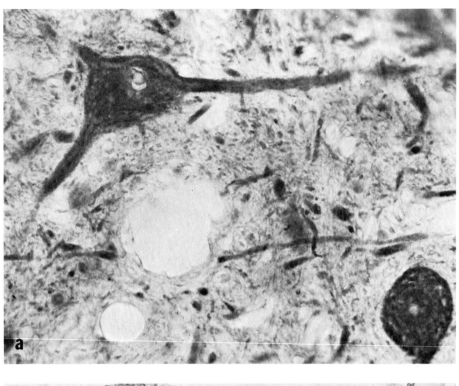

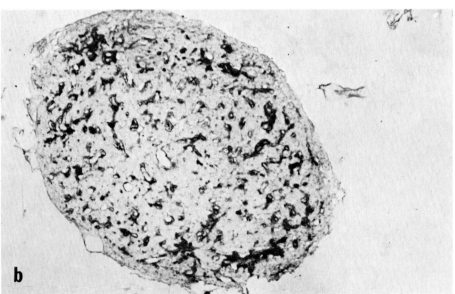

Fig. 3.   Staining of cells in (a) the nucleus reticularis gigantocellularis; and (b) the pineal gland, using antibodies raised against α-MSH; (a) 4394 9/4 and (b) 4372 26/3. (For details see Swaab & Fisser, 1978, and Pévet et al., 1980.)

nucleus reticularis have also been reported to be immunoreactive to enkephalins (Watson & Barchas, 1979). Enkephalins have, on the basis of their immunocytochemical distribution, been considered to be distinct from the ACTH–α-MSH–β-LPH-stained cells and fibers (Watson & Barchas, 1979), which is puzzling, taking the common precursor model into consideration. Immunocytochemical observations point either to an additional, separate pathway for enkephalin synthesis (cf. Rossier, 1981) or to a rapid degradation, release, or masking of the remaining fragments of the opiomelanocortin family. Larsson (1978) found ACTH-positive cells in the supraoptic nucleus, an observation that was extended to the paraventricular and suprachiasmatic nucleus by Sternberger (1979).* Watson and Akil (1979, 1980) described a group of α-MSH-positive cells arching over the top of the third ventricle, continuing laterally into the zona incerta, down toward the fornix, the lateral hypothalamic sulcus, and the insertion of the optic tract, whereas α-MSH-positive fibers, which did not stain for β-endorphin or β-LPH, could be demonstrated in the hippocampus.

Although radioimmunoassayable α-MSH was found in the pineal gland (Table 2), no positive immunocytochemical reaction with anti-α-MSH was reported until recently (Swaab & Fisser, 1978). Pévet et al. (1980) extended this observation and found a positively-staining population of cells in the rat pineal using one antibody against α-MSH, whereas a second antibody stained less well, and a third antibody did not stain these cells at all. By means of pineal cultures, it was demonstrated that this staining remained present during a culture period of 6 days, suggesting that the immunoreactive substance is synthetized in the pineal gland. Because the three antibodies against α-MSH stained α-MSH-containing agarose beads and sections of the intermediate lobe of the rat pituitary equally well, and showed different staining potencies for the pineal and the rest of the brain (see Fig. 3), Pévet et al. (1980) suggested that the stained compound is different from α-MSH. The finding of Dubé et al. (1978) that the pineal did not stain with an antibody against α-MSH, whereas other brain sites were found to be positive, is consistent with this idea. Cross-reaction of antibodies raised against α-MSH with as yet unidentified compounds in the brain that resemble α-MSH immunologically, was also inferred on the same grounds by Swaab and Fisser (1978) and by Dubé, Côte, and Pelletier (1979), and is fully consistent with the presence of α-MSH-like peptides as demonstrated by means of the separation and assay techniques of Loh and Gainer (1977), Loh et al. (1979), O'Donohue et al. (1979a), and Rudman et al. (1979; see Section III-B).

In the *human* brain, neurons in the infundibular nucleus stain with antisera raised against α- and β-endorphin, ACTH$_{17-39}$ and ACTH$_{1-24}$, α- and β-MSH,

---

*Recently, this staining appeared to be due to neurophysin II reactivity in the antiserum. However, the localization of β-LPH-like reactivity in this area was in all probability specific (Joseph & Sternberger, 1979).

and β-LPH. Their axonal fibers either appear to terminate around the blood vessels in the neurovascular zone and in the pituitary stalk, or to establish contacts with nonimmunoreactive perikarya of the infundibular nucleus (Bugnon, Bloch, Lenys, & Fellmann, 1979b). Pelletier and Désy (1979) localized ACTH-positive cell bodies in the arcuate nucleus of the human hypothalamus. These authors found ACTH-containing fibers throughout the hypothalamus, with the greatest density in the paraventricular nucleus. They could not, however, confirm Bugnon et al.'s (1979b) observation of fibers positive for ACTH in the neurovascular zone of the pituitary stalk. Although less numerous and less extensively distributed throughout the hypothalamus, a similar distribution of cells and fibers was found in the human hypothalamus using antibodies against α-MSH (Désy & Pelletier, 1978) and β-LPH (Pelletier, Désy, Lissitzky, Labrie, & Li, 1978).

The crucial question of immunocytochemistry, namely, exactly which compound in the tissue is stained, can only be solved by combining immunocytochemical procedures with separation techniques. Although most authors claim to use "specific antibodies" for their immunocytochemical work, this claim is generally based on false assumptions (cf. Swaab, Pool, & van Leeuwen, 1977; van Leeuwen, 1980). Until now, no immunocytochemical study on the localization of α-MSH–ACTH in the brain has fulfilled the criteria for a specificity test as formulated by C. W. Pool (see Swaab & Boer, 1980; van Leeuwen, 1981). Procedures that will fulfill these criteria are currently under development in our institute (Boer, 1979). Thus, at present, extreme care is necessary in interpreting data on α-MSH localization in the brain, including such unexpected findings as the presence of Met-enkephalin-like material in pituitary somatotrophs (Weber, Voigt, & Martin, 1978), of an ACTH-like substance in hypothalamic LHRH neurons (Tramu, Leonardelli, & Dubois, 1977), and of the simultaneous presence of ACTH-like and prolactin-like material in arcuate and ventromedial neurons (Toubeau, Desclin, Parmentier, & Pasteels, 1979).

V. THE OPIOMELANOCORTIN FAMILY IN DEVELOPMENT

In various species, peptides of the opiomelanocortin family in fetal brain seem to determine, to an important degree, normal intrauterine growth and the initiation of labor. Because the various peptides of this family are thought to influence these processes in different ways, increasing attention is now being paid to developmental aspects of this group (for recent reviews, see Boer, Swaab, Uylings, Boer, Buijs, & Velis, 1980; Swaab, 1980; Swaab, Visser, & Dogterom, 1979).

A. The Rat Pituitary and Brain

Endogenous fetal α-MSH was found to stimulate intrauterine body and brain growth at the time of the intrauterine growth spurt that takes place in the Day 19 fetal rat (Honnebier & Swaab, 1974; Swaab, Boer, & Visser, 1978; Swaab, et al., 1976). By means of immunofluorescence microscopy, α-MSH-containing cells were observed from Day 18 onward in the intermediate lobe of the fetal pituitary and in a few scattered cells in the pars distalis (Chatelain, Dubois, & Dupouy, 1976; Swaab et al., 1976). Radioimmunoassayable α-MSH was even present 2 days earlier (Fig. 4). A remarkably rapid rise in fetal pituitary α-MSH was found around Day 19 of pregnancy—the day of the intrauterine growth spurt—both by bioassay (Swaab et al., 1976) and by radioimmunoassay (Fig. 4). The fact that the absolute values of the radioimmunoassay were some fivefold lower than those obtained by means of the bioassay suggests that melanocyte-stimulating substances, other than α-MSH are also increasing rapidly at this time.

Radioimmunoassayable α-MSH is also present early in the fetal rat brain. The content and concentration of α-MSH in this organ is described by a U-shaped curve, between Day 16 of pregnancy and birth. The high levels at Day 16 of pregnancy are within the same range as the adult brain concentrations, whereas the absolute amount in the fetal brain on Day 16 is of the same magnitude as that of the fetal pituitary (Fig. 4). In view of the possible function of peptides of the opiomelanocortin family in brain development (for review, see Swaab, 1980), knowledge of the developmental course of the other members of this family would be of great interest. This is particularly true because opiate receptor binding has been found to be present in the fetal rat brain as early as Day 14 (Clendeninn, Petraitis, & Simon, 1976). The majority of the developmental observations are, however, based only on qualitative, immunocytochemical studies. Immunofluorescence microscopy has revealed the presence of ACTH and β-MSH in the anterior lobe of the fetal pituitary at Days 16–17 and in the pars intermedia at Days 17–18 (Chatelain et al., 1976). The α-MSH-like material stained in the rat pineal gland by immunocytochemistry is present in higher amounts in young (21 days postnatally) than in adult animals (Pévet et al., 1980). The only other data on the developmental pattern of the opiomelanocortin family comes from Bayon, Shoemaker, Bloom, Mauss, and Guillemin's (1979) study on the developmental pattern of radioimmunoassayable endorphin and enkephalins in the fetal brain. Endorphin levels revealed that the adult regional distribution is present at fetal Day 16. In contrast, enkephalin systems were not as developed, supporting the hypothesis that enkephalin occurs in interneurons that usually differentiate later than the endorphin-containing neurons that form long pathways. A decline was found in endorphin levels in the telencephalon, hippocampus, and striatum after Day

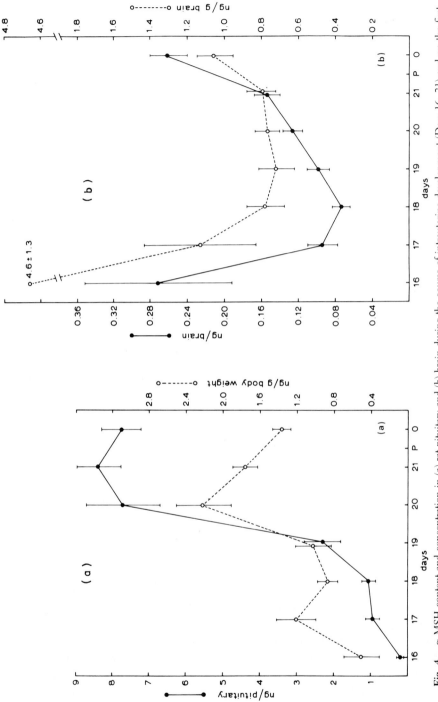

Fig. 4. α-MSH content and concentration in (a) rat pituitary and (b) brain during the course of intrauterine development (Days 16–21) and on the first postnatal day (0) as measured by radioimmunoassay. (P means parturition.) The day on which spermatozoa were found in the vaginal smear was taken as Day 0 of pregnancy. Note the rapid rise in pituitary α-MSH content on Day 19 of pregnancy and the relatively high levels and concentrations of α-MSH in the fetal brain early in development. Male adult pituitary levels in (a) are 2.6 (±.08) μg/pituitary and 13.0 (±4.1) ng/g body, and male adult brain levels in (b) are 12.3 (±2.1) ng/brain and 5.7 (±0.9) ng/g brain. (From Visser & Swaab, unpublished observations.)

16 of pregnancy. A relationship to α-MSH, which also decreases in the brain after Day 16 of pregnancy, is plausible, and underscores the desirability of obtaining a more complete picture of the different components of the opio-melanocortin family during development.

B. The Human Pituitary and Brain

Because the human fetus has a distinct pars intermedia in the pituitary (Romeis, 1940), it might be expected that α-MSH is present during human development. However, the evidence is conflicting. It has been stated that α-MSH

1. is not present in the fetal pituitary (Dubois, Vargues-Regairaz, & Dubois, 1973);
2. is present only in the fetal pituitary (Silman, Chard, Lowry, Smith, & Young, 1976);
3. is not present in the adult pituitary (Scott et al., 1973);
4. is present in 30% of the adult pituitaries (Nieuwenhuyzen Kruseman & Schröder-van der Elst, 1976);
5. is present in only a very few cells in adult pituitaries (Dubois, 1973); and
6. is present in all adult pituitary cells that contain ACTH (Phifer, Orth, & Spicer, 1974).

Because these discrepancies seemed at least partially explainable by the technical pitfalls of immunocytochemistry, we have studied the pituitary distribution of α-MSH and ACTH using antisera that were purified and characterized for their staining capabilities on a number of peptides of the opiomelanocortin family (Visser & Swaab, 1979).

In all fetal pituitaries, the pars intermedia was found to consist mainly of α-MSH-containing cells and only a few ACTH-containing cells. In the pars distalis, the opposite was observed. In the youngest fetus (15 weeks) the pars intermedia contained many α-MSH cells, but no ACTH cells, whereas ACTH cells were present in the pars distalis (Visser & Swaab, 1979). These observations agree with the fact that melanocyte-stimulating activity has been detected by bioassay in the human fetal pituitary from 11 weeks of pregnancy onward (Kastin, Gennser, Arimura, Miller, & Schally, 1968; Levina, 1968). During development the pars intermedia was found to develop gradually into a "zona intermedia," in which there was a large amount of cyst formation. This remnant of the pars intermedia contains α-MSH and ACTH immunoreactive cells, which also invade the posterior lobe. In our study, α-MSH immunoreactive cells were observed in the zona intermedia and pars distalis of all pituitaries throughout life, irrespective of age, sex, pregnancy, therapy, or cause of death.

However, from birth to 19 years of age, progressively fewer α-MSH cells and more ACTH cells were detected in the pars distalis and zona intermedia (Visser & Swaab, 1979). The gradually decreasing ratio of α-MSH–ACTH-containing cells in the human pituitary during development agrees fully with the biochemical work of Silman et al. (1976) and is considered to be of importance, both for intrauterine development and for timing the moment of labor onset (Swaab et al., 1979). The few α-MSH-positive cells found in adulthood is consistent with the low levels of α-MSH in human pituitaries found by radioimmunoassay (Abe et al., 1967). Earlier data (Phifer et al., 1974), indicating that each ACTH cell in the human adult also contained α-MSH can be explained by cross-reactivity, as the antibodies used in these studies were not purified prior to use in immunocytochemistry (Swaab et al., 1976).

The total absence of the pars intermedia and of any α-MSH-containing cells in the pars distalis of the seven anencephalics in our study supports the hypothesis of a stimulatory effect of the hypothalamus on the development of the pars intermedia (Gaillard, 1937). Our study is consistent with the suggestion of Celis (1977) that oxytocin, which we did not find in the anencephalic pituitaries (Swaab & Visser, 1977), might be a prohormone for an α-MSH-releasing factor. This is of particular interest because the influence of the adult mammalian brain on MSH production is generally thought to be inhibitory (Tilders & Smelik, 1978).

Previously, hypothalamic stimulation was not considered to be essential for the development of fetal ACTH cells in the pars distalis (Bégeot, Dubois, & Dubois, 1977). However, we found very few fluorescent cells in the pars distalis of one anencephalic and no ACTH cells in the other six pituitaries when using purified anti-$ACTH_{1-24}$ plasma (Visser & Swaab, 1979). Thus, the human fetal pituitary seems to be dependent on the hypothalamus for the development of ACTH. The staining of "ACTH cells" as reported by Bégeot et al. (1977), might be explained by the staining of a precursor molecule, because when we used anti-$ACTH_{1-39}$ plasma, which cross-reacted with β- and γ-lipotropic hormone, fluorescent cells were found in anterior lobes of *all* the anencephalics (Visser & Swaab, 1979).

Fluorescent cells have been found in the anterior and intermediate lobes of the pituitary of the human fetus as early as the eighth week of fetal life by the use of antibodies to $ACTH_{17-19}$, β-MSH, β-LPH, and α- and β-endorphins (for review see Bégeot, Dubois, & Dubois, 1978). The specificity of these immunocytochemical observations is not sufficiently elaborated (see preceding), but the various studies show that fragments of the opiomelanocortin family are present very early in human fetal development. The same is true for fetal brain; in the infundibular nucleus of the hypothalamus, neurons stained with antibodies against β-endorphin or $ACTH_{17-39}$ from the eleventh week of fetal life onward (Bugnon et al., 1979b).

Various compounds of the opiomelanocortin family have been demonstrated during pregnancy in the human amniotic fluid and maternal blood. Generally, it is unknown, however, what the relative contributions of the fetus, placenta, and mother are to the levels of these peptides in these fluids. In the amniotic fluid, ACTH levels reach a peak between the twenty-sixth and thirty-fourth week of pregnancy and then level off (Schollberg, Seiler, & Schmidt, 1978; Tuimala, Kauppila, & Haapalahti, 1976). However, β-MSH levels in maternal serum rise progressively throughout pregnancy. Because fetal umbilical cord serum and amniotic levels were found to be higher than maternal levels, synthesis by the fetus was suggested (Ances & Pomerantz, 1974). However, data indicate that postpartum maternal serum β-MSH levels are not different from the levels in midpregnancy (Thody, Plummer, Burton, & Hytten, 1974), and, thus, the fetal contribution to the β-MSH levels seems unimportant. β-endorphin in the amniotic fluid probably does not originate in the fetal brain or pituitary either, because anencephalic amniotic fluid contains a high level of this material (Gautray, Vielh, Levaillant, & Rotten, 1979). Corresponding data on the measurement of circulating fetal levels of α-MSH are lacking. Reported levels of more than 0.1 ng of α-MSH/ml in maternal and cord plasma (Clark, Thody, Shuster, & Bowers, 1978) are, in our opinion, at least tenfold too high. This might be explained by the fact that these authors used unextracted plasma in their radioimmunoassay which, with our antisera, yields high nonspecific values.

VI. THE SYNTHESIS OF MSH AND ACTH BY THE BRAIN

Although it is now half a century since Zondek and Krohn (1932) proposed that brain "intermedin" is synthesized in the pituitary, and 20 years since Guillemin et al. (1962) proposed that MSH- and ACTH-like peptides are also synthesized in the brain, the relative contributions of both potential peptide sources is still a matter of intense controversy.

The fact that concentrations of brain peptides, which are identical to pituitary hormones, are unchanged after hypophysectomy (Krieger et al., 1977a,b; Liotta et al., 1978; O'Donohue et al., 1979a,b; Oliver & Porter, 1978; Swaab, 1980; Vaudry et al., 1978; van Dijk, 1979) has not been accepted as proof of their central origin by Moldow and Yalow (1978a,b). These authors proposed that pituitary remnants found in the sella after commercial hypophysectomy could still be the origin of brain peptides. In addition, as concentrations of ACTH and MSH in various brain sites have been found to be inversely related to the distance of these areas from the hypophysis, they suggested that the pituitary may be the sole site of synthesis of these hormones. This idea is, of course, consistent with observations that behavioral deficits after hypophysectomy can be restored by peripheral injections of ACTH or MSH (de Wied, 1964, 1965). Although others have found much less ACTH in the sella after hypophysectomy, and blood levels are apparently affected by "pituitary remnants" (van Dijk, 1979), it is still possible that the pars tuberalis remains functional after hypophysectomy (Ordronneau, 1979). Also, pituitary tissue, which is present in the pharyngeal canal and even the pharynx itself (McGrath, 1976, 1978) might be an additional source of hormones after hypophysectomy. Retrograde transport of peptides from the pituitary to the brain or CSF, as has been proposed by Popa (1938), is also a realistic possibility (Mezey, Kivovics, & Palkovits, 1979).

Nevertheless, evidence for production of "pituitary" hormones by the nervous system is accumulating:

1. Pituitary hormones remain present in the brain after hypophysectomy as shown by means of bioassay, radioimmunoassay, and immunocytochemistry (for references see Section III-B and preceding).

2. α-MSH disappears from the brain sites that are innervated (cf. Section IV) by arcuate nucleus neurons after electrolytic, knife, or sodium glutamate lesions of this area, whereas the pituitary store of this hormone does not diminish (Eskay, Brownstein, & Long, 1979a; Eskay, Giraud, Oliver, & Brownstein, 1979b; O'Donohue et al., 1979a). Sodium glutamate lesions of the arcuate nucleus cause a four- or fivefold decrease of ACTH and β-endorphin in the hypothalamus and a twofold decrease of these peptides in the amygdala, whereas pituitary ACTH content remains the same (Krieger, Liotta, Nicholson, & Kizer, 1979). These observations argue strongly for the arcuate nucleus as a site of peptide synthesis.

3. Enkephalins have been found in cultured spinal cord neurons (Neale, Barker, Uhl, & Snyder, 1978) by immunofluorescence microscopy, whereas in explants of dorsal root ganglia (van Leeuwen, unpublished observations) and organ cultures of the pineal gland (Pévet et al., 1980) α-MSH-like material can be stained immunocytochemically. For the latter experiment, a synthetic culture medium was used, which excludes uptake of exogenous peptides.

4. The localization of "pituitary" peptides in the nervous system is not restricted to site that are immediately connected to the hypophyseal portal vessels as has been claimed by Moldow and Yalow (1978a,b). For example, α-MSH immunoreactivity has been found in neurons of the spinal cord, dorsal root ganglia (Swaab & Fisser, 1978), and the pineal gland (Table 2).

5. Animals that do not have a pituitary have ACTH- and α-MSH immunoreactivity in their nervous systems (Boer, Schot, Roubos, ter Maat, Lodder, Reichelt, & Swaab, 1979).

6. The intracellular localization of "pituitary" hormones observed by light microscopy also supports a central site for their origin (see preceding). Immunoelectron microscopy has shown that immunoreactivity for α-MSH and

ACTH is present in vesicles of arcuate nucleus neurons (Pelletier & Dubé, 1977; Pelletier & Leclerc, 1979).

7. α-MSH is releasable by potassium from hypothalamic synaptosomes (Oliver, Barnea, Warberg, Eskay, & Porter, 1977; Warberg, Oliver, Barnea, Parker, & Porter, 1979).

8. Hippocampal ACTH levels change independently of hypophyseal release (van Dijk, 1979).

From these observations and from the fact that a significant transfer of ACTH from the blood to the brain is very unlikely in view of the existence of an effective blood–brain barrier for such peptides (for references see van Dijk, 1979), it is most probable that opiomelanocortic peptides are produced by the nervous system. Any additional contribution of peptides of pituitary origin to the physiology of the various brain sites remains to be determined.

ACKNOWLEDGMENTS

We thank Dr. H. M. Greven of the Scientific Development Group (Organon) for providing the peptides, and Anke van der Velden for her secretarial work.

REFERENCES

Abe, K., Island, D. P., Liddle, G. W., Fleischer, N., & Nicholson, W. E. (1967). Radioimmunologic evidence for α-MSH (melanocyte stimulating hormone) in human pituitary and tumor tissues. *Journal of Clinical Endocrinology,* 27, 46–52.

Ances, I. G., & Pomerantz, S. H. (1974). Serum concentrations of β-melanocyte-stimulating hormone in human pregnancy. *American Journal of Obstetrics and Gynecology,* 119, 1062–1068.

Applezweig, M. H., & Moeller, G. (1959). The pituitary–adrenocortical system and anxiety in avoidance learning. *Acta Psychologica,* 15, 602–603.

Austen, B. M., Smyth, D. G., & Snell, C. R. (1977). γ-Endorphin, α-endorphin and Met-enkephalin are formed extracellularly from lipotropin C fragment. *Nature,* 269, 619–621.

Bayon, A., Shoemaker, W. J., Bloom, F. E., Mauss, A., & Guillemin, R. (1979). Perinatal development of the endorphin- and enkephalin-containing systems in the rat brain. *Brain Research,* 179, 93–101.

Bégeot, M., Dubois, M. P., & Dubois, P. M. (1977). Growth hormone and ACTH in the pituitary of normal and anencephalic human fetuses: Immunocytochemical evidence for hypothalamic influences during development. *Neuroendocrinology,* 24, 208–220.

Bégeot, M., Dubois, M. P., & Dubois, P. M. (1978). Immunologic localization of α- and β-endorphins and β-lipotropin in corticotropic cells of the normal and anencephalic fetal pituitaries. *Cell and Tissue Research,* 193, 413–422.

Bertagna, X. Y., Nicholson, W. E., Sorenson, G. D., Pettengill, O. S., Mount, C. D., & Orth, D. N. (1978). Corticotropin, lipotropin, and β-endorphin production by a human nonpituitary tumor in culture: Evidence for a common precursor. *Proceedings of the National Academy of Sciences,* 75, 5160–5164.

Bloch, B., Bugnon, C., Fellmann, D., Lenys, D., & Gouget, A. (1979). Neurons of the rat hypothalamus reactive with antisera against endorphins, ACTH, MSH, and β-LPH. *Cell and Tissue Research*, **204**, 1–15.

Boer, G. J. (1979). Development of a specificity test for immunocytochemistry of neuropeptides using high-voltage electrofocusing in polyacrylamide micro slab gels. *Endocrinology*, **80**, 54–55.

Boer, G. J., Swaab, D. F., Uylings, H. B. M., Boer, K., Buijs, R. M., & Velis, D. N. (1980). Neuropeptides in rat brain development. *In* P. S. McConnell, G. J. Boer, H. J. Romijn, N. E. Van Den Poll, & M. A. Corner (Eds.), *Adaptive Capabilities of the Nervous System. Progress in Brain Research* (Vol. 53), pp. 207–227. Amsterdam: Elsevier.

Boer, H. H., Schot, L. P. C., Roubos, E. W., ter Maat, A., Lodder, J. C., Reichelt, D., & Swaab, D. F. (1979). ACTH-like immunoreactivity in two electrotonically coupled giant neurons in the pond snail *Lymnaea stagnalis*. *Cell and Tissue Research*, **202**, 231–240.

Bradbury, A. F., Smyth, D. G., & Snell, C. R. (1976). Prohormones of α-melanotropin (α-melanocyte-stimulating hormone, α-MSH) and corticotropin (adrenocorticotropic hormone, ACTH): Structure and activation. *In* R. Porter & D. W. Fitzsimons (Eds.), *Polypeptide Hormones: Molecular and Cellular Aspects. Ciba Foundation Symposium* (Vol. 41), pp. 61–75. Amsterdam: Elsevier.

Bugnon, C., Bloch, B., Lenys, D., Gouget, A., & Fellmann, D. (1979a). Comparative study of the neuronal populations containing β-endorphin, corticotropin and dopamine in the arcuate nucleus of the rat hypothalamus. *Neuroscience Letters*, **14**, 43–48.

Bugnon, C., Bloch, B., Lenys, D., & Fellmann, D. (1979b). Infundibular neurons of the human hypothalamus simultaneously reactive with antisera against endorphins, ACTH, MSH and β-LPH. *Cell and Tissue Research*, **199**, 177–196.

Buijs, R. M. (1978). Intra- and extrahypothalamic vasopressin and oxytocin pathways in the rat. Pathways to the limbic system, medulla oblongata and spinal cord. *Cell and Tissue Research*, **192**, 423–435.

Buijs, R. M., Swaab, D. F., Dogterom, J., & van Leeuwen, F. W. (1978). Intra- and extra-hypothalamic vasopressin and oxytocin pathways in the rat. *Cell and Tissue Research*, **186**, 423–433.

Celis, M. E. (1975). Serum MSH levels and the hypothalamic enzymes involved in the formation of MSH-RF during the estrous cycle in the rat. *Neuroendocrinology*, **18**, 256–262.

Celis, M. E. (1977). Hypothalamic peptides involved in the control of MSH secretion. Identity, biosynthesis and regulation of their release. *In* F. J. H. Tilders, D. F. Swaab, & Tj. B. van Wimersma Greidanus (Eds.), *Melanocyte-Stimulating Hormone: Control, Chemistry, and Effects*, pp. 69–79. Basel: Karger.

Chatelain, A., Dubois, M. P., & Dupouy, J. P. (1976). Hypothalamus and cytodifferentiation of the foetal pituitary gland. Study in vivo. *Cell and Tissue Research*, **169**, 335–344.

Chrétien, M., Benjannet, S., Gossard, F., Gianoulakis, C., Crine, P., Lis, M., & Seidah, N. G. (1979). From β-lipotropin to β-endorphin and "pro-opio-melanocortin." *Canadian Journal of Biochemistry*, **57**, 1111–1121.

Clark, D., Thody, A. J., Shuster, S., & Bowers, H. (1978). Immunoreactive α-MSH in human plasma in pregnancy. *Nature*, **273**, 163–164.

Clendeninn, N. J., Petraitis, M., & Simon, E. J. (1976). Ontological development of opiate receptors in rodent brain. *Brain Research*, **118**, 157–160.

Désy, L., & Pelletier, G. (1978). Immunohistochemical localization of alpha-melanocyte-stimulating hormone (α-MSH) in the human hypothalamus. *Brain Research*, **154**, 377–381.

de Wied, D. (1964). Influence of anterior pituitary on avoidance learning and escape behavior. *American Journal of Physiology*, **207**, 255–259.

de Wied, D. (1965). The influence of the posterior and intermediate lobe of the pituitary and pituitary peptides on the maintenance of a conditioned avoidance response in rats. *International Journal of Neuropharmacology*, **4**, 157–167.

Dubé, D., Côte, J., & Pelletier, G. (1979). Further studies on the immunohistochemical localization of α-MSH in the rat brain. *Neuroscience Letters*, **12**, 171–176.

Dubé, D., Lissitzky, J. C., Leclerc, R., & Pelletier, G. (1978). Localization of α-melanocyte-stimulating hormone in rat brain and pituitary. *Endocrinology*, **102**, 1283–1291.

Dubois, M. P. (1973). Recherche par immunofluorescence des cellules adénohypophysaires élaborant les hormones polypeptidiques: ACTH, α-MSH, β-MSH. *Bulletin de l'Association des Anatomistes*, **57**, 63–76.

Dubois, P., Vargues-Regairaz, H., & Dubois, M. P. (1973). Human foetal anterior pituitary. Immunofluorescent evidence for corticotropin and melanotropin activities. *Zeitschrift für Zellforschung*, **145**, 131–143.

Eskay, R. L., Brownstein, M. J., & Long, R. T. (1979a). α-Melanocyte-stimulating hormone: Reduction in adult rat brain after monosodium glutamate treatment of neonates. *Science*, **205**, 827–829.

Eskay, R. L., Giraud, P., Oliver, C., & Brownstein, M. J. (1979b). Distribution of α-melanocyte-stimulating hormone in the rat brain: Evidence that α-MSH-containing cells in the arcuate region send projections to extrahypothalamic areas. *Brain Research*, **178**, 55–67.

Ferrari, W., Gessa, G. L., & Vargiu, L. (1963). Behavioral effects induced by intracisternally injected ACTH and MSH. *Annals of the New York Academy of Sciences*, **104**, 330–345.

Gaillard, P. J. (1937). An experimental contribution to the origin of the pars intermedia of the hypophysis. *Acta Neerlandica Morphologica*, **1**, 3–11.

Gautray, J. P., Vielh, J. P., Levaillant, J. M., & Rotten, D. (1979). Immunoassayable beta endorphin in human amniotic fluid. In L. Zichella & P. Pancheri (Eds.), *Psychoneuroendocrinology in Reproduction*, pp. 459–467. Amsterdam: Elsevier.

Guillemin, R., & Krivoy, W. A. (1960). L'hormone mélanophorétique β-MSH jouet-elle un rôle dans les fonctions du système nerveux central chez les mammifères supérieurs? *Comptes Rendus de l'Académie des Sciences Paris*, **250**, 1117–1119.

Guillemin, R., Schally, A. V., Lipscomb, H. S., Andersen, R. N., & Long, J. M. (1962). On the presence in hog hypothalamus of β-corticotropin releasing factor, α- and β-melanocyte stimulating hormones, adrenocorticotropin, lysine-vasopressin and oxytocin. *Endocrinology*, **70**, 471–477.

Guillemin, R., Ling, N., & Burgus, R. (1976). Endorphines, peptides, d'origine hypothalamique et neurohypophysaire à activité morphinomimétique. Isolement et structure moléculaire de l'α-endorphine. *Comptes Rendus de l'Académie des Sciences Paris*, **282**, 783–785.

Harris, I. (1960). The chemistry of pituitary polypeptide hormones. *British Medical Bulletin*, **16**, 189–195.

Honnebier, W. J., & Swaab, D. F. (1974). Influence of α-melanocyte-stimulating hormone (α-MSH), growth hormone (GH) and fetal brain extracts on intrauterine growth of fetus and placenta in the rat. *Journal of Obstetrics and Gynaecology of the British Commonwealth*, **81**, 439–447.

Hughes, J., Smith, T. W., Kosterlitz, H. W., Fothergill, L. A., Morgan, B. A., & Morris, H. R. (1975). Identification of two related pentapeptides from the brain with potent opiate agonist activity. *Nature*, **258**, 577–579.

Jacobowitz, D. M., & O'Donohue, T. L. (1978). α-Melanocyte stimulating hormone: Immunohistochemical identification and mapping in neurons of rat brain. *Proceedings of the National Academy of Sciences*, **75**, 6300–6304.

Joseph, S. A., & Sternberger, L. A. (1979). The unlabeled antibody method: Contrasting color staining of β-lipotropin and ACTH-associated hypothalamic peptides. *Journal of Histochemistry and Cytochemistry*, **27**, 1430–1437.

Kastin, A. J., Gennser, G., Arimura, A., Miller, M. C., & Schally, A. V. (1968). Melanocyte-stimulating and corticotrophic activities in human foetal pituitary glands. *Acta endocrinologica*, **58**, 6–10.

Kovács, G. L., Bohus, B., & Versteeg, D. H. G. (1979). The effects of vasopressin on memory processes: The role of noradrenergic neurotransmission. *Neuroscience*, **4**, 1529–1537.

Kraicer, J., Beraud, G., & Lynwood, D. W. (1977). Pars intermedia ACTH and MSH content: Effect of adrenalectomy, gonadectomy and a neurotropic (noise) stress. *Neuroendocrinology*, **23**, 253–267.

Krieger, D. T., Liotta, A., & Brownstein, M. J. (1977a). Presence of corticotropin in brain of normal and hypophysectomized rats. *Proceedings of the National Academy of Sciences*, **74**, 648–652.

Krieger, D. T., Liotta, A., & Brownstein, M. J. (1977b). Presence of corticotropin in limbic system of normal and hypophysectomized rats. *Brain Research*, **128**, 575–579.

Krieger, D. T., Liotta, A., Nicholson, G., & Kizer, J. S. (1979). Brain ACTH and endorphin reduced in rats with monosodium glutamate-induced arcuate nuclear lesions. *Nature*, **278**, 562–563.

Larsson, L.-I. (1978). Distribution of ACTH-like immunoreactivity in rat brain and gastrointestinal tract. *Histochemistry*, **55**, 225–233.

Levina, S. E. (1968). Endocrine features in development of human hypothalamus, hypophysis, and placenta. *General and Comparative Endocrinology*, **11**, 151–159.

Li, C. H., Barnafi, L., Chrétien, M., Chung, D. (1965). Isolation and amino-acid sequence of β-LPH from sheep pituitary glands. *Nature*, **208**, 1093–1094.

Liotta, A. S., Hauser, H., Brownstein, M. J., & Krieger, D. T. (1978). Effect of long-term hypophysectomy (hypox) or adrenalectomy (adx) on rat brain ACTH. *Federation Proceedings*, **37**, 665.

Loh, Y. P., & Gainer, H. (1977). Heterogeneity of melanotropic peptides in the pars intermedia and brain. *Brain Research*, **130**, 169–175.

Loh, Y. P., Zucker, L., Verspaget, H., & van Wimersma Greidanus, Tj. B. (1979). Melanotropic peptides: presence in brain of normal and hypophysectomized rats and subcellularly localized in synaptosomes. *Journal of Neuroscience Research*, **4**, 147–156.

Mains, R. E., Eipper, B. A., & Ling, N. (1977). Common precursor to corticotropins and endorphins. *Proceedings of the National Academy of Sciences*, **74**, 3014–3018.

Mains, R. E., & Eipper, B. A. (1979). Synthesis and secretion of corticotropins, melanotropins, and endorphins by rat intermediate pituitary cells. *Journal of Biological Chemistry*, **254**, 7885–7894.

McGrath, P. (1976). Further observations on the pharyngeal hypophysis and the postphenoid in the mature male rat. *Journal of Anatomy*, **121**, 193–201.

McGrath, P. (1978). Aspects of the human pharyngeal hypophysis in normal and anencephalic fetuses and neonates and their possible significance in the mechanism of its control. *Journal of Anatomy*, **127**, 65–81.

Mezey, E., Kivovics, P., & Palkovits, M. (1979). Pituitary–brain retrograde transport. *Trends in Neuroscience*, **2**, 57–59.

Mialhe-Voloss, C. (1958). Posthypophyse et activité corticotrope. *Acta Endocrinologica*, Supplement **35**, 9–96.

Moldow, R., & Yalow, R. S. (1978a). Extrahypophysial distribution of corticotropin as a function of brain size. *Proceedings of the National Academy of Sciences*, **75**, 994–998.

Moldow, R. L., & Yalow, R. S. (1978b). Extrahypophysial distribution of thyrotropin as a function of brain size. *Life Sciences*, **22**, 1859–1864.

Moriarty, C. M., & Moriarty, G. C. (1975). Bioactive and immunoactive ACTH in the rat pituitary: Influence of stress and adrenalectomy. *Endocrinology*, **96**, 1419–1425.

Morris, C. J. O. R., Russell, D. S., Landgrebe, F. W., & Mitchell, G. M. (1956). The melanophore-expanding and corticotrophic activity of human pituitary tissue. *Journal of Endocrinology*, **14**, 263–267.

Murphy, J. W., & Miller, R. E. (1955). The effect of adrenocorticotropic hormone (ACTH) on avoidance conditioning in the rat. *Journal of Comparative and Physiological Psychology*, **48**, 47–49.

Nakanishi, S., Inoue, A., Taii, S., & Numa, S. (1977). Cell-free translation product containing corticotropin and β-endorphin encoded by messenger RNA from anterior lobe and intermediate lobe of bovine pituitary. *FEBS Letters*, **84**, 105–109.

Nakanishi, S., Inoue, A., Kita, T., Nakamura, M., Chang, A. C. Y., Cohen, S. N., & Numa, S. (1979). Nucleotide sequence of cloned cDNA for bovine corticotropin-β-lipotropin precursor. *Nature*, **278**, 423–427.

Neale, J. H., Barker, J. L., Uhl, G. R., & Snyder, S. H. (1978). Enkephalin-containing neurons visualized in spinal cord cell cultures. *Science*, **201**, 467–469.

Nieuwenhuyzen Kruseman, A. C., & Schröder-van der Elst, J. P. (1976). The immunolocalization of ACTH and α-MSH in human and rat pituitaries. *Virchows Archiv B Cell Pathology*, **22**, 263–272.

O'Donohue, T. L., Miller, R. L., & Jacobowitz, D. M. (1979a). Identification, characterization and stereotaxic mapping of intraneuronal α-melanocyte stimulating hormone-like immunoreactive peptides in discrete regions of the rat brain. *Brain Research*, **176**, 101–123.

O'Donohue, T. L., Holmquist, G. E., & Jacobowitz, D. M. (1979b). Effect of hypophysectomy on α-melanotropin in discrete regions of the rat brain. *Neuroscience Letters*, **14**, 271–274.

O'Donohue, T. L., Miller, R. L., Pendleton, R. C., & Jacobowitz, D. M. (1979). A diurnal rhythm of immunoreactive α-melanocyte-stimulating hormone in discrete regions of the rat brain. *Neuroendocrinology*, **29**, 281–287.

Oliver, C., Barnea, A., Warberg, J., Eskay, R. L., & Porter, J. C. (1977). Distribution, characterization, and subcellular localization of MSH in the brain. In F. J. H. Tilders, D. F. Swaab, & Tj. B. van Wimersma Greidanus (Eds.), *Melanocyte-Stimulating Hormone: Control, Chemistry, and Effects*, pp. 162–166. Basel: Karger.

Oliver, C., & Porter, J. C. (1978). Distribution and characterization of α-melanocyte-stimulating hormone in the rat brain. *Endocrinology*, **102**, 697–705.

Ordronneau, P. (1979). Pars tuberalis: a possible source of adenohypophyseal hormones in the hypophysectomized rat. *Anatomical Records*, **193**, 641.

Orwoll, E., Kendall, J. W., Lamorena, L., & McGilvra, R. (1979). Adrenocorticotropin and melanocyte-stimulating hormone in the brain. *Endocrinology*, **104**, 1845–1852.

Pelletier, G., & Dubé, D. (1977). Electron microscopic immunohistochemical localization of α-MSH in the rat brain. *American Journal of Anatomy*, **150**, 201–204.

Pelletier, G., Désy, L., Lissitsky, J.-C., Labrie, F., & Li, C. H. (1978). Immunohistochemical localization of β-LPH in the human hypothalamus. *Life Sciences*, **22**, 1799–1804.

Pelletier, G., & Désy, L. (1979). Localization of ACTH in the human hypothalamus. *Cell and Tissue Research*, **196**, 525–530.

Pelletier, G., & Leclerc, R. (1979). Immunohistochemical localization of adrenocorticotropin in the rat brain. *Endocrinology*, **104**, 1426–1433.

Pévet, P., Ebels, I., Swaab, D. F., Mud, M. T., & Arimura, A. (1980). Presence of AVT-, α-MSH-, LHRH- and somatostatin-like compounds in the rat pineal gland and their relationship with the UMO5R pineal fraction. An immunocytochemical study. *Cell and Tissue Research*, **206**, 341–353.

Phifer, R. F., Orth, D. N., & Spicer, S. S. (1974). Specific demonstration of the human hypophyseal adrenocortico–melanotropic (ACTH–MSH) cell. *Journal of Clinical Endocrinology and Metabolism*, **39**, 684–692.

Popa, Gr. T. (1938). Le drainage de l'hypophyse vers l'hypothalamus. *La Presse Médicale*, **34**, 663–666.

Roberts, J. L., & Herbert, E. (1977). Characterization of a common precursor to corticotropin and β-lipotropin: Identification of β-lipotropin peptides and their arrangement relative to corticotropin in the precursor synthesized in a cell-free system. *Proceedings of the National Academy of Sciences*, **74**, 5300–5304.

Romeis, B. (1940). *Handbuch der Mikroskopischen Anatomie des Menschen VI*, Teil 3, Innersekretorische Drüsen II. Hypophyse. Berlin: Springer.

Rossier, J. (1981). Enkephalin biosynthesis. *Trends in Neuroscience*, **4**, 94–97.

Rudman, D., Del Rio, A. E., Hollins, B. M., Houser, D. H., Keeling, M. E., Sutin, J., Scott, J. W., Sears, R. A., & Rosenberg, M. Z. (1973). Melanotropic-lipolytic peptides in various regions of bovine, simian and human brains and in simian and human cerebrospinal fluids. *Endocrinology*, **92**, 372–381.

Rudman, D., Scott, J. W., Del Rio, A. E., Houser, D. H., & Sheen, S. (1974). Melanotropic activity in regions of rodent brain. *American Journal of Physiology*, **226**, 682–686.

Rudman, D., Chawla, R. K., & Hollins, B. M. (1979). N, O-diacetylserine α-melanocyte-stimulating hormone, a naturally occurring melanotropic peptide. *Journal of Biological Chemistry*, **254**, 10102–10108.

Sachs, H., & Takabatake, Y. (1964). Evidence for a precursor in vasopressin biosynthesis. *Endocrinology*, **75**, 943–948.

Sandman, C. A., Denman, P. M., Miller, L. H., Knott, J. R., Schally, A. V., & Kastin, A. J. (1971). Electroencephalographic measures of melanocyte-stimulating hormone activity. *Journal of Comparative and Physiological Psychology*, **76**, 103–109.

Schollberg, K., Seiler, E., & Schmidt, (1978). ACTH levels in late pregnancy. *In* G. Dörner & M. Kawakami (Eds.), *Hormones and Brain Development*, pp. 391–396. Amsterdam: Elsevier.

Schwyzer, R., & Eberle, A. (1977). On the molecular mechanism of α-MSH receptor interactions. *In* F. J. H. Tilders, D. F. Swaab, & Tj. B. van Wimersma Greidanus (Eds.), *Melanocyte Stimulating Hormone: Control, Chemistry and Effects*, pp. 18–25. Basel: Karger.

Scott, A. P., Ratcliffe, J. G., Rees, L. H., Landon, J., Bennett, H. P. J., Lowry, P. J., & McMartin, C. (1973). Pituitary peptide. *Nature: New Biology*, **244**, 65–67.

Scott, A. P., Lowry, P. J., Ratcliffe, J. G., Rees, L. H., & Landon, J. (1974). Corticotrophin-like peptides in the rat pituitary. *Journal of Endocrinology*, **61**, 355–367.

Shapiro, M., Nicholson, W. E., Orth, D. N., Mitchell, W. M., Island, P., & Liddle, G. W. (1972). Preliminary characterization of the pituitary melanocyte stimulating hormones of several vertebrate species. *Endocrinology*, **90**, 249–256.

Silman, R. E., Chard, T., Lowry, P. J., Smith, I., & Young, I. M. (1976). Human foetal pituitary peptides and parturition. *Nature*, **260**, 716–718.

Sofroniew, M. V. (1979). Immunoreactive β-endorphin and ACTH in the same neurons of the hypothalamic arcuate nucleus in the rat. *American Journal of Anatomy*, **154**, 283–289.

Steiner, D. F., Rubenstein, A. H., Peterson, J. D., Kemmler, W., & Tager, H. S. (1973). Proinsulin and polypeptide hormone biosynthesis. *In* R. O. Scott, F. J. G. Ebling, & I. W. Henderson (Eds.), *Endocrinology*, pp. 561–566. New York: Elsevier.

Sternberger, L. A. (1979). The unlabeled antibody method. *Journal of Histochemistry and Cytochemistry*, **27**, 1658–1659.

Swaab, D. F. (1976). Localization of an α-MSH-like compound in the nervous system by immunofluorescence. *International Congress of Endocrinology*, **285**, Hamburg.

Swaab, D. F. (1980) Neuropeptides and brain development—a working hypothesis. *In* C. Di Benedetta, R. Balázas, G. Gombos, & P. Porcellati (Eds.), *A Multidisciplinary Approach to Brain Development*, pp. 181–196. Amsterdam: Elsevier.

Swaab, D. F., & Boer, K. (1980). Technical developments in the study of neuroendocrine mechanisms in rat pregnancy and parturition. *In* P. W. Nathanielsz (Ed.), *Animal Models in Fetal Medicine*, pp. 169–234. Amsterdam: Elsevier.

Swaab, D. F., Boer, G. J., & Visser, M. (1978). The fetal brain and intrauterine growth. *In* D. Barltrop (Ed.), *Paediatrics and Growth*, pp. 63–69. London: Fellowship Postgraduate Medicine.

Swaab, D. F., & Fisser, B. (1978). Immunocytochemical localization of α-melanocyte stimulating hormone (α-MSH)-like compounds in the rat nervous system. *Neuroscience Letters*, 7, 313–317.

Swaab, D. F., Pool, C. W., & van Leeuwen, F. W. (1977). Can specificity ever be proved in immunocytochemical staining? *Journal of Histochemistry and Cytochemistry*, 25, 388–391.

Swaab, D. F., & Visser, M. (1977). A function for α-MSH in fetal development and the presence of an α-MSH-like compound in nervous tissue. In F. J. H. Tilders, D. F. Swaab, & Tj. B. van Wimersma Greidanus (Eds.), *Melanocyte Stimulating Hormone: Control, Chemistry and Effects*, pp. 170–178. Basel: Karger.

Swaab, D. F., Visser, M., & Dogterom, J. (1979). The maturity of the foetal brain and its involvement in labour. *In* L. Zichella, & P. Pancheri (Eds.), *Psychoneuroendocrinology in Reproduction*, pp. 483–495. Amsterdam: Elsevier.

Swaab, D. F., Visser, M., & Tilders, F. J. H. (1976). Stimulation of intrauterine growth in rat by α-MSH. *Journal of Endocrinology*, 70, 445–455.

Thody, A. J. (1974). Plasma and pituitary MSH levels in the rat after lesions of the hypothalamus. *Neuroendocrinology*, 16, 323–331.

Thody, A. J., Plummer, N. A., Burton, J. L., & Hytten, F. E. (1974). Plasma β-melanocyte-stimulating hormone levels in pregnancy. *Journal of Obstetrics and Gynaecology of the British Commonwealth*, 81, 875–877.

Thody, A. J., Penny, R. J., Clark, D., & Taylor, C. (1975). Development of a radioimmunoassay for α-melanocyte-stimulating hormone in the rat. *Journal of Endocrinology*, 67, 385–395.

Tilders, F. J. H. (Ed.). (1975). *Melanofoor Stimulerend Hormoon (MSH) bij Zoogdieren*. Doctoral dissertation, Free University of Amsterdam, 1975.

Tilders, F. J. H., and Smelik, P. G. (1975). A diurnal rhythm in melanocyte-stimulating hormone content of the rat pituitary gland and its independence from the pineal gland. *Neuroendocrinology*, 17, 296–308.

Tilders, F. J. H., & Smelik, P. G. (1978). Effects of hypothalamic lesions and drugs interfering with dopaminergic transmission on pituitary MSH content of rats. *Neuroendocrinology*, 25, 275–290.

Toubeau, G., Desclin, J., Parmentier, M., & Pasteels, J. L. (1979). Compared localizations of prolactin-like and adrenocorticotropin immunoreactivities within the brain of the rat. *Neuroendocrinology*, 29, 374–384.

Tramu, G., Leonardelli, J., & Dubois, M. P. (1977). Immunohistochemical evidence for an ACTH-like substance in hypothalamic LH–RH neurons. *Neuroscience Letters*, 6, 305–309.

Tuimala, R., Kauppila, A., & Haapalahti, J. (1976). ACTH levels in amniotic fluid during pregnancy. *British Journal of Obstetrics and Gynaecology*, 83, 853–856.

Usategui, R., Oliver, C., Vaudry, H., Lombardi, G., Rozenberg, I., & Mourre, A. M. (1976). Immunoreactive α-MSH and ACTH levels in rat plasma and pituitary. *Endocrinology*, 98, 189–196.

van Dijk, A. M. A. (Ed.). (1979). *Determination of Radioimmunoassayable ACTH in Plasma and/or Brain Tissue of Rats as Influenced by Endocrine and Behavioral Manipulations*. Medical Dissertation, University of Utrecht.

van Leeuwen, F. W. (1980). Immunocytochemical specificity for peptides with special reference to oxytocin and arginine-vasopressin. *Journal of Histochemistry and Cytochemistry*, 28, 479–482.

van Leeuwen, F. W. (1981). An introduction to the immunocytochemical localization of neuropeptides and neurotransmitters. *Acta Histochemica*, Suppl. 24, 49–77.

Vaudry, H., Tonon, M. C., Delarue, C., Vaillant, R., & Kraicer, J. (1978). Biological and radioimmunological evidence for melanocyte stimulating hormones (MSH) of extrapituitary origin in the rat brain. *Neuroendocrinology*, **27**, 9–24.

Versteeg, D. H. G., de Kloet, E. R., & de Wied, D. (1979). Effects of α-endorphin, β-endorphin and (Des-Tyr¹)-γ-endorphin on α-MPT-induced catecholamine disappearance in discrete regions of the rat brain. *Brain Research*, **179**, 85–92.

Visser, M., & Swaab, D. F. (1979). Life span changes in the presence of α-melanocyte-stimulating-hormone-containing cells in the human pituitary. *Journal of Developmental Physiology*, **1**, 161–178.

Warberg, J., Oliver, C., Barnea, A., Parker, R., & Porter, J. C. (1979). Release of immunoreactive α-MSH by synaptosome enriched fractions of homogenates of hypothalami. *Brain Research*, **175**, 247–257.

Watson, S. J., & Akil, H. (1979). The presence of two α-MSH positive cell groups in rat hypothalamus. *European Journal of Pharmacology*, **58**, 101–103.

Watson, S. J., & Akil, H. (1980). α-MSH in rat brain: Occurrence within and outside of β-endorphin neurons. *Brain Research*, **182**, 217–223.

Watson, S. J., & Barchas, J. D. (1979). Anatomy of the endogenous opioid peptides and related substances: The enkephalins, β-endorphin, β-lipotropin, and ACTH. *In* R. F. Beers, & E. G. Bassett (Eds.), *Mechanisms of Pain and Analgesic Compounds*, pp. 227–237. New York: Raven.

Weber, E., Voigt, K. H., & Martin, R. (1978). Pituitary somatotrophs contain Met-enkephalin-like immunoreactivity. *Proceedings of the National Academy of Sciences*, **75**, 6134–6138.

Wilkes, M. M., Watkins, W. B., Stewart, R. D., & Yen, S. S. C. (1980). Localization and quantitation of β-endorphin in human brain and pituitary. *Neuroendocrinology*, **30**, 113–121.

Yalow, R. S., & Berson, S. A. (1971). Size heterogeneity of immunoreactive human ACTH in plasma and in extracts of pituitary gland and ACTH-producing thymoma. *Biochemical and Biophysical Research Communications*, **44**, 439–445.

Zondek, B. (Ed.). (1935). *Hormone des Ovariums und des Hypophysenvorderlappens*. Wien: Springer.

Zondek, B., & Krohn, H. (1932). Hormon des Zwischenloppens der Hypophyse (Intermedin). II. Intermedin im Organismus (Hypophyse, Gehirn). *Klinische Wochenschrift*, **11**, 849–853.

CHAPTER 2

Mechanisms of Action of Behaviorally Active ACTH-like Peptides

Albert Witter, Willem H. Gispen, and David de Wied

The behavioral activities of the adrenocorticotropic hormone (ACTH) neuropeptides are complex. They involve processes that serve the adaptive capacities of the organism, such as learning and memory, motivation, arousal, and attention. The essential structural elements are located in the sequence $ACTH_{4-7}$; the phenylalanine in Position 7 plays a key role. The *in vivo* generation of such ACTH neuropeptides, fragments of $ACTH_{1-39}$ that are behaviorally active, but virtually lack peripheral, endocrine activities, can be accomplished through various pathways. Production in the pituitary and subsequent release into the brain by retrograde blood flow from the pituitary to the median eminence, or selective and specific generation in the central nervous system (CNS) are the most likely mechanisms, although, ultimately, these need to be demonstrated.

The neurochemical mechanism(s) underlying the observed behavioral activities appear to be equally complex. An apparent absence of receptors for ACTH neuropeptides could be due to receptor properties that obscure "classical" detection. These properties might be related to very high affinity–very low capacity characteristics or to different physicochemical qualities. Moreover, multiple putative receptors appear to be indicated, functioning via different mechanisms: neurotransmission, neurohormonal, and neuromodulation. Evidence is presented that neuromodulation could be affected by phosphorylation of synaptic membrane constituents (proteins and lipids). Structure–activity studies point to a close correlation between behavioral effects of ACTH and

37

ENDOGENOUS PEPTIDES
AND LEARNING AND MEMORY PROCESSES

congeners and their inhibition of a membrane protein kinase. It is suggested that the modulation is brought about by changes in calcium gating through the membrane. Such changes are thought to be crucial to the efficacy of neurotransmission.

I. INTRODUCTION

The findings of impaired learning behavior in rats after removal of the pituitary gland led to the hypothesis that hypophyseal principles are involved in brain function. This, in turn, led to experiments showing that impaired avoidance behavior as a result of hypophysectomy could be corrected by substitution with ACTH, α-melanocyte-stimulating hormone (α-MSH), or vasopressin (de Wied, 1969). It was found that the behavioral effects of these pituitary hormones resided in certain fragments of the parent hormones devoid of the classical peripheral endocrine effects (de Wied, 1969; see Table 1). These findings suggested that the pituitary manufactures peptide hormones that may function as precursor molecules for peptides involved in various brain functions (de Wied, 1969, 1974). In view of their chemical nature and central effects, such peptides were designated as "neuropeptides" (de Wied, van Wimersma Greidanus, & Bohus, 1974). Remarkably, while the view was emerging that retrograde blood flow from the pituitary to the brain was the most likely route of transport of hormones and/or neuropeptides involved in brain function (Mezey, Kivovics, & Palkovits, 1979; Oliver, Mical, & Porter, 1977), several pituitary hormones, such as ACTH, α-MSH, and β-lipotropic hormone (β-LPH) were discovered in the brain (Krieger, Liotta, & Brownstein, 1977; Orwoll, Kendall, Lamorena, & McGilvra, 1979; Rossier, Vargo, Minick, Ling, Bloom, & Guillemin, 1977). It has been demonstrated that ACTH, β-LPH, and other peptides are derived from the same large precursor molecule (Loh, 1979; Mains, Eipper, & Ling, 1977). Immunohistochemical studies revealed the existence of a widespread and diffuse neuronal system containing the large prohormone pro-opiomelanocortin (Watson & Akil, 1980). This prohormone is probably the parent compound for the hormones ACTH, β-LPH, and related peptides, but these hormones in themselves are "second order" prohormones for neuroactive fragments such as α-MSH and CLIP ($ACTH_{17-39}$), which are formed from ACTH (Scott, Bloomfield, Lowry, Gilke, London, & Rees, 1976); β-endorphin, which is derived from β-LPH (Bradbury, Smyth, & Snell, 1976; Graf, Szekely, Ronai, Dunai-Kovacs, & Bajusz, 1976); and smaller fragments of these respective peptides. It has been suggested that specific enzyme systems present in pituitary and brain control the formation of bioactive peptides from these precursor molecules (Burbach & de Wied, 1980). Environmental stimulation may activate these enzymes to release neuropeptides from pituitary and brain cells, which modulate brain functions involved in learning and memory, motivation, arousal, and attention processes that serve the adaptive capacities

TABLE 1

Amino Acid Sequences of Adrenocorticotropic Hormone and the $ACTH_{4-9}$ Analog, Organon 2766

	1		4		7		10												20		

H-Ser-Tyr-Ser-Met-Glu-His-Phe-Arg-Trp-Gly-Lys-Pro-Val-Gly-Lys-Lys-Arg-Arg-Pro-Val-

25   30   39

Lys-Val-Tyr-Pro-Asn-Gly-Ala-Glu-Asp-Glu-Ser-Ala-Glu-Ala-Phe-Pro-Leu-Glu-Phe-OH   Human $ACTH_{1-39}$

H-Met(O)-Glu-His-Phe-D-Lys-Phe-OH   ORG 2766

of the organism. In view of this complexity, it is not surprising that, as yet, no neurochemical mechanism of action of ACTH and congeners has been found. As has been discussed elsewhere (Gispen, 1980), three models of action have been proposed: namely, ACTH as a neurohormone, as a putative neurotransmitter, and as a neuromodulator. The experimental evidence discussed in this chapter supports the role of ACTH as a neuromodulator.

II. EFFECTS ON CONDITIONED BEHAVIOR OR BEHAVIORAL EFFECTS

Murphy and Miller (1955) first showed that injection of ACTH during shuttlebox training delays subsequent extinction of the avoidance response. A more pronounced effect was found, however, when ACTH was given during the extinction period (de Wied, 1967). This behavioral influence occurs independently of the action of ACTH on the adrenal cortex, for ACTH is also active on extinction of shuttlebox avoidance of adrenalectomized rats (Miller & Ogawa, 1962). Moreover, α-MSH, β-MSH, $ACTH_{1-10}$, and $ACTH_{4-10}$ are as active as $ACTH_{1-24}$ in delaying extinction of pole-jumping avoidance behavior (de Wied, 1966). In addition, ACTH and related peptides improve maze performance (Kastin, Sandman, Stratton, Schally, & Miller, 1975), facilitate passive avoidance behavior (de Wied, 1974; Flood, Jarvik, Bennett, & Orme, 1976; Kastin, Miller, Nockton, Sandman, Schally, & Stratton, 1973; Levine & Jones, 1965; Lissak & Bohus, 1972), delay extinction of food-motivated behavior in hungry rats (Flood et al., 1976; Garrud, Gray, & de Wied, 1974; Gray, 1971; Leonard, 1969; Sandman, Kastin, & Schally, 1969), and delay extinction of conditioned taste aversion (Rigter & Popping, 1976) and sexually motivated behavior (Bohus, Hendrickx, van Kolfschoten, & Krediet, 1975).

The behavioral effects of peptides related to ACTH appear to be of short-term nature. A single injection of $ACTH_{4-10}$ delays extinction of a pole-jumping avoidance response or facilitates passive avoidance retention in intact rats for only a few hours (de Wied, 1974). Similarly, cessation of the administration of $ACTH_{4-10}$ in hypophysectomized rats that normalized the level of performance in these animals, led to a progressive deterioration of avoidance behavior despite shock punishment (Bohus, Gispen, & de Wied, 1973). Electrophysiological findings suggest that $ACTH_{4-10}$ may affect the state of arousal in limbic midbrain structures. $ACTH_{4-10}$ induces a shift in the dominant frequency of hippocampal and posterior thalamic rhythmic slow activity to higher frequencies during paradoxical sleep (de Wied, Bohus, van Ree, & Urban, 1978) or following electrical stimulation of the reticular formation (Urban & de Wied, 1976). Similar shifts are found when the strength of the stimulus is increased. These studies suggest an increase in the state of arousal of the limbic brain by

$ACTH_{4-10}$. Measured at a wide variety of light intensities, the amplitudes of visually evoked responses from cortical Area 17 in rats were significantly diminished following treatment with $ACTH_{4-10}$ or [D-Phe^7] $ACTH_{4-10}$ (Wolthuis & de Wied, 1976). This was interpreted as an effect of $ACTH_{4-10}$ on the CNS vigilance-regulating system. These observations led to the hypothesis that ACTH and related peptides, by temporarily increasing the state of arousal in the limbic brain, enhance the motivational influence of environmental stimuli. This may cause an increase in the probability of generating stimulus-specific responses (de Wied, 1974).

Kastin et al., 1973 and Kastin et al., 1975 proposed actions on learning and attention as the main behavioral effects of ACTH–MSH peptides. This proposal was derived from experiments using a two-choice discrimination problem in which animals avoided shock by running to a white door. After successful learning, the discriminative stimuli were reversed and running to the black door was the correct response (Sandman, Alexander, & Kastin, 1973; Sandman, Beckwith, Gittes, & Kastin, 1974; Sandman, Miller, Kastin, & Schally, 1972). Acquisition of the original response is considered a measure of learning a new response, whereas the reversal stage measured the level of attention. $ACTH_{4-10}$ and $ACTH_{1-24}$ enhanced learning of the original problem but α- or β-MSH were inactive. However, reversal learning was enhanced by α- or β-MSH. These authors suggested that ACTH improves learning and that α- and β-MSH enhance attention (Sandman & Kastin, 1980).

ACTH and related peptides affect memory processes as well. They attenuate carbon dioxide-induced amnesia, or amnesia produced by electroconvulsive shock, in a passive avoidance response task, if they are administered prior to the retention test, but not when given prior to acquisition or to the induction of amnesia following training (Rigter, van Riezen, & de Wied, 1974; Rigter & van Riezen, 1975). Furthermore, they alleviate amnesia produced by intracerebral administration of the protein synthesis inhibitors puromycin (Flexner & Flexner, 1971) or anisomycin (Flood et al., 1976). Rigter et al. (1974) interpreted the effect of $ACTH_{4-10}$ on amnesia as an influence on retrieval performance. Gold and van Buskirk (1976) found that posttrial administration of ACTH can enhance or impair later retention depending upon the dose of the peptide. These authors suggested that ACTH modulates memory storage processing of recent information. Isaacson, Dunn, Rees, and Waldock (1976) demonstrated that $ACTH_{4-10}$ improves the use of information provided on the location of reward in a four-table choice situation. These findings are obviously not in conflict with a motivational or an attentional hypothesis, because motivation and attention are involved in the paradigms used in these studies.

Structure–activity studies to determine the essential elements required for the behavioral effect of ACTH revealed that not more than four amino acid residues are needed. Thus, $ACTH_{4-7}$ is as effective as the whole ACTH molecule in delaying extinction of pole-jumping avoidance behavior (de Wied, Witter, & Greven, 1975; Greven & de Wied, 1973). The amino acid residue phenylalanine in Position 7 plays a key role in this behavioral effect of ACTH. Replacement of this amino acid by the D-enantiomer in $ACTH_{1-10}$ (Bohus & de Wied, 1966), $ACTH_{4-10}$, or $ACTH_{4-7}$ (Greven & de Wied, 1973) causes an effect on extinction of avoidance behavior opposite to that found with the corresponding [L-Phe] ACTH fragments. The [D-Phe^7] ACTH analogs facilitate extinction of active avoidance behavior and approach behavior motivated by food (Garrud et al., 1974). However, [D-Phe^7] $ACTH_{4-10}$, like $ACTH_{4-10}$, facilitates passive avoidance behavior when given prior to the retention test (de Wied, 1974), although it has been shown to attenuate passive avoidance behavior when administered in a relatively high dose immediately following the learning trial (Flood et al., 1976). Also, extinction of conditioned taste aversion is affected by [D-Phe^7] $ACTH_{4-10}$ in the same manner as by $ACTH_{4-10}$ (Rigter & Popping, 1976).

The behavioral activity of ACTH fragments can be completely dissociated from inherent endocrine, metabolic, and opiate-like activities by modification of the molecule. Substitution of -Met^4- by methionine sulfoxide, -Arg^8- by -D-Lys- and -Trp^9- by -Phe- yields a peptide (ORG 2766; see Table 1) which is 1000 times more active than $ACTH_{4-10}$ on extinction of pole-jumping avoidance behavior (Greven & de Wied, 1973). It possesses, however, 1000 times less MSH activity, and its steroidogenic action is markedly reduced. It has no fat-mobilizing activity nor opiate-like effects in the guinea pig ileum preparation. By extending the more active $ACTH_{4-9}$ analog (ORG 2766), with the sequence $ACTH_{10-16}$, and substituting the amino acid residue lysine by its D-enantiomer, a further potentiation of the effect on pole-jumping avoidance behavior is obtained. This peptide is 300,000 times more active than $ACTH_{4-10}$ (Greven & de Wied, 1977).

III. *IN VIVO* FATE

The *in vivo* fate of ACTH has recently been reviewed (Bennett & McMartin, 1979). It appears that ACTH is relatively stable in fresh blood or plasma samples *in vitro*, whereas the half-life of ACTH *in vivo* is short. Evidently, degradation in blood does not substantially contribute to catabolism, and other mechanisms of elimination must be considered.

The disappearance of ACTH from plasma occurs in a biphasic manner (Bennett & McMartin, 1979; Liotta, Li, Schussler, & Krieger, 1978; Normand & Lalonde, 1979). The initial phase proceeds with a half-life on the order of a few minutes and is followed by a slow decay phase with a half-life on the order of 1 hr. Bennett, McMartin, and their associates have provided evidence that the initial clearance is due to uptake of intact peptide and a small, variable amount of -Met^4- sulphoxide in most major tissues. In muscle and skin, extensive degradation by aminopeptidases occurs subsequently, mainly at the H-Ser^1 ↓ Tyr^2 ↓ Ser^3-peptide bonds, because ACTH$_{3-39}$ is the only initial major metabolite in plasma (Hudson, Ambler, Bennett, & McMartin, 1979). However, a portion of the ACTH taken up by muscle and skin survives proteolysis and is stored at sites where it resists further attack. The stored peptide can then return to circulation, and this could be responsible for the second phase of slow decline of plasma ACTH levels. Considerable uptake also takes place in the kidney and in the liver. In the kidney, extensive degradation to smaller fragments or amino acids occurs, but probably only after the peptide and larger fragments have been transferred to lysosomes, where proteolytic fragmentation takes place prior to release. The handling of ACTH by the kidney thus differs essentially from that by muscle and skin.

The initial NH$_2$-terminal cleavage is decreased in peptides with a D-Ser^1-residue. This increases the amount of peptide that survives the initial proteolysis and that subsequently can return to circulation, thus enhancing potency and duration of action of such analogs. A conformational protective effect is also apparent for the ACTH$_{25-39}$ sequence. The circulating fragments after administration of ACTH$_{1-24}$ include, beside ACTH$_{3-24}$, a large range of metabolites resulting from a variety of cleavages in the COOH-terminal region like the -Lys^{15} ↓ Lys^{16} ↓ Arg^{17} ↓ Arg^{18} ↓ Pro^{19} ↓ Val^{20} ↓ Lys^{21} ↓ Val^{22}-bonds (Hudson et al., 1979; Hudson & McMartin, 1980). Similar metabolites were not detected with ACTH$_{1-39}$. Proteolysis is not restricted to the NH$_2$- and COOH-terminal regions. In particular, the peptide bonds -His^6 ↓ Phe^7 ↓ Arg^8 ↓ Trp^9- are susceptible to proteolytic attack, thereby increasing the number of possible circulating fragments. There is evidence that these circulating fragments are generated in muscle and skin. Like ACTH$_{1-39}$, the tetracosapeptide ACTH$_{1-24}$ is distributed to tissue beds and, within 1 min, 70% of the dose is cleared from the circulation by such mechanisms. Liver and kidney contained mainly intact peptide and the initial uptake in these tissues only represents part of the cleared ACTH$_{1-24}$. The greater part was present in muscle, skin, and intestine. Particularly in muscle and skin, extensive degradation occurs extremely rapidly and the fragments present in these tissues are similar to those appearing in the circulation 2 min after intravenous administration. This provides good evidence that these tissues (in particular, muscle) are the sites of generation of circulating fragments in the initial phase following injection of ACTH$_{1-24}$ (Hudson & McMartin, 1980).

The *in vitro* half-life of some synthetic ACTH$_{4-9}$ analogs, as determined from the rate of biotransformation by rat plasma and by the supernatant of a rat brain homogenate, was found to correlate with their behavioral potencies (Witter, Greven, & de Wied, 1975). This indicates that increased behavioral activity can be explained, at least partially, by increased resistance against biotransformation. The structural modifications of the analogs used included substitution of -Arg^8- by -D-Lys^8-, alone or in combination with -Met^4-sulphoxide and -Phe^9-substitutions. The *in vivo* disappearance of the threefold substituted ACTH$_{4-9}$ analog (ORG 2766) from plasma also occurs biphasically, with an initial half-life of 1–3 min followed by a slower phase with a half-life of about 20 min (Verhoef & Witter, 1976).

IV. GENERATION OF NEUROPEPTIDES

Proteolysis of polypeptides serves a twofold function. It is a process of elimination that, by degrading the biologically active peptide moieties, terminates their action. But it is also a process of generation that, by restricted proteolysis of precursor molecules, generates biologically active (oligo) peptides and thus initiates their action. Proteolytic processing allows a fast, efficient physiological regulation of initiation and termination of the responding of an effector system to external and internal stimuli. Large precursor molecules endow the organism with an increased flexibility in that they offer increased possibilities for the generation of a variety of changing amounts of neuropeptides with different activity profiles and/or (partial) synergistic or antagonistic properties. Important features of such a proteolytic on–off response system include the availability of a precursor–neuropeptide at the site of action, either present locally or transported from distant sites, and the location and triggering of enzyme systems responsible for the generation and degradation of neuropeptides.

The production of potential neuropeptide–precursor molecules (like ACTH, α-MSH, β-MSH, and possibly γ-MSH) takes place, at least in the pituitary, from a 30 kdalton proprecursor molecule (pro-opiomelanocortin). This topic is reviewed in Chapter 1.

The presence of ACTH-like peptides in low concentrations in the brain (see Chapter 1) raises the question as to their origin. The aforementioned pituitary–brain transport routes could possibly explain this low brain-pool of ACTH-like peptides, indicating a pituitary origin of this pool. However, there is increasing evidence for an ACTH-like pool in the CNS that is independent of the pituitary (see Chapter 1). This pool could act as a source for ACTH neuropeptides, as the brain appears to be highly active in proteolytic processing of ACTH neuropeptides. Soluble mouse brain preparations preferentially liberate amino acids from the NH$_2$-terminal region of ACTH$_{1-24}$. This initial release is followed by a relatively uniform liberation of amino acids originating throughout the entire sequence in a later stage (Reith, Neidle, & Lajtha, 1979). Although the transient formation of larger intermediate fragments cannot be ruled out, the observed pattern of breakdown makes their accumulation, at least *in vitro*, unlikely. This indicates that, *in vivo* selective, rather than general, mechanisms for the generation of neuropeptides should be operative. Only selective mechanisms, possibly closely connected with ACTH biosynthetic, storage, or release sites, seem physiologically significant, because such mechanisms could prevent complete proteolysis. The presence and the triggering of such selective enzyme systems have been indicated for ACTH neuropeptides. The generation of $ACTH_{1-16}$, which contains the information for all hitherto known CNS actions of ACTH, together with $ACTH_{17-39}$ (CLIP) from $ACTH_{1-39}$, has been demonstrated to occur in the intermediate lobe of the pituitary (Scott et al., 1976). A similar processing of pro-opiomelanocortin in the brain seems probable (Watson & Akil, 1980). Similar evidence exists for other neuropeptide precursor molecules like oxytocin and β-endorphin (cf. de Kloet & de Wied, 1980).

Also, ACTH neuropeptides are subject to proteolytic degradation, illustrating that neuropeptide action cannot only be initiated but can also be terminated by proteolytic processes. Again, free amino acids are released in considerable quantities, possibly by the combined action of endopeptidase(s), with a preference for the $-Phe^7 \downarrow Arg^8-$ peptide bond, and exopeptidases (Marks, Stern, & Kastin, 1976). $ACTH_{4-9}$ neuropeptide analogs with increased behavioral potency, possessing $-Met^4(0)-$ and $-D-Lys^8-$ substitutions, are less susceptible to proteolytic attack. Initial degradation of the analogs might proceed through cleavage of the $-His^6 \downarrow Phe^7-$ bond, followed by further breakdown of the released tripeptide $H-Phe^7 \downarrow D-Lys^8-Phe-OH$ at the position indicated (Marks et al., 1976; Verhoef & Witter, 1976).

In summary, the generation of neuropeptides can be accomplished by various pathways, but the ultimate demonstration of enzymatic mechanisms operating physiologically still needs considerable investigation.

V. RECEPTOR SITES

The location of the site(s) of action of ACTH-like peptides in the CNS has been derived mainly from studies involving the destruction of specific brain regions, the implantation of neuropeptides directly into the brain, and specific uptake studies of intraventricularly administered neuropeptides in defined brain regions. The results of these studies indicate the involvement of multiple sites of action, including posterior thalamic structures as well as the septal–hippocampal complex, in the mediation of behavioral effects of ACTH neuropeptides (van Wimersma Greidanus, Bohus, & de Wied, 1975; Verhoef, Witter, & de Wied, 1977). Furthermore, the results suggest that intact limbic–midbrain circuitry is essential in eliciting behavioral responses; this is substantiated by the presence of an ACTH-containing neuronal system, originating from cell bodies in the arcuate nucleus with nerve terminals in limbic midbrain structures (Watson, Richard, & Barchas, 1978).

Attempts to demonstrate the presence of receptor sites in these brain areas have so far been unsuccessful for $ACTH_{4-9}$, but preliminary work appears to suggest the existence of a high affinity recognition site for $ACTH_{1-39}$ in brain (Akil & Watson, 1980). Although there are a number of possibilities to explain the apparent absence of receptors in brain tissue for smaller ACTH neuropeptides, it has been hypothesized that this apparent absence might have physiological significance (Witter, 1980). The very high potency of ACTH neuropeptides in eliciting behavioral responses could implicate the presence of CNS receptor sites, characterized by very high affinity and very low capacity. Such receptor sites would represent an effector system that could effectively respond at very low CNS neuropeptide concentrations. These receptor properties could also explain the difficulties encountered in demonstrating such receptor sites by direct binding approaches. Similar situations might exist for other neuropeptides, like vasotocin (Pavel, 1979; Pavel, Cristoveanu, Goldstein, & Calb, 1977) and the C-terminal tripeptide of oxytocin, H-Pro-Leu-Gly-NH$_2$ (Chiu, Wong, & Mishra, 1980). Inherent in such a model would be a (very) slow dissociation of the neuropeptide from its putative receptor sites, in accordance with the relatively long-term nature of ACTH neuropeptide effects on extinction of conditioned avoidance behavior (de Wied, 1977).

This model might also have implications for the mechanism of action of ACTH neuropeptides, favoring interactions with the low-capacity sites. Such low-capacity systems could be connected with receptor populations of limited occurrence, for example, locations restricted to nerve terminals indicative of a presynaptic mode of action, rather than presumably more abundant postsynaptic sites of action. A further extension of this hypothesis might be to assign a presynaptic site to a neuromodulating function, rather than a postsynaptic site to a neurotropic function or an action as a neurotransmitter.

Based on the interaction of $ACTH_{1-24}$ at 10^{-9} M or less with planar lipid bilayer membranes, an interesting hypothesis has been put forward (Schwyzer, 1980). According to this hypothesis, the message-carrying part of the $ACTH_{1-24}$ molecule penetrates the membrane, whereas the accumulated positively charged address region ($ACTH_{15-18}$) remains outside. This might mean that ACTH reacts at low concentrations with receptors located in the interior of the membrane or even in the cytoplasm. The properties of such hypothetical receptors could differ essentially from "classical" receptors, usually assumed to be located on the membrane exterior. Differences in physiochemical properties, such as solubility, might have obscured binding data using "classical" approaches, and adaptation of binding methodology to the view presented by Schwyzer could be the key in obtaining positive data.

VI. ACTH AS A NEUROMODULATOR

Assuming that ACTH influences nerve cell metabolism by an interaction with a target cell receptor, the question arises as to what neurochemical events underlie the behavioral activity of the peptide. As pointed out in the preceding section, the literature available to date leaves open the possibility that ACTH can affect brain cell receptors in at least three different ways: as a neurotransmitter, as a neurohormone, or as a neuromodulator. In this chapter, we shall limit ourselves to a neuromodulatory role for ACTH and consider the possible molecular events underlying such action.

Neuromodulation can take place by an altered transmitter release from presynaptic terminals or by an altered responsiveness of postsynaptic elements to incoming information.

Evidence is accumulating that phosphorylation of synaptic constituents (proteins and lipids) may affect the transmission of information between neurons and, thus, could be an ideal target for putative neuromodulators (Greengard, 1979; Michell, 1979). The pioneering studies of Heald (1962) and Rodnight (1979) implied that depolarization of nerve cell membranes was accompanied by changes in membrane protein phosphorylation. In a more refined approach, Browning, Dunwiddie, Bennett, Gispen, and Lynch (1979) and Bär, Schotman, Gispen, Tielen, and Lopes da Silva (1980) demonstrated that synaptic potentiation of specific hippocampal pathways led to an altered calcium-dependent phosphorylation of specific protein bands associated with the synaptic plasma membrane (SPM) fraction. Furthermore, several authors reported calcium-dependent phosphorylation in relation to exocytosis of neurotransmitters. As rapid turnover of membrane phosphoinositides (PI) is often seen in a variety of tissues after effector–receptor coupling, this poly-PI response is also thought to occur after transmitter–synaptic receptor coupling (Michell, 1979). In this context, we shall outline our studies on the effect of ACTH on SPM protein and lipid phosphorylation in more detail.

A. Effects on Phosphoproteins

In a series of experiments, we studied the effect of ACTH on SPM protein phosphorylation *in vitro* (Zwiers, Schotman, & Gispen, 1980; Zwiers, Tonnaer, Wiegant, Schotman, & Gispen, 1979; Zwiers, Veldhuis, Schotman, & Gispen, 1976; Zwiers, Wiegant, Schotman, & Gispen, 1978). Incubation of rat brain SPM in the presence of $ACTH_{1-24}$ resulted in a decrease in phosphorylation of at least five protein bands after polyacrylamide slab gel electrophoresis, as visualized by protein staining and by autoradiography. One of these proteins (molecular weight (MW) of B-50 is 48 kdalton) was especially sensitive to

ACTH both *in vitro* and after introcerebroventricular (i.c.v.) treatment *in vivo*. Evidence was obtained suggesting that in SPM, ACTH interacts with protein kinase rather than a phosphatase. The ACTH-sensitive protein kinase and its substrate protein B-50 could be isolated from SPM as an enzyme–substrate complex. Further purification was obtained by diethyl-diaminoethyl (DEAE) cellulose column chromatography, ammonium sulphate purification (ASP) and two-dimensional separation combining isoelectric focusing and sodium dodecyl sulphate (SDS) gel electrophoresis. The isolated protein kinase had an apparent MW of 71 kdalton and an isoelectric point (IEP) of 5.5. The B-50 substrate protein had a MW of 48 kdalton and an IEP of 4.5. The enzyme was not sensitive to cAMP but the presence of calcium was essential for B-50 protein phosphorylation. The IC_{50} for $ACTH_{1-24}$ was 5×10^{-6} M.

So far, the only other purified and characterized brain membrane-bound protein and its protein kinase have been described by Ueda and Greengard (1977) and by Uno, Ueda, and Greengard (1977). Their protein kinase is sensitive to cAMP and consists of two subunits with apparent MW (in SDS) of 40 and 52 kdalton, respectively, whereas the IEP of the native protein is 5.5. Likewise, the substrate protein I is different from the B-50 protein, as protein I has an IEP of 10.3 and a MW in SDS of 86 kdalton. Thus, the two membrane-associated phosphorylation systems are clearly different. Functionally, the major difference between the two systems may well be their differential sensitivity to cAMP and calcium.

Antibodies were raised against the protein B-50 using a B-50 preparation partially purified by one-dimensional separation on SDS-polyacrylamide slab gels (Oestreicher, Zwiers, Schotman, & Gispen, 1979, 1981). The presence of specific antibodies to B-50 in rabbit serum was demonstrated by an immunoperoxidase staining method. By use of this peroxidase–antiperoxidase (PAP) method, the immunohistochemical localization of B-50 was studied in sections of rat brain cerebellum and hippocampus. In agreement with the presumed synaptic origin of B-50, the antiserum reacted with tissue components in both brain regions rich in synaptic content. In contrast, white matter and cell perikarya did not contain specific immunostaining. The staining pattern was similar to that found by others using synaptic antigens (Matus, Jones, & Mughal, 1976) and suggests that, at the cellular level, there is a restricted localization of the B-50 protein in the synaptic region but that, at the brain regional level, the protein seems ubiquitous.

Structure–activity studies indicated that the interaction of ACTH with endogenous phosphorylation is rather complex. It appeared that the capability of $ACTH_{1-24}$ to inhibit phosphorylation of B-50 is confined to the NH_2-terminal part of the molecule. The shortest active sequence with the NH_2-terminus intact is $ACTH_{1-13}$. The sequence $ACTH_{5-18}$ seems as active as $ACTH_{1-16}$, and it was therefore concluded that the active site is in the region $ACTH_{5-13}$.

Possibly, C-terminal elongation of this sequence is necessary for expression of the activity since $ACTH_{5-16}$ was inactive, whereas $ACTH_{5-18}$ was active.

With respect to the effect of ACTH on avoidance behavior, there seems to be one active site in $ACTH_{4-7}$ and a second active site within the sequence $ACTH_{7-24}$ (Greven & de Wied, 1977). Also, the requirements necessary to displace dihydromorphine from its binding site in rat brain SPM and the counteraction of morphine *in vivo* suggest a site in the sequence $ACTH_{4-10}$ along with that extraactive site (Gispen, Buitelaar, Wiegant, Terenius, & de Wied, 1976).

The presently known ACTH structure–central nervous system (CNS) activity relationships are not completely identical, but the general principle of dormant activity and induction of such activity by chain elongation seems to apply. Apparently, information is encoded in a multiple form, making comparison among peptides on the basis of primary structures alone, hazardous (de Wied & Gispen, 1977).

Interestingly, the effect of ACTH fragments on the phosphorylation of B-50 is very similar to that found for the induction of excessive grooming (Gispen, Wiegant, Greven, & de Wied, 1975) (see Fig. 1). The ACTH sequences 1–24, 1–16, 1–13, 5–18, and, to some extent, 5–16 induce the display of excessive grooming in the rat after intraventricular administration, whereas 1–10, 4–10, 11–24, 7–16 and the combination of 1–10 plus 11–24 are ineffective. This indicates that phosphorylation is related to excessive grooming rather than to learning and memory.

B. Effects on Membrane Phospholipid Phosphorylation

Although evidence has been presented to suggest a regulatory role of synaptic phosphoprotein in synaptic transmission (see Section VI-A), similar ideas were formulated with respect to a special class of membrane phospholipids, the (poly)phosphoinositides (Michell, 1979). The rapid metabolism of these phospholipids in various membranes, especially in brain membranes, led to a search for their roles in membrane function, because poly-PI is present in relatively high amounts. Based on a variety of studies, Michell (1975, 1979) concludes that there is a correlation between membrane poly-PI metabolism and calcium influx into the cell. The binding of a variety of agonists (hormones, neurotransmitters, etc.) to their respective receptors might initiate the hydrolysis of membrane poly-PI, followed by the influx of calcium, which in turn is thought to activate the target cell in its role as second messenger (see Fig. 2).

To investigate whether or not the modulatory interaction of ACTH with synaptic membranes could involve such a mechanism, Jolles, Wirtz, Schotman, and Gispen (1979) incubated synaptosomal fractions of rat brain (previously

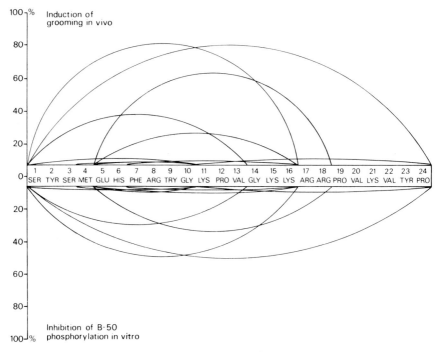

Fig. 1. Similarity in structural requirements of ACTH fragments (top) for induction of excessive grooming and (bottom) for inhibition of B-50 protein phosphorylation. Example: i.c.v. injection of $ACTH_{1-16}$ (2 μg/3 μl) induces grooming behavior in the rat to a level of 80% of the maximal grooming score. *In vitro*, $ACTH_{1-16}$ (10^{-4} M) inhibits B-50 protein phosphorylation in SPM of rat brain by 50%.

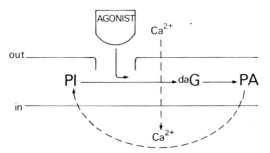

Fig. 2. The Michell hypothesis. Agonist–receptor interaction triggers Phosphatidyl-myo-inositol hydrolysis to diacylglycerol (daG) thereby opening Ca^{2+} gates. The daG is quickly phosphorylated to phosphatidic acid; phosphatidyl-myo-inositol is resynthesized (dashed line) in the endoplasmatic reticulum.

labeled with inorganic orthophosphate) with $ACTH_{1-24}$, and measured the amount of label recovered in MPI, DPI, TPI, and phosphatidic acid (PA). In a time- and dose-dependent manner, the peptide decreased the amount of label in the polyphosphoinositides and PA. Most sensitive to the addition of the peptide were DPI and TPI. Recently, it was shown in a lysed synaptosomal fraction that calcium-dependent mechanisms are involved in the effect of the peptide (Jolles, Zwiers, van Dongen, Schotman, Wirtz, & Gispen, 1980b), in that ACTH was only active in the absence of calcium.

In an effort to relate these peptide-induced changes in poly-PI metabolism to the previously reported effect of the peptide on membrane phosphoproteins, exogenous DPI and $[\gamma\text{-}^{32}P]ATP$ were added to a partially purified membrane fraction containing the B-50 protein kinase/B-50 substrate protein complex (ASP_{55-80}) (Jolles, Zwiers, Schotman, & Gispen, 1980a; Jolles, Zwiers, Dekker, Wirtz, & Gispen, 1981b). The only labeled entities appeared to be the phosphoprotein B-50 and TPI. Apparently, this fraction, in addition to protein kinase activity, contains DPI kinase activity. An inverse relationship was observed between the degree of phosphorylation of B-50 and the amount of TPI produced. These and other findings strongly suggest that B-50 protein phosphorylation and membrane poly-PI metabolism are interrelated (Jolles et al., 1980a). Such a notion was underscored by data on the simultaneous effect of ACTH on B-50 protein kinase and DPI kinase activity in the same ASP_{55-80} fraction. Interestingly, a dose-dependent inhibition of B-50 phosphorylation was observed, again accompanied by a stimulation of TPI production (Jolles et al., 1980a; Jolles et al., 1981b).

DPI and TPI are very potent chelators of calcium and magnesium ions (Hawthorne & Kai, 1970; Michell, 1975); however, they interact strongly with proteins, and, in fact, it has been proposed that the poly-PIs may carry the negative potential of the membrane (Torda, 1974). Thus, a change in the relative amounts of MPI–DPI–TPI in the synaptic membrane, brought about by an ACTH-induced inhibition of B-50 protein phosphorylation, may affect the conformation of membrane proteins lining the presumed ion channels and change the amount of calcium and magnesium bound to the membrane (see Fig. 1). Needless to say, such a hypothetical model needs further experimental support. However, the striking similarity in structure activity of ACTH in the induction of excessive grooming *in vivo*, and the inhibition of B-50 protein kinase activity *in vitro* suggest to us an important role for membrane phosphorylation in the central mechanism of action of ACTH.

C. Mechanism of Action

This chapter has reviewed literature indicating a neuromodulatory mechanism of action of ACTH. The neurochemical data on the interaction of ACTH with synaptic membranes make it entirely possible that ACTH may regulate the transmission process in certain types of synapses.

The experimental evidence that has been provided in the preceding part supports the notion that neuropeptides such as ACTH modulate nerve cell function by virtue of their influence on poly-PI metabolism. A change in protein phosphorylation may mediate this influence. The experimental support comes from

1. the similar structure–activity relationship on the *in vivo* and *in vitro* parameters;

2. the effect of ACTH administered *in vivo* on phosphorylation of B-50 *in vitro*;

3. the causal relation between B-50 protein phosphorylation and poly-PI metabolism;

4. the co-purification of the ACTH-sensitive B-50–B-50 kinase and the lipid kinases; and

5. the regulation of the dynamic properties of the membrane by Ca^{2+} and anionic lipids.

The following working hypothesis for the action of ACTH-like neuropeptides has been proposed. First, the peptide (e.g., $ACTH_{1-24}$) interacts with a mixed protein-lipid receptor (Jolles, Aloyo, & Gispen, 1981a). As a result of the peptide–receptor interaction, the autophosphorylation of the regulatory subunit (B-50) is inhibited, resulting in activation of the DPI kinase possibly by disassociation of the holo enzyme; TPI is formed, and the breakdown of PI is inhibited, resulting in an inhibited Ca^{2+} influx, and probably inhibition of secondary Ca^{2+} dependent processes. However, even if changes in membrane phosphorylation in certain synapses account for some of the behavioral effects brought about by ACTH, there is considerable doubt that this is the only mechanism of action. As pointed out earlier, the presently known neurophysiological, neurochemical, and behavioral effects of ACTH suggest multiple ACTH receptors on several neural substrates (Gispen, van Ree, & de Wied, 1977). In neurochemical terms, this could imply that three possible mechanisms (Gispen, 1980) (neurotransmitter, neurohormone, or neuromodulator) could be functional, depending on the brain system involved in the adaptation to the changing environment.

REFERENCES

Akil, H., & Watson, S. J. (1980). Neuromodulatory functions of the brain pro-opiocortin system. *In* E. Costa & M. Trabucchi (Eds.), *Neural Peptides and Neuronal Communication.* (*Advances in Biochemical Psychopharmacology* 22), pp. 435–445. New York: Raven.

Bär, P. R., Schotman, P., Gispen, W. H., Tielen, A. M., & Lopes da Silva, F. H. (1980). Changes in synaptic membrane phosphorylation after tetanic stimulation in the dentate area of the rat hippocampal slice. *Brain Research*, **198**, 478–484.

Bennett, H.J.P., & McMartin, C. (1979). Peptide hormones and their analogues: Distribution, clearance from the circulation, and inactivation *in vivo*. *Pharmacological Reviews*, **30**, 247–292.

Bohus, B., & de Wied, D. (1966). Inhibitory and facilitatory effect of two related peptides on extinction of avoidance behavior. *Science*, **153**, 318–320.

Bohus, B., Gispen, W. H., & de Wied, D. (1973). Effect of lysine vasopressin and $ACTH_{4-10}$ on conditioned avoidance behavior of hypophysectomized rats. *Neuroendocrinology*, **11**, 137–143.

Bohus, B., Hendrickx, H. H. L., van Kolfschoten, A., & Krediet, T. G. (1975). Effect of $ACTH_{4-10}$ on copulatory and sexually motivated approach behavior in the male rat. *In* M. Sandler & G. L. Gessa (Eds.), *Sexual Behavior: Pharmacology and Biochemistry*, pp. 269–275. New York: Raven.

Bradbury, A. F., Smyth, D. G., & Snell, C. R. (1976). Lipotropin: A precursor to two biologically active peptides. *Biochemical and Biophysical Research Communications*, **69**, 950–956.

Browning, M., Dunwiddie, T., Bennett, W., Gispen, W. H., & Lynch, G. (1979). Synaptic phosphoproteins: Specific changes after repetitive stimulation of the hippocampal slice. *Science*, **203**, 60–62.

Burbach, J. P. H., & de Wied, D. (1980). Adaptive behavior and endorphin biotransformation. *In* E. Usdin, T. Sourkes, & H. Youdim (Eds.), *Enzymes and Neurotransmitters in Mental Diseases*. Baffinsdale: Wiley.

Chiu, S., Wong, Y.-W., & Mishra, R. K. (1980). Direct evidence for L-prolyl-leudyl-glycinamide (PLG) receptor in mammalian brain. *Federation Proceedings*, **39**, 387.

Cornford, E. M., Braun, L. D., Crane, P. D., & Oldendorf, W. H. (1978). Blood–brain barrier restriction of peptides and the low uptake of enkephalins. *Endocrinology*, **103**, 1297–1303.

de Kloet, E. R., & de Wied, D. (1980). The brain as target-tissue for hormones of pituitary origin: Behavioral and biochemical studies. *In* L. Martini & W. F. Ganong (Eds.), *Frontiers in Neuroendocrinology* (Vol. 6), pp. 157–201. New York: Raven.

de Wied, D. (1966). Inhibitory effect of ACTH and related peptides on extinction of conditioned avoidance behavior in rats. *Proceedings of the Society for Experimental Biology and Medicine*, **122**, 28–32.

de Wied, D. (1967). Opposite effects of ACTH and glucocorticosteroids on extinction of conditioned avoidance behavior. *Excerpta Medica International Congress Series*, **132**, 945.

de Wied, D. (1969). Effects of peptide hormones on behavior. *In* W. F. Ganong & L. Martini (Eds.), *Frontiers in Neuroendocrinology*, pp. 97–140. London–New York: Oxford University Press.

de Wied, D. (1974). Pituitary–adrenal system hormones and behavior. *In* F. O. Schmitt & F. G. Worden (Eds.), *The Neurosciences, Third Study Program*, pp. 653–666. Cambridge: MIT Press.

de Wied, D. (1977). Behavioral effects of neuropeptides related to ACTH, MSH and β-LPH. *Annals of the New York Academy of Sciences*, **297**, 263–274.

de Wied, D., Bohus, B., van Ree, G. M., & Urban, G. (1978). Behavioral and electrophysiological effects of peptides related to lipotropin (β-LPH). *Journal of Pharmacology and Experimental Therapeutics*, **204**, 570–580.

de Wied, D., & Gispen, W. H. (1977). Behavioral effects of peptides. *In* H. Gainer (Ed.), *Peptides in Neurobiology*, pp. 397–448. New York: Plenum.

de Wied, D., van Wimersma Greidanus, Tj. B., & Bohus, B. (1974). Pituitary peptides and behavior: Influence on motivational, learning and memory processes. *Excerpta Medica International Congress Series*, **359**, 653–658.

de Wied, D., Witter, A., & Greven, H. M. (1975). Behaviorally active ACTH analogues. *Biochemical Pharmacology*, **24**, 1463–1468.

Dorsa, D. M., de Kloet, E. R., Mezey, E., & de Wied, D. (1979). Pituitary–brain transport of neurotensin: Functional significance of retrograde transport. *Endocrinology*, **104**, 1663–1666.

Flexner, J. B., & Flexner, L. B. (1971). Pituitary peptides and the suppression of memory by puromycin. *Proceedings of the National Academy of Sciences*, **68**, 2519–2521.

Flood, J. F., Jarvik, M. E., Bennett, E. L., & Orme, A. E. (1976). Effects of ACTH peptide fragments on memory formation. *Pharmacology, Biochemistry and Behavior*, **5**, (Suppl. 1), 41–51.

Garrud, P., Gray, J. A., & de Wied, D. (1974). Pituitary–adrenal hormones and extinction of rewarded behavior in the rat. *Physiology and Behavior*, **12**, 109–119.

Gispen, W. H. (1980). On the neurochemical mechanism of action of ACTH. *Progress in Brain Research*, **53**, 193–206.

Gispen, W. H., Buitelaar, J., Wiegant, V. M., Terenius, L., & de Wied, D. (1976). Interaction between ACTH fragments, brain opiate receptors and morphine-induced analgesia. *European Journal of Pharmacology*, **39**, 393–397.

Gispen, W. H., van Ree, J. M., & de Wied, D. (1977). Lipotropin and the central nervous system. *International Review of Neurobiology*, **20**, 209–250.

Gispen, W. H., Wiegant, V. M., Greven, H. M., & de Wied, D. (1975). The induction of excessive grooming in the rat by intraventricular application of peptides derived from ACTH: Structure activity studies. *Life Sciences*, **17**, 645–652.

Gold, P. E., & van Buskirk, R. (1976). Enhancement and impairment of memory processes with post-trial injections of adrenocorticotrophic hormone. *Behavioral Biology*, **16**, 387–400.

Graf, L., Szekely, J. I., Ronai, A. Z., Dunai-Kovacs, Z., & Bajusz, S. (1976). Comparative study on analgesic effect of Met^5-enkephalin and related lipotropin fragments. *Nature*, **263**, 240–241.

Gray, J. A. (1971). Effect of ACTH on extinction of rewarded behavior is blocked by previous administration of ACTH. *Nature*, **229**, 52–54.

Greengard, P. (1979). *Cyclic Nucleotides, Phosphorylated Proteins, and Neuronal Function*. New York: Raven.

Greven, H. M., & de Wied, D. (1973). The influence of peptides derived from corticotropin (ACTH) on performance. Structure activity studies. *Progress in Brain Research*, **39**, 429–442.

Greven, H. M., & de Wied, D. (1977). Influence of peptides structurally related to ACTH and MSH on active avoidance behavior in rats. A structure–activity relationship study. *Frontiers of Hormone Research*, **4**, 140–152.

Hawthorne, J. N., & Kai, M. (1970). Metabolism of phosphoinositides. *In* A. Lajtha (Eds.), *Handbook of Neurochemistry*, pp. 491–508. New York: Plenum.

Heald, P. J. (1962). Phosphoprotein metabolism and ion transport in nervous tissue: A suggested connection. *Nature*, **193**, 451–454.

Hudson, A. M., Ambler, L., Bennett, H. P. J., & McMartin, C. (1979). Metabolism of ACTH. *Journal of Endocrinology*, **81**, 129.

Hudson, A. M., & McMartin, C. (1980). Mechanism of catabolism of corticotrophin (1–24)-tetracosapeptide in the rat *in vivo*. *Journal of Endocrinology*, **85**, 93–103.

Isaacson, R. L., Dunn, A. J., Rees, H. D., & Waldock, B. (1976). ACTH$_{4-10}$ and improved use of information in rats. *Physiological Psychology*, **4**, 159–162.

Jolles, J., Aloyo, V., & Gispen, W. H. (1981a). Molecular correlates between pituitary hormones and behavior. *In* I. R. Brown (Ed.), *Molecular Approaches to Neurobiology*. New York: Academic Press.

Jolles, J., Wirtz, K. W., Schotman, P., & Gispen, W. H. (1979). Pituitary hormones influence polyphosphoinositide metabolism in rat brain. *FEBS Letters*, **105**, 110–114.

Jolles, J., Zwiers, H., Dekker, A., Wirtz, K. W., & Gispen, W. H. (1981b). ACTH$_{1-24}$ affects protein phosphorylation and polyphosphoinositide metabolism in the rat brain. *Biochemical Journal*. **194**, 283–291.

Jolles, J., Zwiers, H., Schotman, P., & Gispen, W. H. (1980a). ACTH and the phosphorylation of proteins and phosphatidylinositides. In M. Brzin, D. Sket, & H. Bachelard (Eds.), Synaptic Constituents in Health and Disease. Proceedings of the Third Meeting of the ESN, pp. 269–270. London: Pergamon.

Jolles, J., Zwiers, H., van Dongen, C., Schotman, P., Wirtz, K. W., & Gispen, W. H. (1980b). Modulation of brain polyphosphoinositide metabolism by ACTH-sensitive protein phosphorylation. Nature, 286, 623–625.

Kastin, A. J., Miller, L. H., Nockton, R., Sandman, C. A., Schally, A. V., & Stratton, L. O. (1973). Behavioral aspects of melanocyte stimulating hormones (MSH). Progress in Brain Research, 39, 461–470.

Kastin, A. J., Sandman, C. A., Stratton, L. O., Schally, A. V., & Miller, L. H. (1975). Behavioral and electrographic changes in rat and man after MSH. Progress in Brain Research, 42, 143–150.

Krieger, D. T., Liotta, A., & Brownstein, M. J. (1977). Presence of corticotropin in brain of normal and hypophysectomized rats. Proceedings of the National Academy of Sciences, 74, 648–652.

Leonard, B. E. (1969). The effect of sodium-barbitone alone and together with ACTH and amphetamine on the behavior of the rat in the multiple "T" maze. International Journal of Neuropharmacology, 8, 427–435.

Levine, S., & Jones, L. E. (1975). Adrenocorticotropic hormones (ACTH) and passive avoidance learning. Journal of Comparative and Physiological Psychology, 59, 357–360.

Liotta, A. S., Li, C. H., Schussler, G. C., & Krieger, D. T. (1978). Comparative metabolic clearance rate, volume of distribution and plasma half-life of human β-lipotropin and ACTH. Life Sciences, 23, 2323–2330.

Lissak, K., & Bohus, B. (1972). Pituitary hormones and avoidance behavior of the rat. International Journal of Psychobiology, 2, 103–115.

Loh, Y. P. (1979). Immunological evidence for two common precursors to corticotropins, endorphins and melanotropin in the neurointermediate lobe of the toad pituitary. Proceedings of the National Academy of Sciences, 76, 796–800.

Mains, R., Eipper, B. A., & Ling, N. (1977). Common precursor to corticotropins and endorphins. Proceedings of the National Academy of Sciences, 74, 3014–3018.

Marks, N., Stern, F., & Kastin, A. J. (1976). Biodegradation of α-MSH and derived peptides by rat brain extracts and by rat and human serum. Brain Research Bulletin, 1, 591–593.

Matus, A. L., Jones, D. H., & Mughal, S. (1976). Restricted distribution of synaptic antigens in the neuronal membrane. Brain Research, 103, 171–175.

Mezey, E., Kivovics, P., & Palkovits, M. (1979). Pituitary–brain retrograde transport. Trends in Neuroscience, 2, 57–59.

Mezey, E., Palkovits, M., de Kloet, E. R., Verhoef, J., & de Wied, D. (1978). Evidence for pituitary–brain transport of a behaviorally potent ACTH analog. Life Sciences, 22, 831–838.

Michell, R. H. (1975). Inositol phospholipids and cell surface receptor function. Biochimica Biophysica Acta, 415, 81–148.

Michell, R. H. (1979). Inositol phospholipids in membrane function. Trends in Biochemical Sciences, 4, 128–131.

Miller, R. E., & Ogawa, N. (1962). The effect of adrenocorticotropic hormone (ACTH) on avoidance conditioning in the adrenalectomized rat. Journal of Comparative and Physiological Psychology, 55, 211–213.

Murphy, J. V., & Miller, R. E. (1955). The effect of adrenocorticotropic hormone (ACTH) on avoidance conditioning in the rat. Journal of Comparative and Physiological Psychology, 48, 47–49.

Normand, M., & Lalonde, J. (1979). Distribution and metabolism of adrenocorticotropin in the rat. Canadian Journal of Physiology and Pharmacology, 57, 1024–1027.

Oestreicher, A. B., Zwiers, H., Schotman, P., & Gispen, W. H. (1979). Immunohistochemical studies of a phosphoprotein band (B-50) modulated in rat brain membranes by ACTH. *Abstract Seventh International Meeting ISN*, Jerusalem, 509.

Oestreicher, A. B., Zwiers, H., Schotman, P., & Gispen, W. H. (1981). Immunohistochemical localization of a phosphoprotein (B-50) isolated from rat brain synaptosomal plasma membranes. *Brain Research Bulletin*, **6**, 145–154.

Oliver, C., Mical, R. S., & Porter, J. C. (1977). Hypothalamic–pituitary vasculature: Evidence for retrograde blood flow in the pituitary stalk. *Endocrinology*, **101**, 598–604.

Orwoll, E., Kendall, J. W., Lamorena, L., & McGilvra, R. (1979). Adrenocorticotropin and melanocyte-stimulating hormone in the brain. *Endocrinology*, **104**, 1845–1852.

Pavel, S. (1979). Pineal vasotocin and sleep involvement of serotonin-containing neurons. *Brain Research Bulletin*, **4**, 731–734.

Pavel, S., Cristoveanu, A., Goldstein, R., & Calb, M. (1977). Inhibition of release of corticotropin releasing hormone in cats by extremely small amounts of vasotocin injected into the third ventricle of the brain. Evidence for the involvement of 5-hydroxytryptamine-containing neurons. *Endocrinology*, **101**, 672–678.

Rapoport, S. I., Klee, W. A., Pettigrew, K. D., & Ohno, K. (1980). Entry of opioid peptides into the central nervous system. *Science*, **207**, 84–86.

Reith, M. E. A., Neidle, A., & Lajtha, A. (1979). Breakdown of corticotropin (1–24) by mouse brain extracts. *Archives of Biochemistry and Biophysics*, **195**, 478–484.

Rigter, H., & Popping, A. (1976). Hormonal influences on the extinction of conditioned taste aversion. *Psychopharmacologia*, **46**, 255–261.

Rigter, H., & van Riezen, H. (1975). Antiamnesic effect of $ACTH_{4-10}$: Its dependence of the nature of the amnesic agent and the behavioral test. *Physiology and Behavior*, **14**, 563–566.

Rigter, H., van Riezen, H., & de Wied, D. (1974). The effects of ACTH- and vasopressin-analogues on CO_2-induced retrograde amnesia in rats. *Physiology and Behavior*, **13**, 381–388.

Rodnight, R. (1979). Cyclic nucleotides as second messengers in synaptic transmission. *International Review of Biochemistry*, **26**, 1–80.

Rossier, J., Vargo, T. M., Minick, S., Ling, N., Bloom, F. E., & Guillemin, R. (1977). Regional dissociation of β-endorphin and enkephalin contents in rat brain and pituitary. *Proceedings of the National Academy of Sciences*, **74**, 5162–5165.

Sandman, C. A., Alexander, W. D., & Kastin, A. J. (1973). Neuroendocrine influences on visual discrimination and reversal learning in the albino and hooded rat. *Physiology and Behavior*, **11**, 613–617.

Sandman, C. A., Beckwith, B. C., Gittes, N. M., & Kastin, A. J. (1974). Melanocyte-stimulating hormone (MSH) and overtraining effects on extradimensional shift (EDS) learning. *Physiology and Behavior*, **13**, 163–166.

Sandman, C. A., & Kastin, A. J. (1980). The influence of fragments of the LPH chains on learning, memory and attention in animals and man. *Pharmacology and Therapeutics*.

Sandman, C. A., Kastin, A. J., & Schally, A. V. (1969). Melanocyte-stimulating hormone and learned appetitive behavior. *Experientia*, **25**, 1001–1002.

Sandman, C. A., Miller, L. H., Kastin, A. J., & Schally, A. V. (1972). Neuroendocrine influence on attention and memory. *Journal of Comparative and Physiological Psychology*, **80**, 54–58.

Schwyzer, R. (1980). Organization and transduction of peptide information. *Trends in Pharmacological Sciences*, **1**, 327–331.

Scott, A. P., Bloomfield, G. A., Lowry, Ph. J., Gilke, T. J., London, J., & Rees, L. H. (1976). Pituitary adrenocorticotrophin and melanocyte stimulating hormones. *In* J. A. Parsons (Ed.), *Peptide Hormones*, pp. 247–271. Baltimore: University Park Press.

Torda, C. (1974). A potential mechanism for reserpine and chlorpromazine generation of myasthenia gravis-like easy fatigability and Parkinsonism involving acetylcholine, dopamine, and cyclic AMP. *International Review of Neurobiology*, **16**, 1–66.

Ueda, T., & Greengard, P. (1977). Adenosine 3',5'-monophosphate-regulated phosphoprotein system of neuronal membranes. I. Solubilization, purification and some properties of an endogenous phosphoprotein. *Journal of Biological Chemistry*, **252**, 5155–5163.

Uno, I., Ueda, T., & Greengard, P. (1977). Adenosine 3',5'-monophosphate-regulated phosphoprotein system of neuronal membranes. II. Solubilization, purification and some properties of an endogenous adenosine 3',5'-monophosphate-dependent protein kinase. *Journal of Biological Chemistry*, **252**, 5164–5174.

Urban, I., & de Wied, D., (1976). Changes in excitability of the theta activity generating substrate by $ACTH_{4-10}$ in the rat. *Experimental Brain Research*, **24**, 324–344.

van Wimersma Greidanus, Tj. B., Bohus, B., & de Wied, D. (1975). CNS sites of action of ACTH, MSH and vasopressin in relation to avoidance behavior. *In* W. F. Stumpf & L. D. Grant (Eds.), *Anatomical Neuroendocrinology*, pp. 284–289. Basel: Karger.

Verhoef, J., & Witter, A. (1976). *In vivo* fate of a behaviorally active $ACTH_{4-9}$ analog in rats after systemic administration. *Pharmacology, Biochemistry and Behavior*, **4**, 583–590.

Verhoef, J., Witter, A., & de Wied, D. (1977). Specific uptake of a behaviorally potent [³H]-$ACTH_{4-9}$ analog in the septal area after intraventricular injection in rats. *Brain Research*, **131**, 117–128.

Watson, S. J., & Akil, H. (1980). On the multiplicity of active substances in single neurons: β-endorphin and α-melanocyte stimulating hormone as a model system. *In* D. de Wied & P. A. van Keep (Eds.), *Hormones and the Brain*, pp. 73–86. Lancaster: MTP Press.

Watson, S. J., Richard, C. W., & Barchas, J. D. (1978). Adrenocorticotropin in rat brain: Immunocytochemical localization in cells and axons. *Science*, **200**, 1180–1182.

Witter, A., (1980). On the presence of receptors for ACTH-neuropeptides in the brain. *In* G. Pepeu, M. J. Kuhar, & S. J. Enna (Eds.), *Receptors for Neurotransmitters and Peptide Hormones, Advances in Biochemical Psychopharmacology* (Vol. 21), pp. 407–414. New York: Raven.

Witter, A., Greven, H. M., & de Wied, D. (1975). Correlation between structure, behavioral activity and rate of biotransformation of some $ACTH_{4-9}$ analogs. *Journal of Pharmacology and Experimental Therapeutics*, **193**, 853–860.

Wolthuis, O. L., & de Wied, D. (1976). The effect of ACTH-analogues on motor behavior and visual evoked responses in rats. *Pharmacology, Biochemistry and Behavior*, **4**, 273–278.

Zwiers, H., Schotman, P., & Gispen, W. H. (1980). Purification and some characteristics of an ACTH-sensitive protein kinase and its substrate protein in rat brain membranes. *Journal of Neurochemistry*, **34**, 1689–1699.

Zwiers, H., Tonnaer, J., Wiegant, V. M., Schotman, P., & Gispen, W. H. (1979). ACTH-sensitive protein kinase from rat brain membranes. *Journal of Neurochemistry*, **33**, 247–256.

Zwiers, H., Veldhuis, H. D., Schotman, P., & Gispen, W. H. (1976). ACTH, cyclic nucleotides, and brain protein phosphorylation *in vitro*. *Neurochemical Research*, **1**, 669–677.

Zwiers, H., Wiegant, V. M., Schotman, P., & Gispen, W. H. (1978). ACTH-induced inhibition of endogenous rat brain protein phosphorylation *in vitro*: Structure activity. *Neurochemical Research*, **3**, 455–463.

CHAPTER *3*

Actions of ACTH- and MSH-like Peptides on Learning, Performance, and Retention

Béla Bohus and David de Wied

The peptide sequence found near the NH_2-terminal end of the adrenocorticotropic hormone (ACTH) molecule has strong influences on many forms of behavior. This peptide sequence, shared by both ACTH and α-melanocyte-stimulating hormone (α-MSH), tends to facilitate the acquisition of appetitive and avoidance tasks and to attenuate the rate of habituation. These findings may be interpreted as a peptide-induced improvement in an organism's ability to respond to environmental stimuli. ACTH and α-MSH also affect performance. Peripheral administration of $ACTH_{4-10}$ increases the response rate of rats to rewarding electrical brain stimulation, if the stimulating current is minimal. The peptide fails to affect response rates when higher current intensities are used. $ACTH_{4-10}$ also interferes with the typical characteristics of performance during brain stimulation reward shift. It may be that the peptide prevents behavioral adjustment to changes in the rewarding properties of stimulation or that it stabilizes performance.

ACTH and related peptides also influence measures of retention as reflected in delayed extinction of learned appetitive and avoidance behaviors. These peptide fragments appear to influence memory retrieval but not memory consolidation processes. On the other hand, γ2-MSH suppresses both memory consolidation and memory retrieval.

Some of the effects of the ACTH- and MSH-related peptides may be due to their enhancement or attenuation of the motivational properties of environmental stimuli. Influences on memory mechanisms may explain some others. However, the multiplicity of behavioral effects of these peptides suggests that the peptide sequences may contain more than one bit of behavioral information and that, under normal physiological conditions, ACTH- and MSH-like peptides may originate from distinct sources. Finally, the observations described in this chapter suggest that the behavioral effects of these peptides are due to direct influences on the central nervous system.

59

ENDOGENOUS PEPTIDES
AND LEARNING AND MEMORY PROCESSES

I. INTRODUCTION

It has been more than a quarter of a century since the behavioral effects of ACTH were first reported in animals (Mirsky, Miller, & Stein, 1953). Further investigation of these pilot observations was not done until the early 1960s. De Wied (1964) reported that the removal of the anterior pituitary was followed by an impairment of acquisition of an active avoidance response. This behavioral deficit was corrected by administration of ACTH in the rat. In the intact rat, Bohus and Endröczi (1965) found that ACTH facilitates the acquisition of a one-way avoidance response. Subsequent studies led to at least three important discoveries concerning the behavioral effects of ACTH. First, it appeared that the behavioral effects of ACTH are not mediated through the adrenal cortex, as adrenalectomy does not prevent these influences (Bohus, Nyakas, & Endröczi, 1968; Miller & Ogawa, 1962). Second, peptide hormones that share a part of the ACTH molecule, such as α-MSH and β-MSH, and also shorter fragments such as $ACTH_{4-10}$, mimic the behavioral activities of ACTH (Bohus & de Wied, 1966; de Wied, 1966). These observations indicated that the behavioral activity of the ACTH molecule resides in a short sequence near the NH_2-terminus. Third, resistance to extinction of avoidance responses that followed the administration of ACTH and related peptides (Bohus & de Wied, 1966; de Wied, 1966) suggested a strong influence of these peptides on the retention of learned responses. Evidence collected during the next decade showed that the influence of ACTH-related peptides on learning and retention can be generalized across avoidance and appetitive behaviors (for review see Bohus & de Wied, 1980).

The release of ACTH from the pituitary gland is one of the classic features of the stress response. It is a rather recent discovery that ACTH and β-lipotropic hormone (β-LPH), the precursor molecule of opiate-like neuropeptides, exist within the same cells in the anterior pituitary (Phifer, Orth, & Spicer, 1974) and originate from the same high molecular weight precursor protein (pro-opiomelanocortin; Mains, Eipper & Ling, 1977; Roberts & Herbert, 1977). This 31 kdalton precursor molecule contains several pairs of basic amino acid residues that may be attacked by proteolytic enzymes, thereby yielding several smaller peptides such as ACTH, β-LPH, α-MSH ($ACTH_{1-13}$), CLIP ($ACTH_{18-39}$), β-MSH ($β-LPH_{41-58}$), β-endorphin ($β-LPH_{61-91}$), and other fragments (Herbert, Roberts, Phillips, Allen, Hinman, Budarf, Policastro, & Rosa, 1980; Lowry, Silman, Jackson, & Estivariz, 1979). In addition, the pituitary gland may not be the only source of pro-opiomelanocortin-related peptides. Immuno- and bioreactive ACTH, α-MSH, β-LPH, and endorphins all have been found in the brains of intact and hypophysectomized rats within the same neuronal system. The cell bodies of this system are located in the region of the hypothalamus and its neurons terminate in the limbic and midbrain areas (Kendall

& Orwoll, 1980; Krieger & Liotta, 1979; Watson & Akil, 1980). It is an interesting characteristic of the pro-opiomelanocortin precursor molecule that the amino acid sequence $ACTH_{4-10}$ (Met-Glu-His-Phe-Arg-Try-Gly) occurs in several "end products" such as ACTH, α-MSH, and β-MSH. This fragment contains two sequences that are the behaviorally active sites of ACTH-related peptides: $ACTH_{4-7}$ and $ACTH_{7-9}$ (Greven & de Wied, 1973; de Wied, Witter, & Greven, 1975). Recently, Nakanishi, Inoue, Kita, Nakamura, Chang, Cohen, and Numa (1979) reported the nucleotide sequence of cloned cDNA for bovine pro-opiomelanocortin. The corresponding amino acid sequence indicated a further melanotropic-like moiety in the NH_2-terminal part of the precursor molecule. This segment of the precursor contains tyrosine and methionine residues and the His-Phe-Arg-Try tetrapeptide sequence that is characteristic of MSH peptides. The authors have proposed the designation γ-MSH for this putative peptide.

In this chapter, we present some of our recent observations on the effects of ACTH- and MSH-like peptides on learning, performance, and retention. These observations suggest that the NH_2-terminus of the pro-opiocortin precursor molecule contains a sequence (γ-MSH) that may serve as an antagonist of ACTH and/or β-endorphin-related peptides. In addition, it is likely that ACTH and α-MSH carry some behavioral information that, despite redundancy of amino acid sequences, may have different characteristics.

II. ACTH- AND MSH-LIKE PEPTIDES AND LEARNING

Based on the observation that active avoidance performance in the acquisition phase is positively correlated with the pituitary–adrenal response to stress in the same rats (Bohus, Endröczi, & Lissák, 1963), it was assumed that increased release of ACTH and corticosteroids because of the stressful nature of avoidance conditioning may improve learning. To test this hypothesis, Bohus and Endröczi (1965) investigated the influence of ACTH on the acquisition of a one-way avoidance response. Facilitation of avoidance responding was found during the early phase of conditioning when the tendency to avoid was low. This facilitatory effect of the peptide did not occur unless at least one avoidance response was scored before treatment. This observation suggested that a certain degree of conditioning had to occur before ACTH could modulate acquisition behavior. Clear indications of an extra-adrenal effect of ACTH were subsequently found because ACTH facilitates active avoidance learning in adrenalectomized rats (Bohus et al., 1968). Subsequent observations were in accord with these findings and, in addition, suggested the importance of variables such as shock intensity and circadian rhythmicity (Beatty, Beatty, Bowman, & Gilchrist, 1970; Ley & Corson, 1972; Pagano & Lovely, 1972; Stratton & Kastin, 1974). That

ACTH (Guth, Seward, & Levine, 1971; Levine & Jones, 1965; Lissák & Bohus, 1972), α-MSH (Sandman, Kastin, & Schally, 1971) and an ACTH analog (ACTH$_{4-9}$[8-D-Lys, 9-Phe]; Martinez, Vasquez, Jensen, Soumireu-Mourat, & McGaugh, 1979) facilitate passive (inhibitory) avoidance learning provided further support of the hypothesis that ACTH and related peptides facilitate acquisition processes.

Recently, we have adopted a massed trial shuttlebox avoidance paradigm to investigate the influences of neuropeptides on avoidance learning. A short CS–US interval of 3 sec and low footshock intensity (0.15 mA, dc, scrambled) results in a slow acquisition rate during a 30-trial single session. Subcutaneous administration of ACTH$_{4-10}$ in a dose of 110 μg/kg facilitates avoidance acquisition (Table 1). The same dose of γ2-MSH failed to affect avoidance behavior, but doses 10–100 times less led to a marked suppression of acquisition behavior. Similar to γ2-MSH, ACTH$_{4-10}$[7-D-Phe] also suppresses shuttlebox avoidance learning, but this peptide is less potent.

Earlier observations on the influence of ACTH fragments suggested that the amino acid residue phenylalanine in Position 7 plays a key role in the effects on extinction behavior. Substitution of 7-Phe of ACTH$_{1-10}$ by its D-enatiomer

TABLE 1

The Influence of ACTH- and MSH-like Peptides on Avoidance Learning in the Rat

Treatment[a]	Dose[b]	Number of avoidance responses[c,d]	
Saline		16.4 ± 1.2	(5)
ACTH$_{4-10}$	110.0	20.6 ± 0.8**	(5)
Saline		16.9 ± 0.8	(6)
γ2-MSH	110.0	17.6 ± 1.6	(6)
γ2-MSH	11.0	6.5 ± 1.2***	(6)
Saline		17.6 ± 1.6	(8)
γ2-MSH	1.1	13.0 ± 2.6**	(6)
Saline		19.8 ± 0.4	(6)
γ2-MSH	0.11	16.7 ± 1.6	(7)
Saline		16.8 ± 1.9	(6)
ACTH$_{4-10}$[7-D-Phe]	110.0	12.5 ± 2.6*	(6)
Saline		16.8 ± 1.9	(6)
ACTH$_{4-10}$[7-D-Phe]	11.0	19.5 ± 1.9	(6)

[a] Treatments were given 1 hr before the onset of shuttlebox avoidance training.
[b] μg/kg body weight, subcutaneously.
[c] Mean ± SEM of conditioned avoidance responses scored during 30 trials.
[d] Numbers in parentheses indicate the number of rats.
* $p < .05$.
** $p < .01$.
*** $p < .001$ (two-tailed t-test).

($ACTH_{1-10}$[7-D-Phe]) resulted in a reversal of the effect of the peptide; whereas $ACTH_{4-10}$ delays avoidance extinction, the 7-D-Phe-analog accelerated it (Bohus & de Wied, 1966). Shorter fragments of this peptide such as $ACTH_{4-10}$[7-D-Phe] have a similar effect (Greven & de Wied, 1973). Accordingly, recent observations suggest that the configuration in Position 7 also plays an important role in the influence of ACTH-related peptides on avoidance learning. The suppressive effect of γ2-MSH is, however, surprising. This peptide shares the sequence 6–9 with ACTH fragments including 7-Phe, but the elongation of the fragment at both the NH_2- and the C-terminus is different.

Facilitation of acquisition by ACTH-like peptides is also apparent. In appetitive paradigms, α-MSH improves the acquisition of a complex maze response for food reward (Stratton & Kastin, 1975), and ACTH also facilitates operant learning motivated by a water reward provided that the motivation level is low (Guth, Levine, & Seward, 1971). These observations suggest that the effect of ACTH-related peptides on learning is a general phenomenon. Observations of the influences of these peptides on habituation, however, call into question the generality of a facilitatory effect. Habituation is interpreted as one form of learning (Harlow, McGaugh, & Thompson, 1971). Habituation represents a decrease in responding to repeated presentation of a specific stimulus, which first directs attention. Habituation of behavioral responses, such as exploration and orientation, also occurs in novel environments both within and between test sessions.

Endröczi, Lissák, Fekete, and de Wied (1970) were the first to report that $ACTH_{1-24}$ and $ACTH_{1-10}$ attenuate electrocortical habituation to repeated sound stimuli in man. File (1978) found that $ACTH_{1-24}$, $ACTH_{4-10}$, and $ACTH_{4-10}$[7-D-Phe] delay habituation of an orienting response. Intracerebroventricular administration of $ACTH_{4-10}$ diminished the decline of open-field ambulatory activity of rats upon repeated testing (Bohus & van Wolfswinkel, unpublished observations); $ACTH_{1-24}$ failed to affect a hole-board exploration response within and between test sessions (File, 1978), and $ACTH_{4-10}$ and $ACTH_{4-9}$[8-D-Lys-9-Phe] had no effect on habituation of ambulation and rearing within a test session in an open-field situation (Bohus, unpublished observations). Accordingly, ACTH-related peptides influence only certain forms of habituation and, if a peptide effect is present, it is an impairment rather than an improvement of habituation.

The obvious difference in the influences of ACTH-related peptides on avoidance or appetitive learning and habituation are not easy to explain unless one accepts the view that these peptides improve an organism's ability to respond to environmental stimuli. According to Tolman (1932), sensory experiences, including those environmental stimuli that signal reward or punishment, determine what an animal will learn. The rewarding or punishing characteristics of the environmental stimuli, the motivation in a particular situation, and what has been learned, are the factors that determine performance. Training provides the possibility of learning about the environment and of selecting the most appropriate behavioral response. In 1973, Bohus, Gispen, and de Wied found that $ACTH_{4-10}$ normalizes impaired avoidance acquisition behavior of hypophysectomized rats. Avoidance performance, however, deteriorates in these animals despite footshock punishment if the daily administration of the peptide is terminated. If the hypophysectomized rats had learned the avoidance response better under the influence of the peptide, then maintenance of responding would be expected. Accordingly, the peptide influenced some processes that enabled the rats to respond appropriately, but without consequences on later behavior. This temporary nature of the influences of $ACTH_{4-10}$ in hypophysectomized rats is in accord with observations in intact rats. De Wied and Bohus (1966) observed that administration of α-MSH during active avoidance acquisition fails to influence later retention of this response. Bohus et al. (1968) found that termination of ACTH treatment during avoidance extinction results in an abrupt decrease in responding. Bohus et al. (1973) suggested that $ACTH_{4-10}$ may have increased the motivation of the hypophysectomized rat to avoid punishment. Because ACTH-like peptides also maintained responding during avoidance (de Wied, 1969) and appetitive extinction (Bohus, Hendrickx, van Kolfschoten, & Krediet, 1975; Garrud, Gray, & de Wied, 1974) in the absence of punishment or reward, it seemed likely that the motivational properties of environmental stimuli are affected by the peptides rather than by the motivational characteristics of punishment and reward. A higher probability of generating stimulus-specific behavioral responses would then follow the administration of these peptides. Although punishment and reward are absent in habituation paradigms, preserved responding during certain types of habituation training in peptide-treated rats may be explained by the maintenance of exploratory motivation.

III. ACTH-LIKE PEPTIDES AND PERFORMANCE

The influence of peptides on acquisition of avoidance and appetitive responses manifests itself in improved performance. The acquisition period may, however, be confounded by a number of factors that make it difficult to precisely investigate performance in relation to the motivational properties of stimuli. Intracranial self-stimulation behavior based upon brain-stimulation reward (Olds, 1976) is often used to study performance. Because a number of the behavioral effects of ACTH-like peptides appear to depend on motivational variables (for review, see Bohus & de Wied, 1980), Nyakas, Bohus, and de Wied (1980) investigated the influence of $ACTH_{4-10}$ on self-stimulation behavior in the rat using a wide range of stimulation current intensities to induce different rates of performance. In addition, two distinct neuroanatomical and functional regions, the medial forebrain bundle (MFB) and the medial septal area, were selected to investigate the importance of the locus of stimulation. It has been suggested that these two areas involve different neuronal circuits, either of which support self-stimulation behavior (Rolls, 1977).

Subcutaneous administration of $ACTH_{4-10}$ increases response rate if the threshold current (i.e., the minimal current that elicits self-stimulation) or 1.2 times this current is used in the medial septum. The peptide is also effective if the MFB is stimulated by the threshold current. $ACTH_{4-10}$ fails to affect response rate when higher current intensities are used in either the medial septum or MFB (Table 2).

In these experiments the rats were tested with one current intensity on each day. The current intensities were varied randomly and peptide and placebo treatment days were alternated. The fact that facilitation is more obvious in the case of the medial septum suggests that the site of the stimulation is important in the effect of $ACTH_{4-10}$ on self-stimulation behavior. Closer inspection of the data, however, indicated that $ACTH_{4-10}$ treatment increases response rate from both the medial septum and the MFB when baseline responding is low. Accordingly, the effect of $ACTH_{4-10}$ depends on the response rate rather than on the site of stimulation. Facilitation of self-stimulation behavior performance by $ACTH_{4-10}$ may be interpreted to mean that the peptide increased the motivational properties of rewarding brain stimulation. The absence of an effect of $ACTH_{4-10}$ at higher baseline levels could be due to ceiling performance. This may be true for the highest current intensities. It is, however, more probable that the modulatory influence of the peptide is absent during high motivational levels because the stimulus exerts a stronger control on the behavior.

$ACTH_{4-10}$ also interferes with characteristics of the performance during brain stimulation reward shift. In this experiment, the intensity of the stimulation current was shifted from 1.2 to 1.5 every 8 min within one session, which is 2–3 times the threshold level. Stimulation electrodes were implanted in the medial septum. $ACTH_{4-10}$ treatment enhanced the response rate at lower currents and attenuated performance at the highest current for self-stimulation (Fig. 1). The most likely explanation is that $ACTH_{4-10}$ may attenuate the contrast in the motivational characteristics of the various current intensities. Attenuation by $ACTH_{4-10}$ of the effects of reinforcement shifts has been reported by Gray and Garrud (1977). $ACTH_{4-10}$ treatment attenuated the alterations in response rate of bar pressing for food reward that rats normally showed when reward conditions changed within a session. Attenuation occurred whether a high reward rate was introduced on a low rate baseline or a low reward rate on a high rate baseline. Accordingly, the peptide attenuated both "favorable" and "unfavorable" effects of reward shifts. It also attenuated the consequences

TABLE 2

Effects of $ACTH_{4-10}$ on Self-Stimulation Behavior in the Medial Septum and the Medial Forebrain Bundle

TMs[a]	Medial septum			Medial forebrain bundle		
	Response rate[b]			Response rate[b]		
	Saline[c]	$ACTH_{4-10}$[c]	Change in %[d]	Saline[c]	$ACTH_{4-10}$[c]	Change in %[d]
1.0	27.4 ± 4.5	38.5 ± 5.6**	40.4 (8)	24.5 ± 4.2	38.9 ± 6.9*	58.9 (6)
1.2	47.7 ± 6.5	59.3 ± 5.1**	24.5 (6)	107.3 ± 19.8	116.5 ± 23.3	8.5 (6)
2.0	104.9 ± 11.6	109.3 ± 10.4	4.2 (7)			
3.0	138.6 ± 5.2	147.1 ± 8.1	6.2 (8)			

[a] Threshold current multiples; each current intensity was used on separate days.

[b] Rate/6 min. Treatments ($ACTH_{4-10}$, 50 μg/rat or saline) were given 1 hr prior to test (from Nyakas, Bohus, & de Wied, 1980).

[c] Mean ± SEM.

[d] Numbers in parentheses indicate the number of observations.

* $p < .05$ (paired t-test).

** $p < .01$.

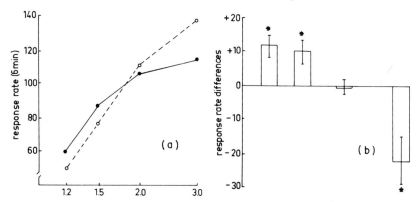

Fig. 1.   Changes in the characteristics of self-stimulation behavior using an ascending sequence of stimulation threshold current multiples in the medial septum (MS) following $ACTH_{4-10}$ administration (n = 7). The threshold multiples are given on the horizontal axes on a logarithmic scale. (a) The performance pattern of the same rats after $ACTH_{4-10}$ (●——●) or saline administration (○---○). (b) Mean differences ± SEM between peptide and saline sessions. $ACTH_{4-10}$ or saline was given subcutaneously 1 hr prior to the test. Probability level of tests of significance: $^*p < .05$ (paired t - test).

of partial reinforcement on performance, or resistance to extinction of a runway response for food reward, when the peptide was administered during acquisition training (Garrud, Gray, Rickwood, & Coen, 1977).

These effects of $ACTH_{4-10}$ may be interpreted in several ways. The peptide may have stabilized performance during changes in the rewarding properties of stimuli preventing either an overshoot or an undershoot of performance. Alternatively, the peptide may have prevented behavioral adjustment to changes in the rewarding properties of stimuli. The available information does not permit one to decide which of these alternative explanations is correct and what kinds of mechanisms are involved in these peptide effects.

IV. ACTH- AND MSH-LIKE PEPTIDES AND RETENTION

Performance of a learned response depends upon the consolidation (establishment of the memory trace), storage (passive maintenance of memory traces), and retrieval (recall) processes. Spear (1973) emphasized that the measurement of retention may profoundly affect retrieval processes. When some relearning criteria are used, the contingencies that were present during learning have to be reinstated on the first test trial. However, when performance measures other than relearning are used, it is not necessary to exactly reinstate the previous reinforcement contingencies. Extinction tests in which reinforcement contingencies are omitted or recall tests in passive (inhibitory) avoidance paradigms may be representative of such retention tests.

Studies of the influence of ACTH and related peptides on extinction behavior represented the early phase of neuropeptide research on behavior; however, the use of the extinction paradigm is still a more useful technique in determining psychoactive properties of various neuropeptides (de Wied, 1980). The finding that ACTH, α-MSH, and shorter fragments of ACTH such as $ACTH_{1-10}$ delay extinction of a conditioned avoidance response (de Wied, 1966) first suggested that retention of a behavioral response is enhanced by these peptides. Subsequently, it was found that the amino acid phenylalanine in Position 7 plays a key role in mediating this behavioral effect of ACTH-related peptides. Replacement of L-phenylalanine by its D-enantiomer in $ACTH_{1-10}$ produces an opposite effect on avoidance extinction (Bohus & de Wied, 1966). Replacement of amino acid residues other than the phenylalanine in ACTH fragments does not affect the direction of action (Greven & de Wied, 1973). The shortest fragment that has full behavioral activity is the tetrapeptide $ACTH_{4-7}$, but other activity sites are also present in the whole ACTH molecule. For example, $ACTH_{4-7}$ and $ACTH_{11-24}$ also contain some activity. Information may be present in a latent form in these molecules and the activity of $ACTH_{4-7}$ is highly potentiated by chain elongation at the carboxyl terminus (de Wied, Witter, & Greven, 1975).

Observations of the influence of ACTH-related peptides on active avoidance extinction provided a clear picture of the structural requirements for an effect on retention performance. Subsequent studies suggested that effects on retention can be generalized across avoidance and appetitive behaviors. Garrud et al. (1974) found that ACTH and $ACTH_{4-10}$ delay extinction of a runway response for food reward. $ACTH_{4-10}$[7-D-Phe], as in avoidance extinction, had the opposite effect. Bohus et al. (1975) showed that $ACTH_{4-10}$ delays extinction of a sexually motivated runway response in the male rat.

These observations suggest that ACTH peptides affect retention, although it is not clear whether consolidation or retrieval processes are affected. Effects on extinction behavior have been observed when the peptides were given after acquisition training, prior to, or right after the first extinction session.

One-trial learning, passive (inhibitory) avoidance paradigms, are often used to study memory processes. McGaugh (1961) proposed that treatments given after the learning trial affect brain processes that are related to memory. Effects on later retention of the behavior depend on influencing the consolidation processes that follow the learning experience. However, treatments given shortly before retention testing may influence retrieval processes.

Recently, we have investigated the influence of several ACTH- and MSH-related peptides on consolidation and retrieval of memory. A step-through, one-trial passive avoidance paradigm as described by Ader, Weijnen, and Moleman (1972) was used. Briefly, on Day 1, the rats were adapted to the dark compartment of the conditioning box. Subsequently, they were allowed to enter the dark compartment from a lighted, elevated platform that was attached to the front center of the large dark compartment. Due to a preference for darkness, the rats leave the platform readily. On the next day, three more approach trials were given. The third trial was followed by an inescapable footshock (0.25 mA for 2 sec, scrambled) administered through the grid floor of the dark compartment. Retention of the response was tested 24 hr after the learning trial. The rats were placed on the elevated platform and the latency to reenter the dark area was used as the behavioral measure.

Postlearning administration of $ACTH_{4-10}$ failed to influence passive avoidance behavior. Also, $ACTH_{4-10}[7\text{-}D\text{-}Phe]$ had no significant effect on retention. However, $\gamma2$-MSH led to a significant reduction in avoidance latencies (Table 3), indicating impaired retention.

The influence of these peptides, when given subcutaneously 60 min prior to the retention test, are also depicted in Table 3. Both $ACTH_{4-10}$ and

TABLE 3

Effects of ACTH- and MSH-Related Peptides on the Retention of a Passive Avoidance Response

Treatment schedule	Peptides	Dose[a]	Avoidance latencies[b]			
			Peptide[e]		Saline[e]	
Postlearning[c]	$ACTH_{4-10}$	90.0	69.5	(6)	68.0	(6)
		9.0	74.5	(6)		
	$ACTH_{4-10}$ [7-D-Phe]	90.0	95.0	(6)	75.0	(6)
		9.0	69.0	(6)		
	$\gamma2$-MSH	90.0	33.5*	(6)	80.0	(6)
		9.0	17.5*	(6)		
Preretention[d]	$ACTH_{4-10}$	90.0	164.0*	(6)	79.5	(6)
		9.0	89.5	(6)		
	$ACTH_{4-10}$ [7-D-Phe]	90.0	212.4**	(6)	85.5	(6)
		9.0	105.0	(6)		
	$\gamma2$-MSH	90.0	9.5**	(6)	98.0	(6)
		9.0	24.5*	(6)		
		0.9	101.0	(6)		

[a] Dose in μg/kg given subcutaneously.
[b] Medians in sec; 24 hr after the learning trial.
[c] Immediately after the learning trial.
[d] 1 hr before retention test.
[e] Numbers in parentheses indicate the number of observations.
* $p < .05$.
** $p < .01$ (Mann-Whitney U-test).

ACTH$_{4-10}$[7-D-Phe] increased avoidance latencies. A decrease in passive avoidance retention was observed in rats that received γ2-MSH.

The observations suggest that ACTH$_{4-10}$ and ACTH$_{4-10}$[7-D-Phe] influence retrieval but not consolidation of memory of the passive avoidance response; but γ2-MSH suppresses both consolidation and retrieval. The ineffectiveness of ACTH$_{4-10}$ and ACTH$_{4-10}$[7-D-Phe] following postlearning administration is surprising because Gold and van Buskirk (1976a,b) found that postlearning administration of ACTH causes a dose-dependent effect on the retention of both inhibitory and active avoidance tasks. They found that lower doses of the peptide facilitated later retention, whereas higher doses had an attenuating effect. Enhancement and disruption of retention by posttrial ACTH treatment in an aversively motivated choice task was reported by Sands and Wright (1979). Flood, Jarvik, Bennett, and Orme (1976) found that ACTH$_{4-10}$[7-D-Phe] had an opposite effect. Absence of an effect of ACTH$_{4-10}$ (van Wimersma Greidanus, 1980) and of a potentiated ACTH$_{4-9}$ analog (Martinez et al., 1979) in passive avoidance paradigms have also been reported.

There are at least two alternatives to explain our negative findings. In those experiments in which positive (enhancement or disruption depending on the dose) postlearning effects were found, the whole molecule of ACTH was administered. Shorter ACTH fragments such as ACTH$_{4-10}$ or an ACTH$_{4-9}$ analog were used in the present study and in other studies on rats. It may therefore be that the active sequence in the ACTH molecule that affects consolidation resides outside the 4–10 sequence. It is important to recall that a second active site has been found in the extinction test in the ACTH$_{11-24}$ sequence and acetyl-ACTH$_{11-13}$-NH$_2$ also has some residual activity (Greven & de Wied, 1977). However, this explanation does not agree with the fact that ACTH$_{4-10}$ appears to be active in mice (Flood et al., 1976). Although a species difference may be important, other behavioral factors may be involved. Control animals in the study of Flood et al. (1976) showed very low performance during the retention test and the majority of the mice were considered to be amnestic. Therefore, ACTH$_{4-10}$ improved otherwise pure retention rather than modulating the consolidation of a well-learned response. A high dose of ACTH$_{4-10}$ given after the learning trial attenuated the retention deficit in rats with hereditary hypothalamic diabetes insipidus (Bohus, 1979). Hypophysectomy-induced retention deficits can also be corrected by postlearning administration of ACTH (Gold, Rose, Spanis, & Hankins, 1977). Accordingly, ACTH$_{4-10}$ or related sequences may affect only disturbed or partial consolidation processes.

Passive avoidance retention was also facilitated by ACTH$_{4-10}$ when it was given shortly before the retention test. This observation is a replication of earlier observations showing enhanced retention of passive avoidance (Greven & de Wied, 1973), conditioned suppression (Schneider, Weinberg, & Weissberg, 1974), and conditioned taste aversion behaviors (Levine, Smotherman, & Hennessy, 1977; Rigter & Popping, 1976) by ACTH-related peptides. As ACTH-related peptides affect performance, one may question whether enhanced retention is indeed an effect on memory retrieval. Reversal or attenuation by ACTH-related peptides of amnesias caused by postlearning electroconvulsive shock (Keyes, 1974; Rigter & van Riezen, 1975) or CO_2 inhalation (Rigter, van Riezen, & de Wied, 1974) support the suggestion of influences on retrieval. In addition, $ACTH_{4-10}$ enhances the use of information in a four-table choice test, but only during the first trial (Isaacson, Dunn, Rees, & Waldock, 1976).

The fact that $ACTH_{4-10}$ and $ACTH_{4-10}$[7-D-Phe] had similar effects on retention following pretest administration is unexpected on the basis of findings in avoidance acquisition and appetitive and avoidance extinction studies. However, both peptides alleviate CO_2-induced amnesia (Ramaekers, Rigter, & Leonard, 1978) and enhance retention of conditioned taste aversion (Rigter & Popping, 1976). Accordingly, the amino acid residue phenylalanine (Phe) in Position 7 does not play a key role in the influence of ACTH-related peptides on retrieval processes. The opposite effect of the 7-D-Phe analog on acquisition and extinction behavior may be due to disruption of performance.

V. CONCLUDING REMARKS

Many observations from our and other laboratories suggest that ACTH- and MSH-like peptides affect learning, performance, and retention. An influence on performance, probably due to enhancing or attenuating the motivational properties of environmental stimuli, and thereby increasing or decreasing the probability of the occurrence of stimulus-specific behavior, may account for a number of behavioral effects of these peptides. The hypothesis that ACTH influences brain processes involved in consolidation and retrieval of memory cannot explain all the behavioral effects of ACTH- and MSH-like peptides so far described. Accordingly, multiple interactions between brain mechanisms and ACTH- and MSH-like neuropeptides may exist.

One of the possible causes of multiple effects on behavior may be that the peptide sequence contains more than one bit of behavioral information. This possibility has been discussed in relation to the influence of ACTH and ACTH fragments on consolidation and retrieval of memory. Hard evidence supporting this possibility is as yet missing. Another alternative for the multiplicity of behavioral effects is that, under physiological circumstances, ACTH- and MSH-like peptides originate from different sources. As was mentioned in the Introduction, these peptides may be of pituitary or brain origin. It has been proposed that α-MSH is the primary peptide in the neuronal pro-opiomelanocortin system (Watson & Akil, 1980). However, ACTH may be transported retrogradely to the brain via the vessels of the pituitary stalk or through the cerebrospinal fluid.

Distinct functions for neuropeptides of pituitary and brain origin are suggested by observations in hypophysectomized rats. Removal of the pituitary, which does not affect bio- and immunoreactive ACTH and α-MSH levels in the brain (Krieger, Liotta, and Brownstein, 1977a,b), results in impairment of acquisition and retention of avoidance responses (de Wied, 1964; Bohus and de Wied, 1966). These abnormalities can be corrected by the administration of ACTH or related peptides (Bohus & de Wied, 1966; Bohus et al., 1973; de Wied, 1964, 1969; Gold et al., 1977).

A novel finding is the influence of γ2-MSH on learning and retention of avoidance responses. The existence of this or similar peptides has been proposed by the nucleotide code of bovine pro-opiomelanocortin within the amino terminal portion of the molecule (Nakanishi et al., 1979). The amino terminal glycopeptide has been isolated from pig pituitary material, and an amino acid sequence corresponding to the proposed γ-MSH has been identified (Hakanson, Ekman, Sundler, & Nilsson, 1980). Hakanson et al. (1980) failed to isolate γ-MSH or a similar fragment, but Shibasaki, Ling, & Guillemin (1980) described three immunoreactive γ-MSH peaks in whole bovine pituitary extract. Interestingly, the melanotropic bioactivity of γ-MSH peptides appeared to be negligible (Ling, Ying, Minick, & Guillemin, 1979). Although the existence of γ-MSH as an end product of pro-opiomelanocortin biotransformation is questionable, recent findings suggest the presence of a behaviorally active amino acid sequence in the NH_2-terminus of the precursor molecule. The behavioral activity of γ2-MSH is of an opposite nature to ACTH- and α- or β-MSH-like peptides. Inspection of the structure of γ2-MSH (see Table 4) suggests that the peptide shares the amino acid sequence 6–9 with ACTH- and MSH-like peptides. The behavioral activity of this sequence in comparison to $ACTH_{4-10}$ is low but becomes greater with chain elongation (de Wied et al., 1975; Greven & de Wied, 1977). This may have been the case with dogfish β-MSH, which showed behavioral activity comparable to α-MSH-like peptides (van Wimersma Greidanus, Lowry, Scott, Rees, & de Wied, 1975). Whether the opposite character of behavioral activity in the γ2-MSH or a similar molecule resides in the C- or NH_2-terminal elongation of the molecule, remains to be shown.

The observations described here suggest profound behavioral activity of ACTH- and MSH-like peptides upon peripheral administration. Because the blood–brain barrier is rather impermeable to large peptide molecules such as ACTH (Allen, Kendall, McGilvra, & Vancura, 1974) and only a small portion of peripherally injected peptides enter the brain (see Kastin, Olson, Schally, & Coy, 1979), one may question whether behavioral effects are due to a direct action of these peptides in the brain. However, because $ACTH_{4-10}$ injected directly in the brain mimics the influence of peripherally injected $ACTH_{4-10}$ on extinction behavior (van Wimersma Greidanus & de Wied, 1971) and 100 times less $ACTH_{4-10}$ is needed to affect extinction behavior after intracerebroventricular administration (de Wied, Bohus, van Ree, & Urban, 1978), it

TABLE 4

Amino Acid Sequence of Some MSH-like Peptides

	1	2	3	4	5	6	7	8	9	10	11	12	13		
Ac -	SER -	TYR -	SER -	MET -	GLU -	HIS -	PHE -	ARG -	TRP -	GLY -	LYS -	PRO -	VAL -	NH$_2$	α-MSH
		TYR -	VAL -	MET -	GLY -	HIS -	PHE -	ARG -	TRP -	ASP -	ARG -	PHE -	GLY -	OH	γ2-MSH
·· -	ASP -	TYR -	LYS -	PHE -	GLY -	HIS -	PHE -	ARG -	TRP -	SER -	VAL -	PRO -	LEU -	OH Dogfish	β-MSH

is likely that these peptides act directly on brain mechanisms by passing into the brain.

REFERENCES

Ader, R., Weijnen, J. A. W. M., & Moleman, P. (1972). Retention of passive avoidance response as a function of the intensity and duration of electric shock. *Psychonomic Science*, 26, 125–128.

Allen, J. P., Kendall, J. W., McGilvra, R., & Vancura, C. (1974). Immunoreactive ACTH in cerebrospinal fluid. *Journal of Clinical Endocrinology Metabolism*, 38, 586–593.

Beatty, P. A., Beatty, W. M., Bowman, R. E., & Gilchrist, J. C. (1970). The effects of ACTH, adrenalectomy, and dexamethasone on the acquisition of an avoidance response in rats. *Physiology and Behavior*, 5, 939–944.

Bohus, B. (1979). Inappropriate synthesis and release of vasopressin in rats: Behavioral consequences and effects of neuropeptides. *Neuroscience Letters*, Supplement, 3, 329.

Bohus, B., & de Wied, D. (1966). Inhibitory and facilitatory effect of two related peptides on extinction of avoidance behavior. *Science*, 153, 318–320.

Bohus, B., & de Wied, D. (1980). Pituitary–adrenal system hormones and adaptive behaviour. *In* I. Chester-Jones & I. W. Henderson (Eds.), *General, Comparative and Clinical Endocrinology of the Adrenal Cortex* (Vol. 3), pp. 256–347. London: Academic Press.

Bohus, B., & Endröczi, E. (1965). The influence of pituitary–adrenocortical function on the avoiding conditioned reflex activity in rats. *Acta Physiologica Academiae Scientiarum Hungaricae*, 26, 183–189.

Bohus, B., Endröczi, E., & Lissák, K. (1963). Correlations between avoiding conditioned reflex activity and pituitary–adrenocortical function in the rat. *Acta Physiologica Academiae Scientiarum Hungaricae*, 24, 79–83.

Bohus, B., Gispen, W. H., & de Wied, D. (1973). Effect of lysine vasopressin and $ACTH_{4-10}$ on conditioned avoidance behavior of hypophysectomized rats. *Neuroendocrinology*, 11, 137–143.

Bohus, B., Hendrickx, H. H. L., van Kolfschoten, A. A., & Krediet, T. G. (1975). The effect of $ACTH_{4-10}$ on copulatory and sexually motivated approach behavior in the male rat. *In* M. Sandler & G. L. Gessa (Eds.), *Sexual Behavior: Pharmacology and Biochemistry*, pp. 269–275. New York: Raven.

Bohus, B., Nyakas, C. S., & Endröczi, E. (1968). Effects of adrenocorticotropic hormone on avoidance behaviour of intact and adrenalectomized rats. *International Journal of Neuropharmacology*, 7, 307–314.

de Wied, D. (1964). Influence of anterior pituitary on avoidance learning and escape behavior. *American Journal of Physiology*, 207, 255–259.

de Wied, D. (1966). Inhibitory effect of ACTH and related peptides on extinction of conditioned avoidance behavior. *Proceedings of the Society for Experimental Biology*, 122, 28–32.

de Wied, D. (1969). Effects of peptide hormones on behavior. *In* W. F. Ganong & L. Martini (Eds.), *Frontiers in Neuroendocrinology*, pp. 97–140. New York: Oxford University Press.

de Wied, D. (1980). Pituitary neuropeptides and behavior. *In* K. Fuxe, T. Hökfelt, and R. Luft (Eds.), *Central Regulation of the Endocrine System*, pp. 297–314. New York: Plenum.

de Wied, D., & Bohus, B. (1966). Long term and short term effects on retention of a conditioned avoidance response in rats by treatment with long acting pitressin and α-MSH. *Nature*, 212, 1481–1486.

de Wied, D., Bohus, B., van Ree, J. M. & Urban, I. (1978). Behavioral and electrophysiological effects of peptides related to lipotropin (β-LPH). *Journal of Pharmacology and Experimental Therapeutics*, 204, 570–580.

de Wied, D., Witter, A., & Greven, H. M. (1975). Behaviourally active ACTH analogues. *Biochemical Pharmacology*, **24**, 1463–1468.

Endröczi, E., Lissák, K., Fekete, T., & de Wied, D. (1970). Effects of ACTH on EEG habituation in human subjects. *In* D. de Wied & J. A. W. M. Weijnen (Eds.), *Pituitary, Adrenal, and the Brain: Progress in Brain Research* (Vol. 32), pp. 254–261. Amsterdam: Elsevier.

File, S. E. (1978). ACTH, but not corticosterone impairs habituation and reduces exploration. *Pharmacology, Biochemistry, and Behavior*, **9**, 161–166.

Flood, J. F., Jarvik, M. E., Bennett, E. L., & Orme, A. E. (1976). Effects of ACTH peptide fragments on memory formation. *Pharmacology, Biochemistry and Behavior*, **5**, Supplement 1, 41–51.

Garrud, P., Gray, J. A., & de Wied, D. (1974). Pituitary–adrenal hormones and extinction of rewarded behavior in the rat. *Physiology and Behavior*, **12**, 109–119.

Garrud, P., Gray, J. A., Rickwood, L., & Coen, C. (1977). Pituitary–adrenal hormones and effects of partial reinforcement of appetitive behavior in the rat. *Physiology and Behavior*, **18**, 813–818.

Gold, P. E., & van Buskirk, R. (1976a). Effects of posttrial injections on memory processes. *Hormones and Behavior*, **7**, 509–517.

Gold, P. E., & van Buskirk, R. (1976b). Enhancement and impairment of memory processes with post-trial injections of adrenocorticotrophic hormone. *Behavioral Biology*, **16**, 387–400.

Gold, P. E., Rose, R. P., Spanis, C. W., & Hankins, L. L. (1977). Retention deficit for avoidance training in hypophysectomized rats: Time-dependent enhancement of retention performance with ACTH injections. *Hormones and Behavior*, **8**, 363–371.

Gray, J. A., & Garrud, P. (1977). Adrenopituitary hormones and frustrative nonreward. *In* L. H. Miller, C. A. Sandman, & A. J. Kastin (Eds.), *Neuropeptide Influences on the Brain and Behavior*, pp. 201–212. New York: Raven.

Greven, H. M., & de Wied, D. (1973). The influence of peptides derived from corticotrophin (ACTH) on performance. Structure activity studies. *In* E. Zimmermann, W. H. Gispen, B. H. Marks, & D. de Wied (Eds.), *Drug Effects on Neuroendocrine Regulation. Progress in Brain Research* (Vol. 39), pp. 429–441. Amsterdam: Elsevier.

Greven, H. M., & de Wied, D. (1977). Influence of peptides structurally related to ACTH and MSH on active avoidance behaviour in rats. A structure–activity relationship study. In Tj. B. van Wimersma Greidanus (Ed.), *Frontiers of Hormone Research* (Vol. 4), pp. 140–152. Basel: Karger.

Guth, S., Levine, S., & Seward, J. P. (1971). Appetitive acquisition and extinction effects with exogenous ACTH. *Physiology and Behavior*, **7**, 195–200.

Guth, S., Seward, J. P., & Levine, S. (1971). Differential manipulation of passive avoidance by exogenous ACTH. *Hormones and Behavior*, **2**, 127–138.

Hakanson, R., Ekman, R., Sundler, F., & Nilsson, R. (1980). A novel fragment of the corticotrophin-β-lipotropin precursor. *Nature*, **283**, 789–792.

Harlow, H. F., McGaugh, J. L., & Thompson, R. F. (1971). *Psychology*. San Francisco: Albion.

Herbert, E., Roberts, J., Phillips, M., Allen, R., Hinman, M., Budarf, M., Policastro, P., & Rosa, P. (1980). Biosynthesis, processing, and release of corticotropin, β-endorphin, and melanocyte-stimulating hormone in pituitary cell culture systems. *In* L. Martini & W. F. Ganong (Eds.), *Frontiers in Neuroendocrinology* (Vol. 6), pp. 67–101. New York: Raven.

Isaacson, R. L., Dunn, A. J., Rees, H. D., & Waldock, B. (1976). ACTH$_{4-10}$ and improved use of information in rats. *Physiological Psychology*, **4**, 159–162.

Kastin, A. J., Olson, R. D., Schally, A. V., & Coy, D. H. (1979). CNS effects of peripherally administered brain peptides. *Life Sciences*, **25**, 410–414.

Kendall, J., & Orwoll, E. (1980). Anterior pituitary hormones in the brain and other extrapituitary sites. *In* L. Martini & W. F. Ganong (Eds.), *Frontiers in Neuroendocrinology* (Vol. 6), pp. 33–65. New York: Raven.

Keyes, J. B. (1974). Effect of ACTH on ECS produced amnesia of a passive avoidance task. *Physiological Psychology*, **2**, 307–309.

Krieger, D. T., & Liotta, A. S. (1979). Pituitary hormones in brain: Where, how, and why? *Science*, **205**, 366–372.

Krieger, D. T., Liotta, A., & Brownstein, M. J. (1977a). Presence of corticotrophin in brain of normal and hypophysectomized rats. *Proceedings of the National Academy of Sciences*, **74**, 648–652.

Krieger, D. T., Liotta, A., & Brownstein, M. J. (1977b). Presence of corticotrophin in limbic system of normal and hypophysectomized rats. *Brain Research*, **128**, 575–579.

Levine, S., & Jones, L. E. (1965). Adrenocorticotropic hormone (ACTH) and passive avoidance learning. *Journal of Comparative and Physiological Psychology*, **59**, 357–360.

Levine, S., Smotherman, W. P., & Hennessy, J. W. (1977). Pituitary–adrenal hormones and learned taste aversion. In L. H. Miller, C. A. Sandman, & A. J. Kastin (Eds.), *Neuropeptide Influences on the Brain and Behavior*, pp. 163–177. New York: Raven.

Ley, K. F., & Corson, J. A. (1972). Sex differences in behavioral response to ACTH depend on time of testing and UCS intensity. *International Journal of Psychobiology*, **2**, 265–271.

Ling, N., Ying, S., Minick, S., & Guillemin, R. (1979). Synthesis and biological activity of four γ-melanotropin derived from the cryptic region of the adrenocorticotrophin/ β-lipotropin precursor. *Life Sciences*, **25**, 1773–1780.

Lissák, K., & Bohus, B. (1972). Pituitary hormones and avoidance behavior of the rat. *International Journal of Psychobiology*, **2**, 103–115.

Lowry, P. J., Silman, R., Jackson, S., & Estivariz, F. (1979). The lipotropin- and corticotropin-related peptides of the mammalian pituitary. In M. T. Jones, B. Gilham, M. F. Dallman, & S. Chattopadhyay (Eds.), *Interaction within the Brain–Pituitary–Adrenocortical System*, pp. 1–6. London: Academic Press.

Mains, R. E., Eipper, B. A., & Ling, N. (1977). Common precursor to corticotrophins and endorphins. *Proceedings of National Academy of Sciences*, **74**, 3014–3018.

Martinez, J. L., Jr., Vasquez, B. J., Jensen, R. A., Soumireu-Mourat, B., & McGaugh, J. L. (1979). ACTH$_{4-9}$ analog (Org 2766) facilitates acquisition of an inhibitory avoidance response in rats. *Pharmacology, Biochemistry and Behavior*, **10**, 145–147.

McGaugh, J. L. (1961). Facilitative and disruptive effects of strychnine sulphate on maze learning. *Psychological Reports*, **9**, 99–104.

Miller, R. E., & Ogawa, N. (1962). The effect of adrenocorticotrophic hormone (ACTH) on avoidance conditioning in the adrenalectomized rat. *Journal of Comparative and Physiological Psychology*, **55**, 211–213.

Mirsky, I. A., Miller, R., & Stein, M. (1953). Relation of adrenocortical activity and adaptive behavior. *Psychosomatic Medicine*, **15**, 574–584.

Nakanishi, S., Inoue, A., Kita, T., Nakamura, M., Chang, A. C. Y., Cohen, S. N., & Numa, S. (1979). Nucleotide sequence of cloned cDNA for bovine corticotrophin–lipotropin precursor. *Nature*, **278**, 423–427.

Nyakas, C., Bohus, B., & de Wied, D. (1980). Effects of ACTH$_{4-10}$ on self-stimulation behavior in the rat. *Physiology and Behavior*, **24**, 759–764.

Olds, J. (1976). Brain stimulation and the motivation of the behavior. In M. A. Corner & D. F. Swaab (Eds.), *Perspectives in Brain Research: Progress in Brain Research* (Vol. 45), pp. 401–425. Amsterdam: Elsevier.

Pagano, R. R., & Lovely, R. H. (1972). Diurnal cycle and ACTH facilitation of shuttlebox avoidance. *Physiology and Behavior*, **8**, 721–723.

Phifer, R. F., Orth, D. N., & Spicer, S. S. (1974). Specific demonstration of the human hypophyseal adrenocortico–melanotropic (ACTH–MSH) cell. *Journal of Clinical Endocrinology and Metabolism*, **39**, 684–692.

Ramaekers, F., Rigter, H., & Leonard, B. E. (1978). Parallel changes in behavior and hippocampal monoamine metabolism in rats after administration of ACTH-analogues. *Pharmacology, Biochemistry and Behavior*, **8**, 547–551.

Rigter, H., & Popping, A. (1976). Hormonal influences on the extinction of conditioned taste aversion. *Psychopharmacologia*, **46**, 255–261.

Rigter, H., & van Riezen, H. (1975). Anti-amnesic effect of $ACTH_{4-10}$: Its independence of the nature of the amnesic agent and the behavioral test. *Physiology and Behavior*, **14**, 563–566.

Rigter, H., van Riezen, H., & de Wied, D. (1974). The effects of ACTH- and vasopressin-analogues on CO_2-induced retrograde amnesia in rats. *Physiology and Behavior*, **13**, 381–388.

Roberts, J. L., & Herbert, E. (1977). Characterization of a common precursor to corticotropin and β-lipotropin: Cell-free synthesis of the precursor and identification of corticotropin peptides in the molecule. *Proceedings of National Academy of Sciences*, **74**, 4826–4830.

Rolls, E. T. (1977). The neurophysiological basis of brain-stimulation reward. *In* A. Waquier & E. T. Rolls (Eds.), *Brain-Stimulation Reward*, pp. 65–87. Amsterdam: North Holland.

Sandman, C. A., Kastin, A. J., & Schally, A. V. (1971). Behavioral inhibition as modified by melanocyte stimulating hormone (MSH) and light–dark conditions. *Physiology and Behavior*, **6**, 45–48.

Sands, S. F., & Wright, A. A. (1979). Enhancement and disruption of retention performance by ACTH in a choice task. *Behavioral and Neural Biology*, **27**, 413–422.

Schneider, A. M., Weinberg, J., & Weissberg, R. (1974). Effects of ACTH on conditioned suppression: A time and strength of conditioning analysis. *Physiology and Behavior*, **13**, 633–636.

Shibasaki, T., Ling, N., & Guillemin, R. (1980). A radioimmunoassay for γ-melanocyte stimulating hormone. *Life Sciences*, **26**, 1781–1785.

Spear, N. E. (1973). Retrieval of memory in animals. *Psychological Review*, **80**, 163–194.

Stratton, L. O., & Kastin, A. J. (1974). Avoidance learning at two levels of motivation in rats receiving MSH. *Hormones and Behavior*, **5**, 149–155.

Stratton, L. O., & Kastin, A. J. (1975). Increased acquisition of a complex appetitive task after MSH and MIF. *Pharmacology, Biochemistry and Behavior*, **3**, 901–904.

Tolman, E. C. (1932). *Purposive Behavior in Animals and Men*. New York: Appleton-Century-Crofts.

van Wimersma Greidanus, Tj. B. (1980). Effects of MSH and related peptides on avoidance behavior in rats. In Tj. B. van Wimersma Greidanus (Ed.), *Frontiers in Hormone Research* (Vol. 4), pp. 129–139. Basel: Karger.

van Wimersma Greidanus, Tj. B., & de Wied, D. (1971). Effects of systemic and intracerebral administration of two opposite-acting ACTH-related peptides on extinction of conditioned avoidance behavior. *Neuroendocrinology*, **7**, 291–310.

van Wimersma Greidanus, Tj. B., Lowry, P. J., Scott, A. P., Rees, L. H., & de Wied, D. (1975). The effects of dogfish MSH's and of corticotrophin-like intermediate lobe peptides (CLIP's) on avoidance behavior in rats. *Hormones and Behavior*, **6**, 319–327.

Watson, S. J., & Akil, H., (1980). α-MSH in rat brain: Occurrence within and outside of β-endorphin neurons. *Brain Research*, **182**, 217–223.

CHAPTER 4

ACTH Modulation of Memory Storage Processing

Paul E. Gold and Richard L. Delanoy

If administered shortly after training, injections of adrenocorticotropin (ACTH) can enhance or impair later retention of avoidance training. These effects on memory storage are time-dependent; delayed injections (e.g., 2-hr posttraining) do not alter retention performance. It therefore seems possible that endogenous hormonal responses to training, including ACTH release, may modulate memory storage processing for the information provided by the learning experience.

This chapter reviews the effects on memory of posttraining injections of ACTH and describes potential neurobiological mechanisms underlying the role of ACTH in memory storage. Specifically, we suggest that ACTH may act on memory through two parallel central nervous system actions. First, ACTH may act on individual neurons to regulate neurophysiological responses to other inputs and, second, ACTH release may potentiate the release of central norepinephrine. The neurobiological actions of the two agents, ACTH and norepinephrine, may act together to enhance and to maintain the neuronal representation of recent experiences.

I. MODULATION OF MEMORY STORAGE PROCESSES

More than 30 years ago, Duncan (1949) reported that a posttraining treatment, electroconvulsive shock, could impair later retention performance of a recently learned response. Although early studies focused on treatments that

ENDOGENOUS PEPTIDES
AND LEARNING AND MEMORY PROCESSES

impaired memory, McGaugh and Dawson (1971) and McGaugh (1973) established the phenomenon of memory enhancement (i.e., some posttraining treatments result in an improvement in later retention performance). A major feature of both memory impairment and memory enhancement experiments is that the effects are time-dependent. A particular treatment has its largest effect on memory if the treatment is administered shortly after training.

On the basis of these findings, a variety of theories of memory consolidation were proposed, generally using the time-dependent nature of the memory effects to support concepts of short- and long-term memory fixation (Barondes, 1970; Gerard, 1949, 1955; Hebb, 1949; McGaugh, 1966). Although it seems clear that memory susceptibility to modification typically decreases with time after training, it is less clear that the time-dependent effects on memory define a basic property of memory storage processes (Gold & McGaugh, 1975). In particular, it seems quite unlikely that a specific temporal gradient of retrograde amnesia or enhancement defines the time-course of memory consolidation. Primarily, this statement is based on the findings in several laboratories that the temporal gradient varies with the intensity (or dose) of a particular treatment (e.g., Alpern & McGaugh, 1968; Cherkin, 1969; Gold, Macri, & McGaugh, 1973; Mah & Albert, 1973; Miller, 1968; cf. Gold & McGaugh, 1975). Thus, the temporal gradient appears to be a measure of the treatment's effectiveness in altering later retention performance rather than an indicant of time-dependent memory processes per se.

Still, the facts that memory storage can be altered by posttraining treatments and that there is a reliable decrease in memory susceptibility over time after training require explanation. Furthermore, it seems particularly useful to generate an explanation that incorporates the possibility that time-dependent posttrial effects on memory reflect mechanisms that have biologically adaptive significance to the animal. These considerations led to the proposal that there may be a set of endogenous responses to training (e.g., responses in arousal level, autonomic function, or neuroendocrine activity) that, in untreated animals, might enhance or impair later retention performance. When such a proposal is applied to studies of retrograde amnesia and enhancement of memory processes, there are two major points to consider. First, agents that modify memory may do so, not because of actions on those neural systems directly involved in storing information, but by altering the pattern of endogenous physiological responses to training. Second, the time-dependent nature of effects on memory may reflect an interaction between the biological effects of the treatment with the endogenous responses to training; responses that typically follow training and continue for a relatively short time thereafter (Gold & Reigel, 1980). We refer to the effects of posttraining treatments on later retention performance as memory modulation, defined as a modification in retention performance produced either by posttraining treatments, or by posttrial endogenous responses to training. The term *modulation* is more descriptive than others based on theoretical explanations of such findings (e.g., memory consolidation, short- and long-term memory processes) (Gold & McGaugh, 1978a,b).

If posttraining hormonal responses to training are important endogenous modulators of memory storage, retention performance may be poor following training with a relatively weak training footshock, not only because the footshock elicits less pain, but also because the hormonal responses are minimal. Therefore, injections of some set of hormones may mimic the endogenous hormonal response to a more intense footshock and thus facilitate memory processing (as interpreted on the basis of later retention performance). It was within this framework that we examined the effects on memory of several different hormones, including ACTH, injected subcutaneously (s.c.) shortly after training. The findings indicated that posttraining injections of ACTH, epinephrine, or norepinephrine (NE) may modulate memory processing. In this chapter, we shall first review some of the behavioral effects we have obtained with ACTH and then discuss some possible neurobiological bases for these effects.

II. ACTH EFFECTS ON MEMORY STORAGE

In a series of experiments, we examined the possibility that posttraining injection of some hormones may facilitate retention performance following training in a one-trial inhibitory (passive) avoidance task (Gold & van Buskirk, 1976a,b). The results of one such study are shown in Fig. 1. Note that the dose–response relationship of ACTH and memory results in an inverted U-shaped curve. Thus, a moderate ACTH dose (administered s.c.) enhanced retention performance and, surprisingly, a higher dose produced retrograde amnesia. We obtained analogous results using epinephrine as the posttrial agent (Gold & van Buskirk, 1975). Furthermore, both the memory-enhancing and amnesic effects of ACTH and epinephrine were time-dependent; injections delayed by 2 hr or more had no effect on retention performance. ACTH, as well as epinephrine, had effects on memory comparable to those of a wide range of treatments that modulate memory—retrograde enhancement, retrograde amnesia, and time-dependency. These findings are therefore consistent with the possibility that many agents that modulate memory may do so shortly after training by affecting the hormonal state of the animal.

A second feature of hormonal enhancement of memory processing is that the dose–response curve for facilitation and impairment of memory appears to interact strongly with training-related stress. An illustration of these findings is shown in Fig. 2. In this experiment (Gold & van Buskirk, 1976a), each rat received a posttraining ACTH injection of either 3 or 6 IU following a single

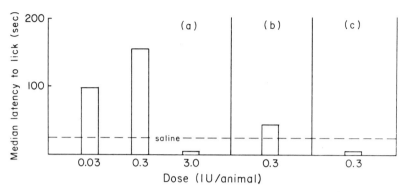

Fig. 1. Median retention latencies of rats trained in a one-trial inhibitory (passive) avoidance task. The animals were pretrained to lick from a waterspout at the far end of a shock compartment. After 4 days of pretraining, the rats received a weak footshock while drinking. Retention was measured as the latency to return to the waterspout on the next day. Note that the effects on memory of immediate posttrial injections of ACTH exhibited an inverted-U dose–response relationship. (a) Animals receiving the two lowest ACTH doses (0.03 or 0.3 IU/animal) immediately after training had significantly enhanced retention performance; those animals that received 3.0 IU/animal immediately after training had significantly impaired retention. (b) ACTH injections delayed by 2 hr after training had no effect on later retention performance. (c) ACTH injections with no footshock had no effect on later retention. (From Gold & van Buskirk, 1976a.)

inhibitory (passive) avoidance training trial at one of four footshock levels. At the lowest footshock level, both ACTH doses enhanced retention performance. At the second slightly higher footshock level, the lower ACTH dose enhanced retention performance, but the higher dose produced amnesia. At the two highest footshock intensities, both doses produced amnesia. Therefore, it appears that the exogenously administered ACTH summates with endogenous responses to training to produce its effects on memory. Other studies have also found interactions between training-related stress and the effects of posttrial treatment on memory. Epinephrine (Gold & van Buskirk, 1976b, 1978a,b), pentylene-tetrazol (Krivanek & McGaugh, 1968; Krivanek, 1971), melanocyte-stimulating hormone (MSH) (Nockton, Kastin, Elder, & Schally, 1972), and electrical stimulation of the midbrain reticular formation (Bloch, 1970) and amygdala (Gold, Hankins, Edwards, Chester, & McGaugh, 1975; Gold, Rose, & Hankins, 1978) each enhance retention of training in low stress situations and impair retention in high stress (strong or multiple footshocks) situations. The inverted U-shaped memory curve resulting from these treatments, as well as the interaction with the severity of the training footshock, suggest that the well-known Yerkes–Dodson Law, which predicts an inverted U relationship between stress and performance (Malmo, 1967), may be mediated by hormonal systems, perhaps including ACTH and epinephrine.

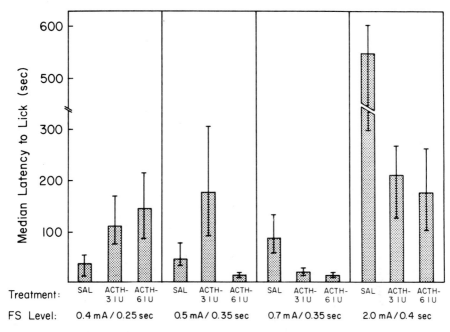

Fig. 2. Median retention latencies (and interquartile ranges) of rats trained in an inhibitory (passive) avoidance task that received posttraining saline or ACTH (3.0 or 6.0 IU) injections. Note that the 3.0 IU ACTH dose enhanced later retention of training with the two lowest footshock levels, but disrupted retention of training with the two highest footshock levels. The 6.0 IU ACTH injection significantly enhanced retention only at the lowest footshock level; this ACTH dose significantly impaired retention of training with all other footshock levels. (From Gold & van Buskirk, 1976a.)

In fact, it is possible that such stress-related responses to training and to memory-modulating agents underlie many classes of memory-enhancing and memory-impairing posttraining treatments. In a recent series of experiments, we found that pretrial injections of several adrenergic antagonists can attenuate the amnesias produced by epinephrine, supraseizure frontal cortex stimulation, pentylenetetrazol, subseizure amygdala stimulation, diethyldithiocarbamate (a NE synthesis inhibitor), and cycloheximide (a protein synthesis inhibitor) (Gold & Sternberg, 1978; Sternberg & Gold, 1980). Memory enhancement produced by electrical stimulation of the amygdala is also blocked by pretreatment with adrenergic antagonists (Sternberg & Gold, 1981). Therefore, it may be the case that memory modulation by agents thought to act by very different neurobiological mechanisms may instead all modulate memory through effects on a restricted set of physiological responses to training and posttraining treatments.

III. MECHANISMS UNDERLYING ACTH MEMORY MODULATION

The effects of ACTH on learned behavior have been variously attributed to mechanisms underlying arousal, dearousal, anxiety, motivation, memory retrieval systems, and memory storage systems. Most of these viewpoints are represented in other chapters, and we can readily accept the view that ACTH has multiple effects on learned behavior. Although hypothetical constructs may offer means of classifying the many effects of ACTH on learned behaviors, psychological terms need not be taken as a classfication scheme for biological modes of action. This seems particularly true when one recognizes that these terms are not mutually exclusive behavioral constructs; motivation, arousal, memory, perception, etc., all interact in most behavioral situations. For this reason, it may be that ACTH appears to fit most or all of the constructs somewhat, but none of them perfectly. Either ACTH is an extremely diverse, multifaceted hormone, or the constructs do not translate in isomorphic fashion to neurobiological mechanisms. Memory modulation may seem at the outset to be an interpretation based on still one more hypothetical construct. However, by restricting the discussion of memory to operational definitions of it (i.e., retention performance), modulating effects of ACTH and other treatments can be pursued entirely at a biological level by relating biological actions of ACTH to performance of learned responses. What follows is a discussion of a very selective (biased) collection of biological events elicted by ACTH that can be evaluated on different levels for their contribution to memory modulation.

A. Peripheral Actions of ACTH

1. Steroidogenesis

As a first guess, one likely mechanism by which ACTH modulates memory might be through its classical steroidogenic properties. This possibility was discounted by the studies of de Wied and others (de Wied, 1967, 1974; de Wied, van Delft, Gispen, Weijnen, & van Wimersma Greidanus, 1972) who found that ACTH analogs devoid of steroidogenic properties retained effects on extinction performance in several avoidance tasks. In terms of memory modulation by posttrial treatments, the question of possible steroidogenic actions mediating changes in retention of shock-avoidance training has not been completely evaluated. Gold and van Buskirk (1976b) failed to find effects of corticosterone on retention of an inhibitory avoidance response. Given the sometimes narrow inverted U-shaped dose–response curves for memory-enhancing agents, as well as important interactions with training-related stress, a more complete evaluation of potential corticosterone effects on memory may be warranted. Furthermore, Flood, Vidal, Bennett, Orme, Vasquez, and Jarvik (1978) reported that injections of hydrocortisone and cortisone, if administered to mice soon after T-maze training, enhanced later retention performance. Although nonsteroidogenic ACTH analogs enhance retention if administered just prior to a retention test trial, posttrial injections of an ACTH$_{4-9}$ analog failed to enhance later retention performance of an inhibitory avoidance response in young adult (Martinez, Vasquez, Jensen, Soumireu-Mourat, & McGaugh, 1979) and aged rats (Gold, Vasquez, & McGaugh, unpublished findings). Thus, the mechanisms mediating the memory modulatory actions of posttraining ACTH on later retention performance may differ from those mediating ACTH effects on extinction performance with pretest injections. The possibility that ACTH memory modulation may involve release of adrenocortical steroids is further supported by the very interesting findings that postexperience injections of ACTH can enhance later submissiveness in mice following an attack, and that the effects appear to be mediated by corticosterone (Leshner, Merkle, & Mixon, Chapter 8; Roche & Leshner, 1979). These behavioral measures appear to share many common features with shock-motivated inhibitory avoidance training, but assess the behavior in a more naturalistic setting. Analogous experiments that independently assess the contributions of ACTH and corticosterone to memory modulation in shock-motivated tasks have not been performed. The possibility remains, therefore, that the effects on memory of posttrial ACTH injections may be mediated by the hormone's steroidogenic properties.

2. Plasma Catecholamines

As described in the previous section, posttrial epinephrine and ACTH injections have similar effects on memory. Both hormones facilitate and impair later retention performance according to an inverted U-shaped dose–response function. A single dose of either hormone will enhance retention of low-level footshock training and impair retention of high-level footshock training. Finally, these hormones modulate memory storage only if they are administered within 2 hr of training. For these reasons, McCarty and Gold (1981) examined the possibility that ACTH injections might result in the release of adrenal medullary epinephrine or NE. Plasma samples were taken from an indwelling tail arterial catheter at various times after injection of ACTH, epinephrine, or footshock. These samples were later assayed for epinephrine and NE concentrations. As shown in Fig. 3, samples taken immediately (within 60 sec) after a footshock indicate that plasma epinephrine concentrations increase dramatically—to a level 15–20 times above basal values. Similar results were obtained for plasma norepinephrine. An injection of epinephrine at a dose that facilitates later

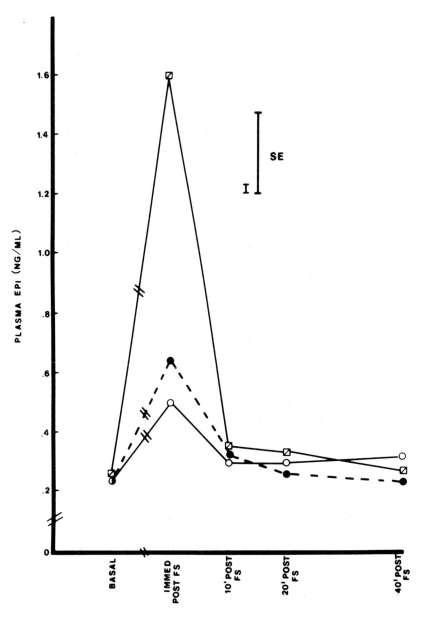

Fig. 3. Plasma epinephrine responses (and range of standard errors) following handling and footshock in an inhibitory (passive) avoidance task. Handled controls indicated by ○, low footshock by ●, and high footshock by ▨. Epinephrine levels in all groups increased in samples taken immediately after placement in the avoidance apparatus (basal versus immediate samples). In the immediate posttraining samples, high footshock training resulted in an increase in plasma epinephrine to a level that was significantly higher than the concentrations seen in either of the other groups. (From McCarty & Gold, 1981.)

retention performance (0.1 mg/kg) increased plasma levels to approximately the same concentration as those seen immediately after training, although the plasma epinephrine concentration remained high for a much longer time after injection than after footshock. In contrast, following injection of either ACTH or ORG 2766 (an ACTH$_{4-9}$ analog), plasma epinephrine and NE concentrations were unchanged at all time intervals up to 80 min (Fig. 4). In addition, if administered immediately after footshock training, ACTH did not alter the epinephrine response to the footshock. Thus, ACTH does not affect memory storage by releasing adrenal medullary catecholamines. It still remains possible that epinephrine causes the release of ACTH. For example in rats, but not in humans and dogs (Ganong, 1963), epinephrine injections result in increased concentrations of plasma glucocorticoids.

B. Central Actions of ACTH

The next section of this chapter deals with ways in which ACTH may affect the brain through mechanisms that could be related to memory processing. We will initially discuss potential access of ACTH to the brain and then will discuss possible mechanisms of action. These portions of the chapter are admittedly speculative. The hypothesis guiding this section is that ACTH effects on memory may be mediated by two parallel neurobiological actions. First, ACTH may act on individual neurons to regulate neurophysiological responses to other inputs. Second, ACTH may result in the release of central NE. One view of the role of NE in memory modulation will be described at the end of the chapter.

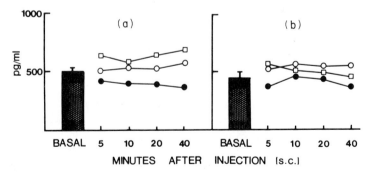

Fig. 4.   Plasma (a) epinephrine and (b) norepinephrine levels measured before (basal) and at various times after low footshock and ACTH (subcutaneous) injections. Saline is indicated by ●, ACTH 0.3 IU by ○, and ACTH 3.0 IU by □. The ordinate is greatly expanded compared to the preceding figure. Note that the ACTH injection did not significantly alter plasma epinephrine concentrations at those intervals tested. (From McCarty & Gold, 1981.)

1. Entry to the Brain

The possibility that ACTH has access to the central nervous system (CNS) now seems far more likely than it did in the recent past. Previously, the major evidence that ACTH effects on memory were mediated by a direct influence on the CNS came from two types of studies (cf. Dunn & Gispen, 1977). First, because ACTH analogs devoid of steroidogenesis have behavioral potency (de Wied, 1969, 1974; de Wied et al., 1972), exclusion of the corticosteroid response left direct action on the brain as a mechanism available by default. Second, ACTH and its analogs could affect behavior if they were administered directly (as crystalline material) into some brain regions (van Wimersma Greidanus & de Wied, 1971). Although quite interesting, the findings of these later studies do not precisely address the issue of whether peripheral injections have analogous behavioral effects mediated by quite different—and perhaps non-physiological—mechanisms.

However, the CNS entry of peripheral peptides from the periphery has recently shifted from being considered impossible to likely. Oliver, Mical, and Porter (1977) demonstrated that retrograde blood flow and high hormone concentrations exist in the hypophyseal portal system. As a result, entry of pituitary hormones into the median eminence appears to be more probable. Consistent with these findings, Mezey, Palkovits, DeKloet, Verhoef, & de Wied (1978) reported brain uptake of an ACTH analog after injection into the pituitary. Additional evidence that peripheral peptides can gain access to the CNS comes from the findings that blood-borne insulin can be shown to label axons and nerve terminals in rat median eminence and arcuate nucleus (van Houten, Posner, Kopriwa, & Brawer, 1980) and that plasma opioid peptides also enter the brain in moderate amounts (Rapoport, Klee, Pettigrew, & Ohno, 1980). Furthermore, there is now evidence that some pituitary hormones are released not only from terminals in the pituitary but also from CNS sites (e.g., vasopressin: Buijs, Swaab, Dogterom, & van Leeuwen, 1978; ACTH: Barchas, Akil, Elliott, Holman, & Watson, 1978) and that endogenous release of ACTH into the cerebrospinal fluid may mediate behavioral responses to mildly stressful stimuli (Dunn, Green, & Isaacson, 1979).

Another potential mechanism for ACTH access to the brain may involve functions of other hormones. For example, vasopressin, a hormone released with ACTH in response to stress, appears to have the capacity to increase brain vascular permeability (Raichle, Hartman, Eichling, & Sharpe, 1975). Also, peripheral epinephrine injections appear to reduce the blood–brain barrier to albumin (Johansson & Martinsson, 1980). Thus, it is possible that some stress-related hormones may facilitate ACTH access into the brain. This evidence supports the view that the behavioral effects of ACTH, in particular those not mediated by steroids, could be the result of direct brain action by ACTH. On the other hand, it may not be necessary for ACTH to enter the brain and to reach widespread regions. There may be receptors at specific brain sites that have particularly high blood–brain permeability, such as the area postrema (Lichtensteiger, Lienhart, & Kopp, 1977) and these receptors may mediate the effects of plasma ACTH on the brain.

2. Neurophysiology

Peripherally administered ACTH can affect such physiological parameters as ketosis, lipolysis, hypoglycemia, and insulin resistance, but these responses seem to be high-dose effects and do not reflect normal physiological actions of the hormone (Engel, 1961). At lower doses, ACTH appears to have the property of altering physiological responses, not by simply increasing or decreasing the rate of neuronal firing, but, by decreasing the variance around the means of several physiological measures (Miller, Fischer, Groves, & Rudrauff, 1977). In addition, ACTH may reduce the sensitivity of some measures to change in response to altered inputs. For example, Bohus (1974) found that $ACTH_{4-10}$ injections resulted in lower susceptibility to increases in blood pressure produced by electrical stimulation of both the posterior hypothalamus and the brainstem reticular formation. There are also reports that $ACTH_{4-10}$ injections result in decreased anxiety (Miller, Kastin, Sandman, Fink, & van Veen, 1974) and that α-melanocyte-stimulating hormone (α-MSH) injections result in an increased occurrence of EEG α-waves (Dyster-Aas & Krakou, 1965; Sandman, Denman, Miller, & Knott, 1971).

These results could be interpreted as stabilization of the activity of physiological systems in response to environmental input. One mechanism by which ACTH might have such an effect is suggested by the findings of a study in which $ACTH_{4-10}$ decreased membrane resistance in the giant dopaminergic neuron (GDN) of the snail (Lichtensteiger & Felix, 1979). A decrease in membrane resistance would reduce current spread down a dendrite from a given synaptic input because of a relatively increased current shunt across the membrane. As a result, the GDN would be less responsive to any synaptic inputs, excitatory or inhibitory. Instead of tight coupling of changes in neuronal firing with changes in synaptic input, a neuron's firing rate would tend to stabilize without necessarily altering the mean rate of firing. In short, decreased membrane resistance should dampen transient changes in firing rates. In addition, although all inputs would be attenuated somewhat, control of neuronal firing would be biased toward proximal dendritic and somatic inputs, because decreased current spread would preferentially tend to electrically isolate the dendritic inputs distal to the cell body. Consequently, ACTH may function to amplify particular synaptic inputs and might thereby potentiate the importance to brain activity of particular neuronal system programs or repertoires.

(This idea brings to mind, perhaps inappropriately, an analogous interpretation of much of the behavioral data described in this chapter and in Leshner et al. (Chapter 8) and Martin (Chapter 5). The suggestion is that ACTH may function to select those behavioral repertoires that are particularly important in the experiential context of the experiments—whether to freeze or to flee, whether to be submissive or aggressive.) Although these interpretations of direct neuronal actions of ACTH are conjectural, the findings do suggest a useful framework within which to attempt to relate ACTH effects on CNS electrophysiology to modulation of memory storage processes.

3. Effects of Central Catecholamines

In contrast to the paucity of information about electrophysiological actions of ACTH, it seems quite clear that peripherally administered ACTH can alter the activity of central catecholaminergic systems (Dunn & Gispen, 1977). For example, Lichtensteiger, et al. (1977) demonstrated that peripherally administered $ACTH_{4-10}$ and α-MSH increased forebrain dopamine fluorescence. This response was abolished by lesions of the area postrema, a brainstem structure with particularly high blood–brain barrier permeability. Many other studies have examined the effects of exogenous ACTH on brain NE metabolism. In experiments that examined norepinephrine turnover, the findings appear to vary with the methods employed to assess such turnover. Those experiments in which turnover was measured by examining the recovery of radioactive catecholamines converted from labeled tyrosine do not indicate an effect of $ACTH_{1-24}$, $ACTH_{4-10}$ (Iuvone, Morasco, Delanoy, & Dunn, 1978) or α-MSH (Iuvone et al., 1978; Leonard, Kafoe, Thody, & Shuster, 1976) on NE metabolism. However, glucocorticoid injections apparently do result in increased NE turnover assessed with this method (Iuvone, Morasco, & Dunn, 1977). In contrast, a rather consistent effect of ACTH on NE turnover has been obtained by quantifying the decline of endogenous NE following synthesis inhibition with α-methyl-p-tyrosine (α-MPT). Using this method of measuring turnover, peripheral administration of $ACTH_{1-24}$ (Hokfelt & Fuxe, 1972, $ACTH_{4-10}$ (Leonard, Ramaekers, & Rigter, 1975; Versteeg, 1973) and α-MSH (Kostrzewa, Kastin, & Spirtes, 1975), each increased the rate of disappearance (turnover) of NE. In addition, hypophysectomy caused a decrease in NE turnover assessed with this method, and the decrease appeared to be mediated by ACTH and not by glucocorticoids (Fuxe, Hokfelt, Jonsson, Levine, Lidbrink, & Lofstrom, 1973; Javoy, Glowinski, & Kordon, 1968).

The seeming inconsistencies of these reports may, in fact, have a ready explanation. Injections of α-MPT are stressful as defined by a steroidogenic response (Scapagnini & Preziousli, 1972) and, most likely, many physiological parameters in addition to steroidogenesis are also affected. Only in those turnover studies using α-MPT was a consistent activation of NE turnover observed following peripheral ACTH injection. The use of α-MPT in studies examining NE turnover may therefore alter the results. In constrast, most studies that measured NE using tracer procedures did not find changes following ACTH injection. Thus, those situations in which ACTH caused obvious changes in NE metabolism all appear to contain an element of stressful stimulation. Therefore, one CNS action of ACTH may be to enhance the release of NE in response to a training footshock or other stressor.

In the rat, footshock and other stressors result in a transient decrease in brain NE content (cf. Stone, 1975), a result generally interpreted as reflecting release of the amine. The extent (typically 20–40% reduction) and duration (min to hr) appear to vary with the severity and duration of the stressor. In the context of memory studies, a fairly intense training footshock (2–3 mA, 1–2 sec) results in a transient 20% decrease in brain NE that is maximal 10 min after training and recovers to baseline over the next 30 min (Gold & van Buskirk, 1978a). A lower footshock intensity results in little change in brain NE content and, as will be discussed in what follows, results in relatively poor retention performance. Several studies have examined the effect of ACTH and α-MSH on brain NE content. Most studies found no effect of the injections of these hormones (Kostrzewa et al., 1975; Leonard et al., 1976). Others found only effects restricted to particular brain areas, such as a decrease in NE content in the ventromedial nucleus of the hypothalamus and the arcuate nucleus and an increase in NE content in the locus coeruleus (Fekete, Stark, Herman, Palkovits, & Kanyicska, 1978; Herman, Fekete, Palkovits, & Stark, 1977).

In preliminary studies, we have found that ACTH injections following low-level footshock training in an inhibitory avoidance task result in a 20–40% decrease in brain NE content when it was measured 10 min after training and injection (Fig. 5) (Gold & McGaugh, 1978b). Brain NE concentrations return to control levels within 30 min (Fig. 5). These results need to be confirmed, but they suggest that an ACTH injection may result in a transient decrease in brain NE content that is comparable to that produced by footshock. Thus, one CNS action of ACTH may be to augment the NE released after a training footshock, a possibility compatible with the results of the NE turnover studies described earlier.

IV. NORADRENERGIC MEMORY MODULATION

In a series of recent experiments, we found that the extent of a posttraining decrease in brain NE is an excellent predictor, over a wide range of experimental conditions, of later retention performance in comparably trained and treated animals (Gold & Murphy, 1980; Gold & van Buskirk, 1978a,b). Briefly, the

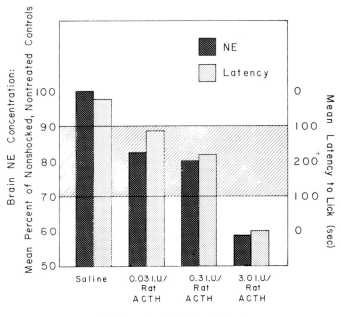

Fig. 5. Forebrain (telencephalon-diencephalon) norepinephrine concentration and retention performance of rats trained with low footshock in an inhibitory (passive) avoidance task. Immediately following the footshock, the animals received a subcutaneous injection of saline or ACTH (0.03, 0.3, or 3.0 IU/rat). The retention performance scale is bidirectional in order to distinguish low latencies resulting from poor retention following weak footshock training itself, with that following an amnestic ACTH dose. Note that retention performance is optimal when forebrain NE concentration exhibits a transient 20% decrease 10 min after training.

results indicate that optimal retention performance is seen under conditions in which brain NE decreases by approximately 20% as measured 10 min after training. A decrease of this extent can be produced by a strong training footshock or by a weak footshock followed by some memory-enhancing treatments including ACTH, as just described. Treatments that increase the noradrenergic response to relatively intense footshock result in amnesia. These findings suggest that a central noradrenergic response to training may modulate the storage of the new information provided by the training experience (cf. Gold & McGaugh, 1978b). There are many neurobiological properties of central noradrenergic systems that seem to be compatible with this view. For example, forebrain axon terminals labeled with ^{3}H-NE seldom appear to make "classic" postsynaptic contacts; most often, the terminals are not in close apposition to other cells (Beaudet & Descarries, 1978). When combined with the diffuse projections of individual locus coeruleus neurons (cf. Amaral & Sinnamon, 1977; Moore

& Bloom, 1979), the anatomical evidence suggests that forebrain NE may mediate a widespread, relatively slow, tropic effect on a very large population of neurons.

Neurophysiological examination of noradrenergic actions indicate that iontophoretically applied NE generally inhibits the action potential frequency in most neurons (Cooper, Bloom, & Roth, 1978). However, considerable evidence suggests that NE does not merely serve to inhibit receptive neurons (cf. Dismukes, 1979). For example, Segal and Bloom (1976) found that auditory stimulus-induced suppressions of hippocampal pyramidal cell activity were correlated with increases in the activity of locus coeruleus neurons. Segal and Bloom also found that, if the same auditory stimulus was first associated with a reward in a conditioning paradigm, the tone elicited a facilitation of hippocampal pyramidal cell activity. In addition, the facilitation was potentiated by locus coeruleus stimulation. Thus, the results suggest that NE released from locus coeruleus cell axon terminals in the hippocampus enhanced the extent of change in firing pattern, whether that change was an increase or a decrease from baseline. On the basis of this evidence, Segal and Bloom suggested that NE may suppress the response to nonrelevant stimuli and to potentiate the responses to significant stimuli, resulting in an improved signal–noise ratio. Other recent findings seem to be consistent with this view. Norepinephrine appears, not simply to inhibit neuronal activity, but to potentiate either excitatory or inhibitory inputs to cortex (Waterhouse & Woodward, 1980; Woodward, Moises, Waterhouse, Hoffer, & Freedman, 1979), hippocampus (Segal & Bloom, 1976), and cerebellum (Freedman, Hoffer, Woodward, & Puro, 1977).

In terms of possible memory modulatory roles for NE, these anatomical and physiological properties of the system would appear to fit well with a function that enables the nervous system to promote the physiological significance (extent and duration) of important events and may well enhance the storage of that information.

V. INTEGRATION OF ACTH AND NE MEMORY MODULATION

In the last section, we reviewed evidence suggesting that both ACTH and central NE modulate the retention of recent experiences. Furthermore, many of the neurobiological actions of these modulators appear to be consistent with a role in the events responsible for memory storage. ACTH may have two CNS actions that are complementary in terms of modulating memory storage. One action may be to potentiate the release of NE. The importance to memory modulation of this response may lie in those anatomical and physiological properties of noradrenergic systems suggesting that norepinephrine may enhance the neurobiological significance of salient experiences. A second, and apparently contradictory, action of ACTH may be to stabilize the physiological activity of neuronal systems in the face of other inputs, perhaps biasing neurons to accept preferentially some synaptic inputs. When combined, these two actions of ACTH may result in a rather interesting scenario for brain activity following a training event. After a training footshock, for example, the release of both ACTH and brain NE would result in an amplification of the enhanced activity of a specific set of both excitatory and inhibitory neuronal systems appropriate to the environmental input, and would also result in a decrease in the activity of less relevant systems. It is also interesting to note that the function of a mechanism such as this would be impaired by either inadequate or excessive release of ACTH and/or NE. Thus, it becomes easier to explain the inverted U-shaped dose–response curves seen with posttraining ACTH effects on memory (cf. Figs. 1, 2) as well as the fact that there is an optimal decrease in posttraining brain NE content that is correlated with good retention performance.

Therefore, ACTH and NE may act in conjunction to enhance and to maintain the neuronal representation of recent experiences. This hypothesis is consistent with both behavioral and neurobiological information about the actions of these modulating agents. Importantly, the proposed functions do not imply that memory will necessarily be impaired in the absence of ACTH and/or NE. First, the roles suggested here imply that memory can still be formed in the absence of either ACTH or NE. Second, we have explicitly excluded from discussion many other neurobiological systems that may have properties important to memory storage such as other hormonal, neurotransmitter, and neuromodulating systems. Although ACTH and NE are emphasized in this review, it seems inappropriate in any case to consider individual neurohumoral systems to have a function in isolation from other such systems. On the contrary, one way in which our understanding of the relationship between neurobiological mechanisms and memory may progress is to determine the experiential conditions under which various systems are activated or inactivated as well as the parallel and sequential nature of interactions between the systems.

ACKNOWLEDGMENTS

Preparation of this manuscript and research described within supported by research grants MH31141 and AG01642 from USPHS and BNS-76-80007 from the National Science Foundation.

REFERENCES

Alpern, H. P., & McGaugh, J. L. (1968). Retrograde amnesia as a function of duration of electroshock stimulation. *Journal of Comparative and Physiological Psychology*, **65**, 265–269.

Amaral, D. G., & Sinnamon, H. M. (1977). The locus coeruleus: Neurobiology of a central noradrenergic nucleus. *Progress in Neurobiology*, 9, 147–196.

Barchas, J. D., Akil, H., Elliott, G. R., Holman, R. B., & Watson, S. J. (1978). Behavioral neurochemistry, neuroregulators, and behavioral states. *Science*, 200, 964–973.

Barondes, S. H. (1970). Multiple steps in the biology of memory. *In* F. O. Schmitt (Ed.), *The Neurosciences: Second Study Program*, pp. 272–278. New York: Rockefeller Press.

Beaudet, A., & Descarries, L. (1978). The monoamine innervation of rat cerebral cortex: Synaptic and nonsynaptic axon terminals. *Neuroscience*, 3, 851–860.

Bloch, V. (1970). Facts and hypotheses concerning memory consolidation. *Brain Research*, 24, 561–575.

Bohus, B. (1974). The influence of pituitary peptides on brain centers controlling autonomic responses. *Progress in Brain Research*, 41, 175–183.

Buijs, R. M., Swaab, D. F., Dogterom, J., & van Leeuwen, F. W. (1978). Intra- and extra-hypothalamic vasopressin and oxytocin pathways in the rat. *Cell and Tissue Research*, 186, 423–433.

Cherkin, A. (1969). Kinetics of memory consolidation: Role of amnesic treatment parameters. *Proceedings of the National Academy of Science*, 63, 1094–1101.

Cooper, J. R., Bloom, F. E., & Roth, R. H. (1978). *The Biochemical Basis of Neuropharmacology*, 3rd ed., pp. 154. London and New York: Oxford University Press.

de Wied, D. (1967). Opposite effects of ACTH and glucocorticosteroids on extinction of conditioned avoidance behavior. *Proceedings of the 2nd International Congress on Hormonal Steroids, Milan*, 945–951, Excerpta Medica International Congress Series, No. 132.

de Wied, D. (1969). Effects of peptide hormones on behavior. *In* W. F. Ganong & L. Martini (Eds.), *Frontiers in Neuroendocrinology*, pp. 97–140. London and New York: Oxford University Press.

de Wied, D. (1974). Pituitary-adrenal system hormones and behavior. *In* F. O. Schmitt and F. G. Worden (Eds.), *The Neurosciences: Third Study Program*, pp. 653–666. Cambridge, Massachusetts: MIT Press.

de Wied, D., van Delft, A. M. L., Gispen, W. H., Weijnen, J. A. W. M., & van Wimersma Greidanus, Tj. B. (1972). The role of pituitary-adrenal system hormones in active avoidance conditioning. *In* S. Levine (Ed.), *Hormones and Behavior*, pp. 135–171. New York: Academic Press.

Dismukes, R. K. (1979). New concepts of molecular communication among neurons. *The Behavioral and Brain Sciences*, 2, 409–448.

Duncan, C. P. (1949). The retroactive effect of electroshock on learning. *Journal of Comparative and Physiological Psychology*, 42, 132–144.

Dunn, A. J., & Gispen, W. H. (1977). How ACTH acts on the brain. *Biobehavioral Reviews*, 1, 15–23.

Dunn, A. J., Green, E., & Isaacson, R. L. (1979). Intracerebral ACTH mediates novelty-induced grooming in the rat. *Science*, 203, 281–283.

Dyster-Aas, H. K. & Krakau, C.E.T. (1965). General effects of α-melanocyte stimulating hormone in the rabbit. *Acta Endocrinologia*, 48, 609–618.

Engel, F. L. (1961). Extra-adrenal actions of adrenocorticotropin. *Vitamins and Hormones*, 19, 189–227.

Fekete, M. I., Stark, E., Herman, J. P., Palkovits, M., & Kanyicska, B. (1978). Catecholamine concentration of various brain nuclei of the rat as affected by ACTH and corticosterone. *Neuroscience Letters*, 10, 153–158.

Flood, J. F., Vidal, D., Bennett, E. L., Orme, A. E., Vasquez, S., & Jarvik, M. E. (1978). Memory facilitating and anti-amnesic effects of corticosteroids. *Pharmacology, Biochemistry and Behavior*, 8, 81–87.

Freedman, R., Hoffer, B. J., Woodward, D. J., & Puro, D. (1977). Interaction of norepinephrine with cerebellar activity evoked by mossy and climbing fibers. *Experimental Neurology*, **55**, 269–288.

Fuxe, K., Hökfelt, T., Jonsson, G., Levine, S., Lidbrink, P., & Lofstrom, A. (1973). Brain and pituitary–adrenal interaction studies on central monoamine neurons. *In* A. Brodish & E. S. Redgate (Eds.), *Brain-Pituitary-Adrenal Interrelationships*, pp. 239–269. Basel: Karger.

Ganong, W. F. (1963). The central nervous system and the synthesis and release of adrenocorticotropic hormone. *In* A. V. Nalbanov (Ed.), *Advances in Neuroendocrinology*, pp. 92–157. Urbana, Illinois: University of Illinois Press.

Gerard, R. W. (1949). Physiology and psychiatry. *American Journal of Psychiatry*, **106**, 161–173.

Gerard, R. W. (1955). Biological roots of psychiatry. *Science*, **122**, 225–230.

Gold, P. E., Hankins, L., Edwards, R. M., Chester, J., & McGaugh, J. L. (1975). Memory interference and facilitation with posttrial amygdala stimulation: Effect on memory varies with footshock level. *Brain Research*, **86**, 509–513.

Gold, P. E., Macri, J., & McGaugh, J. L. (1973). Retrograde amnesia gradients: Effects of direct cortical stimulation. *Science*, **179**, 1343–1345.

Gold, P. E., & McGaugh, J. L. (1975). A single-trace, two process view of memory storage processes. *In* D. Deutsch & J. A. Deutsch (Eds.), *Short-Term Memory*, pp. 355–378. New York: Academic Press.

Gold, P. E., & McGaugh, J. L. (1978a). Neurobiology and memory: Modulators, correlates and assumptions. *In* T. Teyler (Ed.), *Brain and Learning*, pp. 93–103. Stamford, Connecticut: Greylock Publishers.

Gold, P. E., & McGaugh, J. L. (1978b). Endogenous modulators of memory storage processes. *In* L. Carenza, P. Pancheri, & L. Zichella (Eds.), *Clinical Psychoneuroendocrinology in Reproduction*, pp. 25–46. London: Academic Press.

Gold, P. E., & Murphy, J. M. (1980). Brain noradrenergic responses to training and to amnestic frontal cortex stimulation. *Pharmacology, Biochemistry and Behavior*, **13**, 257–263.

Gold, P. E., & Reigel, J. A. (1980). Extended retrograde amnesia gradients produced by epinephrine injections administered at the time of cortical stimulation. *Physiology and Behavior*, **24**, 1101–1106.

Gold, P. E., Rose, R. P., & Hankins, L. L. (1978). Retention impairment produced by unilateral amygdala implantation: Reduction by posttrial amygdala stimulation. *Behavioral Biology*, **22**, 515–523.

Gold, P. E., & Sternberg, D. B. (1978). Retrograde amnesia produced by several treatments. Evidence for a common neurobiological mechanism. *Science*, **201**, 367–369.

Gold, P. E., & van Buskirk, R. B. (1975). Facilitation of time-dependent memory processes with posttrial epinephrine injections. *Behavioral Biology*, **13**, 145–153.

Gold, P. E., & van Buskirk, R. B. (1976a). Enhancement and impairment of memory processes with posttrial injections of adrenocorticotrophic hormone. *Behavioral Biology*, **16**, 387–400.

Gold, P. E., & van Buskirk, R. B. (1976b). Effects of posttrial hormone injections on memory processes. *Hormones and Behavior*, **7**, 509–517.

Gold, P. E., & van Buskirk, R. (1978a). Posttraining brain norepinephrine concentrations: Correlation with retention performance of avoidance training and with peripheral epinephrine modulation of memory processing. *Behavioral Biology*, **23**, 509–520.

Gold, P. E., & van Buskirk, R. B. (1978b). Effects of α- and β-adrenergic receptor antagonists on post-trial epinephrine modulation of memory: Relationship to post-training brain norepinephrine concentrations. *Behavioral Biology*, **24**, 168–184.

Hebb, D. O. (1949). *The Organization of Behavior*. New York: Wiley.

Herman, J. P., Fekete, M., Palkovits, M., & Stark, E. (1977). Catecholamine turnover measurement and ACTH-induced short term changes of catecholamine levels in individual brain nuclei. *Polish Journal of Pharmacology and Pharmacy*, **29**, 323–331.

Hökfelt, T., & Fuxe, K. (1972). On the morphology and neuroendocrine role of the hypothalamic catecholamine neurons. *In* K. M. Knigge, D. E. Scott, & A. Weindl (Eds.), *Brain-Endocrine Interaction. Median Eminence: Structure and Function*, Pp. 181–228. Basel: Karger.

Iuvone, P. M., Morasco, J., Delanoy, R. L., & Dunn, A. J. (1978). Peptides and the conversion of [^{3}H]tyrosine to catecholamines: Effects of ACTH analogs, melanocyte stimulating hormones and lysine vasopressin. *Brain Research*, **139**,131–139.

Iuvone, P. M., Morasco, J., & Dunn, A. J. (1977). Effect of corticosterone on the synthesis of catecholamines in the brains of CD-1 mice. *Brain Research*, **120**, 571–576.

Javoy, F., Glowinski, J., & Kordon, C. (1968). Effect of adrenalectomy on the turnover of norepinephrine in the rat brain. *European Journal of Pharmacology*, **4**, 103–104.

Johansson, B. B. & Martinsson, L. (1980). β-Adrenergic antagonists and the dysfunction of the blood–brain barrier induced by adrenaline. *Brain Research*, **181**, 219–222.

Kostrzewa, R. M., Kastin, A. J., & Spirtes, M. A. (1975). α-MSH and MIF-I effects on catecholamine levels and synthesis in various rat brain areas. *Pharmacology, Biochemistry and Behavior*, **3**, 1017–1023.

Krivanek, J. (1971). Facilitation of avoidance learning by pentylenetetrazol as a function of task difficulty, deprivation and shock levels. *Psychopharmacologia*, **20**, 213–229.

Krivanek, J., & McGaugh, J. L. (1968). Effects of pentylenetetrazol on memory storage in mice. *Psychopharmacologia*, **12**, 303–321.

Leonard, B. E., Kafoe, W. F., Thody, A. J., & Shuster, S. (1976). The effect of α-melanocyte stimulating hormone (α-MSH) on the metabolism of biogenic amines in the rat brain. *Journal of Neuroscience Research*, **2**, 39–45.

Leonard, B. E., Ramaekers, F., & Rigter, H. (1975). Effects of adrenocorticotrophin (4-10)-heptapeptide on changes in brain monoamine metabolism associated with retrograde amnesia. *Biochemical Society Transactions*, **3**, 113–115.

Lichtensteiger, W., & Felix, D. (1979). Peptide action on individual giant dopamine neurons: Electrophysiological and microfluorimetric analysis. *Society for Neuroscience Abstracts*, **5**, 532.

Lichtensteiger, W., Lienhart, R., & Kopp, H. G. (1977). Peptide hormones and central dopamine neuron systems. *Psychoneuroendocrinology*, **2**, 237–248.

Mah, C. J., & Albert, D. J. (1973). Electroconvulsive shock-induced retrograde amnesia: Analysis of the variation in the length of the amnesia gradient. *Behavioral Biology*, **9**, 517–540.

Malmo, R. B. (1967). Motivation. *In* A. M. Freedman & H. I. Kaplan (Eds.), *Comprehensive Textbook of Psychiatry*. Baltimore, Maryland: Williams & Wilkins.

Martinez, J. L., Vasquez, B. J., Jensen, R. A., Soumireu-Mourat, B., & McGaugh, J. L. (1979). ACTH$_{4-9}$ analog (ORG 2766) facilitates acquisition of an inhibitory avoidance response in rats. *Pharmacology, Biochemistry and Behavior*, **10**, 145–147.

McCarty, R., & Gold, P. E. (1981). Plasma catecholamines: Effects of footshock level and hormonal modulators of memory storage. *Hormones and Behavior*.

McGaugh, J. L. (1966). Time-dependent processes in memory storage. *Science*, **153**, 1351–1358.

McGaugh, J. L. (1973). Drug facilitation of learning and memory. *Annual Review of Pharmacology*, **13**, 229–241.

McGaugh, J. L., & Dawson, R. G. (1971). Modification of memory storage processes. *Behavioral Science*, **16**, 45–63.

Mezey, E., Palkovits, M., DeKloet, E. R., Verhoef, J., & de Wied, D. (1978). Evidence for pituitary-brain transport of a behaviorally potent ACTH analog. *Life Sciences*, **22**, 831–833.

Miller, A. J. (1968). Variations in retrograde amnesia parameters of electroconvulsive shock and time of testing. *Journal of Comparative and Physiological Psychology*, **66**, 40–47.

Miller, L. H., Fischer, S. C., Groves, G. A., & Rudrauff, M. E. (1977). MSH/ACTH$_{4-10}$ influences on the CAR in human subjects: A negative finding. *Pharmacology, Biochemistry and Behavior*, **7**, 417–419.

Miller, L. H., Kastin, A. J., Sandman, C. A., Fink, M., & van Veen, W. J. (1974). Polypeptide influences on attention, memory and anxiety in man. *Pharmacology, Biochemistry and Behavior*, **2**, 663–668.

Moore, R. Y., & Bloom, F. E. (1979). Central catecholamine neuron systems: Anatomy and physiology of the norepinephrine and epinephrine systems. *Annual Review of Neuroscience*, **2**, 113–168.

Nockton, R., Kastin, A. J., Elder, S. T., & Schally, A. V. (1972). Passive and active avoidance responses at two levels of shock after injection of melanocyte-stimulating hormone. *Hormones and Behavior*, **3**, 339–344.

Oliver, C., Mical, R. S., & Porter, J. C. (1977). Hypothalamic-pituitary vasculature: Evidence for retrograde blood flow in the pituitary stalk. *Endocrinology*, **101**, 598–604.

Raichle, M., Hartman, B. K., Eichling, J. O., & Sharpe, L. G. (1975). Central noradrenergic regulation of cerebral blood flow and vascular permeability. *Proceedings of the National Academy of Sciences, U.S.A.*, **72**, 3726–3730.

Rapoport, S. I., Klee, W. A., Pettigrew, K. D., & Ohno, K. (1980). Entry of opioid peptides into the central nervous system. *Science*, **207**, 84–86.

Roche, K. E., & Leshner, A. I. (1979). ACTH and vasopressin treatments immediately after a defeat increase future submissiveness in male mice. *Science*, **204**, 1343–1344.

Sandman, C. A., Denman, P. M., Miller, L. H., & Knott, J. R. (1971). Electroencephalographic measures of melanocyte-stimulating hormone activity. *Journal of Comparative and Physiological Psychology*, **76**, 103–109.

Scapagnini, U., & Preziousli, P. (1972). Role of brain norepinephrine and serotonin in the tonic and phasic regulation of hypothalamic-hypophyseal axis. *Archives Internationales de Pharmacodynamie et de Therapie, Suppl. 1*, **196**, 205–220.

Segal, M., & Bloom, F. E. (1976). The action of norepinephrine in the rat hippocampus. IV. The effects of locus coeruleus stimulation on evoked hippocampal unit activity. *Brain Research*, **107**, 513–525.

Sternberg, D. B., & Gold, P. E. (1980). Effects of α- and β-adrenergic receptor antagonists on retrograde amnesia produced by frontal cortex stimulation. *Behavioral and Neural Biology*, **29**, 289–302.

Sternberg, D. B., & Gold, P. E. (1981). Retrograde amnesia produced by electrical stimulation of the amygdala: Attenuation with adrenergic antagonists. *Brain Research*, **211**, 59–65.

Stone, E. A. (1975). Stress and catecholamines. *In* A. J. Friedhoff (Ed.), *Catecholamines and Behavior*, Vol. 2, Neuropsychopharmacology, pp. 31–72. New York: Plenum.

van Houten, M., Posner, B. I., Kopriwa, B. M., & Brawer, J. R. (1980). Insulin binding sites localized to nerve terminals in rat median eminence and arcuate nucleus. *Science*, **207**, 1081–1083.

van Wimersma Greidanus, Tj. B. & de Wied, D. (1971). Effects of systemic and intracerebral administration of two opposite acting ACTH-related peptides on extinction of conditioned avoidance behavior. *Neuroendocrinology*, **7**, 291–301.

Versteeg, D. H. G. (1973). Effect of two ACTH-analogs on noradrenaline metabolism in rat brain. *Brain Research*, **49**, 483–485.

Waterhouse, B. D., & Woodward, D. J. (1980). Interaction of norepinephrine with cerebrocortical activity evoked by stimulation of somatosensory afferent pathways in the rat. *Experimental Neurology*, **67**, 11–34.

Woodward, D. J., Moises, H. C., Waterhouse, B. D., Hoffer, B. J., & Freedman, R. (1979). Modulatory actions of norepinephrine in the central nervous system. *Federation Proceedings*, **38**, 2109–2116.

CHAPTER 5

ACTH and Brain Mechanisms Controlling Approach–Avoidance and Imprinting in Birds

James T. Martin

Adrenocorticotropin (ACTH)-like peptides influence the imprinting process in ducklings through an action on unlearned approach responsiveness. Peripheral administration of 10 µg $ACTH_{1-10}$ facilitates approach to a novel moving mother figure in newly hatched ducklings (*Anas Platyrhynchos*). Adrenocorticotropic hormone or segments such as $ACTH_{1-24}$ have no apparent effect on approach, which is probably due to their action in releasing corticosterone. Corticosterone inhibits approach responsiveness when given intraperitoneally in nanogram amounts. Injections of Metyrapone, a drug that blocks corticosterone synthesis and elevates plasma ACTH levels, has no stimulatory effect on approach, but does cause a deficit in retention of the imprinting experience as measured in a subsequent choice test. This deficit is apparently not due to the drug's action on the pituitary–adrenal system, because its action is not reversed by corticosterone, or simulated by the injection of ACTH or antibodies to corticosterone. Peripheral administration of 5 µg $ACTH_{1-10}$ or 10 µg $ACTH_{1-24}$ before exposure to the imprinting model or of 20 µg $ACTH_{1-24}$ immediately after exposure to the model is ineffective in changing the later preference of the duckling for a model. Pretraining corticosterone treatment (500 ng, i.p.) causes deficits in retention that are probably due to decreased following of the model during training. Taken together, these experiments suggest that ACTH-like peptides and corticosteroids act on the imprinting process through their influence on brain structures subserving unlearned approach and avoidance responses. Evidence that three separate regions of the avian brain are involved in imprinting is discussed, and a new hypothesis is presented suggesting possible action of ACTH-like peptides on the brain to produce the observed behavioral changes in ducklings.

99

ENDOGENOUS PEPTIDES
AND LEARNING AND MEMORY PROCESSES

I. INTRODUCTION

Memory exists in two forms. Nature has provided all higher animals with an inherited memory, representing information the species has assimilated during its evolution, and with a developmental memory for recording information of significance in the animal's lifetime. Inherited memory, long ago encoded in molecular sequences in the genome, is decoded and consolidated in the individual by an essentially unknown process in early development that produces various species-specific connections and pathways in the central nervous system (CNS). These pathways are the physical representation or basis of this evolutionary memory, and their activation by specific external stimuli represents retrieval of this memory. New information acquired by the individual must be added into the framework and superstructure of this inherited memory. It may be unwise to carry the comparison between these two types of memory too far, but new perspective and insight may arise from a judicious exposure to the relationship between the two processes. Mark (1979) aptly pointed out that observers depend on changes in behavior to infer the presence of stored information or memory in an animal and that to change behavior the learning experience must be either very strong or easily linked to existing motor patterns. Lower animals have only a limited capacity to make use of information not closely meshed with the neuronal network representing the inherited memory. The animal may be able to acquire considerable information in the form of cues about the external environment but be unable to integrate this information into a meaningful behavioral output. Physiological processes that normally mediate neuronal and behavioral events and promote adaptation may actually impede or derange CNS function when the animal finds itself facing an unnatural or "stressful" task. Therefore, studies of learning and memory should utilize tasks adapted to the animal's natural proclivities and should measure parameters from the animal's natural behavior repertoire. In this respect, birds, particularly young precocial birds, offer important advantages as experimental models. In the remainder of this chapter, studies about the effects of neuropeptides on learning and approach–avoidance behavior in birds will be reviewed, and the relationship of this information to work on lower mammals will be discussed.

II. APPROACH–AVOIDANCE BEHAVIOR AND IMPRINTING IN BIRDS

Approach and avoidance tendencies are involved in most, if not all, behaviors directed toward objects in the animal's environment. Schneirla (1959) described how approach and withdrawal responses are integrated into more complex functional patterns as the young animal develops. In the neonate, these response tendencies are present in a form that is unconstrained by experience, and hence, studies on young animals may more readily reveal the neural and neuroendocrine systems mediating avoidance responses. Researchers in the past have most often examined approach behavior in birds within the context of imprinting. The process of imprinting first described by Lorenz (1935) and more recently reviewed by Bateson (1966), Fischer (1969), Hess (1973), and Sluckin (1972) concerns, in its narrower sense, the development of a filial preference or attachment by the neonate for a mother figure. The development of this attachment takes place during a "sensitive period" shortly after the animal hatches, and the onset of this sensitive period appears to depend on developmental changes in the animal's behavioral machinery (Bateson, 1979; Gottlieb, 1961). The sensitive period for imprinting has been characterized in two ways. Hess (1959) saw it as a sensitive period for learning the characteristics of the mother; whereas, others viewed it as the period when approach responses could be elicited by objects in the environment (Fabricius, 1951; Fabricius & Boyd, 1954). A distinction between these views is crucial for understanding the nature of the process and the physiological events controlling it.

A. Evidence that Neuropeptides Contribute to the Imprinting Process

Volumes have been written on the effects of neuropeptides on learning processes in mammals, but few studies exist implicating neuropeptides in bird behavior. The existing studies include reports of effects of ACTH or ACTH-like peptides on displacement activity in pigeons (Delius, Craig, & Chaudoir, 1976), on wing flapping and precopulatory displays in domestic ducks (Deviche, 1976, 1979), on vocalizations in chicks (DeLanerolle, Elde, Sparber, & Frick, 1981; Panksepp, Vilberg, Bean, ·Coy, & Kastin, 1978) and on imprinting responses in ducks (Landsberg & Weiss, 1976; Martin, 1975, 1978a, Martin & van Wimersma Greidanus, 1978). The first evidence that ACTH-like peptides might be involved in imprinting came from experiments showing that the level of plasma corticosterone rises during the early posthatch period in Pekin ducklings and then drops near the end of the sensitive period for imprinting (Martin, 1975). The rise in plasma corticosteroids was confirmed by Weiss, Kohler, and Landsberg (1977), and they indicated that a significant circadian periodicity accompanied this rise.

To explore the relationship between imprinting and pituitary adrenal hormones, we have conducted a series of experiments with mallard and Pekin ducklings. Normally, the naive duckling is placed near an imprinting model when it is 18–24 hr old and its response to the model is recorded during this "training period." We have used either a moving vocalizing red cube or a blue ball as the imprinting stimulus (Martin & Schutz, 1975) or, in one case, a flashing light (Martin & van Wimersma Greidanus, 1978). During this training period the latency to approach and the time spent near the model are recorded. To see if the bird has developed a preference for the model and to obtain a measure of how much the bird has learned, 2 days later the duckling is given a choice test, and the amount of time with familiar and unfamiliar models is recorded. Because the initial response of the duckling is not conditioned, it is possible to study the effects of neuropeptides on preprogrammed or unlearned approach responses as well as on the learning process that takes place during and after the initial exposure to the model.

1. ACTH Effects on Unlearned Approach

To alter the pituitary–adrenal axis, we have used exogenous hormone injections, treatment with enzyme inhibitors or with antibodies, as well as genetic strains, and developmental groups with different pituitary–adrenal dynamics. Injection of ducklings with 5 μg ACTH$_{1-10}$ intraperitoneally (i.p.) 30 min prior to the initial training with the model reduces the latency to approach the model and increases the amount of time the duckling spends following the moving model in the 10 min training period (Fig. 1) (Martin & van Wimersma Greidanus, 1978). This peptide is not corticotropic in ducklings, and a shorter noncorticotropic molecule, ACTH$_{4-10}$, also has a similar effect in stimulating approach. However, the larger corticotropic peptides ACTH$_{1-24}$ and ACTH$_{1-39}$ do not facilitate approach. This may be caused by their action in releasing corticosteroids and/or androgens from the adrenal gland, because exogenous corticosterone in nanogram amounts (500 ng, i.p.) given 15 min before training significantly inhibits approach in ducklings (Martin, 1978a), and both testosterone propionate and 5α-dihydrotestosterone in microgram amounts reduce following in chicks (Balthazart & deRycker, 1979). Alternatively, the large ACTH molecule may have more difficulty penetrating the blood–brain barrier, although conflicting data obtained after peripheral administration of peptides suggests some penetration of large peptides does occur in adult rats (Cornford, Braun, Crane, & Oldendorf, 1978; Kastin, Olson, Schally, & Coy, 1979). It is generally thought that the blood–brain barrier is poorly developed in young animals, although in precocial animals it appears to be functional at birth (Bradbury, 1979). Studies in the chick embryo indicate that the brain excludes large amounts of proteins by the end of incubation and that there is a well-developed blood–brain barrier in newly hatched chicks (Birge, Rose, Haywood, & Doolin, 1974). Injection (i.p.) of 0.1 ml of rabbit plasma containing antibodies to corticosterone also facilitates approach, presumably by its action in reducing plasma corticosterone levels. The fact that fragments of ACTH and

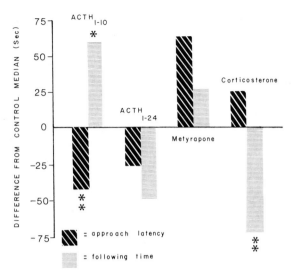

Fig. 1. Effects of ACTH and related substances on unlearned imprinting approach responses in naive ducklings. Both 5 μg $ACTH_{1-10}$ ($N = 36$) and 10 μg $ACTH_{1-24}$ ($N = 14$) were given i.p. 30 min before testing. Metyrapone ($N = 16$, 1.25 mg, i.p.) was given 4 hr before testing and corticosterone ($N = 18$, 0.5 μg, i.p.) 15 min before testing. Vehicle controls were 0.1 ml 0.75% saline ($N = 38$; median latency = 59.5 sec; median following time = 376.5 sec) for $ACTH_{1-10}$; 0.1 ml $ZnCl_2$-phosphate buffer ($N = 11$; median latency = 114 sec; median following time = 296 sec) for $ACTH_{1-24}$; 0.1 ml propylene glycol-saline (3:2) ($N = 15$; median latency = 20 sec; median following time = 434 sec) for metyrapone, or 0.1 ml propyleneglycol-saline-ethanol (5:9:2) ($N = 24$; median latency = 35 sec; median following time = 431 sec) for corticosterone. Mann-Whitney U Test: Corticosterone $U_{following} = 228$, $p < .01$; $ACTH_{1-10}$ $U_{latency} = 923.5$, $p < .01$, $U_{following} = 851.5$, $p < .05$. (Modified from Martin, 1978a; copyright © 1978, American Association for the Advancement of Science.)

corticosterone affect approach in opposite ways in reminiscent of the opposite effects of these substances on extinction of conditioned avoidance behavior (de Wied, 1974).

We have also studied the effect on imprinting of drugs known to inhibit steroid synthesis and increase ACTH levels. Metyrapone (Metopirone, Ciba-Geigy) is an 11-β-hydroxylase inhibitor that blocks the conversion of 11-deoxy-corticosterone to corticosterone. This drug was injected (1.25 mg, i.p.) 4 hr before initial exposure to the model, and although complete cessation of cortisol synthesis occurs within 30 min of metyrapone infusion no stimulatory effect on approach responding occurred (Fig. 1). On the contrary, there was a tendency ($p < .15$) for *longer* latencies to approach in the drug group. Failure to facilitate approach behavior may be due to increased plasma levels of 11-deoxycorti-costerone or to extra-adrenal or CNS effects of metyrapone, a compound originally developed as an anticonvulsant drug. The influence of deoxycorticosterone on imprinting has not been studied, but, like corticosterone, it is known to impair passive avoidance behavior in rats (van Wimersma Greidanus, 1977).

The functional dynamics of the pituitary adrenal system can be permanently affected by experimentally changing hormone levels in early ontogeny (Erskine, Geller, & Yuwiler, 1979; Krieger, 1972, 1974; Lorenz, Branch, & Taylor, 1974; Martin, 1975). Injection of corticosterone into the chorioallantoic membrane of developing Pekin duckling eggs causes dose-dependent changes in posthatch approach behavior and in adrenocortical response to immobilization stress (Martin, 1975). Embryos treated with 40 μg of corticosterone in three injections during the second half of incubation and tested later, approached the imprinting model more quickly and followed it more than did controls, but significantly fewer of these embryonically injected ducklings approached the imprinting model. Embryonic treatment with $ACTH_{1-39}$ seemed to have no effect on later approach latency, but the lowest dose (0.005 IU) significantly stimulated following (Martin, 1978b). Embryonic treatment with metyrapone in four increasing doses (0.25, 0.25, 0.5, and 1.0 ng) during the last half of incubation failed to influence posthatch approach latency or following time. Another steroid enzyme inhibitor, aminoglutethimide phosphate, which blocks the conversion of cholesterol to pregnenolone, also failed to influence approach behavior when it was administered during the latter half of incubation (Martin, 1978b).

2. ACTH Effects on Imprinting Retention

To determine retention of the imprinting experience, we gave ducklings a choice test 2 to 9 days following the initial 10 min exposure to the imprinting model. The latency to approach and the time spent following both familiar and unfamiliar models provides an indication as to whether the treatments influenced some component of learning, such as consolidation, storage, or retrieval. One may assume that ducklings that follow the unfamiliar model more, or the familiar model less than do controls, show deficits in retention. Conversely, those that follow the familiar model more and the unfamiliar one less, show improved retention. This assumption could be questioned if decreased preference for the familiar model signifies reduced anxiety or fear in the test situation. If this were the case, however, one would predict reduced following of both models.

In only one instance have we reported that an ACTH-like peptide, administered prior to training with the model, had any significant effect on the behavior in a subsequent choice test (Martin, 1975). Long-acting porcine $ACTH_{1-39}$ (1 IU) given 20 min prior to training, along with a blue ball model, reduced the amount of time spent with the unfamiliar red cube among those birds that actually followed the strange model in the choice test, 8 days after training. However, because more controls (35%) than ACTH-treated birds (21%) failed to follow the strange model, a conclusion that ACTH improved retention seemed unwarranted. Certainly, $ACTH_{1-10}$ (5 μg, i.p., 20 min prior to training) failed to influence the preference scores in the choice tests (Fig. 2), even though it had a strong stimulating effect on approach behavior during training (Fig. 1). Also, 20 μg of $ACTH_{1-24}$ given i.p. immediately after the training experience had no effect on later preference for the model as measured by the choice test (Fig. 2). Embryonic injections of long-acting porcine $ACTH_{1-39}$ or corticosterone, as described earlier, also failed to influence choice test scores. Metyrapone, however, produced consistent effects on choice test preference scores when given either just prior to imprinting (Fig. 2) or during the latter stages of incubation (Martin, 1978b). In all cases, metyrapone significantly increased the time spent with the unfamiliar model, or interfered with the formation of a preference for the imprinting model. It seems unlikely that this deficit in retention was due to the action of metyrapone in elevating ACTH, as similar effects were not found by injecting ACTH. Because metyrapone blocks corticosterone synthesis, we attempted to correct the deficit in retention by injecting both corticosterone and metyrapone prior to training with the model. The result was a greater deficit in retention in the subsequent choice test than with metyrapone alone (Martin, unpublished observations). Therefore, it seems unlikely that the retention deficit is related directly to the drug's action on the pituitary–adrenocortical system, although this possibility cannot be totally excluded.

Landsberg and Weiss (1976) reported imprinting retention deficits in ducklings that were either stressed by exposure to cold or injected with porcine ACTH prior to training. They used a different method for assessing retention and found significant retention deficits only with ACTH doses that produced elevation of plasma corticosterone levels. It seems likely, therefore, that the observed deficits in retention performance are due to elevated corticosterone levels. As indicated, corticosterone reduces approach responsiveness; furthermore, plasma corticosterone levels are inversely related to the amount of following during training (Martin, 1978a). Earlier studies have documented a direct relationship between retention scores and time spent following the model during training (Martin & Schutz, 1975). As Landsberg and Weiss (1976) do not report data on their ducklings' performance during the *training period*, it is not possible to determine whether the cause of the reported retention deficit is a hormone-induced reduction of following time during training, or if it is an effect on retention per se. The more parsimonious conclusion is that ACTH-like peptides and corticosterone act not on the learning processes of imprinting, but on the approach and avoidance machinery associated with the learning tasks. It is not necessary to postulate an effect on memory or learning to explain the available data in birds. In the remaining sections, possible mechanisms of action of ACTH-like peptides and corticosteroids on structures in the avian brain involved in imprinting will be discussed.

B. A Hypothesis on How the Avian Brain Controls Imprinting

Lorenz (1973) points out that it is easier to drive an animal out of a flower bed than into a cage. Whatever the mechanisms in the vertebrate brain that control approach, they are usually overshadowed by those commanding escape or withdrawal. The imprinting approach response is an example of a behavior in which no deprivation is necessary to induce approach to a novel object. Because approach occurs only during a restricted or sensitive period, the neural systems mediating escape and withdrawal may be either blocked or poorly developed at this time. There is some evidence to support both possibilities (Bateson, 1979; Hoffman & Ratner, 1973), but the fact that young ducklings will avoid novel objects literally minutes after initial exposure to a mother figure suggests that neural programs mediating avoidance are functionally mature in the neonate. It thus seems likely that the avoidance program is blocked until a specific link is established between the neuronal elements encoding the physical characteristics of the mother figure and those controlling approach. After this connection is established, the blockade of the avoidance system can be removed and the sensitive period ends. Subsequent approach can then only be elicited by stimulation of the neuronal set containing the memory trace, the encoded representation of the mother figure. Such a conception of imprinting postulates three interacting and interconnected but separate neuronal systems: (1) an approach system; (2) an avoidance system (both of which are laid down as part of the inherited memory); and (3) a storage system that forms the template of the mother figure and that includes a means for matching or comparing continuing sensory input with the template. There are, in addition, at least three processes involving these neural systems that occur simultaneously during imprinting: (1) a generalized suppression of avoidance with accompanying activation of approach; (2) construction of a template of the mother figure; and (3) intermeshing of this template with the approach and/or the avoidance system(s). What evidence exists to indicate that these three postulated neuronal systems are indeed physically separate entities or that these hypothetical processes actually occur? There are several lines of evidence (see Fig. 3). For example, Gervai and Csanyi (1973) demonstrated that puromycin injected into the medial dorsal forebrain (Wulst) prevented the development of a preference for the imprinting model but did not impair the approach response. In another study,

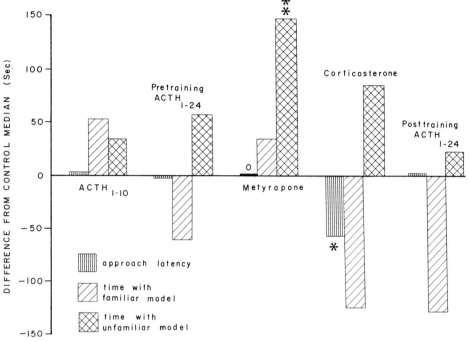

Fig. 2. Effects of ACTH and related substances on behavior in a choice test 2 days after imprinting to a moving model. Doses as in Fig. 1 except posttraining ACTH$_{1-24}$ was 20 μg i.p. immediately after training. ACTH$_{1-10}$ (N_{exp} = 15; $N_{control}$ = 12; median control latency = 61.5 sec, follow familiar model = 247 sec, follow unfamilar model = 114 sec); Pre-ACTH$_{1-24}$ (N_{exp} = 4; $N_{control}$ = 6; median control latency = 37 sec, follow familiar model = 320 sec, follow unfamiliar model = 51 sec); Metyrapone (N_{exp} = 17, $N_{control}$ = 16; median control latency = 27 sec, follow familiar model = 250 sec, follow unfamiliar model = 39 sec; $U_{unfamiliar model}$ = 48, p < .001); corticosterone (N_{exp} = 9, $N_{control}$ = 14; median control latency = 82 sec, follow familiar model = 241 sec, follow unfamiliar model = 56 sec; $U_{latency}$ = 35.5, p < .05); Post ACTH$_{1-24}$ (N_{exp} = 8, $N_{control}$ = 7; median control latency = 45 sec, follow familiar model = 339 sec, follow unfamiliar model = 142 sec).

the details of which were not published, Hess (1973) indicated that decerebrate chicks were able to follow a moving model but did not develop a clear-cut preference for the model. Salzen and Parker (1975), and Salzen, Parker, and Williamson (1975, 1978) reported that, unlike controls, chicks with lateral forebrain lesions that were exposed to objects in their cages did not subsequently prefer the stimulus object. These lesioned birds usually suffered bilateral damage to the dorsal part of the archistriatum, a region in the ventral lateral forebrain thought to be homologous to the amygdala. Two other lesion groups with damage to the dorsal forebrain or to the posterior forebrain showed less severe deficits, and an anterior forebrain lesion group did not differ from controls. Interestingly, the dorsal and posterior groups sustained damage to pathways projecting to the archistriatum. The major deficit occurring in these groups was a decreased likelihood that the chick would approach either of the imprinting models. In other words, these lesioned birds showed less approach responsiveness.

1. The Archistriatum

There is no doubt that the archistriatum has a major function in controlling avoidance or escape responses. Phillips (1964) clearly showed that ablation of the wild mallard's archistriatum caused apparent loss of fearfulness characterized by reduced escape reactions. Electrical stimulation of this area, moreover, led to flight or escape responses in chickens (Phillips & Youngran, 1971), pigeons (Goodman & Brown, 1966), peach-faced lovebirds (Phillips, 1968) and mallards (Maley, 1969; Phillips, 1964). Efferent projections from the archistriatum leave this region in the occipitomesencephalic tract (OM) (Fig. 3). Damage to this tract with electrolytic lesions in ducks (Phillips, 1964) or with knife cuts in ducks (Martin, DeLanerolle, & Phillips, 1979) or in Barbary doves (Wright, 1975) causes a dramatic reduction in escape and avoidance responding and a facilitation of approach toward food and water in a novel setting. Similarly, Hess (1973) mentioned that his chicks with "amygdaloid" lesions approached and followed imprinting objects that were ineffective in eliciting approach from controls.

The archistriatum and its efferents may well represent the neuronal avoidance system just postulated to be involved in the imprinting process. This region, which receives input from more rostral forebrain structures, may function to orient and to organize escape from, or avoidance of unfamiliar objects or situations. Suppression or inhibition of the output of this region may be necessary for imprinting approach and following responses to take place. Inhibition of the output from this region may be the basis of the sensitive period for imprinting. Although there is no direct evidence for suppression of neuronal activity in this region during the sensitive period, there is some circumstantial evidence that ACTH suppresses archistriatal output. For example, rising levels of corticosterone after hatching indicate that ACTH is being released from the pituitary; also, levels of α-MSH are increasing at this time in the duckling midbrain–hypothalamus region (Martin, Dogterom, & Swaab, unpublished data). The increasing amounts of ACTH-like peptides in the brain may serve to suppress archistriatal output. This seems plausible since Korányi, Endröczi, and Tamásy (1969) found that intravenous (i.v.) injections of $ACTH_{1-10}$ in young chicks resulted in a rapid diminution of potentials evoked in the cortical area after electrical stimulation of the reticular nuclei of the mesencephalon. This is particularly significant in that they observed these effects in young chicks but not in adult chickens. Furthermore, the archistriatum may normally inhibit ACTH secretion, as bilateral section of the output from this region (OM) at the point where it leaves the archistriatum causes basal plasma corticosterone levels to rise (Martin et al., 1979), suggesting that low output from this structure is correlated with high ACTH levels. Moreover, this region is extremely rich in opiate binding sites in the chick (Pert et al., 1974) (Fig. 3), and opiate-like peptides suppress distress vocalizations in young chicks (Panksepp et al., 1978). Radioimmunoassay and immunocytochemical studies with pigeons have found β-endorphin and Leu- and Met-enkephalin in the archistriatum (Bayon, Koda, Battenberg, Azad, & Bloom, 1980). This information, coupled with the findings of others that β-endorphin, ACTH, and α-MSH are localized within the same neurons (Sofroniew, 1979; Watson & Akil, 1979), and that the rodent amygdala contains ACTH (Pacold, Kirsteins, Hojvat, Lawrence, & Hagen, 1978) and α-MSH (O'Donahue, Miller, & Jocobowitz, 1979), suggests that posthatching increases in levels of ACTH-like peptides act on the archistriatum to suppress instinctive avoidance or escape reactions and, thus, produce a sensitive period for approach in the young bird. This hypothesis would be strengthened if one could cite evidence from other vertebrates that peptides act on the amygdala to suppress avoidance responding. A number of studies have implicated the amygdala in avoidance conditioning in rodents (Gallagher & Kapp, 1978; Slotnick, 1973; van Wimersma Greidanus, Croiset, Bakker, & Bouman, 1979). van Wimersma Greidanus et al. (1979) reported that the effect of $ACTH_{4-10}$ in prolonging extinction of conditioned avoidance responses is absent in rats with lesions of the amygdala. ACTH may act on the rat amygdala to suppress certain inherited escape programs, and this may explain increased response latency effects of the hormone using dark-box passive avoidance tests (de Wied, 1974; Lissak & Bohus, 1972). If ACTH treatment reduces escape tendencies, longer latencies would be expected in a passive avoidance task. Most passive avoidance studies have failed to include controls for this possibility; for example, in studies of ACTH affects on "learning" by Ader and de Wied (1972) and de Wied (1974), shocked saline-treated controls failed to show an increased latency to enter the shock box suggesting that no "learning" took place in any group. In a study by Lissak and Bohus (1972), shock levels that produced retention in saline-treated controls were those in which ACTH effects were absent. These results could be interpreted as evidence that ACTH blocks escape reactions. Similarly, an ACTH effect in prolonging extinction of active avoidance responses might be attributable to a suppression of escape programs that normally interfere with attention to the conditioning cues. Not all of the effects of ACTH

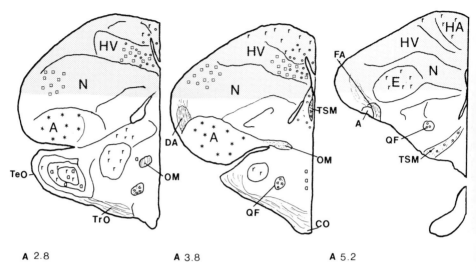

A 2.8   **A** 3.8   **A** 5.2

Fig. 3. Neuroanatomical and neurochemical correlates of imprinting in the chick brain. Stippled area is site of imprinting-specific protein synthesis of Bateson, Horn, and Rose (1971). Uptake sites for corticosterone in duck brain are indicated by ○ (Rhees, Abel, & Haack, 1972). Area of imprinting-specific deoxyglucose metabolism are designated by □ (Kohsaka, Takamatsu, Aoki, and Tsukada, 1979). Asterisks include areas of abundant opiate receptor binding of Pert, Aposhian, and Snyder (1974). Small a's show efferent projections of the archistriatum (A). Input to the archistriatum includes the tractus archistriatalis dorsalis (DA) and the tractus archistriatalis frontalis (FA). Output is via tractus occipitomesencephalicus (OM). Abbreviations are CO, chiasma opticum; E, ectostriatum; HA, hyperstriatum accessorium; N, neostriatum; QF, tractus quintofrontalis; TeO, tectum opticum; TrO, tractus opticus; TSM, tractus septomesencephalicus. (Transverse sections adopted from chick brain atlas of Youngren and Phillips, 1978.)

in rats, however, can be explained in this way. The posttraining memory effects reported by Gold and van Buskirk (1977) and the reduction of social interactions effect (File & Vellucci, 1978) demand an additional explanation and a different mechanism of action. Injection of ACTH after avoidance training can influence a rat's performance several days later. This can be attributed to an effect on memory processes but not to an effect on escape circuitry per se because the peptide is probably metabolized by the time of retesting and because the direction of the effect varies with the dose level (Gold & van Buskirk, 1977). Therefore, no single mechanism seems to account for all of the observed effects of ACTH-like peptides.

2. The Dorsal Forebrain of Hyperstriatum

Whereas a blockade of output from the archistriatum by hormonal signals after hatching may lead to a sensitive period, a more specific long-lasting blockade must remain after the young bird constructs a template of the mother figure. The data of Salzen and Parker (1975) discussed in the previous section suggests that, for proper approach to take place after the imprinting experience, the connections between the rostral forebrain structures and the archistriatum (FA and DA, Fig. 3) must be intact. Such an input from the rostral forebrain could serve to inhibit the output of the archistriatum. We now know from the elegant work of Bateson and his colleagues that only the forebrain roof is specifically the site of changes in RNA and protein synthesis associated with the imprinting experience (Bateson et al., 1971; Horn, Rose, & Bateson, 1973). The specific activity of RNA polymerase (Haywood, Rose, & Bateson, 1975) and acetylcholinesterase (Haywood, Hambley, & Rose, 1975) increases in the forebrain roof only in imprinted birds; and furthermore, deoxyglucose and autoradiographic studies implicate this region as well (Horn, McCabe, & Bateson, 1979. Kohsaka et al., 1979; Fig. 3). This region appears to be involved in the construction of a neural template or memory of the mother figure, and may be the third neural system mentioned earlier and postulated to be involved in the imprinting process. Output from this region of the forebrain projects to the archistriatum via the tractus archistriatalis dorsalis (DA) and the tractus frontoarchistriatalis (FA) (Zeier & Karten, 1971). A specific connection from the hyperstriatum ventrale to the archistriatum has recently been described (Bradley & Horn, 1979). This input to the archistriatum from the dorsal forebrain may possibly suppress archistriatal activity in the presence of the mother figure. Sensory input representing the mother figure would presumably be carried to the forebrain, activate the neural template of the mother figure, and generate a stimulus-specific suppression of the archistriatum and its avoidance system. Other novel objects would fail to activate the template, and therefore fail to suppress the archistriatal avoidance system.

Karten (1969) and Bradley and Horn (1978) described visual projections to the hyperstriatum, and the information flow in these pathways may be modulated by corticosteroid levels. Rhees et al. (1972) studied uptake of corticosterone in the brain of the adult Pekin duck. They found extensive uptake in the medial dorsal forebrain, and large numbers of cells were labeled in the dorsal and ventral hyperstriatum (HV) and hippocampus as well as in the septomesencephalic tract (TSM) (Fig. 3). These labeled cells are in the region of the proposed neural template and in an efferent projection from this region (TSM). Korányi et al. (1969) reported data indicating that corticosteroids increase the amplitude of nerve impulses arriving in the cortex from the midbrain reticular nuclei and of impulses arriving in the optic tectum from the optic chiasm. Because plasma corticosteroid levels are elevated during the sensitive period for imprinting, information flow to the region of template formation may be accelerated at that time. This possibility does not appear to be consistent, however, with the available behavioral data. The data reviewed here do not

support a facilitatory role for corticosterone in approach or in preference formation. Similarly, the observation that imprinted chicks, stressed with shock,
follow the mother figure more intensely (DePaulo, Hoffman, Klein & Gaioni,
1978; Kovách & Hess, 1963) is unlikely to result from a shock-induced elevation
in corticosteroid levels per se. However, the effects of corticosterone on following
in prior imprinted birds have not been studied and may well prove to be
different from those in birds in which no preference exists.

3. The Midbrain and the Optic Tectum

For the duckling to approach a model, it must be able to process information
about the location and about the direction and rate of movement of the model;
it must also coordinate its own movement in relation to the model's movement.
Hess (1973) reported, without histological documentation, that decerebrate
chicks follow a moving model; hence, we assumed that the approach control
system is located in the midbrain or hindbrain. More recently Collias (1980)
found, however, that the paleostriatum, but not its lateral and overlying structures, was necessary for approach behavior. The paleostriatum is thought to
be analogous to the basal ganglia, and destruction of this area causes gross
motor disturbances (Rieke, 1980).

The integrity of the optic tectum seems to be essential for effecting visually
guided behavior. Thus, it may be relevant that a major projection area of the
archistriatum is the optic tectum (Zeier & Karten, 1971) (Fig. 3), and that
visual input to this region can be reduced by ACTH and augmented by corticosterone administration (Koranyi et al., 1969). One may speculate that approach responses are organized and executed by the paleostriatum and by the
optic tectum operating under the inhibitory control of the archistriatum. Further
research is needed to identify brain regions essential for producing approach
responses.

III. CONCLUSIONS

Our studies on approach–avoidance behavior and learning in neonatal ducklings indicate that ACTH-like peptides act on inherited brain structures to
suppress avoidance and possibly to stimulate approach responses to novel moving
objects in the environment. During imprinting, ACTH may act to suppress
escape responses, permitting the young bird to develop a template of the mother
that will direct its subsequent behavior. In this fashion, ACTH-like peptides
may act to facilitate assimilation of new information from the environment.

New information acquired by the animal must be integrated into the relatively
limited motor output available to the animal. The imprinting process involves

such an integration of information and can be viewed as the interaction of several processes controlled by different parts of the avian brain. Peptides similar to ACTH generate new probabilities for particular behaviors by their presence at certain sites in the brain, and the sensitive period for imprinting may arise from elevated levels of these substances after hatching. It is unlikely, however, that the presence of these endogenous peptides is sufficient, or even necessary for imprinting to occur. Rather, their action represents a mechanism for fine tuning in the adaptation of the newly hatched bird to its environment.

ACKNOWLEDGMENTS

Supported in part by NIH Fellowship NICHHD00495 and grants from the Stockton Research and Professional Development Fund. I am indebted to Jon Swalboski, Phyllis Davy, and Patricia Coleman for technical support.

REFERENCES

Ader, R. & de Wied, D. (1972). Effects of lysine vasopressin on passive avoidance learning. *Psychonomic Science*, **29**, 46–49.

Balthazart, J., & deRycker, C. (1979). Effects of androgens and oestrogens on the behaviour of chicks in an imprinting situation. *Zeitschrift für Tierpsychologie*, **49**, 55–64.

Bateson, P. P. G. (1966). The characteristics and context of imprinting. *Biological Reviews*, **41**, 177–220.

Bateson, P. (1979). How do sensitive periods arise and what are they for? *Animal Behaviour*, **27**, 470–486.

Bateson, P. P. G., Horn, G., & Rose, S. P. R. (1971). Effects of early experience on regional incorporation of precursors into RNA and protein in the chick brain. *Brain Research*, **39**, 449–465.

Bayon, A., Koda, L., Battenberg, E., Azad, R., & Bloom, F. E. (1980). Regional distribution of endorphin, met^5-enkephalin and leu^5-enkephalin in the pigeon brain. *Neuroscience Letters*, **16**, 75–80.

Birge, W. J., Rose, A. D., Haywood, J. R., & Doolin, P. F. (1974). Development of the blood–cerebrospinal fluid barrier to proteins and differentiation of cerebrospinal fluid in the chick embryo. *Developmental Biology* **41**, 245–254.

Bradbury, M. (1979). *The Concept of a Blood–Brain Barrier.* New York: Wiley.

Bradley, P., & Horn G. (1978). Afferent connexions of hyperstriatum ventrale in the chick brain. *Journal of Physiology* **278**, 46P.

Bradley, P., & Horn, G. (1979). Efferent connexions of the hyperstriatum ventrale in the chick brain. *Journal of Anatomy*, **128**, 414–415.

Collias, N. E. (1980). Basal telencephalon suffices for early socialization in chicks. *Physiology and Behavior* **24**, 93–97.

Cornford, E. M., Braun, L. D., Crane, P. D., & Oldendorf, W. H. (1978). Blood-brain barrier restriction of peptides and the low uptake of enkephalins. *Endocrinology*, **103**, 1297–1303.

DeLanerolle, N. C., Elde, R. P., Sparber, S. B., and Frick, M. (1981). Distribution of methionine-enkephalin immunoreactivity in the chick brain: An immunohistochemical study. *Journal of Comparative Neurology*, (in press).

Delius, J., Craig, B., & Chaudoir, C. (1976). Adrenocorticotrophic hormone, glucose and displacement activities in pigeons. *Zeitschrift für Tierpsychologie*, **40**, 183–193.

DePaulo, P., Hoffman, H. S., Klein, S., & Gaioni, S. (1978). Effect of response-contingent shock on ducklings' preference for novel imprinting stimuli. *Animal Learning and Behavior*, **6**, 458–462.

Deviche, P. (1976). Behavioural effects of ACTH or corticosterone administration to adult male domestic ducks, *Anas platyrhynchos* L. *Journal of Comparative Physiology*, A, **110**, 357–366.

Deviche, P. (1979). Effects of testosterone propionate and pituitary–adrenal hormones on the social behaviour of male ducklings (Anas platyrhynchos L.) in two test situations. *Zeitschrift für Tierpsychologie*, **49**, 77–86.

de Wied, D. (1974). Pituitary-adrenal system hormones and behavior. In F. O. Schmidt & F. G. Worden (Eds.), *The Neurosciences, Third Study Program*, pp. 653–666. Cambridge, Massachusetts: MIT Press.

Erskine, M. S., Geller, E., & Yuwiler, A. (1979). Effects of neonatal hydrocortisone treatment on pituitary and adrenocortical response to stress in young rats. *Neuroendocrinology*, **29**, 191–199.

Fabricius, E. (1951). Zur Ethologie junger Anatiden. *Acta Zoologica Fennica*, **68**, 1–175.

Fabricius, E., & Boyd, H. (1954). Experiments on the following reactions of ducklings. *Wild Fowl Trust Fund Annual Report, Slimbridge, England*, **6**, 84–89.

File, S., & Vellucci, S. V. (1978). Studies on the role of ACTH and of 5-HT in anxiety, using an animal model. *Journal of Pharmacy and Pharmacology*, **30**, 105–110.

Fischer, G. J. (1969). The behavior of chickens. In E. S. E. Hafez (Ed.), *The Behavior of Domestic Animals*, pp. 515–553. Baltimore, Maryland: Williams and Wilkins.

Gallagher, M., & Kapp, B. S. (1978). Manipulation of opiate activity in the amygdala alters memory processes. *Life Sciences*, **23**, 1973–1978.

Gervai, J., & Csanyi, V. (1973). The effects of protein synthesis inhibitors on imprinting. *Brain Research*, **53**, 151–160.

Gold, P. E., & van Buskirk, R. B. (1977). Enhancement and impairment of memory processes with posttrial injections of adrenocorticotrophic hormone. *Behavioral Biology*, **16**, 387–400.

Goodman, I. J., & Brown, J. L. (1966). Stimulation of positively and negatively reinforcing sites in the avian brain. *Life Sciences*, **5**, 693–704.

Gottlieb, G. (1961). Developmental age as a baseline for determination of the critical period in imprinting. *Journal of Comparative and Physiological Psychology*, **54**, 422–427.

Haywood, J., Hambley, J., & Rose, S. (1975). Effects of exposure to an imprinting stimulus on the activity of enzymes involved in acetylcholine metabolism in chick brain. *Brain Research*, **92**, 219–225.

Haywood, J., Rose, S. P. R., & Bateson, P. (1975). Changes in chick brain RNA polymerase associated with an imprinting procedure. *Brain Research*, **92**, 227–235.

Hess, E. H. (1959). Imprinting. *Science*, **130**, 133–141.

Hess, E. H. (1973). *Imprinting: Early Experience and the Developmental Psychobiology of Attachment*. Princeton, New Jersey: Van Nostrand-Reinhold.

Hoffman, H. S., & Ratner, A. M. (1973). Effects of stimulus and environmental familiarity on visual imprinting in newly hatched ducklings. *Journal of Comparative and Physiological Psychology*, **85**, 11–19.

Horn, G., McCabe, B. J., & Bateson, P. P. G. (1979). An autoradiographic study of the chick brain after imprinting. *Brain Research*, **168**, 361–373.

Horn, G., Rose, S. P. R., & Bateson, P. P. G. (1973). Experience and plasticity in the central nervous system. *Science*, **181**, 506–514.

Karten, H. J. (1969). The organization of the avian telencephalon and some speculations on the phylogeny of the amniote telencephalon. *Annals of the New York Academy of Sciences*, **167**, 164–179.

Kastin, A. J., Olson, R. D., Schally, A. V., & Coy, D. H. (1979). CNS effects of peripherally administered brain peptides. *Life Sciences*, **25**, 401–414.

Kohsaka, S., Takamatsu, K., Aoki, E., & Tsukada, Y. (1979). Metabolic mapping of chick brain after imprinting using [^{14}C]2-deoxyglucose technique. *Brain Research*, **172**, 539–544.

Korányi, L., Endröczi, E., & Tamásy, V. (1969). Influence of pituitary hormones (ACTH and ACTH analogues) and corticosteroid on the central nervous system in rats and chicks. *Acta Physiologica Academiae Scientiarum Hungaricae*, **36**, 73–82.

Kovách, J. K., & Hess, E. H. (1963). Imprinting: Effects of painful stimulation upon the following response. *Journal of Comparative and Physiological Psychology*, **56**, 461–464.

Krieger, D. T. (1972). Circadian corticosteroid periodicity: Critical period for abolition by neonatal injection of corticosteroid. *Science*, **178**, 1205–1207.

Krieger, D. T. (1974). Effect of neonatal hydrocortisone on corticosteroid circadian periodicity, responsiveness to ACTH and stress in prepuberal and adult rats. *Neuroendocrinology*, **16**, 355–363.

Landsberg, J. W., & Weiss, J. (1976). Stress and increase of the corticosterone level prevent imprinting in ducklings. *Behaviour*, **57**, 173–189.

Lissak, K., & Bohus, B. (1972). Pituitary hormones and avoidance behavior of the rat. *International Journal of Psychobiology*, **2**, 103–115.

Lorenz, K. (1935). Der Kumpan in der Umwelt des Vogels. *Journal für Ornithologie*, **83**, 137–214, 289–413.

Lorenz, K. (1973). *Civilized Man's Eight Deadly Sins*. New York: Harcourt Brace Jovanovich.

Lorenz, R. J., Branch, B. J., & Taylor, A. N. (1974). Ontogenesis of circadian pituitary–adrenal periodicity in rats affected by neonatal treatment with ACTH. *Proceedings of the Society for Experimental Biology and Medicine*, **145**, 528–532.

Maley, M. J. (1969). Electrical stimulation of agonistic behavior in the mallard. *Behaviour*, **34**, 138–160.

Mark, R. (1979). Concluding comments. In M.A.B. Brazier (Ed.), *Brain Mechanisms in Memory and Learning: From the Single Neuron to Man*, pp. 375–381. New York: Raven.

Martin, J. T. (1975). Hormonal influences in the evolution and ontogeny of imprinting behavior in the duck. In W. Gispen, Tj. B. van Wimersma Greidanus, B. Bohus, and D. de Wied (Eds.), *Hormones, Homeostasis and the Brain*, pp. 357–366. Amsterdam: Elsevier.

Martin, J. T. (1978a) Imprinting behavior: Pituitary-adrenal control of the approach response. *Science*, **200**, 565–567.

Martin, J. T. (1978b) Embryonic pituitary adrenal axis, behavior development, and domestication in birds. *American Zoologist*, **18**, 489–499.

Martin, J. T., DeLanerolle, N., & Phillips, R. E. (1979). Avian archistriatal control of fear-motivated behavior and adrenocortical function. *Behavioural Processes*, **4**, 283–293.

Martin, J. T., & Schutz, F. (1975). Arousal and temporal factors in imprinting in mallards. *Developmental Psychobiology*, **8**, 69–78.

Martin, J. T., & van Wimersma Greidanus, Tj. B. (1978). Imprinting behavior: Influence of vasopressin and ACTH analogues. *Psychoneuroendocrinology*, **3**, 261–269.

O'Donohue, T. L., Miller, R. L., & Jacobowitz, D. M. (1979). Identification, characterization and stereotaxic mapping of intraneuronal α-melanocyte stimulating hormone-like immuno-reactive peptides in discrete regions of the rat brain. *Brain Research*, **176**, 101–123.

Pacold, S. T., Kirsteins, L., Hojvat, S., Lawrence, A. M., & Hagan, T. C. (1978). Biologically active pituitary hormones in the rat brain amygdaloid nucleus. *Science*, **199**, 804–805.

Panksepp, J., Vilberg, T., Bean, N. J., Coy, D. H., & Kastin, A. J. (1978). Reduction of distress vocalization in chicks by opiate-like peptides. *Brain Research Bulletin*, **3**, 663–667.

Pert, C. B., Aposhian, D., & Snyder, S. H. (1974). Phylogenetic distribution of opiate receptor binding. *Brain Research*, **75**, 356–361.

Phillips, R. E. (1964). "Wildness" in the mallard duck: Effects of brain lesions and stimulation on "escape behavior" and reproduction. *Journal of Comparative Neurology*, **122**, 139–156.

Phillips, R. E. (1968). Approach-withdrawal behavior of peach-faced lovebirds, *Agapornis roseicolis*, and its modification by brain lesions. *Behaviour*, **31**, 163–184.

Phillips, R. E., & Youngren, O. M. (1971). Brain stimulation and species typical behaviour: Activities evoked by electrical stimulation of the brains of chickens (*Gallus gallus*). *Animal Behaviour*, **19**, 757–779.

Rieke, G. K. (1980) Kainic acid lesions of pigeon paleostriatum: A model for study of movement disorders. *Physiology and Behavior*, **24**, 683–687.

Rhees, R. W., Abel, J. H., & Haack, D. W. (1972). Uptake of tritiated steroids in the brain of the duck (*Anas platyrhynchos*). An autoradiographic study. *General and Comparative Endocrinology*, **18**, 292–300.

Salzen, E. A., & Parker, D. M. (1975). Arousal and orientation functions of the avian telencephalon. *In* P. Wright, P. G. Caryl, & D. M. Vowles (Eds.), *Neural and Endocrine Aspects of Behaviour in Birds*, pp. 205–242. Amsterdam: Elsevier.

Salzen, E. A., Parker, D. M., & Williamson, A. J. (1975). A forebrain lesion preventing imprinting in domestic chicks. *Experimental Brain Research* **24**, 145–157.

Salzen, E. A., Parker, D. M., & Williamson, A. J. (1978). Forebrain lesions and retention of imprinting in domestic chicks. *Experimental Brain Research* **31**, 107–116.

Schneirla, T. C. (1959). An evolutionary and developmental theory of biphasic processes underlying approach and withdrawal. *In* M. R. Jones (Ed.), *Current Theory and Research on Motivation*, Vol. 7, pp. 1–42. Lincoln, Nebraska: University of Nebraska Press.

Slotnick, B. M. (1973). Fear behavior and passive avoidance deficits in mice with amygdala lesions. *Physiology and Behavior*, **11**, 717–720.

Sluckin, W. (1972). *Imprinting and Early Learning* (2nd ed.). London: Methuen.

Sofroniew, M. V. (1979). Immunoreactive β-endorphin and ACTH in the same neurons of the hypothalamic arcuate nucleus in the rat. *American Journal of Anatomy*, **154**, 283–289.

Watson, S. J., & Akil, H. (1979). β-endorphin and α-MSH: Common cells of origin and binding properties. *Society for Neuroscience Abstracts*, **5**, 543.

Weiss, J., Kohler, W., & Landsberg, J. W. (1977). Increase of the corticosterone level in ducklings during the sensitive period of the following response. *Developmental Psychobiology*, **10**, 59–64.

van Wimersma Greidanus, Tj. B. (1977). Pregnene-type steroids and impairment of passive avoidance behavior in rats. *Hormones and Behavior*, **9**, 49–56.

van Wimersma, Greidanus, Tj. B., Croiset, G., Bakker, E., & Bouman, H. (1979). Amygdaloid lesions block the effect of neuropeptides (vasopressin and ACTH$_{4-10}$) on avoidance behavior. *Physiology and Behavior*, **22**, 291–295.

Wright, P. (1975). The neural substrate of feeding behaviour in birds. *In* P. Wright, P. G. Caryl, and D. M. Vowles (Eds.), *Neural and Endocrine Aspects of Behaviour in Birds*, pp. 319–349. Amsterdam: Elsevier.

Youngren, O. M., & Phillips, R. E. (1978). A stereotaxic atlas of the brain of the three-day-old domestic chick. *The Journal of Comparative Neurology*, **181**, 567–599.

Zeier, H., & Karten, H. J. (1971). The archistriatum of the pigeon: Organization of afferent and efferent connections. *Brain Research*, **31**, 313–326.

CHAPTER 6

ACTH and the Reminder Phenomena

David C. Riccio and James T. Concannon

Evidence that forgetting can be alleviated by exogenous adrenocorticotropin (ACTH) is reviewed. Data indicate that memory losses produced by amnesic treatment, ontogenetic change, or the Kamin effect can be attenuated through administration of ACTH or related peptide fragments prior to testing. ACTH also tends to increase resistance to extinction of both active and passive avoidance, as well as of conditioned taste aversion, an effect that may be viewed as a form of memory reactivation. Two special types of reminder phenomena induced by ACTH are considered: In one case, there is some evidence that the hormone may serve as a stimulus capable of producing state-dependent retention, but information about its role as a discriminative cue is lacking; in the second paradigm, data from recent experiments on "redintegration" indicate that ACTH administration can produce associative learning to new cues in previously stressed animals. Apparently, the hormonal state reactivates a representation of the earlier aversive episode, and this memory then becomes linked with environmental cues present at the time of reactivation. It is suggested that ACTH administration may produce these various reminder effects by making the internal context at testing more similar to that present during the original encoding episode, thus facilitating memory retrieval.

I. INTRODUCTION

It has become increasingly recognized that ACTH may play an important role in modulating learned behavior, either directly or through its classic en-

117

ENDOGENOUS PEPTIDES
AND LEARNING AND MEMORY PROCESSES

docrine function as a tropic hormone acting on the adrenal cortex (see Beckwith & Sandman, 1978; de Wied & Gispen, 1977, for reviews). A particularly intriguing aspect of ACTH is its capacity to reverse or alleviate certain memory deficits. In this chapter we attempt to identify several types of memory loss phenomena, and to review evidence of recovery produced by ACTH or related peptide fragments. Our coverage will be restricted, by and large, to situations where the hormone is administered after the retention loss has occurred.

Although ACTH may well influence memory storage processes, our focus on recovery from memory loss naturally leads to an emphasis on the role of retrieval deficits at testing. However, as Spear (1973, 1978) and Tulving (1974) have persuasively argued, retrieval is intimately linked with storage, that is, the way in which the original learning was encoded. Indeed, our general conceptual framework for the reminder phenomena to be described here rests upon the importance of congruence between the conditions, including both external and internal stimuli, present at training and at testing (cf. Spear, 1978). We would include stimuli and events occurring shortly following a nominal learning episode as part of the context of training; there is ample documentation that information continues to be processed following a learning trial (e.g., McGaugh, 1966; Wagner, Rudy, & Whitlow, 1973). Thus, our working assumption is that many manipulations act on the encoding features of memory input, rather than change the strength of storage.

Finally, a caution may be in order in interpreting some instances of hormonally induced memory recovery. Just as there is a distinction between learning and performance, a point recently reiterated by Kimble (1977) in his evaluation of peptide effects on behavior, there is a comparable difference between memory and performance. Put most simply, not all improvements in test performance at retention necessarily reflect recovery of the target memory. Admittedly, disentangling memory from performance is not always achieved as easily as it might seem to be, but the failure to recognize the problem can lead to unwarranted conclusions.

II. ACTH-INDUCED ALLEVIATION OF RETROGRADE AMNESIA

Experimentally induced retrograde amnesia (RA) has provided a particularly attractive paradigm to investigators interested in memory processes. Early concerns that the apparent memory loss might be an artifact of punishment-like effects produced by the amnesic agent have been virtually eliminated by the use of the single-trial passive avoidance procedure. In this paradigm, training typically consists of inescapable shock administered contingent upon the subject making a response such as entering a distinctive compartment. For some animals, training is followed by delivery of the amnesic agent. At a later time the retention test is given: Each subject is placed in the safe area and its tendency to refrain from moving into the shocked area is measured in terms of latency or probability (hence, the term passive avoidance). Retrograde amnesia is reflected in poorer passive avoidance in experimental animals than in controls. By reentering the fear compartment, subjects return to the stimuli temporally contiguous with both footshock and the amnesic treatment; thus, it seems reasonable to conclude that short latencies are based upon a genuine memory loss. However, whether the amnesia represents a disruption of storage–consolidation, or a deficit in retrieval, has been the topic of vigorous debate. Proponents of retrieval failure in explanations of amnesia have pointed to demonstrations of induced memory recovery as evidence par excellence that storage or consolidation was not disrupted and that the "engram" remained intact (Lewis, 1976; Meyer & Meyer, in press; Miller & Springer, 1973). Although the theoretical contention has not gone unchallenged (e.g., Gold & King, 1974), the fact that recovery can be induced by a variety of manipulations, including ACTH administration, has become increasingly well substantiated.

Rigter and his colleagues provided what appears to be the first clear demonstration that exogenous hormone administration at testing attenuates retrograde amnesia. In an initial study (Rigter, van Riezen, & de Wied, 1974), $ACTH_{4-10}$ given prior to testing reversed the CO_2-induced amnesia for a step-through passive avoidance response in rats. The failure of the same manipulation prior to acquisition to protect the subjects from amnesia suggested that the peptide effect was upon a retrieval process. The generality of the phenomenon was confirmed employing an appetitively reinforced task and electroconvulsive shock (ECS) as the amnesic agent (Rigter & van Riezen, 1975). Again, the administration of $ACTH_{4-10}$ before testing, but not before training, alleviated or reversed the memory deficit. The use of the peptide fragment in both of these studies was intended to rule out any influence of corticosteroid hormones. Responses from control groups, as well as the differing response requirements of the two tasks, indicated that the improved performance in the $ACTH_{4-10}$ condition was not an artifact of locomotor or activity changes. In keeping with these considerations, it should be noted that reversal of amnesia can also be achieved in situations where retention is evidenced by physiological changes rather than by the occurrence of a specific behavioral response. For example, when CO_2 treatment is presented following punishment training, the elevated corticosteroid response to the training cues seen in control subjects is eliminated. However, this amnesia effect is reversed by administration of $ACTH_{4-10}$ shortly before testing (Rigter, 1975a).

Although the reversal of amnesia by ACTH is clearly established, it should be noted that many studies lack a condition in which noncontingent footshock is substituted for the training trial to control for performance changes that might result from cumulative effects of systemic stress (e.g., footshock, amnesic treatment, and hormone administration). However, research examining other reactivation agents has commonly included a similar type of control group and found no evidence of artifactual changes in performance (e.g., Hinderliter, Webster, & Riccio, 1975; Mactutus & Riccio, 1978; Miller & Springer, 1972).

An important question about the effectiveness of reactivation treatments concerns the persistence of recovery of the memory. Administration of footshock in a neutral area (noncontingent footshock) seems to result in a relatively permanent attenuation of ECS-induced amnesia (Miller & Springer, 1972). However, reactivation in which subjects are exposed to the original amnesic agent (e.g., ECS or hypothermia) prior to testing produces recovery of a much more limited duration, several hours at most (Mactutus & Riccio, 1978; Thompson & Neely, 1970).

With respect to ACTH administration, recovery of memory also appears to be transient. An extensive parametric study (Rigter, Elbertse, & van Riezen, 1975) using independent groups of subjects found a time-dependent persistence of memory following $ACTH_{4-10}$ administration. Reversal of amnesia was nearly complete when testing occurred 1 or 2 hr after hormone administration, but little recovery was seen after an 8 hr delay (cf. Rigter, Janssens-Elbertse, & van Riezen, 1976). More recently, Mactutus, Smith, and Riccio (1980) investigated the effect of ACTH on memory loss induced when deep body cooling served as the amnesic agent. They observed that ACTH given 30 min prior to testing was highly effective in alleviating RA, but had no discernible effect when testing was delayed for 24 hr, a finding that was consonant with those of Rigter et al. (1976).

The transient nature of the hormonal reactivation of memory is consistent with an interpretation emphasizing that the treatment reestablishes certain internal attributes of the target memory that permit retrieval. As the hormonally elicited state diminishes, memory failure (amnesia) reappears. But are there ways that the reactivation might be made to outlast the immediate effects of exogenous ACTH? If memory is momentarily available, then reassociating it with the original fear cues might provide a more durable reminder effect. This notion was examined in a second experiment by Mactutus et al. (1980). Subjects received an ACTH injection 24 hr after passive avoidance training and an amnesia-producing hypothermia treatment. However, rather than receiving a test 30 min later, the rats were returned to the fear cues of the passive avoidance chamber for either 0 (control), 30, or 60 sec. No footshock was given during the exposure to the conditioned stimuli. As a control for possible reactivating influence of cue exposure per se, a fourth group, not given ACTH, was placed for 60 sec in the fear compartment. All groups were tested 1 day later.

As can be seen in Fig. 1, administration of ACTH in conjunction with brief exposure to the training cues led to a reactivation effect that extended over the 24-hr interval. Consistent with Rigter's data (Rigter et al., 1975a) and our initial experiment, there was little, if any, attenuation of amnesia in the group given ACTH without cue exposure and tested 24 hr later. Interestingly, mere exposure to fear cues also produced some attenuation of amnesia (cf. Hinderliter, et al., 1975), but the effect was significantly weaker than that seen when hormones are paired with cues.

Having found conditions under which ACTH-induced recovery from amnesia could be extended to 24 hr, we thought it of interest to determine whether the recovery could be sustained over more prolonged intervals. Would reactivation remain relatively permanent, or would it, like other responses, undergo retention loss? This experiment on the permanence of induced recovery is the converse of those that have varied the interval between amnesic treatment and the reactivation episode. The 60-sec cue exposure with ACTH that provided strong reactivation in the preceding experiment was employed as the major experimental condition here. However, the retention intervals between the reactivation and testing were either 1, 3, 7, or 14 days. To determine whether the use of the reactivated memory increased or decreased its durability, the 1-day interval group received a second test after a 7-day interval.

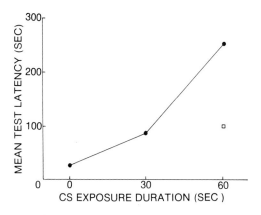

Fig. 1. Mean response latency on a passive avoidance test trial following reexposure to fear cues (CS) in conjunction with ACTH administration (●——●). Group given fear cues but not ACTH is indicated by □. All groups had previously received an amnesic treatment immediately following training. Testing occurred 24 hr after hormone–cue reactivation exposure. (Data reprinted with permission from Mactutus, Smith, & Riccio, 1980.)

The data confirmed that the reactivation effect could be extended with hormone–cue exposure, but also showed that the recovery underwent a temporally graded decline. There was some evidence of continued reactivation at the 7-day interval but, by 14 days, memory recovery was totally absent. Under comparable training conditions without amnesic treatment, the passive avoidance response is well-retained after more than 3 weeks (Schulenburg, Riccio, & Stikes, 1971); therefore, the lack of a reactivation effect after 14 days appears to represent an actual return of amnesia. Loss of the reactivated memory also was seen in the group tested a second time, although this decrement probably also reflects the effect of extinction from the earlier test. Thus, the transient quality of ACTH-induced recovery from retrograde amnesia was extended by our manipulation.

These findings with exogenously administered ACTH and ACTH analogs raise the interesting question of whether endogenous increases in ACTH level can attenuate amnesia. This notion implies that a variety of stressors eliciting ACTH, including different amnesic agents, might be capable of reversing amnesia if they were given prior to testing. However, there is some reason to expect that the success of "cross-modal" induced recovery may be limited by the role of specific attributes encoded during the training–treatment episode (Mactutus & Riccio, 1978).

Finally, our emphasis on retrieval processes should not be taken to suggest that this is the only mechanism through which ACTH (or its analogs) can alleviate amnesia. Indeed, there is considerable evidence that posttraining administration of the peptide fragment $ACTH_{4-10}$ not only improves retention of passive or active avoidance tasks, but also markedly reduces the severity of anisomycin-induced amnesia (Flood, Jarvik, Bennett, & Orme, 1976). Generally similar results have also been obtained with corticosteroids (Flood, Vidal, Bennett, Orme, Vasquez, & Jarvik, 1978). The time-dependent nature of these findings suggests the hormones were influencing storage or encoding mechanisms.

III. ONTOGENETIC MEMORY LOSS: REINSTATEMENT AND ACTH

Retention loss occurring during development represents an important and pervasive form of memory deficit. Immature animals show more rapid forgetting than do adults, even when care has been taken to equate the strength of original learning (Campbell & Campbell, 1962). The similarity of infantile amnesia in humans (Freud, 1938, pp. 580–603) to maturationally related forgetting in animals suggests that, although Freud's observations probably were well-founded, basic biological mechanisms rather than Oedipal conflicts are responsible for the forgetting (Campbell & Coulter, 1976). However, the relative contributions of experiential factors such as interference and neurological changes as sources of ontogenetic memory loss remain unclear (cf. Coulter, 1979; Miller & Berk, 1979).

Relatively rapid retention loss in young organisms appears to pose a problem in explaining the importance of early experience as a determinant of adult behavior. In this connection, Campbell and Jaynes (1966) suggested that in the natural environment, repetitions of the event may occur that serve as reminders to maintain the original association. In support of this notion, Campbell and Jaynes (1966) demonstrated that the retention loss of early learned fear could be alleviated by periodically exposing the subjects to an abridged form of the training conditions, or reinstatement. Weanling rats, which received Pavlovian discriminative fear conditioning followed by a single re-pairing of the fear cues and shock (CS–UCS) once per week, showed significantly better retention after a month than groups given only training or "reinstatement" episodes without training. Thus, reinstatement alone was not sufficient to establish new learning but did improve retention of a previously learned response.

Might a reminder effect be achieved if subjects were merely exposed to the fear cues (without further shocks) during the retention interval? An initial study in our laboratory (Silvestri, Rohrbaugh, & Riccio, 1970) found a small but significant improvement in retention when the cues only were presented briefly and periodically during the retention interval. However, this "conditioned reinstatement" outcome was substantially inferior to the retention performance of a group receiving cues with footshock. A probable explanation for these differences is that degree of memory reactivation depends upon the degree of overlap of attributes between the original conditioning and later reinstatement episodes (Spear, 1978) and that the cues-only condition is less adequate in this regard. If so, perhaps the conditioned reinstatement effect could be bolstered by inducing arousal (cf. Kety, 1970) and mimicking some of the attributes of the internal state that accompanies painful stimulation without actually administering such painful stimulation (Riccio & Haroutunian, 1979). Several experiments now seem to support this possibility. In one study (Haroutunian & Riccio, 1977), rats were fear conditioned to apparatus cues (brightness) at 21 days of age and tested for retention 2 weeks later. During a single reminder episode introduced 7 days after training, groups received one of two manipulations to modify the organism's state (attributes). Epinephrine was administered to some animals shortly before they were placed for a brief period in the compartment previously associated with shock; others first received a footshock in an apparatus very different from the one used in conditioning ("noncontingent footshock"). In the latter case, we assumed that arousal and the internal state elicited by the shock would not cease immediately upon shock termination, but would persist through subsequent cue exposure. Both treatments proved to be highly effective in improving retention as compared with saline injected or untreated groups. Interestingly, epinephrine was ineffective unless it was paired with the training cues. The importance of contiguity between the hormonal state and external fear cues seemed to be consistent with evidence from humans that attributions induced by epinephrine are dependent upon particular environmental conditions rather than on the presence of the drug per se (Schacter & Singer, 1962). In contrast, noncontingent footshock in the reinstatement paradigm enhanced retention with or without cue exposure. Although we initially suggested that the effectiveness of footshock even without fear cues was related to the high level of nonspecific arousal it produced, subsequent work (see following sections) led us to favor the interpretation that footshock is a particularly effective agent because it arouses a larger proportion of the attributes of the original fear conditioning event (Riccio & Haroutunian, 1979).

To explore further the role of hormonally induced changes at the time of reinstatement, we next conducted an experiment using ACTH (Haroutunian & Riccio, 1979). The intimate role of ACTH in the organism's response to stress suggested that this hormone would effectively mimic many of the internal changes that occur during fear conditioning. The procedure paralleled that used in the epinephrine portion of our earlier reinstatement experiment. One week after training, two groups received injections of either saline or ACTH followed by exposure to conditioned fear stimuli. In a third group, ACTH was followed by handling but not fear cue exposure. A fourth group, neither injected nor exposed to cues, provided a baseline for retention. The results were strikingly similar to those obtained with epinephrine. In conjunction with fear cues, ACTH produced a potent reinstatement of memory. In an attempt to delineate further the processes involved in reinstatement, Pesselman (in our laboratory) has begun to assess memory reactivation produced at the time of hormone administration. The paradigm, analogous to the reversal of RA studies described earlier, also yields a memory recovery effect. A comparison of male and female rats suggested the presence of a sex-related difference in reactivation in which the ACTH-induced recovery was greater in females.

Was the enhancement of conditioned reinstatement related specifically to stress-induced stimuli or to a more general state of arousal from the hormones? Although these components are not totally separable, the use of central nervous system (CNS) stimulants might permit central arousal with only minimal mimicking of peripheral stress cues. Accordingly, an ancillary experiment was carried out by Haroutunian and Riccio (1979) using the same methodology but with amphetamine (0.5, 1.0, or 3.0 mg/kg) or strychnine (0.5 or 0.75 mg/kg) as the reinstatement agents. Whereas these dose levels spanned a reasonable range for producing many behavioral effects, the results provided no hint that pharmacologically induced arousal of the CNS influenced reinstatement under conditions in which the exogenous hormones were effective. This outcome does not necessarily conflict with other evidence that analeptic agents can facilitate memory recovery (Gordon & Spear, 1973; Sara & Remacle, 1977), as the latter studies have typically involved paradigms in which the drug is present at testing.

That ACTH in conjunction with conditioned fear cues can maintain or increase fear behavior at later testing is also suggested by a finding obtained under very different circumstances (Korányi, Endröczi, Lissak, & Szepes, 1967). The effects of ACTH on fear motivated behavior in mice were studied in a situation where the hormone was administered prior to passive avoidance retention testing. Because most of the subjects reentered the previously shocked area on the initial test, they essentially self-administered a pairing of fear cues with an exogenously induced ACTH state. Interestingly, on subsequent daily trials, when ACTH was no longer present, most subjects repeatedly displayed enhanced passive avoidance responding.

Taken together, these studies support the supposition that exogenous administration of hormones associated with stressful experience can reestablish important internal attributes of the memory produced by fear conditioning. When this constellation of internal conditions is linked with the appropriate external cues, a powerful and enduring reminder effect is obtained. Although analeptic agents are effective sources of arousal, they may fail to reproduce adequately the totality of stress state changes (and attributes) elicited by electric footshock, or by administration of epinephrine and ACTH. It appears that administration of ACTH produces a state similar to that present during Pavlovian fear conditioning and, in the presence of the conditioned stimuli, results both in more effective retrieval and perhaps further strengthening of the association.

IV. THE KAMIN EFFECT AND ACTH-INDUCED REACTIVATION

Retention of an active avoidance response has often been found to follow a peculiar time course: When testing occurs either shortly after training, or 24 hr later, subjects perform very well, but when testing is given at intermediate retention intervals in the range of 1–6 hours, responding is impaired (Brush, 1971; Kamin, 1957). Because performance recovers at the long intervals, the phenomenon cannot represent simple retention loss. Experimental analysis of this U-shaped retention function, or Kamin effect, has focused on the source of the decline at the intermediate intervals and a variety of explanations have been offered, including such possibilities as motivational changes, alterations in physiological status, or memory impairment.

A seminal study by Levine and Brush (1967) proposed that changes in functioning of the pituitary–adrenal axis might mediate the Kamin effect. To examine the idea that manipulations that effectively elevated corticosteroid levels at the intermediate intervals might improve performance, they administered ACTH or hydrocortisone prior to a retention test given 4 hr after training. As expected, ACTH, and, to a lesser extent, hydrocortisone, alleviated the usual deficit. The authors concluded that motivational states based upon similarity of cues played a critical role in modulating retention.

The contribution of internal states as cues to retrieval of the original memory was demonstrated more extensively in an elegant series of experiments by Klein (1972). This research incorporated a strategy previously used by Spear and his colleagues (Klein & Spear, 1970; Spear, Klein, & Riley, 1971) to establish a particularly compelling case for the retrieval deficit explanation of the Kamin effect. Briefly, Spear's approach has been to use a "negative transfer" paradigm. Although the experimental designs become rather complex, the logic is straightforward. If there is memory failure at the intermediate intervals, acquisition of a conflicting task should be relatively unimpaired; therefore, subjects should be comparable to naive controls. Conversely, conditions with a strong representation of the original learning should result in interference and, therefore, slower learning of an opposing response. To control for locomotor and activity influences on performance, these studies typically included conditions in which some subjects were switched from active to passive avoidance tasks and vice versa. By and large, a *lack* of negative transfer has been found. This is consistent with a memory deficit view.

In his first experiments, Klein (1972) showed that a stressful experience (exposure to cold water), given prior to the intermediate retention interval test, resulted in negative transfer. Presumably, ACTH release during stress reactivated the original target memory, resulting in interference with acquisition of the new (conflicting) task. More direct evidence for the role of ACTH was then provided by experiments in which the hormone level was altered by physiological manipulations. Animals that were trained on one avoidance task and, after an intermediate interval, given electrical stimulation in the anterior hypothalamus (a region previously implicated in regulating ACTH release) required more trials to acquire a conflicting response than did controls receiving brain stimulation in another region. Assuming that the critical internal state also might be achieved by the presence of the pituitary hormone in the CNS, Klein then examined the effect of directly implanting crystals of ACTH into the anterior hypothalamus prior to testing at the intermediate retention interval. An increase in negative transfer, presumably related to enhanced retrieval of the original memory, was again obtained. Because the effects obtained with these manipulations were independent of the order of the avoidance tasks, a number of potential performance artifacts can be ruled out. This ACTH-induced alleviation of the Kamin effect supports the interpretation that reestablishing the internal state produced during the original aversively motivated task provides the necessary cues for more efficient retrieval of that memory. A later study (Klein, 1975) showed that systemic administration of ACTH, but not of corticosterone or saline, would reactivate memory of active avoidance training in hypophysectomized, adrenalectomized, or sham operated rats. This finding suggests that ACTH may exert its reactivation effect by inducing some of the attributes of arousal associated with original training.

One puzzling aspect of memory reactivation by induced ACTH release at the intermediate intervals is why the same effect is not achieved by stress-elicited hormone activity during retention tests per se, as testing typically involves relearning in which painful footshock stimuli are used. Thus, the conditions for ACTH release also seem inherent in the paradigm demonstrating the Kamin effect. Perhaps the administration of noncontingent aversive treatments works as a reactivation because the time interval between stress and testing, or the organism's lack of control over aversive events, are important factors in providing cues for memory retrieval.

V. EFFECTS OF ACTH ON EXTINCTION OF AVERSIVELY MOTIVATED BEHAVIOR

Investigation of extinction, in which reinforcement is withheld for previously reinforced responses, is traditionally viewed as separate from the study of retention. Extinction research focuses on how the change in contingencies alters performance; memory research tends to examine the effects of manipulations (time, traumatic events, etc.) orthogonal to the original training conditions. Nevertheless, it may be of interest to examine the influence of ACTH upon extinction, especially in situations such as avoidance learning, where the nominal change in contingencies (e.g., aversive stimulus is turned off) initially may not alter the subject's perception of the situation (response → no shock). From this perspective, the persistence of responding during extinction may partially reflect the strength of retrieval of the target memory.

Miller and Ogawa (1962) reported that administration of ACTH during training increased the number of shuttle avoidance responses made in an extinction test, although the hormone had no apparent effect on rate of initial acquisition. Because their subjects were adrenalectomized rats, the investigators suggested that the increased resistance to extinction was an extraadrenal consequence of ACTH administration. Although their study does not qualify as a case of reactivation, as the hormone was discontinued at the end of training, it was historically important in calling attention to behavioral effects of ACTH not mediated by the adrenal cortex.

De Wied (1966) confirmed an effect of ACTH on learned behavior by showing that periodic administration of ACTH and related peptide fragments during extinction sessions increased resistance to extinction in two types of avoidance tasks, the shuttlebox and pole-jumping situations. A related study (de Wied & Bohus, 1966) used α-MSH, a peptide consisting of part of the amino acid sequence of ACTH, but lacking corticotropic activity. This hormone, administered to intact rats during acquisition of a shuttle-avoidance response had no effect on trials to criterion or subsequent extinction performance. In contrast, when periodic injections of the α-MSH peptide were given during the 14 days of extinction testing, the number of conditioned avoidance responses made was significantly greater than it was in placebo treated controls. Although Murphy and Miller (1955) failed to obtain an effect of ACTH administered during extinction, other evidence, both direct and indirect, supports the conclusion that high levels of ACTH during testing retard extinction of avoidance responding (de Wied & Gispen, 1977; see also Leshner, 1978).

These outcomes might be considered to be a form of memory reactivation in which the hormonal treatment during extinction enhances retrieval by providing or maintaining cues that were originally produced during the stress of avoidance learning. Alternatively, these treatments may impair the acquisition of new learning in nonreward (extinction) conditions when avoidance behaviors no longer have programmed consequences (cf. Bohus & de Wied, 1966; Gray, 1971; Gray & Garrud, 1977). These data could also reflect performance changes induced by energizing or arousing properties of the hormones. In this context, evidence that ACTH also increased the persistence of punishment-induced suppression is particularly instructive (Levine & Jones, 1965). Although the design did not include a group receiving only posttraining administration of ACTH, rats receiving hormone during the brief punishment training and throughout the 7-day extinction test showed profound suppression of the baseline response of bar pressing for water reward. Also, rats punished for entering a dark compartment are reported to display increased test latencies when $ACTH_{4-10}$ is administered prior to the retention trial (van Wimersma Greidanus, 1979). This latter finding suggests that ACTH induces memory reactivation and that the hormone can facilitate performance during extinction in a task requiring withholding of a motor response. The similarity of outcomes across experiments with opposite response requirements (go versus no-go) implies that simple performance factors alone cannot account for the effect of ACTH.

A somewhat different form of the extinction of aversively motivated learning is reflected in conditioned taste aversion (CTA) learning, in which an animal learns to avoid ingesting a novel taste such as sweetened water associated with malaise induced by lithium chloride (LiCl) or other agents. Levine and his associates (Levine, Smotherman, & Hennessey, 1977; Smotherman & Levine, 1978) proposed that CTA may provide a sensitive paradigm for assessing the influence of peptides on behavior. For example, in certain types of extinction testing measuring recovery from CTA, endogenous levels of ACTH and glucocorticoids remain low, permitting the influence of an exogenously administered peptide to be seen.

A series of studies summarized by Levine et al. (1977) and Smotherman and Levine (1978) carefully examined ACTH modulation of CTA relative to the level of activation of the adrenal–pituitary axis. Preliminary findings (Levine et al., 1977) revealed that corticosterone levels are elevated following a LiCl injection, and that elevations could be blocked by dexamethasone (DEX) pretreatment. When subjects are tested in a two-choice situation, in which plain water and the taste solution previously paired with poisoning are present, conditioned plasma corticosterone levels during recovery are similar for LiCl and vehicle injected animals, even though behavioral aversion to the taste is displayed. However, during forced extinction, in which thirsty subjects were presented with only the flavored (aversive) substance, corticosterone levels in animals that received LiCl were elevated relative to those levels in animals that drank a neutral solution under forced-choice conditions (cf. Ader, 1976; Smotherman, Hennessey, & Levine, 1976). Furthermore, the effects of either ACTH or DEX state-dependent retention paradigms differed in a manner related to the baseline of corticosterone. That is, animals pretreated with DEX prior to LiCl conditioning showed attenuated taste aversion relative to those animals conditioned under the influence of saline. Dexamethasone was ineffective when it was administered during free-extinction testing, when corticosterone levels were found to be low. However, animals treated with ACTH prior to testing in free extinction conditions, when adrenal pituitary levels were low, showed prolonged recovery from CTA relative to animals receiving saline during recovery. In summary, DEX manipulations were effective only when adrenal–pituitary activity was high (near the time of LiCl injection), whereas ACTH was effective only when adrenal–pituitary activity was low (during free extinction).

Because ACTH was most effective during recovery from CTA, it might be concluded that its principal site of action is on a retrieval process, rather than on conditioning processes. To test this notion, animals were given CNS implants of hydrocorticosterone, a procedure that blocks ACTH release and reduces peripheral steroid levels, prior to LiCl injections. This manipulation was found not to affect conditioning. Hence, the effects of DEX on conditioning were probably not due to reduced ACTH levels, as ACTH can be reduced with no effects on taste aversion conditioning. ACTH, therefore, appears to be affecting retrieval processes (van Wimersma Greidanus, van Dijk, de Rotte, Goedemans, Croiset, & Thody, 1978).

Smotherman and Levine (1978) subsequently conducted a series of studies to compare the capacity of $ACTH_{1-39}$ and $ACTH_{4-10}$, a peptide fragment that is devoid of corticotropic activity, to either induce CTA or to alter recovery from CTA. Results from these studies showed that both $ACTH_{1-39}$ and $ACTH_{4-10}$ delayed the extinction of LiCl-induced CTA when the peptides were administered during free-extinction recovery days, although the parent peptide had a significantly stronger effect. These results are consistent with findings presented by Rigter (1975b) and Rigter and Popping (1976), who found that $ACTH_{4-10}$ delayed recovery from CTA when it was administered prior to extinction sessions using a two-bottle preference test. ACTH administered prior to forced-drinking extinction tests was not effective. Hence, the results of Levine and associates and Rigter and associates are consistent in showing that ACTH fragments prolong the rate of recovery from CTA; results that add generality to the effects found for extinction of other forms of aversively and appetitively motivated behavior.

VI. STATE-DEPENDENT RETENTION

Numerous psychopharmacological studies have shown that training and testing an organism in the same drug state (drug versus placebo) leads to superior retention relative to organisms trained in one drug state and tested in the alternative state—a phenomenon known as state-dependent retention (for reviews, see Bliss, 1974; Overton, 1971, 1978). The traditional state-dependent retention paradigm includes a 2×2 factorial combination of the presence of drug in training and in testing. Presence of a statistical interaction between agent present in training and agent present in testing is indicative that retention is state-dependent. That is, optimal retention *depends on* having the same state present during training and testing. Although the mechanism responsible for state-dependent retention has not been precisely specified, memory theorists such as Spear (1978) have suggested that state-dependent performance reflects cue-dependent retention, with the interoceptive cues experienced while in a drug state serving as contextual cues for retrieval. Keeping the drug state constant between training and testing ensures that appropriate retrieval–contextual cues are available.

From our perspective, the issue is whether hormonal states may provide cues similar to those produced by psychotropic drugs. In contrast to the enormous amount of research conducted on state-dependent retention with psychotropic drugs, relatively few studies have been done using hormones as state-dependent agents. One of the few studies was done by Gray (1975) who directly examined the possibility that stress-released hormones produce state-dependent retention. Gray (1975) trained and tested rats in a passive avoidance task under each of the four conditions resulting from a 2×2 factorial combination of exogenous ACTH or vehicle given before training and before testing. His results indicated that retention was indeed state-dependent, because the same-state groups performed better than the changed-state groups, and no main effects of ACTH on conditioning and testing were found. Similar effects were reported for a state-dependent manipulation using DEX, a synthetic glucocorticoid that blocks ACTH release (Pappas & Gray, 1971).

Although the findings of Gray (1975) and Pappas and Gray (1971) concerning ACTH- and DEX-induced state-dependent retention seem to be straightforward, they are not always readily reproducible in other types of learning paradigms (cf., Guth, Seward, & Levine, 1971; Levine et al., 1977; Mormede & Dantzer, 1977). The findings by Levine et al. (1977) imply that inconsistencies in the different studies may be related to uncontrolled variations in the baseline of adrenal–pituitary activity.

Research reported by Gold and van Buskirk (1976a,b) may represent an unusual posttrial example of hormone-induced state-dependent retention. In general, these studies showed that postconditioning administration of low doses of ACTH facilitated retention of a passive avoidance response, whereas high doses of ACTH led to impairment of memory. Retention deficits produced by high doses of ACTH may not have been due to impairment of memory formation (cf. Sands & Wright, 1979), but rather may be due to state-dependent retention, with the hormonal state being altered between memory processing and retention testing (cf. Chute & Wright, 1973). This state-dependent effect might be expected to be particularly profound when both high-intensity footshock is given and a high dose of ACTH is given posttrial (Gold & van Buskirk, 1976b). Readministration of these high doses of ACTH prior to testing would help to determine whether these animals suffered impaired storage or state-dependent retention. Admittedly, *enhancement* of retention with low doses of ACTH following training cannot readily be placed into a state-dependent framework. Indeed, this outcome not only represents a strong point of evidence in favor of a storage view of the action of ACTH, but also offers an intriguing challenge to state-dependent interpretations involving ACTH.

Another apparent inconsistency concerning hormone-induced state-dependent retention is reflected in the previously described research of Rigter and associates concerning hormonal reversal of ECS- or CO_2-induced retrograde amnesia (e.g., Rigter, et al., 1974; Rigter & van Riezen, 1975; Rigter, Schuster, & Thody, 1977). Each of these carefully controlled studies included animals receiving passive avoidance training without amnesic treatment and hormone administration at training, testing, or both. Neither systemically administered $ACTH_{4-10}$, MSH, vasopressin, nor β-lipotropin—in doses that reversed retrograde amnesia—produced state-dependent retention. That is, animals trained in one hormonal condition and tested in the other state performed as well as animals trained and tested in the same state. These results suggest at least two possibilities for the lack of state-dependent or dissociated learning: (1) the doses used are too low to produce dissociation, or (2) dissociation only comes about as a result of the classical tropic effects of these hormones (cf. Gray, 1975). Both possibilities are empirically testable.

Given the substantial body of evidence for the influence of ACTH on memory processing, it might be asked if an ACTH-induced state might be sufficiently salient to function as a cue or discriminative stimulus. One approach to such drug discrimination learning (DDL) is to train an animal to make one response, such as turn left in a T-maze, in one drug state, and another response, such as turn right, in a different drug state, when the presence of the drug is the only reliable predictor of which of the alternative responses will be reinforced. Unlike state-dependent retention paradigms, drug discrimination paradigms are not used to assess the presence or absence of the *same* response, but rather to assess the relative availability of two well-learned responses. Furthermore, DDL is readily produced by low drug doses that will produce dissociation only with great difficulty, and is obtained almost exclusively with drugs that affect CNS functioning. There is some evidence indicating that DDL can be produced by exogenous administration of hormones, including progesterone (Stewart, Krebs, & Kaczender, 1967) and epinephrine (Schuster & Brady, 1964) but we know of no evidence for ACTH-induced DDL. This possibility is currently being investigated by Bellush (1980) in our laboratory.

VII. REDINTEGRATION

We have recently become interested in the possibility that hormonal states might result in what might be termed *redintegration* of prior stress experience. The major question is whether administration of a stress-related hormone can activate a representation of a previous traumatic episode in such a way that this memory becomes associated with contiguous neutral cues in the environment. The assumption is that the internal state associated with endogenous hormone release during stress provides an attribute of aversiveness that can be reinduced by appropriate exogenous hormones. The behavioral consequence of such a transfer would be the presence of fear responses to the environmental cues at a later test. The issue was provoked partly by hunch and partly by hints in the literature. With respect to the latter, Anderson, Crowell, Koehn, and Lupo (1976) found that high-intensity footshock administered in a different noncontingent apparatus decreased locomotor activity in subsequent open-field testing in rats. The authors suggested that the source of this effect was the novelty of the open-field situation eliciting unconditioned emotional reactions to prior shock, resulting in freezing behavior. Hence, exposure to novelty amplified the importance of the internal cues associated with prior shock. Anderson et al. (1976) proposed that generalized fear based on exteroceptive cues was unlikely because of marked dissimilarity in the shock and testing apparatuses. They inferred that the increased freezing was based on similarity of interoceptive stress states between the two situations. Thus, it appears that a novel environment elicited an emotional response that, with its interoceptive feedback, served to remind subjects of earlier shock episodes.

Our emphasis has been upon the hormonally mediated acquisition of new learning, rather than on the amplification of an elicited response. In this approach, redintegrative conditioning should occur in animals subjected to prior shock stress, and later given hormone administration in conjunction with cue exposure, but not in animals exposed to only part of the conditioning contingency. This associative process may depend upon, among other variables, the dose of the hormone used, and it should be weakened when a blocker of the hormone action is utilized. In the following paragraphs we shall briefly describe a Pavlovian conditioning paradigm that we have utilized to explore some of these notions.

In all experiments, adult Holtzman rats were utilized as subjects. The basic experimental design, which has been described more extensively elsewhere (Concannon, Riccio, & McKelvey, 1980), consisted of three phases: (1) initial noncontingent footshock (NCFS) experience; (2) hormone injection and cue exposure; and (3) a test for spatial avoidance of environmental cues, with each phase separated by 24 hr. The NCFS experience consisted of delivery of three brief shocks for 2 consecutive days; some animals received exposure to the NCFS apparatus but were not shocked. Phase 2 (injection plus cue exposure) consisted of injecting animals subcutaneously with either saline (control), epinephrine (Experiments 1–3) or $ACTH_{1-39}$ (8 IU) 20 min prior to a brief (nonshock) exposure to either (1) the black side of a black–white shuttlebox (CUES) or a pine box (PINE) placed on the shuttlebox floor; or (2) being returned to their home cages without further treatment (HOME). Phase 3, the testing period, consisted of placing the animal on the white side of the shuttlebox and measuring the rat's tendency to avoid the black side during a 30 min (no shock) test.

In the initial experiment, four groups of rats that earlier had received NCFS were given either saline or epinephrine (0.01, 0.05, or 0.10 mg/kg) 20 min prior to black cue exposure. Results showed an elevation of crossover latencies as a function of dose level. Saline control scores averaged 42 sec, whereas epinephrine scores were 84, 512, and 428 sec, respectively. Analysis of these data showed that the 0.05 mg/kg group differed significantly from all other groups except the 0.10 mg/kg group, and that no other between-group differences were statistically reliable. The finding cannot be explained simply as stimulus generalization, as the saline group had the same sequence of noncontingent footshock and cue exposure. However, to ensure that the aversion seen in the 0.05 mg/kg group was not due to place aversion in animals without prior shock (PL–AV) or to sensitization to shock (SENS), we ran two additional groups of animals. Place aversion animals received epinephrine injection and black cue exposure, but did not have prior shock. Sensitization animals received prior shock and epinephrine injection prior to exposure to the pine box. Mean crossover latencies for these two groups (66 and 43 sec, respectively) showed total lack of aversion resulting from these manipulations. Hence, neither simple place aversion nor sensitization appear to contribute to the aversion.

The animals in the initial experiment previously had served as subjects in an unrelated but mildly stressful study. Therefore, a replication of the phenomenon was undertaken using experimentally naive rats and the intermediate dose of epinephrine. Two basic groups received NCFS and subsequently epinephrine or saline (control) at time of cue exposure. Two additional control groups that did not receive NCFS were administered epinephrine followed by either exposure to cues (place aversion control) or to the apparatus used for NCFS. Again, rats with prior shock experience in one apparatus showed fear (avoidance) to a new set of cues after being exposed to the new cues while in the epinephrine state. Animals in this condition differed reliably from the other three groups, which did not differ from each other.

A third experiment incorporating additional control groups was conducted to demonstrate that the redintegration phenomenon was based upon new associative learning, rather than performance artifacts. As in the other experiments, the major finding was that animals with irrelevant shock experience given epinephrine–cue exposure avoided the fear cues significantly more than did rats in other conditions. These results suggest that epinephrine can mediate a specific association between prior footshock in an irrelevant context and later exposure to environmental stimuli in a manner not easily explained on the basis of nonassociative variables.

In a subsequent study, we attempted to extend these findings by employing ACTH as the stress-released hormone (Concannon, Riccio, Maloney, & McKelvey, 1980). Of particular interest is that, to assess the possible involvement of opiate systems in mediating the effects of ACTH (Gispen, Reith, Schotman, Wiegant, Zwiers, & de Wied, 1977; Gispen, Wiegant, Bradbury, Hulme, Smyth, Snell, & de Wied, 1976; Wiegant, Gispen, Terenius, & de Wied, 1977), we administered the opiate receptor blockers naloxone and naltrexone prior to ACTH plus cue exposure. Thus, the major condition consisted of NCFS groups given saline (or no injection), naloxone (1.0 mg/kg), or naltrexone (1.0 mg/kg) intraperitoneally 5 min prior to the ACTH injection. Subjects were briefly exposed to the black cues 20 min later. Several control groups were also included; however, naloxone or naltrexone only conditions were deliberately omitted, because there was no reason to believe that learning would occur without the presence of the ACTH state. As can be seen in Fig. 2, the group receiving ACTH prior to cue exposure spent significantly more time on the

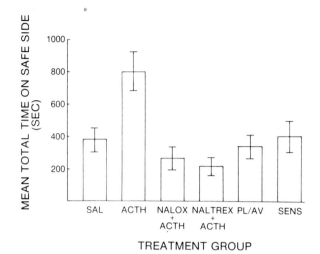

Fig. 2. Mean total time ($\pm$ SEM) spent avoiding the apparatus cues that earlier had been paired with saline or ACTH. For two groups, naloxone or naltrexone injection preceded ACTH administration. Two other groups represent controls for place aversion (PL/AV) and sensitization (SENS) effects. (Data reprinted with permission from Concannon, Riccio, Maloney, & McKelvey, 1980.)

safe side of the apparatus than did all other groups. (The results for groups receiving saline or no injection prior to ACTH were pooled, as their performance did not differ.) An analysis of variance confirmed the presence of a reliable treatment effect for the six groups, and subsequent analytic comparisons showed significantly less avoidance in the naloxone and naltrexone with ACTH groups than in the ACTH only condition. Hence, exogenous administration of ACTH in previously stressed rats can mediate an aversion to novel cues, and this redintegration effect was blocked by reasonably low doses of the specific narcotic antagonists. Thus, it appears that opiate receptors may be involved in this type of learning, as has also been suggested for conditioned hyperthermia, conditioned alleviation of morphine withdrawal hypothermia (Lal, Miksic, & Drawbaugh, 1978), and enkephalin-mediated conditioned antinociception (Chance, White, Krynock, & Rosecrans, 1978).

One implication of these findings on redintegration is that epinephrine and ACTH administered to previously shock-stressed rats can produce acquired aversiveness to new stimuli. The psychological processes by which this learning is achieved remain to be specified. A possible mechanism might be Pavlovian higher-order conditioning, in which the endogenous hormones released during shock serve as an internal cue (CS_1) by virtue of being paired with subsequent shock. Subsequent exogenous administration of a hormone (CS_1) might then elicit fear, which is associated with environmental cues (CS_2: black cues), resulting in transfer of fear to these environmental cues. The unique finding from our point of view is that hormonal states provide the putative CS_1.

A related interpretation can be made in terms of memory reactivation. That is, the hormonal treatment redintegrates the state of prior stress and allows this state to be linked with environmental cues. Conversely, hormones in the absence of prior shock, or prior shock not followed by hormone–cue exposure, should be ineffective, as was the case. Or perhaps hormone administration reactivates a memorial representation of the shock UCS (Rescorla & Heth, 1975) via their shared interoceptive attributes. In either case, the hormonal state results in an old experience becoming associated with new cues.

VIII. CLINICAL IMPLICATIONS OF THE ACTH-INDUCED REMINDER PHENOMENA

Demonstrations that diverse types of retention loss can be ameliorated by ACTH administered subsequent to the original learning episode promote the hope that the hormone might prove to be clinically valuable in treating memory disorders. However, examination of the clinical literature failed to uncover direct empirical evidence that ACTH influences learning or memory processes (e.g., Rapoport, Quinn, Copeland, & Burg, 1976), although several studies with retarded subjects (Sandman, George, Walker, Nolan, & Kastin, 1976; Walker & Sandman, 1979) or elderly subjects (Branconnier, Cole, & Gardos, 1979; Ferris, Sathananthan, Gershon, Clark, & Mohinsky, 1976) offer cautious encouragement in this regard. It seems clear that the pertinent data base at this time is extremely limited. Meyer (1972) has pointed out the importance of motivational context as a determinant of memory reactivation, and has considered the implications for electroconvulsive therapy. Similarly, if there is any merit to our interpretation that the role of ACTH in the reminder phenomena is based upon correspondence of internal states, then one would not expect the hormone to provide a panacea for memory deficits. Selective recovery of memories established under circumstances involving stress or high arousal levels would seem more likely to occur.

IX. CONCLUSIONS AND MUSINGS

The preceding empirical studies document the ACTH-induced reminder phenomena and support the view that the hormone can facilitate memory retrieval.

The topics discussed in Sections II through V of this chapter were primarily focused on direct memory reactivation, in which a hormonal treatment given during the retention interval or around the time of testing served as a reminder. The success of hormone-induced memory reactivation is believed to depend on the reinstatement of a sufficient number (or kind) of retrieval attributes that were present at, or immediately following, original learning (cf. Spear, 1973). Although the latter sections included studies based upon paradigms rather different from reactivation, we believe that they bear an important relationship to reminder phenomena.

Whereas we have proposed that ACTH may exert its effect by increasing the similarity of the retrieval context to the original encoding episode, the specific nature of the stimuli or states involved is not yet clear. Are they central (cf. Dunn & Gispen, 1977), peripheral, or both? Work with $ACTH_{4-10}$ is analytically powerful in demonstrating hormonally induced behavioral changes not mediated by adrenal cortical hormones, but can we presume that $ACTH_{1-39}$ would exert its effects in the same way as the peptide fragment? It is appealing to think in terms of altered CNS states as the common denominator for the memory effects of these peptides, but that notion may not be justified. Indeed, McGaugh, Martinez, Jensen, Messing, & Vasquez (1980) recently reported evidence implicating peripheral processes as important determinants of pharmacological effects on memory, and a similar conclusion is implicit in some of our reinstatement work (Haroutunian & Riccio, 1977). Thus, it may be premature to discount the contribution of peripheral changes, and associated feedback, resulting from ACTH.

Finally, it seems to us that situations involving memory loss provide an important general paradigm for detecting an influence of ACTH on behavior. In contrast, effects of ACTH on acquisition of fear may be obscured by high endogenous levels of hormone resulting from aversive stimulation, whereas, in other situations, it can be shown that ACTH is not necessary to the establishment of fear conditioning (and memory). Also, long-term retention of fear is often reflected on the initial test trial, even though endogenous ACTH levels should have remained near baseline level until *after* the situation is recognized as fearful. Indeed, unless one appeals to the ACTH release attendant upon transporting and handling subjects at testing, most studies of fear retention inherently involve mismatched states, because ACTH levels are high during fear conditioning and the subsequent memory registration period, but low at onset of testing. These outcomes indicated that various other attributes of the training episode are often sufficient for establishing and retrieving memory (cf. Spear, 1978). But in cases where certain cues are minimized, or perhaps altered, ACTH can have a pronounced influence on responding, as was noted some years ago (Weiss, McEwen, Silva, & Kalkut, 1969). It seems likely that, in these cases, ACTH enhances memory retrieval by increasing the proportion or quality of available cues.

ACKNOWLEDGMENTS

Preparation of this chapter was supported in part by NIMH grant MH30223 to the first author and NIMH postdoctoral fellowship MH07358 to the second author. A portion of the contribution by James Concannon was written while he was serving as Visiting Assistant Professor in the Department of Psychology at the University of Rhode Island. The authors gratefully acknowledge the yeomanship typing job performed by Michelle Iannetta, who had to decode two notoriously illegible handwritings.

REFERENCES

Ader, R. (1976). Conditioned adrenocortical steroid elevations in the rat. *Journal of Comparative and Physiological Psychology*, **90**, 1156–1163.

Anderson, D. C., Crowell, C., Koehn, D., & Lupo, J. V. (1976). Different intensities of unsignalled inescapable shock treatments as determinants of non-shock-motivated open field behavior: A resolution of disparate results. *Physiology and Behavior*, **17**, 391–394.

Beckwith, B. E., & Sandman, C. A. (1978). Behavioral influences of the neuropeptides ACTH and MSH: A methodological review. *Neuroscience and Biobehavioral Reviews*, **2**, 311–338.

Bellush, L. (1980). Investigation of discriminative stimulus properties of ACTH and dexamethasone. Unpublished Master's Thesis, Kent State University, Kent, Ohio.

Bliss, D. K. (1974). Theoretical explanations of drug-dissociated behaviors. *Federation Proceedings*, **33**, 1787–1796.

Bohus, B., & de Wied, D. (1966). Inhibitory and facilitatory effect of two related peptides on extinction of avoidance behavior. *Science*, **155**, 318–320.

Branconnier, R. J., Cole, J. O., & Gardos, G. (1979). ACTH$_{4-10}$ in the amelioration of neuropsychological symptomology associated with senile organic brain syndrome. *Psychopharmacology*, **61**, 161–165.

Brush, F. R. (1971). Retention of aversively motivated behavior. *In* F. R. Brush (Ed.), *Aversive Conditioning and Learning*. New York: Academic Press.

Campbell, B. A., & Campbell, E. H. (1962). Retention and extinction of learned fear in infant and adult rats. *Journal of Comparative and Physiological Psychology*, **55**, 1–8.

Campbell, B. A., & Coulter, X. (1976). The ontogenesis of learning and memory. *In* M. R. Rosenzweig & E. L. Bennett (Eds.), *Neural Mechanisms of Learning and Memory*, pp. 209–235. Cambridge, Massachusetts: MIT Press.

Campbell, B. A., & Jaynes, J. (1966). Reinstatement. *Psychological Review*, **73**, 478–480.

Chance, W. T., White, A. C., Krynock, G. M., & Rosecrans, J. A. (1978). Conditional fear-induced antinociception and decreased binding of (^{3}H)N-Leu-enkephalin to rat brain. *Brain Research*, **141**, 371–374.

Chute, D. L., & Wright, D. C. (1973). Retrograde state-dependent learning. *Science*, **180**, 878–880.

Concannon, J. T., Riccio, D. C., Maloney, R., & McKelvey, J. (1980). ACTH mediation of learned fear: Blockage by naloxone and naltrexone. *Physiology and Behavior*, **25**, 977–979.

Concannon, J. T., Riccio, D. C., & McKelvey, J. (1980). Pavlovian conditioning of fear based upon hormonal mediation of prior aversive experience. *Animal Learning and Behavior*, **8**, 75–80.

Coulter, X. (1979). The determinants of infantile amnesia. *In* N. E. Spear & B. A. Campbell (Eds.), *The Ontogeny of Learning and Memory*. Hillsdale, New Jersey: Lawrence Erlbaum and Associates.

de Wied, D. (1966). Inhibitory effect of ACTH and related peptides on extinction of conditioned avoidance behavior in rats. *Proceedings of the Society for Experimental Biology and Medicine*, **122**, 28–32.

de Wied, D., & Bohus, B. (1966). Long-term and short-term effects on retention of a conditioned avoidance response in rats by treatment with long acting pitressin and α-MSH. *Nature (London)*, **212**, 1484–1488.

de Wied, D., & Gispen, W. H. (1977). Behavioral effects of peptides. *In* H. Gainer (Ed.), *Peptides in Neurobiology*. New York: Plenum.

Dunn, A. J., & Gispen, W. H. (1977). How ACTH acts on the brain. *Biobehavioral Reviews*, **1**, 15–23.

Ferris, S. H., Sathananthan, G., Gershon, S., Clark, C., & Moshinsky, J. (1976). Cognitive effects of ACTH$_{4-10}$ in the elderly. *Pharmacology, Biochemistry and Behavior*, **5**, 73–78.

Flood, J. F., Jarvik, M. E., Bennett, E. L., & Orme, A. E. (1976). Effects of ACTH peptide fragments on memory formation. *Pharmacology, Biochemistry and Behavior* (Supplement 1), **5**, 41–51.

Flood, J. F., Vidal, D., Bennett, E. L., Orme, A. E., Vasquez, S., & Jarvik, M. E. (1978). Memory facilitating and anti-amnestic effects of corticosteroids. *Pharmacology, Biochemistry and Behavior*, **8**, 81–87.

Freud, S. (1938). Infantile sexuality. *In* A. A. Brill (Ed.), *The Basic Writings of Sigmund Freud*, pp. 580–603. New York: The Modern Library.

Gispen, W. H., Reith, M. E. A., Schotman, P., Wiegant, V. M., Zwiers, H., & de Wied, D. (1977). CNS and ACTH-like peptides: Neurochemical response and interaction with opiates. *In* L. H. Miller, C. A. Sandman, & A. J. Kastin (Eds.), *Neuropeptide Influences on the Brain and Behavior*, pp. 61–80. New York: Raven.

Gispen, W. H., Wiegant, V. H., Bradbury, A. F., Hulme, E. C., Smyth, D. G., Snell, C. R., & de Wied, D. (1976). Induction of excessive grooming in the rat by fragments of lipotropin. *Nature (London)*, **264**, 794–795.

Gold, P. E., & King, R. D. (1974). Retrograde amnesia: Storage failure versus retrieval failure. *Psychological Review*, **81**, 465–469.

Gold, P. E., & van Buskirk, R. B. (1976a). Effect of post-trial hormone injections on memory processes. *Hormones and Behavior*, **7**, 509–517.

Gold, P. E., & van Buskirk, R. B. (1976b). Enhancement and impairment of memory processes with post-trial injections of adrenocorticotropic hormone. *Behavioral Biology*, **16**, 387–400.

Gordon, W. C., & Spear, N. E. (1973). The effects of strychnine on recently acquired and reactivated passive avoidance memories. *Physiology and Behavior*, **10**, 1071–1075.

Gray, J. A. (1971). Effect of ACTH on extinction of rewarded behavior is blocked by previous administration of ACTH. *Nature (London)*, **229**, 52–54.

Gray, J. A., & Garrud, P. (1977). Adrenopituitary hormones and frustrative nonreward. *In* L. H. Miller, C. A. Sandman, & A. J. Kastin (Eds.), *Neuropeptide Influences on the Brain and Behavior*, pp. 201–212. New York: Raven.

Gray, P. (1975). Effect of adrenocorticotropic hormone on conditioned avoidance in rats interpreted as state dependent learning. *Journal of Comparative and Physiological Psychology*, **88**, 281–284.

Guth, S., Seward, J. P., & Levine, S. (1971). Differential manipulation of passive avoidance by exogenous ACTH. *Hormones and Behavior*, **2**, 127–138.

Haroutunian, V., & Riccio, D. C. (1977). Effect of arousal conditions during reinstatement treatment upon learned fear in young rats. *Developmental Psychobiology*, **10**, 25–32.

Haroutunian, V., & Riccio, D. C. (1979). Drug-induced "arousal" and the effectiveness of CS exposure in the reinstatement of memory. *Behavioral and Neural Biology*, **26**, 115–120.

Hinderliter, C. F., Webster, T., & Riccio, D. C. (1975). Amnesia induced by hypothermia as a function of treatment-test interval and recooling in rats. *Animal Learning and Behavior*, **3**, 257–263.

Kamin, L. J. (1957). Retention of an incompletely learned avoidance response. *Journal of Comparative and Physiological Psychology*, **50**, 457–460.

Kety, S. S. (1970). The biogenic amines in the central nervous system: Their possible roles in arousal, emotion, and learning. In F. O. Schmitt (Ed.), *The Neurosciences: Second Study Program*, pp. 324–336. New York: Rockefeller University Press.

Kimble, G. A. (1977). Is learning involved in neuropeptide effects on behavior? In L. H. Miller, C. A. Sandman, & A. J. Kastin (Eds.), *Neuropeptide Influences on the Brain and Behavior*, pp. 189–200. New York: Raven.

Klein, S. B. (1972). Adrenal–pituitary influence in reactivation of avoidance-learning memory in the rat after intermediate intervals. *Journal of Comparative and Physiological Psychology*, **79**, 341–359.

Klein, S. B. (1975). ACTH-induced reactivation of prior active avoidance training after intermediate intervals in hypophysectomized, adrenalectomized, and sham-operated rats. *Physiological Psychology*, **3**, 395–399.

Klein, S. B., & Spear, N. E. (1970). Forgetting by the rat after intermediate intervals ("Kamin effect") as a retrieval failure. *Journal of Comparative and Physiological Psychology*, **71**, 165–170.

Korányi, L., Endröczi, E., Lissak, K., & Szepes, E. (1967). The effect of ACTH on behavioral processes motivated by fear in mice. *Physiology and Behavior*, **2**, 439–445.

Lal, H., Miksic, S., & Drawbaugh, R. (1978). Influence of environmental stimuli associated with narcotic administration on narcotic actions and dependence. In M. L. Adler, L. Manara, & Samanin, R. (Eds.), *Factors Affecting the Action of Narcotics*, pp. 643–668. New York: Raven.

Leshner, A. I. (1978). *An Introduction to Behavioral Endocrinology*. London and New York: Oxford University Press.

Levine, S., & Brush, F. R. (1967). Adrenocortical activity and avoidance learning as a function of time after avoidance training. *Physiology and Behavior*, **2**, 385–388.

Levine, S., & Jones, L. E. (1965). Adrenocorticotropic hormone (ACTH) and passive avoidance learning. *Journal of Comparative and Physiological Psychology*, **59**, 357–360.

Levine, S., Smotherman, W. P., & Hennessy, J. W. (1977). Pituitary-adrenal hormones and learned taste aversion. In L. H. Miller, C. A. Sandman, & A. J. Kastin (Eds.), *Neuropeptide Influences on the Brain and Behavior*, pp. 163–177. New York: Raven.

Lewis, D. J. (1976). A cognitive approach to experimental amnesia. *American Journal of Psychology*, **89**, 51–80.

Mactutus, C. F., & Riccio, D. C. (1978). Hypothermia-induced retrograde amnesia: Role of body temperature in memory retrieval. *Physiological Psychology*, **6**, 18–22.

Mactutus, C. F., Smith, R. L., & Riccio, D. C. (1980). Extending the ACTH-induced memory reactivation in an amnestic paradigm. *Physiology and Behavior*, **24**, 541–546.

McGaugh, J. L. (1966). Time dependent processes in memory storage. *Science*, **153**, 1351–1358.

McGaugh, J. L., Martinez, Jr., J. L., Jensen, R. A., Messing, R. B., & Vasquez, B. J. (1980). Central and peripheral catecholamine function in learning and memory processes. In R. F. Thompson & U. B. Shvyrkov (Eds.), *Neural Mechanisms of Goal Directed Behavior and Learning*, pp. 75–91. New York: Academic Press.

Meyer, D. R. (1972). Access to engrams. *American Psychologist*, **27**, 124–133.

Meyer, D. R., & Meyer, P. M. (in press). Induction of recoveries from amnesias. In J. L. McGaugh & R. F. Thompson (Eds.), *Handbook of Behavioral Neurobiology*. New York: Plenum.

Miller, R. R., & Berk, A. M. (1979). Sources of infantile amnesia. In N. E. Spear & B. A. Campbell (Eds.), *Ontogeny of Learning and Memory*. Hillsdale, New Jersey: Lawrence Erlbaum Associates.

Miller, R. R., & Ogawa, N. (1962). The effect of adrenocorticotrophic hormone (ACTH) on avoidance conditioning in the adrenalectomized rat. *Journal of Comparative and Physiological Psychology*, **55**, 211–213.

Miller, R. R., & Springer, A. D. (1972). Induced recovery of memory in rats following electro-convulsive shock. *Physiology and Behavior*, 8, 645–651.

Miller, R. R., & Springer, A. D. (1973). Amnesia, consolidation and retrieval. *Psychological Review*, 80, 69–79.

Mormede, P., & Dantzer, R. (1977). Effects of dexamethasone on fear conditioning in pigs. *Behavioral Biology*, 21, 225–235.

Murphy, J. V., & Miller, R. E. (1955). The effect of adrenocorticotrophic hormone (ACTH) on avoidance conditioning in the rat. *Journal of Comparative and Physiological Psychology*, 48, 47–49.

Overton, D. A. (1971). Discriminative control of behavior by drug states. In G. Thompson & R. Pickens (Eds.), *Stimulus Properties of Drugs*, pp. 87–110. New York: Appleton.

Overton, D. A. (1978). Major theories of state-dependent learning. In B. Ho, D. Chute, & D. Richards (Eds.), *Drug Discrimination and State-Dependent Learning*. New York: Academic Press.

Pappas, B. A., & Gray, P. (1971). Cue value of dexamethasone for fear-motivated behavior. *Physiology and Behavior*, 6, 127–130.

Rapoport, J. L., Quinn, P. P., Copeland, A. P., & Burg, C. (1976). $ACTH_{4-10}$: Cognitive and behavioral effects in hyperactive, learning disabled children. *Neuropsychobiology*, 2, 291–296.

Rescorla, R. A., & Heth, C. D. (1975). Reinstatement of fear to an extinguished conditioned stimulus. *Journal of Experimental Psychology: Animal Behavior Processes*, 104, 88–96.

Riccio, D. C., & Haroutunian, V. (1979). Some approaches to the alleviation of ontogenetic memory loss. In N. E. Spear & B. A. Campbell (Eds.), *Ontogeny of Learning and Memory*. Hillsdale, New Jersey: Lawrence Erlbaum Associates.

Rigter, H. (1975a). Plasma corticosterone levels as an index of $ACTH_{4-10}$ induced attenuation of amnesia. *Behavioral Biology*, 15, 207–211.

Rigter, H. (1975b). Peptide hormones and the extinction of conditioned taste aversion. *British Journal of Pharmacology*, 55, 270–271.

Rigter, H., Elbertse, R., & van Riezen, H. (1975). Time-dependent anti-amnestic effect of $ACTH_{4-10}$ and desglycinamide-lysine vasopressin. In W. H. Gispen, Tj. B. van Wimersma Greidanus, B. Bohus, & D. de Wied (Eds.), *Progress in Brain Research: Hormones Homeostasis and the Brain*, Volume 42, pp. 163–171. Amsterdam: Elsevier.

Rigter, H., Janssens-Elbertse, R., & van Riezen, H. (1976). Reversal of amnesia by an orally active $ACTH_{4-9}$ analog (Org 2766). *Pharmacology, Biochemistry and Behavior* (Supplement 1), 5, 53–58.

Rigter, H., & Popping, A. (1976). Hormonal influences on the extinction of conditioned taste aversion. *Psychopharmacologia*, 46, 255–261.

Rigter, H., Shuster, S., & Thody, A. J. (1977). ACTH, α-MSH and β-LPH: Pituitary hormones with similar activity in an amnesia test· in rats. *Journal of Pharmacy and Pharmacology*, 29, 110–111.

Rigter, H., & van Riezen, H. (1975). Anti-amnestic effect of $ACTH_{4-10}$: Its independence of the nature of the amnestic agent and the behavioral test. *Physiology and Behavior*, 14, 563–566.

Rigter, H., van Riezen, H., & de Wied, D. (1974). The effects of ACTH- and vasopressin-analogues on CO_2-induced retrograde amnesia in rats. *Physiology and Behavior*, 13, 381–388.

Sandman, C. A., George, J., Walker, B. B., & Nolan, J. D. (1976). Neuropeptide MSH/$ACTH_{4-10}$ enhances attention in the mentally retarded. *Pharmacology, Biochemistry and Behavior* (Supplement 1), 5, 23–28.

Sands, S. F., & Wright, A. A. (1979). Enhancement and disruption of retention performance by ACTH in a choice task. *Behavioral and Neural Biology*, 27, 413–422.

Sara, S. J., & Remacle, J. F. (1977). Strychnine-induced passive avoidance facilitation after electroconvulsive shock or undertraining: A retrieval effect. *Behavioral Biology*, 19, 465–475.

Schacter, S., & Singer, J. E. (1962). Cognitive, social, and physiological determinants of emotional state. *Psychological Review*, **69**, 379–399.

Schulenburg, C. J., Riccio, D. C., & Stikes, E. R. (1971). Acquisition and retention of a passive-avoidance response as a function of age in rats. *Journal of Comparative and Physiological Psychology*, **74**, 75–83.

Schuster, C. R., & Brady, J. V. (1964). The discriminative control of a food reinforced operant by interoceptive stimulation. *Pavlovian Journal of Higher Nervous Activity*, **14**, 448–458.

Silvestri, R., Rohrbaugh, M., & Riccio, D. C. (1970). Conditions influencing the retention of learned fear in young rats. *Developmental Psychology*, **2**, 389–395.

Smotherman, W. P., Hennessy, J. W., & Levine, S. (1976). Plasma corticosterone levels during recovery from LiCl produced taste aversions. *Behavioral Biology*, **16**, 401–412.

Smotherman, W. P., & Levine, S. (1978). ACTH and ACTH$_{4-10}$ modification of neophobia and taste aversion responses in the rat. *Journal of Comparative and Physiological Psychology*, **92**, 22–33.

Spear, N. E. (1973). Retrieval of memory in animals. *Psychological Review*, 80, 163–194.

Spear, N. E. (1978). *The Processing of Memories: Forgetting and Retention*. Hillsdale, New Jersey: Lawrence Erlbaum Associates.

Spear, N. E., Klein, S. B., & Riley, E. P. (1971). The Kamin effect as "state-dependent" learning: Memory retrieval failure in the rat. *Journal of Comparative and Physiological Psychology*, **74**, 416–425.

Stewart, J., Krebs, W.H., & Kaczender, E. (1967). State-dependent learning produced with steroids. *Nature (London)*, **216**, 1223–1224.

Thompson, C.I., & Neely, J.E. (1970). Dissociated learning in rats produced by electroconvulsive shock. *Physiology and Behavior*, **5**, 783–786.

Tulving, E. (1974). Cue-dependent forgetting. *American Scientist*, **62**, 74–82.

van Wimersma Greidanus, Tj.B. (1979). Neuropeptides and avoidance behavior; with special reference to the effects of vasopressin, ACTH, and MSH on memory processes. *In* R. Collu, A. Barbeau, J. R. Ducharme, & J. Rochefort (Eds.), *Central Nervous System Effects of Hypothalamic Hormones and Other Peptides*. New York: Raven.

van Wimersma Greidanus, Tj.B., van Dijk, A. M. A., de Rotte, A. A., Goedemans, J. H. J., Croiset, G., & Thody, A. J. (1978). Involvement of ACTH and MSH in active and passive avoidance behavior. *Brain Research Bulletin*, **3**, 227–230.

Wagner, A. R., Rudy, J. W., & Whitlow, Jr., J. W. (1973). Rehearsal in animal conditioning. *Journal of Experimental Psychology Monograph*, 97, 407–426.

Walker, B. B., & Sandman, C. A. (1979). Influences of an analog of the neuropeptide ACTH$_{4-9}$ on mentally retarded adults. *American Journal of Mental Deficiency*, **83**, 346–352.

Weiss, J. M., McEwen, B. S., Silva, M. T. A., & Kalkut, M. F. (1969). Pituitary–adrenal influence on fear responding. *Science*, **163**, 197–199.

Wiegant, V. M., Gispen, W. H., Terenius, L., & de Wied, D. (1977). ACTH-like peptides and morphine: Interaction at the level of the CNS. *Psychoneuroendocrinology*, **2**, 63–69.

CHAPTER 7

ACTH$_{4-9}$ Analog (ORG 2766)
and Memory Processes in Mice

Bernard Soumireu-Mourat, Jacques Micheau,
and Cécile Franc

These experiments investigated the effects on memory consolidation of a low dose of the adrenocorticotropic hormone (ACTH) sequence 4–9 analog, ORG 2766, injected either subcutaneously or into the cerebral ventricles. In one-way active avoidance conditioning, the peptide injected immediately after the acquisition session had no effect on retention 1 day later. However, when injected after the retention session, the peptide induced, on the following days, a facilitation of extinction in all treatment conditions. In contrast, when given after a partial acquisition session in a continuously reinforced operant conditioning task, the posttraining treatment induced an impairment in retention after peripheral administration, but an improvement after central administration.

I. INTRODUCTION

Over the past several years, a growing body of evidence has shown that neuropeptides are involved in behavioral processes, particularly in learning and memory, largely through the work of de Wied and his colleagues. This work has been summarized in many reviews (de Wied, 1974, 1977; de Wied & Gispen, 1977). This area of research originated with the observation that re-

143

moval of the pituitary gland markedly impaired acquisition of a shuttlebox active avoidance response (Applezweig & Baudry, 1955; de Wied, 1964). This finding was later confirmed, using other active and passive avoidance tasks. The deficiency was found to be due to the disappearance of anterior pituitary hormones, because posterior lobectomy did not induce a disturbance in learning (de Wied, 1965). Avoidance learning performance in hypophysectomized animals can be restored by administration of natural ACTH or structurally related peptides (see the following), such as α-melanocyte-stimulating hormone (α-MSH)' and $ACTH_{4-10}$, which are virtually devoid of corticotropic activity (de Wied, 1969).

The behavioral effects of ACTH and related peptides on intact animals have also been studied. Different effects were observed, mainly on acquisition and on extinction but only rarely on retention. In most studies, shock-motivated tasks were used. Depending on the training conditions, ACTH and other peptides were found to be capable of altering the acquisition of an active avoidance task (Beatty, Beatty, Bowman, & Gilchrist, 1970; Bohus, Nyakas, & Endröczi, 1968; Stratton & Kastin, 1974). Acquisition of appetitively motivated responses has also been found to be facilitated by ACTH (Guth, Levine, & Seward, 1971) and α-MSH (Kastin, Sandman, Stratton, Schally, & Miller, 1975), and modulates imprinting behavior (Martin & van Wimersma Greidanus, 1978). Passive avoidance conditioning is also affected by ACTH or ACTH fragments when they are given either prior to acquisition (Levine & Jones, 1965; Martinez, Vasquez, Jensen, Soumireu-Mourat, & McGaugh, 1979) or prior to the retention test (Ader, Weijnen, & Moleman, 1972). Gold and McGaugh (1977) observed that immediate posttraining administration of ACTH affected retention of a passive avoidance task measured on the next day, depending on dose and footshock level. Flood, Jarvik, Bennett, and Orme (1976) also reported that posttraining administration of $ACTH_{4-10}$ influenced retention of an inhibitory avoidance task. However, other authors have reported no effect of $ACTH_{4-10}$ (van Wimersma Greidanus, 1977) or an $ACTH_{4-9}$ analog (Martinez et al., 1979) given after training. The most pronounced effect of ACTH or ACTH analogs is on extinction of learned behaviors. Given during training (Murphy & Miller, 1955) or during extinction (de Wied, 1969), ACTH treatment delays subsequent extinction of the behavior. This finding has been observed mostly with shock-motivated responses such as pole-jumping avoidance behavior (de Wied & Bohus, 1979) or shuttlebox avoidance (Greven & de Wied, 1973). In addition, ACTH effects on extinction were found using behaviors not motivated by electric shock, such as food-motivated tasks (Garrud, Gray, & de Wied, 1974; Gray, Mayes, & Wilson, 1971; Guth et al., 1971; Kastin, Dempsey, Leblanc, Dyster-Aas, & Schally, 1974; Leonard, 1969; Sandman, Kastin, & Schally, 1969), conditioned taste aversion (Rigter & Popping, 1966), and sexually motivated behavior (Bohus, Hendrickx, van Kolfschoten,

& Krediet, 1975). Another effect of the peptides was found by Rigter, Janssens-Elbertse, and van Riezen (1976). They found that treatment with the ACTH analog (ORG 2766) attenuated carbon dioxide-induced amnesia of a passive avoidance response when the drug was administered prior to the retrieval test but not when given prior to acquisition.

Finally, there is now a great deal of evidence that other neuropeptides related to ACTH seem to be involved in modulating learned behaviors (de Wied & Gispen, 1977). These molecules are generated from lipotropic hormone (β-LPH). The 47–53 sequence of this molecule is common to ACTH($_{4-10}$ fragment) and α- and β-MSH. Moreover, de Wied and colleagues were able to manipulate amino acids of the ACTH sequence and found several artificial fragments that are active on learned behaviors. If ACTH$_{4-10}$ is considered to be the cortico-tropically active center of the ACTH molecule, even though it is 10^{-6} times less active than the full molecule, then ACTH$_{4-7}$ would be the shortest peptide that produces the typical behavioral effects of ACTH (Greven & de Wied, 1973, 1977).

ORG 2766 is an ACTH$_{4-9}$ analog in which three modifications have been introduced: an oxidized methionine residue in Position 4, a D-lysine residue in Position 8 instead of arginine, and replacement of tryptophan by phenyl-alanine in Position 9. This analog has been found to produce the same types of effects as ACTH fragments (Greven & de Wied, 1977; Martinez et al., 1979; Rigter et al., 1976), but the three substitutions in the sequence induce a thousandfold potentiation of the behavioral activity of the ACTH$_{4-9}$ fragment (Greven & de Wied, 1977). In addition, it has been reported that an antiserum to ORG 2766, injected intracerebroventricularly 1 hr before retention testing, reduces the number of conditioned avoidance responses to 45% (Loeber, van Wimersma Greidanus, & de Wied, 1979).

Three characteristics of this experimental literature are notable. First, most of the cited experiments used shock-motivated avoidance responses, thereby limiting the generality of the results. Moreover, except for a few experiments studying retention of avoidance conditioning (Flood et al., 1976; Gold & McGaugh, 1977; Gold & van Buskirk, 1976; Martinez et al., 1979), the clearest effects were observed on extinction. The interpretation of delayed extinction is difficult in terms of memory processes. Indeed, delayed extinction can be considered to be either better retention of the previously learned task, or less adaptation and slower learning of a new situation when reinforcement is not present. Many experiments have used the de Wied paradigm in which several extinction sessions are given 2 hr apart, and the experimental treatment is given after the first extinction session (van Wimersma Greidanus & de Wied, 1969). In some other cases already summarized, the peptide was administered before the test session. Finally, except in a few experiments, no clear findings are available using the classical memory consolidation paradigm (posttraining treatment) (Kesner & Wilburn, 1974) with the retention test given 1 day later (cf. Bohus, Kovacs, & de Wied, 1978a), or using appetitively motivated tasks.

The present series of experiments was designed to study the effects on retention of ORG 2766, one of the most behaviorally potent of the ACTH analogs, administered immediately after either shock-motivated or food-motivated training. The subjects were 10-wk-old male mice of the BALB/c strain. In all the experiments, the dose of $ACTH_{4-9}$ analog was constant (10.0 ng), but subcutaneous or intraventricular routes of administration were used. The peptide was dissolved in saline (0.5 ml for peripheral injections and 1.0 μl for central injections, for each animal) and kept in polyethylene containers. For intraventricular treatment, a cannula was implanted in a lateral ventricle under anesthesia 10 days before the learning experiment (Micheau, Destrade, & Soumireu-Mourat, 1981). At the conclusion of the experiments, placement of the cannula tip was determined by injecting blue dye. In experiments with intraventricular injections, one-half of the control group was injected with saline; the other half consisted of nonimplanted mice. All of the experiments were run blind.

II. PRELIMINARY EXPERIMENT: EFFECTS ON RETENTION AND EXTINCTION OF ONE-WAY ACTIVE AVOIDANCE CONDITIONING

Our first experiment consisted of testing the effects of the peptide on retention and extinction of a one-way active avoidance task. Each training trial consisted of placing the mouse in the dark shock-compartment of a trough-shaped alleyway, facing away from the door. The door to the smaller white compartment was then immediately opened and the animal was allowed 10 sec to step through. On the tenth second, if the subject had not made an avoidance response by entering the white compartment, a 0.3 mA/30 sec sinusoidal footshock was delivered to the mouse. The trial was terminated when the animal escaped to the white safe compartment. If the subject failed to cross within 30 sec, the door was closed and the animal was placed into the white compartment. In this case, a score of 40 sec (10 sec no shock period followed by 30 sec of shock) was assigned to the animal. The intertrial interval was 30 sec. On both the first day (Acquisition) and the second day (Retention), eight trials were run. On the 4 following days, an extinction session was run, consisting of two consecutive blocks of eight trials. In the first experiment, 10.0 ng of $ACTH_{4-9}$ was administered subcutaneously twice, 1 min after the acquisition session and 1 min after the retention test session. The results are summarized in Fig. 1. No effect of the first posttraining injection was observed during the retention test session on the next day. Surprisingly, during the successive extinction

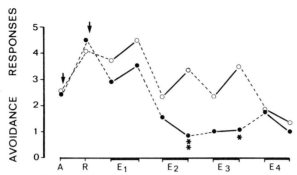

Fig. 1. Mean number of avoidance responses of mice trained in a one-way active avoidance task receiving saline (control, N = 8) or ACTH$_{4-9}$ analog (N = 10) subcutaneous injections (see arrows). Control is designated by ○, ACTH$_{4-9}$ s.c. by ●. A = acquisition session (8 trials); R = retention session (8 trials); E$_1$, E$_2$, E$_3$, E$_4$ = daily successive extinction sessions (2 blocks of 8 trials for each). Probability levels of tests of significance: * $p < .05$, ** $p < .02$.

sessions, the treated animals exhibited facilitation of extinction. The mean number of avoidance responses (i.e., the number of latencies to cross that were shorter than 10 sec) decreased faster in the treated group than in the control group. This tendency should be considered to be a weak effect, because statistical significance was observed only on one block of the second and one block of the third extinction session. However, as discussed later, the mean number of avoidance responses may not be the best measure of extinction in this one-way active avoidance paradigm. Nevertheless, the main finding of this study is the long-term effect of ORG 2766, as the largest difference between groups was observed 2 and 3 days after injection.

III. PEPTIDE EFFECTS
AND TRAINING LEVEL

This result seemed paradoxical compared to other results in the literature. To clarify this finding, we decided to test the effects of the peptide in a similar paradigm with the same treatment and dose, but with altered training conditions. It could be that the effect of the peptide may be different according to the level of training (Beatty et al., 1970; Stratton & Kastin, 1974). Three training conditions were used: (1) 8 trials with a 0.3 mA footshock, (2) 16 trials with a 0.3 mA footshock, or (3) 16 trials with a 0.6 mA footshock. A single subcutaneous injection of saline or ACTH$_{4-9}$ was given immediately after the training session. In this experiment, we chose to use the mean latency to cross into the safety compartment ($\leq$ 40 sec) as our behavioral measure because it appeared to be a more sensitive measure than the number of avoidance responses. Figure 2

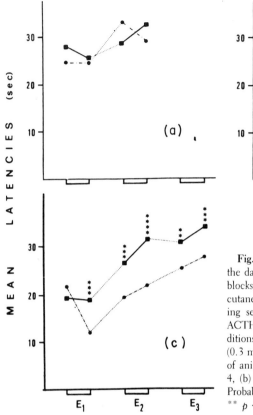

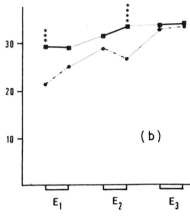

Fig. 2. Mean latencies (sec) on each of the daily extinction sessions (E_1, E_2, E_3 = 2 blocks of 8 trials) for saline and ACTH$_{4-9}$ subcutaneously injected mice after a single training session. Control is denoted by ●—●, ACTH$_{4-9}$ s.c. by ■—■. The training conditions are (a) 8 trials (0.3 mA), (b) 16 trials (0.3 mA), (c) 16 trials (0.6 mA). The number of animals in each condition are: control (a) 4, (b) 9, (c) 10; treated (a) 5, (b) 10, (c) 10. Probability levels of tests of significance: ** $p < .02$, *** $p < .01$, **** $p < .001$.

shows the mean latencies for each training condition for treated and control animals during the 3 days of extinction. For the lowest level of training, ACTH$_{4-9}$ was without effect. But for the intermediate and highest level of training, the facilitation of extinction by the peptide (i.e., an increase in latencies) appeared again as in the first experiment. Thus, if an effect of the peptide was observed, it was facilitation of extinction, contrary to the initial hypothesis.

IV. CENTRAL VERSUS PERIPHERAL EFFECTS ON ACTIVE AVOIDANCE BEHAVIOR

We then compared the effects of the ACTH$_{4-9}$ analog under the same conditions as used in the first experiment (eight training trials with a 0.3 mA footshock on two sessions, 1 day apart) with subcutaneous (s.c.) and intraventricular (i.c.v.) administration of 10.0 ng of the peptide. We also decided to study the effects of either one injection (after the retention session on the second day) or two injections (after both the first and second sessions). Under these conditions, we observed a facilitation of extinction in all four situations. A close examination of the results revealed that this effect was strongly shown by an increase in the mean latencies and not so much by a decrease in the mean number of avoidance responses. For instance, Table 1 shows both measures for each block of eight extinction trials in which each animal was injected once after the retention session into the lateral ventricle. The facilitation of extinction reached statistical significance only at one point when the mean number of avoidances is used as the dependent measure; however, when mean latencies are examined, facilitation of extinction by ORG 2766 is highly significant throughout the extinction sessions. In fact, it can be seen that the increase in the mean latencies in the treated animals is related to a large increase in the number of ceiling latencies (i.e., the trials in which the mice did not escape to the white compartment within 40 sec). Finally, we chose to use this criterion, the mean number of ceiling latencies, for the comparison of s.c. versus i.c.v. administration and one versus two injections. These results are summarized in Fig. 3. In each of the four situations, ORG 2766 resulted in a facilitation of extinction. No clear difference appeared between the four conditions, except that the animals that received two injections may have had more ceiling latencies with both the s.c. and i.c.v. routes of administration.

Overall, the results show that ORG 2766, a potent ACTH$_{4-9}$ analog, administered peripherally or centrally at a low dose (10.0 ng) immediately after training results in facilitation of extinction of a one-way active avoidance task. This effect is related mainly to an increase in the number of the longest latencies rather than just to a decrease in avoidance responses (latencies less than 10 sec).

V. PERIPHERAL VERSUS CENTRAL ADMINISTRATION OF ORG 2766 IN APPETITIVE OPERANT CONDITIONING

To better understand the effects of ORG 2766 on memory consolidation, a study was designed using appetitive operant conditioning and posttraining central or peripheral administration of the peptide. The same dose (10 ng) used in the avoidance experiments was employed in this study. Ten days after cannula implantation, a food deprivation schedule was initiated. Three days later, on the first day of the experiment, without pretraining or shaping, the animals underwent a 15-min session in a Skinner box, with a continuous reinforcement schedule (6 mg pellets). The implanted mice were injected with the drug or

TABLE 1

Effects of ACTH$_{4-9}$ Analog on Responses of Mice in One-Way Active Avoidance Task

	First extinction		Second extinction		Third extinction		Fourth extinction	
	8 trials	8 trials	8 trials	8 trials	8 trials	8 trials	8 trials	8 trials
Mean number of avoidance responses								
CONTROLS[a] (N = 10)	3.7 ± 0.7	4.5 ± 1.0	3.0 ± 0.8	3.5 ± 0.9	3.0 ± 0.6	3.1 ± 0.9	2.9 ± 0.8	3.3 ± 0.6
ORG 2766[b] (N = 9)	2.1 ± 0.7	3.6 ± 1.2	2.7 ± 0.8	2.3 ± 0.9	1.8 ± 0.5	1.7 ± 0.7	1.4 ± 0.4	0.8 ± 0.4
	n.s.	n.s.	n.s.	n.s.	n.s.	n.s.	n.s.	$p < .001$
Mean latency (sec)								
CONTROLS	15.4 ± 1.4	12.5 ± 1.3	16.5 ± 1.4	15.8 ± 1.3	16.3 ± 1.3	16.5 ± 1.4	20.1 ± 1.7	16.7 ± 1.4
ORG 2766	21.4 ± 1.7	17.6 ± 1.7	22.7 ± 1.8	25.0 ± 2.1	22.1 ± 1.6	24.4 ± 1.7	27.6 ± 1.7	29.7 ± 1.6
	$p < .01$	$p < .05$	$p < .01$	$p < .01$	$p < .01$	$p < .001$	$p < .01$	$p < .001$

[a] Saline was intraventricularly injected after the retention session.
[b] ACTH$_{4-9}$ was intraventricularly injected after the retention session.

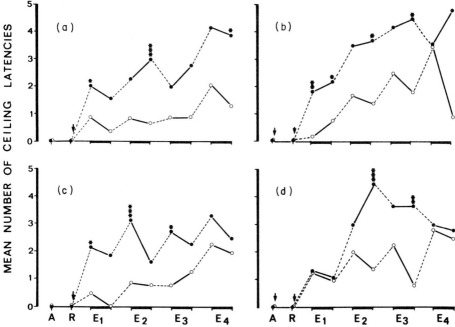

Fig. 3.   Mean number of ceiling latencies (= 40 sec) in each 8-trial block of mice trained in a one-way active avoidance task and receiving one (a and c, see arrow) or two (b and d) injections (see arrows) of saline or ACTH$_{4-9}$ analog, administered intraventricularly (a and b) or subcutaneously (c and d). Control is indicated by ○, ACTH$_{4-9}$ by ●. The number of animals for each group is 10, except for the one-injection s.c. or i.c.v. treated group (N = 9). A = acquisition session (8 trials); R = retention session (9 trials); E$_1$, E$_2$, E$_3$, E$_4$ = daily successive extinction sessions (2 blocks of 8 trials for each). Probability levels of tests of significance: * $p < .05$, ** $p < .02$, *** $p < .01$.

saline immediately after the end of the session and fed 30 min later. The next day the animals were again placed into the Skinner box for a 30-min retention session. As can be seen in Fig. 4, at the beginning of the retention session (first 5-min block), all of the control animals displayed an increase in the number of reinforced responses compared to their performance during the first session. We called this improvement the reminiscence phenomenon (Jaffard, Destrade, Soumireu-Mourat, & Cardo, 1974). After subcutaneous administration, ORG 2766 produced an impairment of the reminiscence phenomenon during the first 5-min blocks of the retention session. In contrast, intraventricular injection of the peptide facilitated the reminiscence phenomenon; however, this facilitation was weak, as the improvement just reached statistical significance on the first two 5-min blocks. In conclusion, with central and peripheral routes of administration of the same dose of the peptide, we observed opposite effects on retention of an appetitive operant behavior.

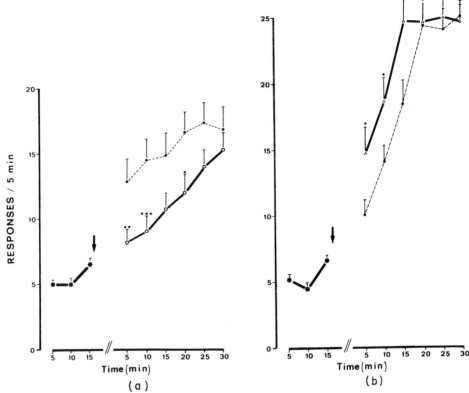

Fig. 4.   Performances for each 5-min block during first and second sessions of the continuously reinforced operant appetitive conditioning for (a) subcutaneously or (b) intraventricularly injected mice, either drug-injected (respectively, N = 18 and N = 9) or control (s.c.: N = 19; i.c.v.: N = 5 saline + 8 noninjected mice). Control is designated by ●---●, ACTH$_{4-9}$ (a) s.c. and (b) i.c.v. by ○—○. Probability levels of tests of significance: * $p < .05$, ** $p < .02$, *** $p < .01$.

VI. DISCUSSION

The aim of these experiments was to study possible long-lasting effects of small doses of this ACTH$_{4-9}$ analog. At first glance, our results seem to be ambiguous and contradictory to the literature. In fact, close examination of the literature suggests that the effects of this analog, and of the ACTH fragments in general, may be dependent on the task and its parameters, the exact fragment used, dose, and the route of injection.

Considering the issue of task, Beatty et al. (1970) reported ACTH facilitation of shuttlebox avoidance acquisition for a high but not a moderate footshock intensity. Stratton and Kastin (1974) observed an opposite effect using α-MSH. Similarly, Gold and McGaugh (1977) reported that the same dose of ACTH enhanced retention of an inhibitory avoidance task with a low footshock level, but disrupted retention with a high footshock level. Similarly, Micco, McEwen, and Shein (1979) report results of either facilitated or delayed extinction after corticosterone treatment. This is interesting, as Bohus (1973) emphasized that ACTH (or ACTH-like peptides) and corticosteroids have opposite effects on different kinds of behaviors. The pole-jumping test (used by de Wied and other authors) and the present avoidance task are both one-way active avoidance conditioning paradigms, but they are structurally very different. In our task, the increase in mean latencies after ACTH$_{4-9}$ treatment is much more significant than the decrease in the number of avoidances. If we examine the behavior of the treated animals, they often ran faster to the partition between compartments, but they hesitated more than the controls to cross into the white safe compartment. In other words, it is possible that treated subjects feared the footshock as much as did the controls, but they seemed to return to their natural preference (the black compartment) faster than the untreated animals. If we consider the results with the appetitive task, our data are difficult to compare with the literature, as no previous experiment has been done with posttraining administration of ACTH, and later retention testing (and not extinction testing). Generally, the effects of peptides on appetitively motivated responses have been studied less extensively.

The exact ACTH fragment used appears to be important. Opposite effects have sometimes been observed with only small structural alterations. For instance, replacement of the 7-L phenylalanine by 7-D phenylalanine in the ACTH$_{4-10}$ fragment facilitates the extinction of the conditioned response in contrast to the delayed extinction observed with the L-isomer (de Wied, Witter, & Greven, 1975). Moreover, ACTH and ACTH fragments generally delay extinction of learned behavior, but sometimes opposite effects are reported. For example, de Wied and Gispen (1977) found that newly synthesized material facilitated extinction. Also, Weijnen and Slangen (1970) reported negative findings.

The next point to be considered is the route of administration and dose of the drug. We observed the same effects with both sites of injection in the avoidance task, but opposite effects in the appetitive operant conditioning task. Such opposite effects have been reported for other drugs, e.g., morphine (Jensen, Martinez, Messing, Spiehler, Vasquez, Soumireu-Mourat, Liang, & McGaugh, 1978) and oxytocin. Oxytocin increased resistance to extinction using peripheral administration (de Wied & Bohus, 1979), but tends to facilitate extinction of the pole-jumping test when given intraventricularly (Bohus, Urban, van Wimersma Greidanus, & de Wied, 1978b). In addition, Gold and McGaugh (1977) reported opposite effects by increasing the dose of ACTH.

Thus, several inconsistencies in the literature could be related to use of various doses of the drug. For instance, Flood et al. (1976), after posttrial administration of $ACTH_{4-10}$ observed effects on retention of both inhibitory and active avoidance tasks. With ORG 2766, a more potent analog of the $ACTH_{4-9}$, we did not find any influence on consolidation of an inhibitory avoidance task (Martinez et al., 1979). The dose of ACTH fragments generally used is in the μg range (see de Wied & Gispen, 1977, p. 403; Flood et al., 1976); as ORG 2766 is a potent analog, the dose has been reduced to the nanogram range (Rigter et al., 1976). Comparing the effects of $ACTH_{4-10}$, de Wied, Bohus, van Ree, and Urban (1978) found the same level of delayed extinction with microgram doses given by the s.c. route and nanogram doses given by the i.c.v. route. This thousandfold difference may explain our opposite results with the same dose using both central and peripheral routes of administration in the appetitive operant task.

Finally, we must consider the time of administration relative to the problem of memory effects of ORG 2766. De Wied (1977) postulated that peptides related to ACTH have a short-term behavioral effect compared to the long-term behavioral effect of vasopressin. He tested the effects of ACTH-related peptides several hours following injection (Bohus, 1979; de Wied & Bohus, 1979). In our paradigm, we tested the effects of the peptide on the days following training. So, whether ACTH is considered to have a long-term effect is dependent on how it is studied. In the de Wied paradigm, it is difficult to say whether the peptide has a retrograde effect on the preceding session or an anterograde effect on the next session a few hours later. Thus, the nature of the short-term effect is at best relative because, in our experiments, administration of the peptide after training affected retention or extinction 1 day later. This finding confirms a few other studies that used posttraining administration on avoidance tasks (Flood et al., 1976; Gold & McGaugh, 1977). To determine whether the peptide directly affects memory consolidation, one would need a study of time-gradient processes with delayed administration of the drug after the learning session. Moreover, it is not clear that the peptide mechanisms are normally a part of the memory consolidation process. Indeed, we found opposite effects in Skinner box with s.c. and i.c.v. routes of administration and also different retention effects using the i.c.v. route with negative and positive reinforcements.

Despite the results of many studies, the mechanisms underlying the behavioral effects of peptides are not known. Several hypotheses have been developed, but it is not presently possible to choose among them. It is difficult to extrapolate results between different behavioral models, species, molecules, and doses (Meyerson, 1979). Peptide effects on fear (Bohus, 1973) seem to be untenable with positively reinforced tasks. One interesting hypothesis was advanced by de Wied. He suggested that the peptides "may alter the motivational value of environmental stimuli" (de Wied et al., 1978). That hypothesis could explain the results of our second experiment. However, whether the peptides increase (de Wied, 1977) or attenuate (Bohus, 1973) the motivational properties of the environmental cues is not clear. The direction of the effects could depend on the task and on its conditions. A slightly different hypothesis that has support in both animal and human studies postulates that the peptides improve arousal, vigilance, and selective attention processes (Meyerson, 1979; Oades, 1979). Finally, all these interpretations fit with the more general hypothesis put forward by Kastin et al. (1975) and developed by de Wied; namely, that ACTH fragments play an important role in adaptive behavioral mechanisms. The neuropeptides might modulate a kind of behavioral homeostasis, as the hormonal state may modulate brain reactions to environmental stimuli during spontaneous or learned behaviors (Gold & McGaugh, 1977; Leshner, 1978). Thus, the neuropeptides may act by modulating memory processes rather than by participating directly in memory consolidation, even though they would have a preferential time for their effects in the posttraining period (on consolidation processes) or in the pretest period (on retrieval processes). In fact, all these interpretations are still presently at a speculative and theoretical level (Kimble, 1977). We are still at the beginning of gaining an understanding of how neuropeptides act on brain functions. It will be of great interest to determine the physiological sites and the cellular mechanisms of action. Several papers deal with the purported molecular level of action of ACTH-like peptides (de Wied & Gispen, 1977; Gispen, Reith, Schotman, Wiegant, Zwiers, & de Wied, 1977). Anatomically, it seems that limbic structures, including nonspecific thalamic nuclei and dorsal hippocampus, need to be intact for behavioral effects of the peptides to be evident (van Wimersma Greidanus & de Wied, 1976; reviewed in de Wied & Gispen, 1977 and in Oades, 1979). Also, these peptides influence theta rhythm (Bohus et al., 1978b; Urban & de Wied, 1978). This is of considerable interest, as the hippocampus appears to be largely involved in attention and memory processes (Kesner & Wilburn, 1974; Soumireu-Mourat, Destrade, & Cardo, 1975).

In summary, there is much evidence that the ACTH-related peptides may modulate memory consolidation as well as retrieval. In this chapter we have shown that postlearning administration of an ACTH$_{4-9}$ analog, ORG 2766, can produce long-term effects. Second, these effects are dependent on the training conditions. Finally, subcutaneous and intraventricular administration of this peptide produces opposite effects on retention of a food-motivated operant-conditioning task.

ACKNOWLEDGMENTS

This work was supported by the CNRS (ERA 416 and grant ATP 4189). The authors would like to thank Dr. H. Rigter and Organon International for the donation of ORG 2766. We thank

A. M. Perret and Chr. Baquerin for technical assistance. We are grateful to Dr. J. L. Martinez, Jr. for help in the preparation of the manuscript.

REFERENCES

Ader, R., Weijnen, J.A.W.M., & Moleman, P. (1972). Retention of a passive avoidance response as a function of the intensity and duration of electric shock. *Psychonomic Science*, 26, 125–128.

Applezweig, M. H., & Baudry, F. D. (1955). The pituitary–adrenocortical system in avoidance learning. *Psychological Reports*, 1, 417–420.

Beatty, D. A., Beatty, W. A., Bowman, R. E., & Gilchrist, J. C. (1970). The effects of ACTH, adrenalectomy and dexamethasone on the acquisition of an avoidance response in rats. *Physiology and Behavior*, 5, 939–944.

Bohus, B. (1973). Pituitary–adrenal influences on avoidance and approach behavior of the rat. *In* E. Zimmerman, W. H. Gispen, B. H. Marks, & D. de Wied (Eds.), *Drug Effects on Neuroendocrine Regulation: Progress in Brain Research*, 39, pp. 407–419, Amsterdam: Elsevier.

Bohus, B. (1979). Effects of ACTH-like neuropeptides on animal behavior and man. *Pharmacology*, 18, 113–122.

Bohus, B., Hendrickx, H.H.L., van Kolfschoten, A. A., & Krediet, T. G. (1975). The effect of $ACTH_{4-10}$ on copulatory and sexually motivated approach behavior in the male rat. *In* M. Sandler & G. L. Gessa (Eds.), *Sexual Behavior: Pharmacology and Biochemistry*, pp. 269–275. New York: Raven.

Bohus, B., Kovacs, G. L., & de Wied, D. (1978a). Oxytocin, vasopressin and memory: Opposite effects on consolidation and retrieval processes. *Brain Research*, 157, 414–417.

Bohus, B., Nyakas, C., & Endröczi, E. (1968). Effects of adrenocorticotrophic hormone on avoidance behavior of intact and adrenalectomized rats. *International Journal of Neuropharmacology*, 7, 307–314.

Bohus, B., Urban, I., van Wimersma Greidanus, Tj. B., & de Wied, D. (1978b). Opposite effects of oxytocin and vasopressin on avoidance behavior and hippocampal theta rhythm in the rat. *Neuropharmacology*, 17, 239–247.

de Wied, D. (1964). Influence of anterior pituitary on avoidance learning and escape behavior. *American Journal of Physiology*, 207, 255–259.

de Wied, D. (1965). The influence of the posterior and intermediate lobe of the pituitary and pituitary peptides on the maintenance of a conditioned avoidance response in rats. *International Journal of Neuropharmacology*, 4, 157–167.

de Wied, D. (1969). Effects of peptide hormones on behavior. *In* W. F. Ganong & L. Martini (Eds.), *Frontiers in Neuroendocrinology*, pp. 97–140. Oxford: Oxford University Press.

de Wied, D. (1974). Pituitary–adrenal system hormones and behavior. *In* F. O. Schmitt & F. G. Worden (Eds.), *The Neurosciences: Third Study Program*, pp. 653–666. Cambridge, Massachusetts: MIT Press.

de Wied, D. (1977). Peptides and behavior. *Life Sciences*, 20, 195–204.

de Wied, D., & Bohus, B. (1979). Modulation of memory processes by neuropeptides of hypothalamic–neurohypophyseal origin. *In* M.A.B. Brazier (Ed.), *Brain Mechanisms in Memory and Learning: From the Single Neuron to Man*, pp. 139–149. New York: Raven.

de Wied, D., Bohus, B., van Ree, J. M., & Urban, I. (1978). Behavioral and electrophysiological effects of peptides related to lipotropin (β-LPH). *Journal of Pharmacology and Experimental Therapeutics*, 204, 570–580.

de Wied, D., & Gispen, W. H. (1977). Behavioral effects of peptides. *In* H. Gainer (Ed.), *Peptides in Neurobiology*, pp. 397–448. New York: Plenum.

de Wied, D., Witter, A., & Greven, H. M. (1975). Behaviorally active ACTH analogs. *Biochemical Pharmacology*, 24, 1463–1468.

Flood, J. F., Jarvik, M. E., Bennett, E. L., & Orme, A. E. (1976). Effects of ACTH peptide fragments on memory formation. *Pharmacology, Biochemistry and Behavior*, (Supplement 1), 5, 41–51.

Garrud, P., Gray, J. A., & de Wied, D. (1974). Pituitary–adrenal hormones and extinction of rewarded behavior in the rat. *Physiology and Behavior*, 12, 109–119.

Gispen, W. H., Reith, M.E.A., Schotman, P., Wiegant, V. M., Zwiers, H., & de Wied, D. (1977). CNS and ACTH-like peptides: Neurochemical response and interaction with opiates. *In* L. H. Miller, C. A. Sandman, & A. J. Kastin (Eds.), *Neuropeptide Influences on the Brain and Behavior*, pp. 61–80. New York: Raven.

Gold, P. E., & McGaugh, J. L. (1977). Hormones and Memory. *In* L. H. Miller, C. A. Sandman, & A. J. Kastin (Eds.), *Neuropeptide Influences on the Brain and Behavior*, pp. 127–143. New York: Raven.

Gold, P. E., & van Buskirk, R. B. (1976). Enhancement and impairment of memory processes with post-trial injections of adrenocorticotrophic hormone. *Behavioral Biology*, 16, 387–400.

Gray, J. A., Mayes, A. R., & Wilson, M. (1971). A barbiturate-like effect of adrenocorticotrophic hormone on the partial reinforcement acquisition and extinction effects. *Neuropharmacology*, 10, 223–230.

Greven, H. M., & de Wied, D. (1973). The influence of peptides derived from corticotrophin (ACTH) on performance. Structure activity studies. *In* E. Zimmermann, W. H. Gispen, B. H. Marks, & D. de Wied (Eds.), *Drug Effects on Neuroendocrine Regulation: Progress in Brain Research*, pp. 429–441. Amsterdam: Elsevier.

Greven, H. M., & de Wied, D. (1977). Influence of peptides structurally related to ACTH and MSH on active avoidance behavior in rats. *Frontiers of Hormone Research*, 4, 140–152.

Guth, S., Levine, S., & Seward, J. P. (1971). Appetitive acquisition and extinction effects with exogenous ACTH. *Physiology and Behavior*, 7, 195–200.

Jaffard, R., Destrade, C., Soumireu-Mourat, B., & Cardo, B. (1974). Time-dependent improvement of performance on appetitive tasks in mice. *Behavioral Biology*, 11, 89–100.

Jensen, R. A., Martinez, J. L., Jr., Messing, R. B., Spiehler, V., Vasquez, B. J., Soumireu-Mourat, B., Liang, K. C., & McGaugh, J. L. (1978). Morphine and naloxone alter memory in the rat. *Society for Neuroscience Abstracts*, 4, 260.

Kastin, A. J., Dempsey, G. L., Leblanc, B., Dyster-Aas, K., & Schally, A. V. (1974). Extinction of an appetitive operant response after administration of MSH. *Hormones and Behavior*, 5, 135–139.

Kastin, A. J., Sandman, C. A., Stratton, L. O., Schally, A. V., & Miller, L. H. (1975). Behavioral and electrographic changes in rat and man after MSH. *Progress in Brain Research*, 42, 143–150.

Kesner, R. P., & Wilburn, M. W. (1974). Review of electrical stimulation of the brain in context of learning and retention. *Behavioral Biology*, 10, 259–293.

Kimble, G. A. (1977). Is learning involved in neuropeptide effects on behavior? *In* L. H. Miller, C. A. Sandman, & A. J. Kastin (Eds.), *Neuropeptide Influences on the Brain and Behavior*, pp. 189–200. New York: Raven.

Leonard, B. E. (1969). The effect of sodium-barbitone, alone and together with ACTH and amphetamine, on the behavior of the rat in the multiple "T" maze. *International Journal of Neuropharmacology*, 8, 427–435.

Leshner, A. I. (1978). *An Introduction to Behavioral Endocrinology*. New York–London: Oxford University Press.

Levine, S., & Jones, L. E. (1965). Adrenocorticotrophic hormone (ACTH) and passive avoidance learning. *Journal of Comparative and Physiological Psychology*, 59, 357–360.

Loeber, J. G., van Wimersma Greidanus, Tj. B., & de Wied, D. (1979). Evidence for the existence of highly specific neuropeptides which affect the maintenance of avoidance behavior. *Journal of Endocrinology*, 80, 9P.

Martin, J. T., & van Wimersma Greidanus, Tj. B. (1978). Imprinting behavior: Influence of vasopressin and ACTH analogs. *Psychoneuroendocrinology*, **3**, 261–270.

Martinez, J. L., Jr., Vasquez, B. J., Jensen, R. A., Soumireu-Mourat, B., & McGaugh, J.L. (1979). $ACTH_{4-9}$ analog (ORG 2766) facilitates acquisition of an inhibitory avoidance response in rats. *Pharmacology, Biochemistry and Behavior*, **10**, 145–147.

Meyerson, B. J. (1979). Hypothalamic hormones and behavior. *Medical Biology*, **57**, 69–83.

Micco, D. J., Jr., McEwen, B. S., & Shein, W. (1979). Modulation of behavioral inhibition in appetitive extinction following manipulation of adrenal steroids in rats: Implications for involvement of the hippocampus. *Journal of Comparative and Physiological Psychology*, **93**, 323–329.

Micheau, J., Destrade, C., & Soumireu-Mourat, B. (1981). Memory facilitation by intraventricular corticosterone injection on an appetitive discriminative task in mice. *Behavioral and Neural Biology*, **31**, 100–104.

Murphy, J. V., & Miller, R. E. (1955). The effect of adrenocorticotrophic hormone (ACTH) on avoidance conditioning in the rat. *Journal of Comparative and Physiological Psychology*, **48**, 47–49.

Oades, R. D. (1979). Search and attention: Interactions of the hippocampal–septal axis, adrenocortical and gonadal hormones. *Neurosciences and Biobehavioral Reviews*, **3**, 31–48.

Rigter, H., Janssens-Elbertse, R., & van Riezen, H. (1976). Reversal of amnesia by an orally active $ACTH_{4-9}$ analog (ORG 2766). *Pharmacology, Biochemistry and Behavior* (Supplement 1), **5**, 53–58.

Rigter, H., & Popping, A. (1966). Hormonal influence on the extinction of conditioned taste aversion. *Psychopharmacologia*, **46**, 255–261.

Sandman, C. A., Kastin, A. J., & Schally, A. V. (1969). Melanocyte-stimulating hormone and learned appetitive behavior. *Experientia*, (Basel), **25**, 1001–1002.

Soumireu-Mourat, B., Destrade, C., & Cardo, B. (1975). Effects of seizure and subseizure posttrial hippocampal stimulation on appetitive operant behavior in mice. *Behavioral Biology*, **15**, 303–316.

Stratton, L. O., & Kastin, A. J. (1974). Avoidance learning at two levels of motivation in rats receiving MSH. *Hormones and Behavior*, **5**, 149–155.

Urban, I., & de Wied, D. (1978). Neuropeptides: Effects on paradoxical sleep and theta rhythm in rats. *Pharmacology, Biochemistry and Behavior*, **8**, 51–59.

van Wimersma Greidanus, Tj. B. (1977). Effects of MSH and related peptides on avoidance behavior in rats. *Frontiers of Hormone Research*, **4**, 129–139.

van Wimersma Greidanus, Tj. B., & de Wied, D. (1969). Effects of intracerebral implantation of corticosteroids on extinction of an avoidance response in rats. *Physiology and Behavior*, **4**, 365–370.

van Wimersma Greidanus, Tj. B., & de Wied, D. (1976). Dorsal hippocampus: A site of action of neuropeptides on avoidance behavior? *Pharmacology, Biochemistry and Behavior* (Supplement 1), **5**, 29–34.

Weijnen, J.A.W.M., & Slangen, J. L. (1970). Effects of ACTH-analogs on extinction of conditioned behavior. *In* D. de Wied & J.A.W.M. Weijnen (Eds.), *Pituitary, Adrenal and the Brain: Progress in Brain Research*, **32**, pp. 221–233. Amsterdam: Elsevier.

CHAPTER 8

Pituitary–Adrenocortical Effects on Learning and Memory in Social Situations

Alan I. Leshner, Dennis A. Merkle, and James F. Mixon

To test further the generality of the effects of the pituitary–adrenocortical hormones on learning and memory, a series of studies was conducted to examine the effects of these hormones on avoidance behavior in social situations. One situation, called "avoidance-of-attack," is a standard passive avoidance paradigm, but the aversive stimulus in this case is attack by a trained fighter mouse, rather than electric shock. The second situation studied is a series of repeated tests for submissiveness, wherein the effects of these hormones on learning and memory can be evaluated by comparing the change in submissiveness over repeated encounters of normal mice with those of mice with altered pituitary–adrenocortical hormone levels.

In both cases, the pituitary–adrenocortical hormones were found to affect social avoidance responding differently from the way they affect avoidance in shock-mediated situations. Specifically, it was found that corticosterone is the primary pituitary–adrenal hormone in the control of social avoidance responding, whereas in shock-mediated situations, the critical hormone of this axis has always been found to be adrenocorticotropin (ACTH). This basic finding imposes some limitation on the generality of the findings of other studies of pituitary–adrenocortical effects on learning and memory processes.

159

ENDOGENOUS PEPTIDES
AND LEARNING AND MEMORY PROCESSES

The second part of this chapter concerns the ways by which these hormones affect responding in social learning or memory situations. A variety of alternative mechanisms are considered, including direct effects of these hormones on the quality of submissiveness, effects on learning or memory processes per se, and state-dependent effects. It is concluded that corticosterone probably does not affect responding in social learning–memory situations because this hormone directly modifies some general behavioral process. Rather, it is suggested that corticosterone acts as a part of the normal stimulus complex of the defeat experience. That is, because changes in the level of this hormone are a normal part of the stimulus complex of the experience of defeat, any change in the magnitude or intensity of that stimulus complex should increase the magnitude of the experience, or its "meaningfulness." This change in meaningfulness may be expressed in later performance tests, appearing as an increase in the memory of the initial experience. The possibility that this interpretation might be appropriate in other learning situations is also considered.

I. INTRODUCTION

As is discussed in detail elsewhere in this book, there is by now a vast literature relating the hormones of the pituitary–adrenocortical axis—ACTH and corticosterone—to the behaviors observed in learning and memory situations. Despite some occasional bits of contradictory evidence, the literature leaves little doubt that these hormones, particularly the pituitary peptide ACTH, have some important behavioral effects. However, it is not yet clear just how these hormones exert many of their effects, or on what processes they are acting (see the extensive discussions in Beckwith & Sandman, 1978; Bohus, 1975; de Wied, 1976; Oades, 1979; Rigter & van Riezen, 1978; and other chapters in this book).

One limitation in this extensive literature is that the large majority of the evidence consistently pointing to effects of ACTH and corticosterone on learning and memory has come from studies of a single class of learning–memory situations, avoidance responding. Furthermore, within the class of avoidance behavior, virtually all studies have used a single kind of aversive stimulus, electric shock. This limitation raises questions about the generality of these hormonal effects.

In the process of studying the effects of the pituitary–adrenal hormones on submissive behaviors, we came to the realization that the test paradigms we use provide a mechanism for studying avoidance responding in situations where the aversive stimulus is different from electric shock, specifically, the threat of being attacked or actual defeat by an opponent. We have carried out a variety of studies, and our findings suggest that the pituitary–adrenocortical hormones affect avoidance behavior in social situations quite differently from the way in which they affect avoidance-of-shock.

II. PITUITARY–ADRENOCORTICAL EFFECTS ON AVOIDANCE-OF-ATTACK

The first paradigm to be discussed is called *avoidance-of-attack*. In this situation, we use a standard passive avoidance paradigm, but substitute attack by a trained fighter mouse for the more typically used aversive stimulus of electric shock. The test apparatus and procedures have been described in detail elsewhere (Leshner & Moyer, 1975; Leshner, Moyer, & Walker, 1975), but it may be useful to summarize them here. The test apparatus consists of two connected chambers, one large and one small, separated by a guillotine door. The smaller start chamber is well lit, whereas the larger attack chamber is kept dark. A highly aggressive, trained fighter mouse lives in the attack chamber for at least 1 wk prior to behavioral testing.

Passive avoidance conditioning is begun by placing the experimental animal into the start chamber, opening the guillotine door, and recording the latency of the test animal to enter the attack chamber. Immediately upon entering the attack chamber, the test animal is subjected to 5 sec of physical attack by the trained fighter. The animals then are separated, and the experimental animal is returned to the start chamber. This procedure is repeated until the experimental animal remains in the start chamber without entering the attack chamber for 300 sec (the passive avoidance criterion). Twenty-four hours following acquisition of this avoidance response, the experimental animal is returned to the start chamber, the guillotine door is opened, and the latency to enter the attack chamber is recorded (retention test latency). The number of trials needed to achieve the passive avoidance criterion and the retention test latency are used as the measures of avoidance responding.

A. Simultaneous Manipulation of ACTH and Corticosterone Levels

Our first study (Leshner, et al., 1975) was an attempt to establish, simply, whether there is an effect of manipulating ACTH and corticosterone levels on the tendency to avoid being attacked. In that study, mice were either (1) hypophysectomized; (2) adrenalectomized; (3) treated for a week with 325 μg/day corticosterone; or (4) treated for a week with 2 IU ACTH (Cortrophin-Zinc, Organon), and their avoidance behavior was compared with that of their appropriate controls.

The results of that study, reproduced from Leshner et al. (1975), can be seen in Table 1. All of the manipulations significantly increased the tendency to avoid attack relative to controls. These findings established that there is an

TABLE 1

Mean (± Standard Error) Trials to Criterion and Retention Test Latencies

Group	n	Trials to criterion	Retention latency (sec)
Hypophysectomized	20	3.8 ± 0.4**	257.3 ± 19.7***
Sham-hypophysectomized	20	5.4 ± 0.4	139.3 ± 16.5
Adrenalectomized	20	4.4 ± 0.3**	278.5 ± 7.0***
Sham-adrenalectomized	20	6.0 ± 0.4	167.6 ± 25.3
Corticosterone-treated	20	4.2 ± 0.2*	202.6 ± 21.4
Placebo for corticosterone	20	5.6 ± 0.8	154.0 ± 18.9
ACTH-treated	24	4.2 ± 0.3**	239.1 ± 16.3**
Placebo for ACTH	22	5.6 ± 0.3	159.9 ± 21.3

*$p < .05$ (by t-tests).
**$p < .01$.
***$p < .001$.

effect and suggested a bimodal relationship between the pituitary–adrenocortical hormones and avoidance-of-attack, in which either very low or very high levels of these hormones lead to increased avoidance responding.

To test the authenticity of that bimodal relationship, we turned to a direct test of the dose–response relationship between ACTH and avoidance respond-ing. In this study (Leshner et al., 1975), we treated hypophysectomized mice with a range of dosages of ACTH and examined their avoidance performance.

As shown in Fig. 1, ACTH treatment did not affect the number of trials needed by hypophysectomized mice to reach the avoidance criterion. However, the expected results were observed on the retention test latency measure. As shown in Fig. 2, a sham-operated control group exhibited a shorter retention test latency than did all other groups, except for a group given an intermediate dosage, 0.75 IU/day, of ACTH. Furthermore, that group receiving the 0.75 IU/day dosage had a shorter retention test latency than did all other groups, except the one receiving the highest dosage (1.75 IU/day), which still differed from the sham-operated controls. Thus, this second study provided results consistent with those of the first study. It showed that mice with either very low or very high pituitary–adrenocortical hormone levels display a greater retention of the avoidance response than do mice with intermediate levels of these hormones. (We continue to use the general term, pituitary–adrenocortical hormones, when discussing the effects of ACTH in this study, because ACTH treatment leads to a consequent increase in corticosterone levels.)

It may be useful to mention at this time that the effects of the pituitary–adrenocortical hormones on avoidance acquisition, the trials to cri-terion measure, are not always consistent across our studies. We have no simple

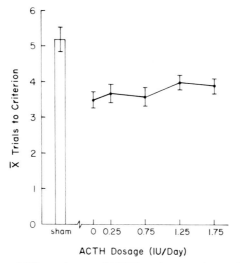

Fig. 1.   The effects of different dosages of ACTH on the mean (± standard error) number of trials needed to achieve the passive avoidance criterion in hypophysectomized mice. Difference between controls and all hypophysectomized groups: $p < .001$, by analysis of variance. (From Leshner, Moyer, & Walker, 1975.)

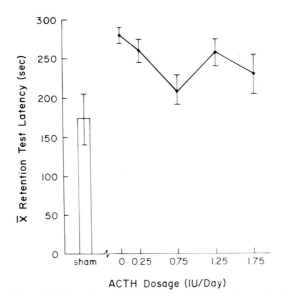

Fig. 2.   The effects of different dosages of ACTH on the mean (± standard error) retention test latency scores in hypophysectomized mice. Differences as discussed in the text: $p < .005$, by analysis of variance and Scheffé's tests. (From Leshner et al., 1975.)

explanation for that inconsistency, but the same problem occurs in avoidance-of-shock situations (e.g., Bohus, 1975; DiGiusto, Cairncross, & King, 1971; Miller & Ogawa, 1962). This inconsistency suggests that we focus on what appear to be major effects of these hormones, those on retention, and, although we shall present the data on both measures, we shall limit our discussion primarily to the retention measure.

These two studies provided the first indication that the relationship between the pituitary–adrenal hormones and avoidance responding in our avoidance-of-attack situation is different from that observed in more traditional shock-mediated situations. Although there are exceptions (Gold & van Buskirk, 1976), a generally linear relationship between pituitary–adrenal hormone levels and avoidance responding has been observed in most studies of shock-mediated situations; animals with very low levels of these hormones show decreased avoidance, and those with high pituitary–adrenal hormone levels exhibit increased avoidance retention (DiGiusto et al., 1971; Weiss, McEwen, Silva, & Kalkut, 1969). Our studies suggested, however, a different bimodal relationship, with both low and high pituitary–adrenocortical hormone levels, leading to increased avoidance retention. Our later studies supported and extended the view that these two kinds of avoidance responding are affected differently by the pituitary–adrenocortical hormones.

B. Separation of the Effects of ACTH and Corticosterone

We next turned to a separation of the effects of ACTH and corticosterone on avoidance-of-attack. Earlier studies had shown that the effects of altering pituitary–adrenal hormone levels on shock-mediated avoidance were primarily a result of the changes in ACTH and not corticosteroid levels (e.g., de Wied, 1969, 1976); therefore, we attempted a similar separation for the effects of ACTH and corticosterone on avoidance-of-attack.

Earlier studies separating ACTH and corticosterone effects on avoidance-of-shock used two general classes of techniques. First, some studies combined pituitary–adrenal manipulations in such a way that one could compare the effects of having high ACTH and high corticosterone levels with those of having high ACTH and low corticosterone levels, and then with the effects of having low ACTH levels and low corticosterone levels, and so on. This kind of analysis suggested that ACTH is the primary pituitary–adrenal hormone in the control of avoidance responding in shock-mediated situations (DiGiusto et al., 1971). But, if this strategy is applied in the avoidance-of-attack situation, then, as shown in Table 1, this approach does not yield the same kind of separation of effects. All manipulations led to similar increases in the tendency to avoid attack.

Second, some studies separating ACTH and corticosterone effects on avoidance behavior in shock-mediated situations took advantage of the availability of fragments of the ACTH molecule that are devoid of tropic actions on the adrenal cortex (de Wied, 1976). Again, these kinds of studies pointed to ACTH as the critical pituitary—adrenocortical hormone. But, we use mice as subjects, and Brain (1972) had reported a study suggesting that ACTH fragments might have some tropic effects on the adrenal cortices of mice. Therefore, we could not use these fragments with confidence and had to turn to other techniques for separating the effects of ACTH and corticosterone on avoidance-of-attack.

The first study we conducted attempting to separate ACTH and corticosterone effects was concerned with the effects of low levels of these hormones (Moyer & Leshner, 1976). Our earlier studies had shown that hypophysectomy, which leads to low levels of both ACTH and corticosterone, increases avoidance-of-attack retention. Furthermore, replacement therapy with ACTH and corticosterone, was found to restore avoidance performance to normal levels. The question, then, became which hormone is the critical one in this effect. One way to answer that question is to attempt to restore the avoidance performance of hypophysectomized mice by restoring the level of only one hormone.

Figure 3 shows the effects of treating hypophysectomized mice with a range of dosages of corticosterone on their avoidance-of-attack performance. Again, hormone replacement had no effect on the acquisition measure. But, corticosterone treatment did restore the avoidance retention of hypophysectomized mice to normal levels. Thus, treatment with this steroid hormone alone is as effective as is increasing the levels of both ACTH and corticosterone in maintaining avoidance retention at normal levels. Therefore, we concluded that ACTH is not necessary for normal avoidance-of-attack behavior, and that the effects on avoidance retention of simultaneously reducing both ACTH and corticosterone levels probably are due to the withdrawal of the glucocorticoid.

We then turned to the question of the effects of increasing pituitary—adrenal hormone levels. Our earlier studies had shown that generally increasing pituitary—adrenocortical activity levels increases avoidance-of-attack; we then asked which is the critical hormone, ACTH or corticosterone?

The approach we used was to selectively raise the level of one hormone, ACTH, without altering the level of the other, corticosterone. In this paradigm, we compare the effects of ACTH on the avoidance responding of intact mice with the effects of this peptide on the behavior of mice with *controlled corticosterone levels*. Controlled corticosterone levels are produced by subjecting the mice to adrenalectomy, and treating them with a fixed behavioral replacement dosage of corticosterone (200 µg/day) for two weeks prior to testing. Thus, mice

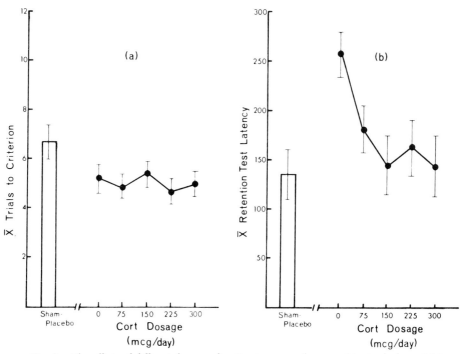

Fig. 3. The effects of different dosages of corticosterone on the mean (± standard error) (a) number of trials needed to achieve the passive avoidance criterion and (b) retention test latency scores in hypophysectomized mice. On the trials measure, there were no significant effects by analysis of variance. On the retention latency measure, differences as discussed in the text: $p <$ .025, by analysis of variance and tests of least significant differences. (From Moyer & Leshner, 1976.)

with controlled corticosterone levels have sufficient amounts of this hormone to maintain normal responding, but they cannot respond to ACTH treatment with further increases in the level of this steroid. If the effects of generally increasing pituitary–adrenal activity levels are due to the rise in ACTH level, it should not matter whether corticosterone levels are concomitantly increased or not; both intact and controlled corticosterone mice should respond to ACTH with increased avoidance performance. But, if the effect is due to corticosterone's actions, ACTH should only affect the avoidance behavior of intact mice, who would have responded to the peptide treatment with increases in corticosterone levels; there should be no effect of ACTH on the avoidance behavior of mice with controlled corticosterone levels. This technique has been used effectively in studying aggressiveness (Leshner, Walker, Johnson, Kelling, Kriesler & Svare, 1973).

As can be seen in Fig. 4, ACTH significantly increased the avoidance behavior of intact mice, in this case again only on the retention measure. But, ACTH had no effect on the avoidance behavior of mice with controlled corticosterone levels. Therefore, ACTH does not affect avoidance-of-attack independently of that peptide's effects on the adrenal secretion of corticosterone, and we conclude that corticosterone is the primary pituitary–adrenal hormone in the control of avoidance-of-attack.

C. Summary Statement: Pituitary–Adrenocortical Effects on Avoidance-of-Attack

The studies discussed here on the effects of ACTH and corticosterone on the tendency to avoid attack argue that these pituitary–adrenal hormones affect

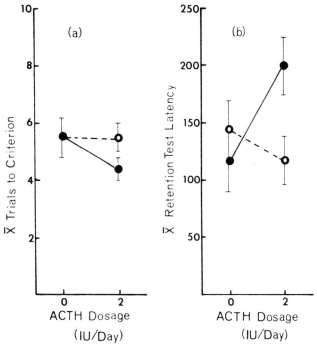

Fig. 4. The effects of ACTH on the mean (± standard error) (a) number of trials needed to achieve the passive avoidance criterion; and (b) retention test latency scores of intact mice (Sham, ●——●) and mice with controlled corticosterone levels (C-C, ○---○). On the trials measure, by analysis of variance, there were no significant differences among the groups. On the retention measure, differences as discussed in the text: $p < .05$ by analysis of variance and tests of least significant differences. (From Moyer & Leshner; 1976.)

avoidance responding in social situations quite differently from the way they affect avoidance behavior in shock-mediated situations. First, the form of the general relationship between these hormones and avoidance retention is different. As mentioned before, the relationship between pituitary–adrenocortical hormone levels and avoidance retention in the shock-mediated situation is approximately linear, whereas here the effect appears to be bimodal; both decreases and increases in hormone levels lead to increased avoidance-of-attack. Second, the critical hormone in the two situations appears to be different. In the shock mediated situation, ACTH exerts an extraadrenal facilitative effect on avoidance responding (de Wied, 1976), whereas in the avoidance-of-attack situation, ACTH levels seem to be unimportant; the critical pituitary–adrenal hormone appears to be corticosterone.

III. PITUITARY–ADRENOCORTICAL EFFECTS ON SUBMISSIVENESS

Another aspect of our work has focused on the role of the pituitary–adrenal hormones on the quality of submissiveness measured directly, rather than indirectly via studying the tendency to avoid attack. The paradigm we use in studying submissiveness provides another example of learning and memory as they occur in social situations.

A. Some Comments about Submissiveness and the Test Procedures

Before going into the results of those studies and their relevance to this discussion, it may be useful to describe briefly the way in which we view submissiveness, and the way in which we study it. More detailed discussions are published elsewhere (Leshner, Korn, Mixon, Rosenthal, & Besser, 1980; Leshner & Politch, 1979). Submissiveness is the readiness of an individual to surrender, or submit, following attack or the threat of attack by an opponent. The most likely submissive response in natural situations is flight, because it is the most effective means of avoiding attack. However, flight is usually impossible in laboratory situations and under some more natural conditions. Therefore, we study the next most likely submissive responses in the laboratory, and these include a set of ritualized appeasement gestures. These gestures seem to be signals of submission that can terminate attack by the opponent, thereby providing mechanisms for avoiding attack (Lorenz, 1966; Scott, 1967).

These ritualized appeasement gestures have been identified in almost all vertebrate species, and they are species-typical in form. It seems likely that little if any learning is demanded before an individual knows *how* to act submissively (Eibl-Eibesfeldt, 1975; Lorenz, 1966), although they do seem to learn *when* to be submissive; that is, they learn the context or cues that demand appeasement behavior. For our discussion here, it is important that, over repeated experiences of defeat, animals will submit more and more readily after the initial attack or threat of attack, showing the learned quality of this behavior (Leshner, 1980).

A rather strict criterion of submission is used in our studies, to provide an accurate measure of the readiness to surrender. We require that, to demonstrate submission, a mouse must (1) exhibit the classic upright submissive posture; and (2) not fight back in response to a subsequent attack by the opponent (cf. Ginsburg & Allee, 1942; Nock & Leshner, 1976; Scott, 1946). The submission tests are carried out by pairing two mice, an experimental animal and an opponent, in a novel arena, and allowing them to interact until one submits. The opponents almost always win, as they have been housed in isolation for at least 6 weeks prior to their use, and, therefore, are much more aggressive than the test animals, who never have been isolated for longer than 2 weeks. During the encounter, we record a variety of measures, but the primary measure of submissiveness that is used, and the one that will be discussed here, is the number of aggressive bouts (called *aggressions*) by the opponent needed to induce submission in the test animal. The smaller the number of aggressions by the opponent needed to induce submission in the test animal, the more submissive that test animal is considered to be.

The experimental paradigm used in the studies to be discussed here is a repeated testing one, in which the test animal is subjected to two defeats, separated by at least 24 hr. The experimental animal meets a different opponent on each test. We consider the first test to be an acquisition or learning trial, and the second test is viewed as a retention test. That the mice learn during the first encounter to submit more readily in the second can be seen from Fig. 5, which shows the increase in submissiveness from Test 1 to Test 2 of untreated mice. Because, as discussed earlier, submissive responding is a response used to avoid attack, we consider this paradigm to be an analog of that already discussed, avoidance-of-attack, providing a second mechanism for studying pituitary–adrenal effects on learning and memory in social situations.

B. The Effects of Pituitary–Adrenocortical Manipulations on Responding in Repeated Submission Tests

Our first study using this paradigm (Roche & Leshner, 1979) was designed to be comparable to some studies of shock-mediated avoidance and involved subjecting mice to an initial defeat and treating them immediately thereafter with either ACTH (2 IU, Cortrophin-Zinc) or a placebo. The mice received

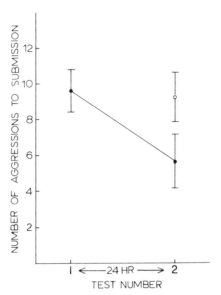

Fig. 5. The mean (± standard error) number of aggressions to submission during an initial defeat (Test 1) and a second defeat 24 hr later (Test 2), by *t*-test, *p* < .01. No prior defeat is shown by ○.

a second submission test either 24 hr, 48 hr, or seven days later. The protocol is the same as one used frequently in avoidance-of-shock situations, where the test animal is treated with the hormone immediately after a passive avoidance acquisition trial, and then tested for retention at some time later (Gold & van Buskirk, 1976).

As can be seen in Fig. 6, immediate postdefeat treatment with ACTH increased subsequent submissiveness relative to controls when the mice were tested either at the 24 hr or at the 48 hr retention interval. This is expressed in the form of a decrease in the number of aggressions by the opponent needed to induce submission in the test animal. There was no effect at the 168 hr interval.

The findings of this initial study are consistent with those of our earlier studies on avoidance-of-attack in showing that generally increasing pituitary–adrenal activity levels increases responding in these social avoidance situations. Furthermore, these findings are similar to what has been observed in shock-mediated passive avoidance situations, where postacquisition ACTH treatment increases retention of that avoidance response (Gold & McGaugh, 1977; Gold & van Buskirk, 1976).

However, our earlier studies of avoidance-of-attack had suggested that the critical pituitary–adrenal hormone in social learning situations is corticosterone;

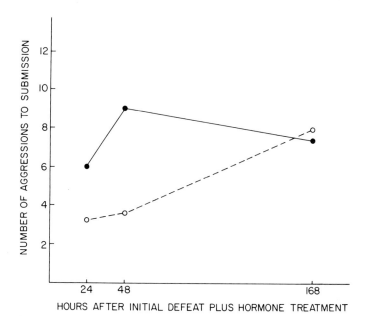

Fig. 6. Mean number of aggressions to submission in mice treated with either a placebo (●——●) or 2 IU ACTH (○---○) immediately after an initial defeat and tested for submissiveness either 24 hr, 48 hr, or 7 days after the initial defeat plus hormone treatment. All points represent independent groups. Differences as discussed in the text: $p < .01$ by analysis of variance and tests of least significant differences. (From Roche & Leshner, 1979.)

whereas, in shock-mediated situations the critical hormone appears to be ACTH (de Wied, 1976). In addition, some of our other studies (e.g., Leshner & Politch, 1979; Leshner, 1980) suggested that in the case of submissiveness, the critical pituitary–adrenal hormone is corticosterone as well. Therefore, we conducted the next logical study in the series, attempting to mimic the observed effects of ACTH by corticosterone treatment.

In this study, we used only one retention interval, 48 hr postdefeat and hormone treatment. Mice were subjected to an initial defeat and treated immediately thereafter with 300 μg corticosterone or a placebo. All were tested again 48 hr later. As can be seen in Fig. 7, postdefeat treatment with corticosterone had the same effect as had postdefeat treatment with ACTH: Both hormone treatments, when coupled with an initial defeat experience, increased subsequent submissiveness. Because ACTH treatment increases both ACTH and corticosterone levels, whereas corticosterone treatment increases only its own levels while, in fact, decreasing ACTH levels (Hodges & Sadow, 1969), we interpret the results of these studies as showing that corticosterone is the critical pituitary–adrenal hormone in this effect; ACTH levels appear to be irrelevant to pituitary–adrenal effects on submissiveness.

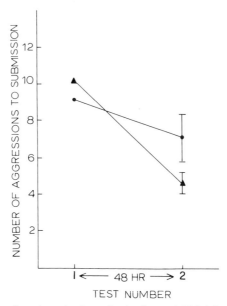

Fig. 7. The effects of treating mice immediately after an initial defeat (Test 1) with either a placebo (●———●) or corticosterone (▲———▲) on the mean (± standard error) level of submissiveness during a second test 48 hr later (Test 2). Differences as discussed in the text: $p < .05$, by analysis of variance and tests of least significant differences.

IV. HOW DOES CORTICOSTERONE AFFECT BEHAVIOR IN SOCIAL LEARNING–MEMORY SITUATIONS?

Having shown that corticosterone has important effects in social learning situations, a reasonable next question might concern the way in which corticosterone is exerting its effects. For example, is corticosterone selectively altering the behaviors that occur in these social situations, such as submissiveness, or is this hormone affecting some more general process, such as learning and/or memory?

A first possibility is that corticosterone directly increases submissiveness, the behavior being studied in these social situations. That explanation, although the simplest, seems unlikely. It is true that in these and other studies (Leshner et al., 1980; Leshner & Politch, 1979) we found corticosterone to increase submissiveness. However, Brain and Evans (1977) and Heller (1978) found that acute increases in corticosterone levels also increase aggressiveness. It seems impossible that if corticosterone directly increases submissiveness, it would increase aggressiveness, a behavior that normally is exclusive to submissiveness.

Therefore, corticosterone must be affecting some more general behavioral process in the situations described in this chapter.

A second possibility is that corticosterone is directly altering some aspect of learning and/or memory processes. We also believe this to be an unlikely explanation. Again, it is true that we are observing effects of this hormone on the learned behaviors displayed in social situations, and it is true that our effects are seen in the same paradigms as have been used to demonstrate hormonal effects on memory in other situations. In this latter case, we see very dramatic effects of postacquisition treatment with corticosterone on later performance (or retention) of these social responses, effects similar to those used to ascribe memory effects to other hormones (Gold & McGaugh, 1977; Gold & van Buskirk, 1976; McGaugh, Gold, Handwerker, Jensen, Martinez, Meligeni, & Vasquez, 1979). However, when corticosterone manipulations are used in *nonsocial* learning–memory situations, such as shock-mediated passive avoidance situations, the results are not consistent. Some studies have found corticosterone to decrease retention (Bohus, 1975; Weiss et al., 1969); other studies have found corticosterone to increase avoidance performance (Flood, Vidal, Bennett, Orme, Vasquez, & Jarvik, 1978; Kovács, Telegdy, & Lissák, 1977), whereas yet others have found no effects (Gold, van Buskirk, & McGaugh, 1975). If corticosterone were working on learning or memory processes per se in our situations, surely they should affect these same processes in other, somewhat similar situations. Therefore, we believe it to be unlikely that corticosterone is exerting its effects in social learning and memory situations by affecting either of these processes directly.

A third explanation that might seem reasonable is that corticosterone affects these behaviors in a state-dependent way. This explanation has been suggested for the case of the effects of ACTH on avoidance responding (Gray, 1975), and therefore, it should also be considered a possibility here. To test this, we conducted a study of the effects of corticosterone on submissiveness using the traditional state-dependency paradigm (Leshner et al., 1980). That is, four groups of mice each received two submission tests. One group (CP) was treated with a single injection of corticosterone (300 μg) before Test 1 and a placebo before Test 2. A second group (PC) received a placebo before Test 1 and corticosterone before Test 2. The third group (CC) received two injections of corticosterone, one before each test, and the fourth group (PP) received a placebo injection before each test. If corticosterone is affecting submissiveness in a state-dependent way, there should be decrements from original to later performance (an increase or no change in submissiveness) in Groups CP and PC, because their states are changing, and only Group CC should show the facilitative effects of corticosterone on future submissiveness.

As can be seen from Fig. 8, the results obtained in this study were inconsistent with a state-dependency explanation. In fact, statistical analysis of the data

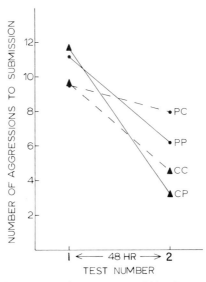

Fig. 8. The effects of treatment with corticosterone (300 μg) or a placebo at different times during the testing sequence on the mean (± standard error) level of submissiveness on each test. See text for explanation of abbreviations. Differences as discussed in the text and as presented further in Fig. 9: $p < .05$, by analysis of variance. (From Leshner et al., 1980.)

showed that only the main effect of treatment with corticosterone before Test 1 was significant, and then only on performance during Test 2. That effect is shown in Fig. 9. This effect on Test 2 performance is identical to that observed following post-Test 1 treatment with corticosterone discussed before (and shown in Fig. 7). Thus, corticosterone's effects on behavior in social learning–memory situations cannot be attributed to state-dependency.

We believe the most likely explanation of corticosterone's effects in social learning–memory situations is one somewhat similar to what Spear (1973) has suggested may be operating in many cases where physiological manipulations appear to affect learning or memory processes, although our view is slightly more phenomenological. All experiences are made up of a variety of stimulus elements, and the physiological responses to those experiences can be viewed as some portion of the stimulus complex of the experience. (There is no question that individuals can sense their physiological reactions to important experiences; perhaps the individual can sense changes in corticosterone levels as well.) Furthermore, it is now well documented that the experience of defeat is accompanied by marked elevations in circulating corticosterone levels (Bronson & Eleftheriou, 1965; Leshner, 1980). Because corticosterone increases are a normal consequence of the defeat experience, they might be viewed as stimulus elements, contributing to the "meaning" or "magnitude" of that experience.

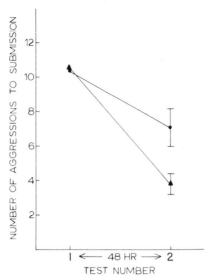

Fig. 9. The effects of placebo (●———●) or corticosterone (▲———▲) treatment before Test 1 on the mean (± standard error) level of submissiveness on each submission test. (From Leshner et al., 1980.)

Then, as the magnitude of the defeat experience is altered, one would expect to find changes in later performance of the behaviors learned during that experience, because more meaningful experiences are remembered better then less meaningful experiences. This is precisely what happens in our studies of both avoidance-of-attack and submissiveness. We increase (or in other cases decrease, e.g., Nock & Leshner, 1976) the normal postdefeat corticosterone responses by exogenous corticosterone treatments. And, these manipulations increase (or decrease) future avoidance–submissive responding.

This hypothesis would predict that in intact animals, the greater the corticosterone responses to defeat, the greater the magnitude of that experience and, therefore, the greater is avoidance retention or future submissiveness. We have conducted a series of studies that show just that. In the avoidance-of-attack situation (Politch & Leshner, 1977), there is no correlation between resting corticosterone levels and either the trials to acquisition measure ($r = 0.19$, $df = 18$) or the retention test latency measure ($r = -0.05$, $df = 18$). However, there is a significant correlation between retention test latencies and the mouse's corticosterone responses to being placed into the start chamber of the apparatus ($r = 0.46$, $df = 19$, $p < .05$). This correlation shows that the greater the corticosterone response to the threat of being defeated, the greater the tendency to avoid attack.

We have conducted a similar kind of analysis in the submission testing situation, and it yielded similar results. One group of mice was first subjected to two submission tests, separated by 48 hr, and then plasma was collected 5 days later to discover if there is a relationship between resting corticosterone levels and levels of submissiveness. As the data on avoidance-of-attack would predict, there was no correlation between resting corticosterone levels and submissiveness (the number of aggressions to submission) either on Test 1 ($r = -0.02$, $df = 27$) or on Test 2 ($r = -0.04$, $df = 27$).

A second group of mice received a single submission test, and, then, 48 hr later, were placed into the submission testing arena across a wire mesh barrier from an opponent for 2 min. Plasma was collected 15 min after the beginning of this exposure period to determine the corticosterone response to the threat of defeat. We then correlated these response levels of corticosterone with the level of submissiveness (number of aggressions to submission) that had been displayed during the earlier submission test, and that correlation was quite significant ($r = -0.47$, $df = 25$, $p = .02$). Thus, there is a positive relationship between an animal's submissiveness and its corticosterone responses to the threat of being attacked.

These correlational data cannot be considered conclusive by themselves, because they only establish that a relationship exists. However, when they are considered in combination with the experimental data discussed in the rest of this chapter, namely, data showing that experimentally increasing the corticosterone responses to initial attack–defeat experiences increases later performance; these data argue in support of our hypothesis. They suggest that the greater the corticosterone response associated with an experience (i.e., the greater the stimulus complex of the initial experience), the greater the later retention, or performance, of the behavior learned during that experience.

V. CONCLUDING COMMENTS

The studies discussed in this chapter clearly show that, although they do not totally determine behavioral responding, the pituitary–adrenocortical hormones have important effects on the behaviors displayed in social learning–memory situations. Two kinds of situations were discussed, avoidance-of-attack and repeated submissiveness testing, and in both cases major effects were demonstrated. Furthermore, these studies show that the critical pituitary–adrenal hormone in these social situations is corticosterone; ACTH levels appear to be relatively unimportant.

The effects observed in these studies differ markedly from those observed in other avoidance situations, such as those in which electric shock is the motivating stimulus. In shock-mediated situations, the critical pituitary–adrenocortical hormone has been shown to be ACTH (e.g., de Wied, 1976), whereas in social situations the critical hormone is corticosterone. An important question then arises: What do these differences mean?

A first suggestion might be that the observed differences simply are a result of species differences. We study mice, whereas most studies of shock-mediated avoidance have used rats as subjects. But, the more usual kinds of results also have been obtained in studies of shock-mediated avoidance in mice (e.g., Korányi, Endröczi, Lissák, & Szepes, 1967). Therefore, that simple explanation of species differences probably cannot be applied in this case.

The more likely interpretation is that our findings demonstrate some limitation in the generality of the effects of pituitary–adrenocortical hormones on learning and memory as observed in other situations. That is not to suggest that those effects are not reliable or important, or that they are to be considered limited only to avoidance-of-shock. In fact, ACTH has been shown to have extraadrenal, facilitative effects on performance in other, nonaversively motivated learning–memory situations (Garrud, Gray, & de Wied, 1974; Guth, Levine, & Seward, 1971; Sandman, George, Nolan, van Riezen, & Kastin, 1975). Therefore, some generality has already been demonstrated. However, the results of our studies do demand some caution when proposing that those effects have very broad and general implications.

We also suggest that some consideration be given to the hypothesis put forth earlier (Section IV) as an explanation for the effects of hormones observed in some other learning and memory situations. For example, changes in pituitary–adrenocortical hormone levels naturally occur during shock-mediated avoidance conditioning procedures. It may well be that, rather than ACTH working directly on learning and/or memory processes per se (cf. Gold & McGaugh, 1977; Klein, 1972), or on other processes such as attention (Beckwith & Sandman, 1978; Oades, 1979), this peptide affects the retention of avoidance responses simply because a rise in its levels is a normal part of the stimulus complex associated with the acquisition phases of training. Then, ACTH treatment would increase later avoidance retention or performance, not because it affects some fundamental behavioral process, but merely because that treatment is increasing the intensity of the normal stimulus characteristics of the acquisition experience and, therefore, its meaningfulness as expressed in future behavioral responses.

ACKNOWLEDGMENTS

The work reported here was supported by Research Grants No. BMS 75-08120 from NSF and No. MH 31086 from NIMH. In addition to those collaborators mentioned in the text and in the references, P. McIntosh, C. Prindle, L. Stein, and T. Whitaker participated in this work.

REFERENCES

Beckwith, B. E., & Sandman, C. A. (1978). Behavioral influences of the neuropeptides ACTH and MSH: A methodological review. *Neuroscience and Biobehavioral Reviews*, 2, 311–338.

Bohus, B. (1975). The hippocampus and the pituitary-adrenal system hormones. *In* R. L. Isaacson, & K. H. Pribram (Eds.), *The Hippocampus*, Vol. I, pp. 323–353. New York: Plenum.

Brain, P. (1972). Study on the effect of the 4-10 ACTH fraction on isolation-induced intermale fighting behaviour in the albino mouse. *Neuroendocrinology*, 10, 371–376.

Brain, P. F., & Evans, A. E. (1977). Acute influences of some ACTH-related peptides on fighting and adrenocortical activity in male laboratory mice. *Pharmacology, Biochemistry and Behavior*, 7, 524–533.

Bronson, F. H., & Eleftheriou, B. E. (1965). Adrenal responses to fighting in mice: Separation of physical and psychological causes. *Science*, 147, 627–628.

de Wied, D. (1969). Effects of peptide hormones on behavior. *In* W. F. Ganong & L. Martini (Eds.), *Frontiers in Neuroendocrinology*, pp. 97–140. London and New York: Oxford University Press.

de Wied, D. (1976). Pituitary-adrenal system hormones and behavior. *In Symposium on Developments in Endocrinology*. Oss: Organon.

DiGiusto, E. L., Cairncross, K., & King, M. G. (1971). Hormonal influences on fear-motivated responses. *Psychological Bulletin*, 75, 432–444.

Eibl-Eibesfeldt, I. (1975). *Ethology: The Biology of Behavior*. New York: Holt.

Flood, J. F., Vidal, D., Bennett, E. L., Orme, A. E., Vasquez, S., & Jarvik, M. E. (1978). Memory facilitating and anti-amnesic effects of corticosteroids. *Pharmacology, Biochemistry and Behavior*, 8, 81–87.

Garrud, P., Gray, J. A., & de Wied, D. (1974). Pituitary–adrenal hormones and extinction of rewarded behaviour in the rat. *Physiology and Behavior*, 12, 109–119.

Ginsburg, B., & Allee, W. C. (1942). Some effects of conditioning on social dominance and subordination in inbred strains of mice. *Physiological Zoology*, 15, 495–506.

Gold, P. E., & McGaugh, J. L. (1977). Hormones and memory. *In* L. H. Miller, C. A. Sandman, & A. J. Kastin (Eds.), *Neuropeptide Influences on the Brain and Behavior*, pp. 127–143. New York: Raven.

Gold, P. E., & van Buskirk, R. (1976). Enhancement and impairment of memory processes with post-trial injections of adrenocorticotrophic hormone. *Behavioral Biology*, 16, 387–400.

Gold, P. E., van Buskirk, R. B., & McGaugh, J. L. (1975). Effects of hormones on time-dependent memory storage processes. *In* W. H. Gispen, Tj. B. van Wimersma Greidanus, B. Bohus, & D. de Wied (Eds.), *Hormones, Homeostasis, and the Brain: Progress in Brain Research* (Vol. 42), pp. 210–211. Amsterdam: Elsevier.

Gray, P. (1975). Effect of adrenocorticotropic hormone on conditioned avoidance behavior in rats interpreted as state-dependent learning. *Journal of Comparative and Physiological Psychology*, 88, 281–284.

Guth, S., Levine, S., & Seward, J. P. (1971). Appetitive acquisition and extinction effects with exogenous ACTH. *Physiology and Behavior*, 7, 195–200.

Heller, K. E. (1978). Role of corticosterone in the control of post-shock fighting behavior in male laboratory mice. *Behavioral Processes*, 3, 211–222.

Hodges, J. R., & Sadow, J. (1969). Hypothalamo–pituitary–adrenal function in the rat after prolonged treatment with cortisol. *British Journal of Pharmacology*, 36, 489–495.

Klein, S. B. (1972). Adrenal–pituitary influence in reactivation of avoidance–learning memory in the rat after intermediate intervals. *Journal of Comparative and Physiological Psychology*, 79, 341–359.

Korányi, L., Endröczi, E., Lissák, K., & Szepes, E. (1967). The effect of ACTH on behavioral processes motivated by fear in mice. *Physiology and Behavior*, 2, 439–445.

Kovács, G. L., Telegdy, G., & Lissák, K. (1977). Dose-dependent action of corticosteroids on brain serotonin content and passive avoidance behavior. *Hormones and Behavior*, 8, 155–165.

Leshner, A. I. (1980). The interaction of experience and neuroendocrine factors in determining behavioral adaptations to aggression. *In* P. S. McConnell, G. J. Boer, H. J. Romijn, N. E. van de Poll, & M. A. Corner (Eds.), *Adaptive Capabilities of the Nervous System: Progress in Brain Research* (Vol. 53), pp. 427–438. Amsterdam: Elsevier.

Leshner, A. I., Korn, S. J., Mixon, J. F., Rosenthal, C., & Besser, A. K. (1980). Effects of corticosterone on submissiveness in mice: Some temporal and theoretical considerations. *Physiology and Behavior*, 24, 283–288.

Leshner, A. I., & Moyer, J. A. (1975). Androgens and agonistic behavior in mice: Relevance to aggression and irrelevance to avoidance-of-attack. *Physiology and Behavior*, 15, 695–699.

Leshner, A. I., Moyer, J. A., & Walker, W. A. (1975). Pituitary—adrenocortical activity and avoidance-of-attack in mice. *Physiology and Behavior*, 15, 689–693.

Leshner, A. I., & Politch, J. A. (1979). Hormonal control of submissiveness in mice: Irrelevance of the androgens and relevance of the pituitary—adrenal hormones. *Physiology and Behavior*, 22, 531–534.

Leshner, A. I., Walker, W. A., Johnson, A. E., Kelling, J. S., Kreisler, S. J., & Svare, B. B. (1973). Pituitary—adrenocortical activity and intermale aggressiveness in isolated mice. *Physiology and Behavior*, 11, 705–711.

Lorenz, K. (1966). *On Aggression*. New York: Bantam Books.

McGaugh, J. L., Gold, P. E., Handwerker, M. J., Jensen, R. A., Martinez, Jr., J. L., Meligeni, J. A., & Vasquez, B. J. (1979). Altering memory by electrical and chemical stimulation of the brain. *In* M. A. B. Brazier (Ed.), *Brain Mechanisms in Learning and Memory from the Single Neuron to Man*, pp. 151–164. New York: Raven.

Miller, R. E., & Ogawa, N. (1962). The effect of adrenocorticotrophic hormone (ACTH) on avoidance conditioning in the adrenalectomized rat. *Journal of Comparative and Physiological Psychology*, 55, 211–213.

Moyer, J. A., & Leshner, A. I. (1976). Pituitary—adrenal effects on avoidance-of-attack in mice: Separation of the effects of ACTH and corticosterone. *Physiology and Behavior*, 17, 297–301.

Nock, B. L., & Leshner, A. I. (1976). Hormonal mediation of the effects of defeat on agonistic responding in mice. *Physiology and Behavior*, 17, 111–119.

Oades, R. D. (1979). Search and attention: Interaction of the hippocampal—septal axis, adrenocortical and gonadal hormones. *Neuroscience and Biobehavioral Reviews*, 3, 31–48.

Politch, J. A., & Leshner, A. I. (1977). Relationship between plasma corticosterone levels and the tendency to avoid attack in mice. *Physiology and Behavior*, 19, 781–785.

Rigter, H., & van Riezen, H. (1978). Hormones and memory. *In* M. A. Lipton, A. DiMascio, & K. F. Killam (Eds.), *Psychopharmacology: A Generation of Progress*, pp. 677–689. New York: Raven.

Roche, K. E., & Leshner, A. I. (1979). ACTH and vasopressin treatments immediately after a defeat increase future submissiveness in mice. *Science*, 204, 1343–1344.

Sandman, C. A., George, J. M., Nolan, J. D., van Riezen, H., & Kastin, A. J. (1975). Enhancement of attention in man with ACTH/MSH 4–10. *Physiology and Behavior*, 15, 427–431.

Scott, J. P. (1946). Incomplete adjustment caused by frustration of untrained fighting mice. *Journal of Comparative Psychology*, 39, 379–390.

Scott, J. P. (1967). Discussion. *In* C. D. Clemente & D. B. Lindsley (Eds.), *Aggression and Defense*, pp. 45–51. Berkeley, California: University of California Press.

Spear, N. E. (1973). Retrieval of memory in animals. *Psychological Review*, 80, 163–194.

Weiss, J. M., McEwen, B. S., Silva, M. T. A., & Kalkut, M. F. (1969). Pituitary—adrenal influences on fear responding. *Science*, 163, 197–199.

CHAPTER 9

ACTH Analogs and Human Performance

Anthony W. K. Gaillard

Studies investigating the behavioral effects of two adrenocorticotropic hormone (ACTH) analogs, ACTH $_{4-10}$ and ORG 2766, are reviewed. Although there is evidence that these neuropeptides are behaviorally active, it is still unclear which psychological processes are involved. Comparison of the available studies is hindered because they differ in several respects, such as route of administration, dose, type of subject, and also the type of test used. In addition, a number of the studies suffer from important methodological difficulties. These are summarized in a separate section. The principal conclusion of this chapter is that there is no evidence for the notion that ACTH analogs have a direct effect on learning or memory. Pronounced effects are found only in long-duration tasks (continuous performance, vigilance tasks). Thus, ACTH analogs seem to help the subject to maintain alertness during long-term performance.

The psychological processes that mediate the behavioral effects of these neuropeptides are of a general character and could be labeled sustained attention or task-oriented motivation. However, the effects are not easily described in terms of general arousal because, on the one hand, such

181

ENDOGENOUS PEPTIDES
AND LEARNING AND MEMORY PROCESSES

basal physiological indices as heart rate, are *not* affected in rest conditions; on the other hand, several studies reported effects on phasic changes in task-related physiological measures.

In conclusion, ACTH analogs seem to be behaviorally active in a way similar to stimulants, except that these neuropeptides do *not* have the adverse effects commonly observed with, for example, amphetamines.

I. INTRODUCTION

This chapter reviews the effects of two ACTH fragments ($ACTH_{4-10}$ and the $ACTH_{4-9}$ analog, ORG 2766) on human performance. In studies with animals, it has been found that these fragments improve acquisition in hypophysectomized rats, but retard extinction performance in intact animals, and attenuate the effects of amnesic agents (see Chapter 3). These neuropeptides, which have virtually no corticotropic effects, seem to affect memory and learning processes. However, the question still remains whether these drugs directly affect learning, or whether they improve performance by increasing attention or motivation. This issue was tested by Sandman, Alexander, and Kastin (1973) in a reversal learning task with rats. It was predicted that improved attention would result in enhanced reversal learning, whereas improved memory would be manifested by impaired reversal learning. The data clearly supported the idea that $ACTH_{4-10}$ enhances attention.

Similarly, in studies with human subjects the question of whether or not these peptides affect general arousal, motivation or attention, as opposed to more specific learning and memory processes, is the main issue in evaluating the behavioral effects of ACTH analogs.

In this chapter, an attempt is made to find out which psychological processes mediate the behavioral effects of ACTH fragments. In Section III, the results of published studies are discussed for each type of task, i.e., perceptual, motor, cognitive, memory, and learning tasks. In reviewing the articles, many methodological difficulties were encountered, so a separate section is dedicated to that issue.

II. METHODOLOGICAL ISSUES

The methodologies used in the 18 studies reviewed here are summarized in Table 1. It is clear that the studies differ in many respects, and one purpose of the present chapter is to show that these factors are important reasons for the inconsistency in findings.

TABLE 1

Studies on the Effects of ACTH Analogs on Human Performance

Author(s)	Dose, route, ACTH analog[a]	TST[b] (min)	Number of subjects[c]	Design[d]	Type of subject
Branconnier et al., 1979	30 mg (s.c.) 4–10	60	18	CO	Cognitively impaired elderly
Bunt & Sanders, 1974	30 mg (s.c.) 4–10	120	9	BS	Students
Dornbush & Nikolovski, 1976	30 mg (s.c.) 4–10	30	10	CO	Male students
Ferris et al., 1976	15, 30 mg (s.c.) 4–10	60	24	CO	Cognitively impaired elderly
Gaillard & Sanders, 1975	30 mg (s.c.) 4–10	30	9	BS	Students
Gaillard & Varey, 1979	5,10,20 mg (or) 2766	90	22	CO	Students
Miller et al., 1974	10 mg (i.v.) 4–10	60	10	BS	Male students
Miller et al., 1976	30 mg (s.c.) 4–10	20	20	CO	Male students
O'Hanlon et al., 1978	40 mg (or) 2766	60	17	CO	Male students
Rapoport et al., 1976	30 mg (s.c.) 4–10	30	10	CO	Hyperactive children
Sanders et al., 1975	30 mg (s.c.) 4–10	60	16	BS	Students
Sandman et al., 1975	15 mg (i.v.) 4–10	0	10	BS	Male students
Sandman et al., 1976	15 mg (i.v.) 4–10	15	10	BS	Mentally retarded males
Sandman et al., 1977	15 mg (i.v.) 4–10	15	11	BS	Male volunteers
Veith et al., 1978	30 mg (s.c.) 4–10	30	12	BS	Female students
Wagenaar, 1977	30 mg (s.c.) 4–10	60	42	BS	Students
Wagenaar et al., 1977	30 mg (s.c) 4–10	60	44	BS	Students
Walker & Sandman, 1979	5, 20 mg (or) 2766	90	8	BS	Mentally retarded adults

[a] s.c. = subcutaneously; or = orally; i.v. = intravenously; 4–10 = ACTH $_{4-10}$; 2766 = ORG 2766.
[b] TST = time since treatment.
[c] Number of subjects per treatment group.
[d] CO = cross-over; BS = between-subject design.

A. Type of ACTH Analog, Route of Administration, and Dose

As indicated in the second column of Table 1, two types of ACTH fragments were used: Met-Glu-His-Phe-Arg-Trp-Gly ($ACTH_{4-10}$; ORG OI 63) and Met(0)-Glu-His-Phe-D-Lys-Phe (ORG 2766). Both peptides were synthesized and provided by Organon (Oss, The Netherlands). In 10 of 19 studies, 30 mg $ACTH_{4-10}$ was administered subcutaneously whereas in three studies, 15 mg was given intravenously by infusion. In one study (Miller, Kastin, Sandman, Fink, & van Veen, 1974), 10 mg was given by intravenous injection. In the other three studies, the longer lasting ORG 2766 was administered in several doses (5–40 mg). At the moment there is no evidence suggesting that route of administration or type of analog is an important determinant of the behavioral effects. In a continuous performance task, the same results were obtained with $ACTH_{4-10}$ (Gaillard & Sanders, 1975) as with ORG 2766 (Gaillard & Varey, 1979). No dosage effects were found in the few studies that examined this issue (Ferris, Sathanathan, Gershon, Clark, & Moskinsky, 1976; Gaillard & Varey, 1979; Walker & Sandman, 1979).

B. Time Since Treatment

As can be seen in the third column of Table 1, testing started in most studies 30–60 min after treatment, when $ACTH_{4-10}$ was administered, and 60–90 min when ORG 2766 was given. When $ACTH_{4-10}$ was given by infusion, testing started either immediately or after 15 min.

C. Type of Subjects

Most studies used students as volunteers, either or both sexes (six studies) or males only (six studies). Only one study was performed with female students. One or two studies were carried out with the following groups: cognitively impaired elderly, mentally retarded subjects, and hyperkinetic children. Although Veith, Sandman, George, and Stevens (1978) suggested that sex-related factors interact with the behavioral effects of ACTH analogs, this notion was not supported by the results of studies that used both males and females within the same experiment (e.g., Branconnier, Cole, & Gordos, 1979; Ferris et al., 1976; Gaillard & Varey, 1979; Wagenaar, Timmers, & Frowein, 1977; Walker & Sandman, 1979). In general, it should be noted that the amount of variability in patient groups is much larger than in student groups (see also Walker & Sandman, 1979). This implies that the number of subjects in patient studies should be much larger to obtain the same results as with students.

D. Type of Test

The variety of tests used is so large that a summary is impossible within the context of the present chapter. For example, to test memory functions, more tests have been used than there are studies. Moreover, there are several types of learning (e.g., concept, functional, verbal, skill learning) and a distinction might also be made between "absent-mindedness" and impaired short-term or long-term memory. Drug research in human subjects usually involves the administration of large batteries of relatively brief tests, designed to assess impairment of a number of psychological functions, such as memory, decision making, and perceptual and motor skills. As has been argued elsewhere (Gaillard, 1980), this approach runs the risk of finding differential effects, which may be interpreted erroneously as specific effects on the psychological mechanisms that the task is presumed to measure. However, tasks not only differ from each other with regard to the psychological functions involved, but in many other aspects as well. Thus, a signal-detection task not only differs from a reaction task in that the subject has to make a judgment in the former and a fast response in the latter task, but tasks may differ in other dimensions as well, and it is difficult to say which of these dimensions is the critical determiner of the treatment effects obtained. Because the effects of ACTH analogs have been investigated in a wide range of tasks without many replications, it is not surprising that so little progress has been made and that the results are often contradictory, or at least inconsistent.

E. Statistical Design

Most investigators are familiar with basic statistical procedures. There are, however, certain issues that have arisen repeatedly in the studies reviewed, and that deserve special mention. The majority of the studies used a between-subject design (see Table 1), but seven studies used a cross-over design. When learning or memory processes are tested in a cross-over design, strong interaction may be found between the first and the second session, a factor that increases the error variance against which the treatment effect is tested. On the other hand the error term in the between-subject design is rather large because of the large variability in test performance of the subjects (which exists prior to treatment). This variance may be reduced by matching the subjects in pairs on the basis of their performance in a pretest (e.g., Gaillard & Varey, 1979; Wagenaar, 1977). Table 1 also indicates the number of subjects in each treatment group. In half of the studies, only 10 subjects or fewer were used. Wagenaar et al. (1977; see also Walker & Sandman, 1979) have pointed out that 10 subjects per treatment group is far too few to be able to find *reliable* effects.

F. Statistics

Another point that hinders the evaluation of the experimental findings is that in about half of the studies, statistics are either used incorrectly or reported insufficiently. Because it is beyond the scope of this review to examine each study in this respect, only the mistakes generally found will be summarized here:

1. In a pre–post and drug–placebo design, only a significant interaction between these two factors can be regarded as a treatment effect. Pre–post comparisons per treatment group, or posttreatment comparisons between treatment groups, are weak tests and, at best, should be regarded as post hoc statistics.

2. Statistical tests (*t*-tests, Mann-Whitney, etc.) that are *not* planned before the experiment, and that are usually carried out when an analysis of variance does not reveal significant effects, can be regarded as predictive, but cannot be taken as evidence of a treatment effect.

3. Similarly, when subjects or conditions are divided into subgroups *after* the experiment, the statistical results are provisional and should be tested in a separate experiment.

4. When a significance level of 5% is used, it should be realized that in one out of 20 statistical tests, the null hypothesis is rejected erroneously. Thus, in most studies, 10 to 20 statistical tests are carried out, making the possibility of finding a significant effect by chance rather high. This picture is aggravated by the general tendency of researchers (and editors) not to publish nonsignificant results.

It is clear from the preceding overview that a systematic review of the results is hardly possible. The variety of tests and experimental procedures used, combined with too few subjects and poor statistics, do not provide sufficient evidence for definitive conclusions. This situation is especially bothersome because only a few tests have been replicated. This was done in most cases by the same research group, although a replication by a different group is to be preferred. Because an effect on a certain test can become significant by chance, only a replication of the effect should be taken as evidence for a behavioral effect of a certain drug.

III. TASK EFFECTS

The behavioral effects of ACTH analogs are discussed separately for the various psychological functions the test or task is presumed to measure. The following summary concentrates on experimental results obtained in tests that were given in more than one study. Tests that were used only once and did not provide significant results are not included.

A. Perceptual Performance

Only a few studies address this issue. Sandman, George, McCanne, Nolan, Kaswan, and Kastin (1977) found that pattern discrimination was improved by ACTH$_{4-10}$, but only with the easier type of stimuli. However, in the same study, performance in a visual detection task was impaired. The value of these observations can be questioned, because the planned interaction between the factors pre–post treatment and drug–control were not reported; nor were the second-order interactions with stimulus intensity (in the detection task) and with easy–difficult stimuli (in the discrimination task). It is possible that perceptual performance of the two treatment groups was already different *before* the subjects were given an infusion with the peptide.

Performance on the rod and frame test (Miller, Kastin, Sandman, Fink, & van Veen, 1974; Miller, Harris, van Riezen, & Kastin, 1976; Sandman et al., 1977; Veith et al., 1978) and on the threshold of critical flicker fusion (O'Hanlon, Fussler, Sancin, & Grandjean, 1978) was not affected by the administration of an ACTH analog.

O'Hanlon et al. (1978) investigated visual detection performance in a vigilance task of long duration (1 hr and 45 min). The effect of ORG 2766 (40 mg) was compared to that of *d*-amphetamine (10 mg) (see Fig. 1). Under placebo treatment, performance deteriorated, whereas the percentage of detected signals could be maintained at the same level when the subjects were given either amphetamine or the ACTH$_{4-9}$ analog.

In summary, there is no evidence for any direct effect of ACTH analogs on

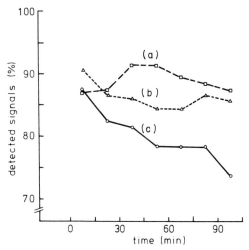

Fig. 1. Mean percentage of detected signals, as a function of time in the vigilance test, separately for (a) *d*-amphetamine, (b) the ACTH$_{4-9}$ analog (ORG 2766), and (c) placebo conditions. (Data derived from O'Hanlon et al. (1978). Reproduced by permission of the authors.)

perceptual processes, except in the O'Hanlon study. As will be discussed in the next section, this convincing result seems to be produced by the long duration of the session. This notion is supported by the fact that no effects of either drug were present in the short pre- and posttests (3.3 min).

B. Motor Performance

We investigated possible beneficial effects of the peptides on motor performance in a skill learning task. This task consists of a six-choice test of self-paced serial reaction where the next stimulus appears immediately after a response to the preceding stimulus has been made. To increase the size of the learning effect, the stimulus and the response key were coupled in an incompatible way. Apart from the effects on learning, we were also interested in possible effects on motivation or sustained attention. Therefore, the subjects worked for 30 min continuously.

This task was used in two separate studies: In the first (Gaillard & Sanders, 1975), the subjects were treated with 30 mg $ACTH_{4-10}$, and in the second (Gaillard & Varey, 1979), with either 5, 10, or 20 mg ORG 2766. Both studies produced essentially the same effects. The results of the second study are presented in Fig. 2. This figure shows the overall average reaction time (RT) and the mean of the fast (first quartile) and the slow (fourth quartile) RTs. Subjects improved more in RT when treated with ORG 2766 than when they received a placebo. The interaction between treatments and the six 5-min periods was significant for the slow RTs, but not for the fast RTs. On the basis of earlier experiments (see Sanders & Hoogenboom, 1970), the fast RTs can be regarded as representing a learning effect and the slow RTs as representing a motivation effect. As is also shown in Fig. 2, the fast RTs become faster during the session because the subjects gain more proficiency in the complex stimulus–response relationships. In contrast, the long RTs become longer due to an increase in the number of lapses of attention. Thus, the data suggest that the peptide does not facilitate skill learning, but counteracts the loss of motivation usually found in continuous performance tasks.

An improvement in RT performance was also found by Miller et al. (1976), using a similar, but simpler continuous performance task. However, a significant effect was found in only one of the four experimental conditions of their task. The results with the long-duration task described previously suggest that Miller and colleagues would have obtained larger effects if they had not introduced 2-min rest periods between successive test periods. As was shown by Sanders & Hoogenboom (1970), such rest periods are sufficient to resolve the mental fatigue induced by the task. Also Branconnier et al. (1979) and Ferris et al. (1976) observed small improvements in the RT performance of the cognitively

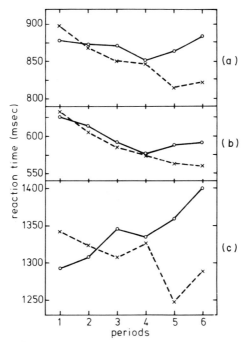

Fig. 2. (a) The overall average reaction times, (b) the mean reaction time of the first quartile, and (c) the mean reaction time of the fourth quartile as a function of periods, separately for the ACTH$_{4-9}$ analog ORG 2766 (x----x), and for the placebo conditions (o——o). (Reproduced from Gaillard & Varey, 1979.)

impaired elderly group. Again, longer test durations may have led to larger effects. This might also be the reason that Rapoport, Quinn, Copeland, & Burg (1976) found no effects on RT performance in hyperactive children.

In summary, as with perceptual performance, motor behavior is only affected in tests of long duration (e.g., 30 min).

C. Cognitive Performance

The results obtained in tests that do not explicitly involve motor, learning, or memory processes are not consistent. On the one hand, a beneficial effect of ACTH$_{4-10}$ was obtained by Miller et al. (1976) in a digit symbol substitution task using male students. On the other hand, Ferris et al. (1976) did not find such an effect in cognitively impaired elderly people.

In other cognitive tests, the results have also been inconsistent (e.g., Ferris et al., 1976; Gaillard & Varey, 1979). Thus, there is no support for the hypothesis that ACTH analogs increase mental capacity or cognitive proficiency.

D. Learning

Tests of verbal learning have generally used a paired-associate paradigm. In such tests, subjects usually learn the relationship between two words or between a letter and a word by rote memorization. Performance in those studies in which this test was used was not affected by $ACTH_{4-10}$ (Ferris et al., 1976; Miller et al., 1976; Rapoport et al., 1976; Sanders, Truijens, & Bunt, 1975; Veith et al., 1978).

Sandman and co-workers investigated the possible effects of $ACTH_{4-10}$ in a concept-learning task. In this task, the subject is trained on a two-choice problem with one relevant dimension (e.g., color) and one irrelevant dimension (e.g., shape). After acquiring this discrimination, the subject learns several shifts in the relevant (and irrelevant) dimensions in succession. Although in an early study (Sandman, George, Nolan, van Riezen, & Kastin, 1975), a trend toward better learning performance was found, this result could not be replicated in later studies, which used either male volunteers (Sandman et al., 1977), female students (Veith et al., 1978), or mentally retarded (Sandman, George, Walker, & Nolan, 1976; Walker & Sandman, 1979). However, a significant effect of ACTH was found in a subproblem analysis of the performance data after the extradimensional shift (Sandman et al., 1976, 1977; Walker & Sandman, 1979). In the extradimensional shift paradigm, one stimulus pair remains the same (unchanged subproblem), whereas the other changes. Subjects treated with an ACTH analog responded to this change in a more adaptive way; that is, the percentage of correct response on the first two trials was larger in the ACTH-treated group than in the placebo group. Sandman and co-workers regard their results as a peptide induced enhancement of attentional processes, and not as an increase in learning capacity.

Wagenaar et al. (1977) studied the effect of $ACTH_{4-10}$ on adaptive learning in a shift-learning task, which is similar to the concept-learning task already discussed. The subject had to learn a mathematical rule. He was presented with a number and had to respond with a corresponding number, which initially he did not know. Feedback was given on each trial by presenting to the subject the difference between his actual response and the correct response. In this way, the subject was able to find the relationship (rule) between stimuli and responses. In the middle of the test (on trial 50 or 51), the rule was changed and the subject had to adapt to the new situation by correcting his response pattern. In three successive experiments, the effect of $ACTH_{4-10}$ was studied on the performance of 88 students. In the first experiment, a facilitating effect was found on adaptation to the shift in mathematical rule. However, this finding could not be replicated in the second and third experiments. According to Wagenaar et al. (1977), the number of subjects (17 per treatment group) in the first study was too small to guarantee a reliable comparison between the groups, given the large variability between subjects.

As is discussed in the section on motor performance, no effects were found on the learning of a perceptual–motor skill in a six-choice RT task (Gaillard & Sanders, 1975; Gaillard & Varey, 1979).

In summary, the results do not provide evidence for the notion that ACTH analogs facilitate learning processes. At best, the results suggest an increase in attention following peptide administration.

E. Memory

Short-term memory has been investigated in a variety of tests; the most widely used have been the Benton visual retention test and the Wechsler memory scale. In the latter test, a beneficial effect of $ACTH_{4-10}$ was found in male students by Miller et al. (1976) and in female students by Veith et al. (1978), although Branconnier et al. (1979) failed to find such an effect in cognitively impaired elderly subjects. The visual retention test, which involves the reproduction of complex geometric figures, was used in six studies. A significant effect was found by Miller et al. (1976), whereas Sandman et al. (1976) found an effect only in a "attentive" and not in a "nonattentive" group of subjects. Because this subdivision in groups was made after the experiment, this result should be regarded as tentative. Negative results were found by Branconnier et al. (1979), Rapoport et al. (1976), Sandman et al. (1977) and Veith et al. (1978).

Dornbusch & Nikolovski (1976) did not find an enhancement of short-term memory in four tests in which the material to be memorized was either auditory, visual–verbal, visual–nonverbal, or bisensory. The issue has also been extensively studied by Ferris et al. (1976) in a sample of 24 cognitively impaired elderly. Using a cross-over design, $ACTH_{4-10}$ administration did not result in an improvement in performance in any of the seven short-term memory tests employed. Negative results were also found by Bunt & Sanders (1974).

Long-term memory was investigated in our Institute (Wagenaar, 1977) in four separate experiments. In all experiments, subjects had to learn two lists of paired associates (letter followed by adjective). In the first experiment, each subject was treated between the learning of the first and the second list and tested on both lists in the same session, and again 1 wk later. Thus, only the second list was acquired under the influence of $ACTH_{4-10}$. There was no treatment effect on either the number of trials needed to reach the criterion or the percentage of items correctly recalled. The aim of the other three experiments was to test long-term retention; therefore an ACTH analog was administered 1 wk after acquisition prior to the recall of both lists. In these experiments, 84 subjects were tested in a between-subject design and no effects were found on the number of items correctly recalled.

In summary, there is no consistent evidence for the notion that ACTH analogs affect either long- or short-term memory.

IV. PHYSIOLOGICAL MEASURES

It has generally been found that basal indices of the autonomic nervous system are not affected by ACTH analogs under resting conditions. This finding has been obtained for heart-rate frequency, cephalic and digital blood pressure, and for plasma control levels (Ferris et al., 1976; Sandman et al., 1975, 1976, 1977; Veith et al., 1978; Walker & Sandman, 1979). However, several studies found that task-induced changes could be modified under the influence of ACTH analogs. Sandman and co-workers (1976, 1977) observed a larger heart-rate deceleration in response to the test stimulus in an orientation sequence. Consistent with this, Miller, Fisher, Groves, Rudeauff, & Kastin (1976), reported a larger reduction in heart-rate variability during extinction in a shock avoidance task. Gaillard & Varey (1979) found, after treatment with ORG 2766, that heart rate remained at a higher frequency in the rest period following 30 min of continuous performance, and furthermore, during task performance, the suppression in heart-rate variability tended to be larger.

A similar effect was obtained by Brunia & van Boxtel (1978) in a two-choice reaction task. In this study, not only heart rate, but also the monosynaptic Achilles tendon reflex, which is assumed to reflect the excitability of moto-neurons, was measured. For both measures, the enhancement induced by the task was larger under $ACTH_{4-10}$ than under the placebo.

Studies that investigated the effects of ACTH analogs on spontaneous EEG under resting conditions have not yielded consistent results. Miller et al. (1974) observed an increase in alpha activity, whereas Branconnier et al. (1979) and Miller et al. (1976) found a decrease. Other studies by Ferris et al. (1976), Fink (cited by Branconnier et al., 1979), and by Sannita, Irwin, & Fink (1976) reported no effect.

Endröczi, Lissák, Fekete, & de Wied (1970) have reported a decrease in the number of synchronized EEG responses to regularly repeated stimuli after treatment with $ACTH_{1-10}$. However, Brunia (personal communication) was unable to replicate this finding with $ACTH_{4-10}$. Miller et al. (1974) observed that $ACTH_{4-10}$ tended to *increase* the amount of alpha activity in the intersti-mulus interval of a disjunctive reaction task. In contrast, O'Hanlon et al. (1978) found that the increase in alpha activity observed under placebo conditions during the 105-min vigilance task was largely attenuated with ORG 2766 and also with an amphetamine.

In summary, the neurophysiological indices seem to yield results similar to the autonomic measures. Only task-induced changes are affected by ACTH analogs. Similar findings have been obtained by Bohus (1975) for phasic changes in the heart rate of rats in an avoidance-learning situation. The increase in heart rate when the rats approach the dark compartment, where they were shocked earlier, is larger after the intake of $ACTH_{4-10}$.

V. SUBJECTIVE RATING SCALES

The effects of ACTH analogs on the subjective state, as indicated by rating scales, have yielded contradictory evidence. Sandman et al. (1975) reported a tendency toward a larger reduction in state anxiety in male students administered $ACTH_{4-10}$, than in the control group. In contrast, Veith et al. (1978) found an increase in trait anxiety in female subjects. Branconnier et al. (1979) suggested that $ACTH_{4-10}$ is capable of improving the moods of cognitively impaired elderly subjects. In their view, this improvement is caused by an increase in nonspecific arousal. Gaillard and Sanders (1975) also obtained evidence for increased arousal as indicated by bipolar rating scales. However, no effects were found on mood scales (Ferris et al., 1976; Gaillard & Varey, 1979) and on trait anxiety (Miller et al., 1974, 1976).

In the vigilance study of O'Hanlon et al. (1978), subjective ratings were collected on a bipolar scale before and after the test session. When ORG 2766 or placebo were administered, subjects felt less aroused after the test session, whereas with d-amphetamine, ratings were not different before and after the session. This is notable because both drugs were equally effective in counteracting a performance decrement and an increase in alpha activity during the vigilance test. In the same study, subjects reported that they were unaware of the fact that ORG 2766 was a psychoactive agent, whereas most of them recognized subjective reactions to the treatment with amphetamine.

In summary, it seems that ACTH analogs, at most, affect the subjective state or the level of arousal only marginally, as measured by subjective rating scales. In contrast to this, when ORG 2766 is administered (sub)chronically it seems to have considerable potential as a mood-lifting agent. It reduces self-rated anxiety and depression and improves observer-rated sociability and ward behavior (see Pigache & Rigter, 1980, for review).

VI. CONCLUSION

The preponderance of the data indicate that ACTH analogs increase the general attention or motivation of human subjects in situations of boredom or mental fatigue. Studies that used prolonged test sessions (Gaillard & Sanders, 1975; Gaillard & Varey, 1979; O'Hanlon et al., 1978) show that ACTH analogs do help the subject to remain concentrated on the task for long periods. However, there is no evidence that these neuropeptides have a direct effect on specific cognitive processes such as memory and learning. The few significant effects found in memory or learning tests could also be caused by increased motivation or attention. The effects on these tests remain small and are therefore only occasionally significant because, in these short-duration tests, motivation and attention play a minor role in determining test performance. Thus, a motivation hypothesis implies that effects are found only when a subject has to work under conditions in which a deterioration in motivation may be expected. However, the behavioral effects are not easily explained by an increase in general arousal, as autonomic and cortical measures are not affected during rest conditions. Nor do rating scales indicate a change in general arousal. Therefore, the psychological process involved could be labeled task-oriented motivation or sustained attention. This terminology indicates that the increase in motivation is goal oriented (i.e., is limited to a specific task situation).

Sustained attention refers to the intensive aspect of attention, i.e., to the observation that subjects are more able to maintain alertness in tasks that normally show a decline in performance as a function of time at work. Just as there are different types of memory and learning, there are also several kinds of attention. Thus, whereas ACTH analogs seem to affect sustained attention, there is no evidence that these neuropeptides facilitate selective attention. This notion has still to be tested in situations with contradictory attentional requirements, as for example, in a dichotic-listening task where messages are simultaneously presented to both ears, while the information coming to only one ear has to be memorized.

Thus, single doses of an ACTH analog seem to have beneficial properties similar to those of stimulants, without the adverse effects commonly observed with, for example, amphetamines: neuropeptides do not affect the adrenal cortex, do not increase motor activity, and there is no increase in autonomic arousal or change in subjective state.

The present results based on acute administration are in general agreement with studies where ORG 2766 is administered daily for 1 or 2 weeks, in that no direct effects on memory and learning were found. However, recent studies using subchronic medication (for review see Pigache & Rigter, 1980) showed that ORG 2766 improves mood, as measured by rating scales. This discrepancy between acute and chronic effects is not surprising, as mood is often slow in responding to drugs, and therefore single doses will not show effects on mood and other dimensions of feeling.

The acute effects reviewed here were interpreted in terms of task-oriented motivation, which is rather similar to the interpretation of animal research data by de Wied (1976). Unfortunately, it is not clear how the acute effects on sustained performance are related to the chronic effects on mood. The only common characteristic seems to be the capacity to tolerate or to cope better with an unpleasant or frustrating environment. On the one hand, motivated people are often better equipped to meet difficult situations and therefore are less likely to become distressed; on the other hand, people in good moods will be better motivated to cope in demanding situations. Further research is necessary to tease out the relations between mood and motivation in this regard.

REFERENCES

Bohus, B. (1975). Pituitary peptides and adaptive automatic responses. *In* W. H. Gispen, Tj. B. van Wimersma Greidanus, B. Bohus, & D. de Wied (Eds.), *Hormones, Homeostasis and the Brain: Progress in Brain Research* (Vol. 42), pp. 275–283. Amsterdam: Elsevier.

Branconnier, R. J., Cole, J. O., & Gardos, G. (1979). $ACTH_{4-10}$ in amelioration of neuropsy-chological symptomatology associated with senile organic syndrome. *Psychopharmacology*, **61**, 161–165.

Brunia, C.H.M., & van Boxtel, A. (1978). $MSH/ACTH_{4-10}$ and task-induced increase in tendon reflexes and heart rate. *Pharmacology, Biochemistry and Behavior*, **9**, 615–618.

Bunt, A. A., & Sanders, A. F. (1974). The effect of $ACTH_{4-10}$ on a serial learning and short term retention task. Soesterberg, Report of the Institute for Perception (TNO), No. 21.

de Wied, D. (1976). Hormone influence on motivation, learning and memory. *Hospital Practice*, 123–131.

Dornbusch, R. L., & Nikolovski, O. (1976). $ACTH_{4-10}$ and short term memory. *Pharmacology, Biochemistry and Behavior*, (Supplement 1), **5**, 69–72.

Endröczi, E., Lissák, K., Fekete, T., & de Wied, D. (1970). Effects of ACTH on EEG habituation in human subjects. *In* D. de Wied & J.A.W.M. Weijnan (Eds.), *Pituitary, Adrenal, and the Brain: Progress in Brain Research*, Vol. 32, pp. 254–262. Amsterdam: Elsevier.

Ferris, S. H., Sathanathan, G., Gershon, S., Clark, C., & Moskinsky, J. (1976). Cognitive effects of $ACTH_{4-10}$ in the elderly. *Pharmacology, Biochemistry and Behavior*, **5**, 73–78.

Gaillard, A.W.K. (1979). The use of task variables and brain potentials in the assessment of cognitive impairment in epileptic patients. *In* B. M. Kulig, H. Meinardi, & G. Stores (Eds.), *Epilepsy and Behavior*, pp. 104–110. Lisse: Swets and Zeitlinger.

Gaillard, A.W.K., & Sanders, A. F. (1975). Some effects of $ACTH_{4-10}$ on performance during a serial reaction task. *Psychopharmacologia*, **42**, 201–208.

Gaillard, A.W.K., & Varey, C. A. (1979). Some effects of an $ACTH_{4-9}$ analog (ORG 2766) on human performance. *Physiology and Behavior*, **23**, 79–84.

Miller, L. H., Fisher, S. C., Groves, G. A., Rudeauff, M. E., & Kastin, A. J. (1977). $MSH/ACTH_{4-10}$ influences of the CAR in human subjects: A negative finding. *Pharmacology, Biochemistry and Behavior*, **7**, 417–419.

Miller, L. H., Harris, L. C., van Riezen, H., & Kastin, A. J. (1976). Neuroheptapeptide influence on attention and memory in man. *Pharmacology, Biochemistry and Behavior*, (Supplement 1), **5**, 17–21.

Miller, L. H., Kastin, A. J., Sandman, C. A., Fink, M., & van Veen, W. J. (1974). Polypeptide influences on attention, memory and anxiety in man. *Pharmacology, Biochemistry and Behavior*, **2**, 663–668.

O'Hanlon, J. F., Fussler, C., Sancin, E., & Grandjean, E. P. (1978). Efficacy of an $ACTH_{4-9}$ analog, relative to that of a standard drug (d-amphetamine), for blocking the vigilance decrement in men. Zürich, Report of the Swiss Federal Institute of Technology.

Pigache, R. M., & Rigter, H. (1980). *In* L. H. Rees & Tj. B. van Wimersma Greidanus (Eds.), *Frontiers in Hormone Research*. Basel: Karger.

Rapoport, J. L., Quinn, P. O., Copeland, A. P. & Burg, C. (1976). $ACTH_{4-10}$: Cognitive and behavioral effects in hyperactive, learning-disabled children. *Neuropsychobiology*, **2**, 291–296.

Sanders, A. F., & Hoogenboom, W. (1970). On the effects of continuous active work on per-formance. *Acta Psychologica*, **33**, 414–431.

Sanders, A. F., Truijens, C. L., & Bunt, A. A. (1975). $ACTH_{4-10}$ and learning. Soesterberg, Report of the Institute for Perception (TNO), No. 4.

Sandman, C. A., Alexander, W. D., & Kastin, A. J. (1973). Neuroendocrine influences on visual discrimination and reversal learning in the albino and hooded rat. *Physiology and Behavior*, **11**, 613–617.

Sandman, C. A., George, J., McCanne, T. R., Nolan, J. D., Kaswan, J., & Kastin, A. J. (1977). MSH/ACTH$_{4-10}$ influences behavioral and physiological measures of attention. *Journal of Clinical Endocrinology and Metabolism*, **44**, 884–890.

Sandman, C. A., George, J. M., Nolan, J., van Riezen, H., & Kastin, A. J. (1975). Enhancement of attention in man with ACTH/MSH$_{4-10}$. *Physiology and Behavior*, **15**, 427–431.

Sandman, C. A., George, J., Walker, B. B., & Nolan, J. D. (1976). Neuropeptide MSH–ACTH$_{4-10}$ enhances attention in the mentally retarded. *Pharmacology, Biochemistry and Behavior*, (Supplement 1), **5**, 23–28.

Sannita, W. G., Irwin, P., & Fink, M. (1976). EEG and task performance after ACTH$_{4-10}$ in man. *Neuropsychobiology*, **2**, 283–290.

Veith, J. L., Sandman, C. A., George, J. M., & Stevens, V. C. (1978). Effects of MSH/ACTH$_{4-10}$ on memory, attention and endogenous hormone levels in women. *Physiology and Behavior*, **20**, 43–50.

Wagenaar, W. A. (1977). ACTH fragments and verbal learning. Soesterberg, Report of the Institute for Perception (TNO), No. 15.

Wagenaar, W. A., Timmers, H., & Frowein, H. (1977). ACTH$_{4-10}$ and adaptive learning. Soesterberg, Report of the Institute for Perception (TNO), No. 13.

Walker, B. B., & Sandman, C. A. (1979). Influences of an analog of the neuropeptide ACTH$_{4-9}$ on mentally retarded adults. *American Journal of Mental Deficiency*, **83**, 346–352.

PART II

Endorphins

CHAPTER 10

Cellular Distribution and Function of Endorphins

Floyd E. Bloom and Jacqueline F. McGinty

Since the discovery of the endorphin peptides, much information about their functional roles and interrelationships has emerged. It now seems that β-endorphin-containing cells in brain exist independently from enkephalins and from pituitary endorphins. Furthermore, there appear to be subclasses of opioid receptors that may be peptide-specific. Studies of central nervous system (CNS) enkephalin content indicate that Met-enkephalin concentrations are 5–10 times higher than Leu-enkephalin concentrations, and that enkephalin concentrations are low in cortical regions, high

199

ENDOGENOUS PEPTIDES
AND LEARNING AND MEMORY PROCESSES

in the diencephalon, and highest of all in the striatum in most common laboratory species. At the present time, five potential enkephalin-containing circuits have been proposed:

1. interneurons in the dorsal horn of the spinal cord;
2. innervation of the globus pallidus from cell bodies in the caudate and putamen;
3. innervation of the bed nucleus of the stria terminalis from cell bodies in the central nucleus of the amygdala;
4. innervation of the neurohypophysis by cell bodies in the paraventricular and supraoptic nuclei;
5. a projection to the spinal cord from paraolivary cells in the medulla.

Unfortunately, there are discrepancies between enkephalin immunoreactivity and autoradiographically detected opiate binding sites in cortex, striatum, amygdala, and spinal cord. β-Endorphin has a different distribution, with cells in the tuberal zone of the hypothalamus innervating the lateral hypothalamus, preoptic area, medial amygdala and the midline of the thalamus and brainstem (periventricular system) as far caudal as the locus coeruleus.

Electrophysiologically, opioid peptides and opiate alkaloids generally naloxone reversibly inhibit single-unit discharge. Important exceptions are naloxone reversible excitatory effects in the hippocampus and Renshaw cells of the spinal cord. Hierarchically, the β-endorphin system may be a component of the endocrine-peptidergic network in the CNS, whereas enkephalins may be local circuit mediators of synaptic events. Biochemically, it may be the case that the function of opioids is to prevent the action of another transmitter substance, such as a monoamine, glutamate, or acetylcholine.

I. INTRODUCTION

An explosion of recent scientific reports has dealt with the endorphin peptides (Childers, Schwarcz, Coyle, & Snyder, 1978; Kosterlitz & Hughes, 1978; Terenius, 1978a,b), the nonpeptidic endorphins (Gintzler, Gershon, & Spector, 1978; Terenius, 1978a,b), the actions of opiates (Bloom, Rossier, & Battenberg, 1978b; Zieglgänsberger, Siggins, French, & Bloom, 1978), and the effects of these materials on the radio-receptor displaceable opioid ligand binding assays (Hollt & Herz, 1978; Simon & Hiller, 1978). These substances have become the subject of one of the most rapidly expanding chapters in the history of chemical neurotransmitters, including earlier explosions that dealt with the discovery of hypothalamic hypophysiotropic peptides (see Guillemin, 1978; Schally, 1978). The following detailed survey indicates the current status of cellular level research on the distribution and function of these peptides. Several detailed reviews are also available for further pursuit (Bloom, Battenberg, Rossier, Ling, & Guillemin, 1978a; Childers et al., 1978; Goldstein, 1976; Simon & Hiller, 1978; Snyder, 1978; Terenius, 1978a,b). Elsewhere in this book we consider possible actions of endorphins and opiates on behavior (see Chapter 12).

At the time of the original molecular identification of Met⁵-enkephalin (M-e) and Leu⁵-enkephalin (L-e) (Hughes, Smith, Kosterlitz, Fothergill, Morgan, & Morris, 1975), the possibility of one or more other endorphins of pituitary origin had already been repeatedly suggested (Cox, Opheim, Tesche-macher, & Goldstein, 1975; Goldstein, 1976; Graf, Ronai, Bajusz, Cseh, & Szekely, 1976a; Ross, Dingledine, Cox, & Goldstein, 1977; Teschemacher, Opheim, Cox, & Goldstein, 1976). When sequencing studies of the purified M-e revealed it to be the NH_2-terminal pentapeptide (Hughes et al., 1975) of the erstwhile pituitary hormone β-lipotropin (β-LPH) (Li, 1964; Li, Barnafi, Chretien, & Chung, 1965; Li & Chung, 1976; see the sequence in Fig. 1), the possibility that β-LPH was the prohormone of pituitary M-e was temporarily viable. That possible relationship appeared to be strengthened by the subsequent isolation, purification, sequencing and synthesis of α-endorphin (A-E) (Guil-lemin, Ling, & Burgus, 1976; Ling, 1977; Ling, Burgus, & Guillemin, 1976), β-endorphin (B-E) called also C-fragment (Bradbury, Feldberg, Smyth, & Snell, 1976a; Bradbury, Smyth, Snell, Birdsall, & Hulme, 1976b; Bradbury, Smyth, Snell, Deaken, & Nendlandt, 1977; Doneen, Chung, Yamashiro, Law, Loh, & Li, 1977; Dragon, Seidah, Lis, Routhier, & Chretien, 1977; Graf et al., 1976a; Li et al., 1976; Li, Yamashiro, Tseng, & Loh, 1977; Ling & Guillemin, 1976; Li & Chung, 1976), γ-endorphin (G-E) (Ling, 1977; Ling et al., 1976; Ling & Guillemin, 1976) and δ-endorphin (D-E) called also C'-fragment (Graf et al., 1976a; Smyth & Snell, 1977). All of these fragments of β-LPH were found in extracts of brain and pituitary, exhibited some action as specific opioid agonists, and contained M-e as their NH_2-terminal pentapeptide (see Fig. 1 for structures).

When subsequent tests *in vitro* (Bradbury et al., 1976b; Doneen et al., 1977; Lazarus, Ling, & Guillemin, 1976) and *in vivo* (Bloom, Segal, Ling, & Guil-lemin, 1976; Bradbury et al., 1976a, 1977; Graf, Szekely, Ronai, Dunai-Kovacs, & Bajusz, 1976b; Loh, Tseng, Wei, & Li, 1976) revealed that B-E was by far the most potent and longest acting of the natural peptides, some

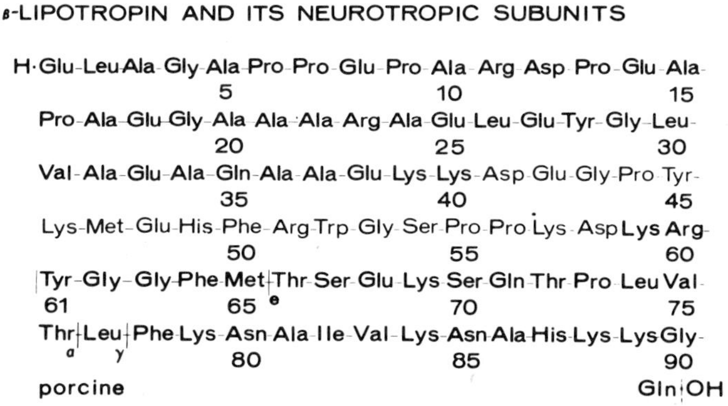

β-LIPOTROPIN AND ITS NEUROTROPIC SUBUNITS

```
H·Glu–Leu Ala–Gly Ala Pro Pro Glu Pro Ala Arg Asp Pro Glu Ala
                5                    10                    15
      Pro Ala Glu Gly Ala Ala Ala Arg Ala Glu Leu Glu Tyr Gly Leu
                20                    25                    30
      Val Ala Glu Ala Gln Ala Ala Glu Lys Lys Asp Glu Gly Pro Tyr
                35                    40                    45
      Lys Met Glu His Phe Arg Trp Gly Ser Pro Pro Lys Asp Lys Arg
                50                    55                    60
     |Tyr Gly Gly Phe Met|Thr Ser Glu Lys Ser Gln Thr Pro Leu Val
      61        65  ᵉ             70                    75
     Thr|Leu|Phe Lys Asn Ala Ile Val Lys Asn Ala His Lys Lys Gly
      ᵟ    ᵞ        80                    85                    90
     porcine                                         Gln|OH
```

Fig. 1. Sequence of "pro-opiocortin," the 31K MW molecule containing lipotropin, endorphin, ACTH, MSH, and enkephalin fragments.

workers concluded that the transient opioid actions of M-e and L-e indicated that these substances were merely weakly active breakdown products of the naturally active hormone, B-E (Bradbury et al., 1977; Feldberg & Smyth, 1977; Goldstein, 1976; Smyth & Snell, 1977). Others interpreted the same data to mean that the natural putative "neurotransmitter" opioid peptides were the succinctly acting M-e and L-e (Urca, Frenk, Leibeskind, & Taylor, 1977; Volavka, Marya, Baig, & Perez-Cruet, 1977), whereas B-E was regarded exclusively as a pituitary product whose longer duration of action arose from proteolytic protection afforded by the greater length of its peptide chain. Curiously, this greater length did not improve the potency or duration of A-E, D-E, or G-E (Geisow, Deakin, Dostrovsky, & Smyth, 1977). Nevertheless, all workers seemed agreeable to the idea that opiate receptors in innervated tissues really represented the natural receptors to the endorphins and enkephalins.

Better definition of the functional roles and relations among these peptides has now emerged following the development of: (1) perfected methods for optimal preservation and extraction (Guillemin, Ling, & Vargo, 1977; Kobayashi, Palkovits, Miller, Chang, & Cuatrecasas, 1978; Krieger, Liotta, Suda, Palkovits & Brownstein, 1977; Rossier, Bayon, Vargo, Ling, Guillemin, & Bloom, 1977a; Rossier, Vargo, Minick, Ling, Bloom, & Guillemin, 1977b; Simantov & Snyder, 1976a; Teschemacher et al., 1976) of the individual peptides; (2) specific antisera for RIA (see Guillemin et al., 1976) and for immunocytochemical localization of their storage sites in brain and pituitary (Bloom, Battenberg, Rossier, Ling, Leppaluoto, Vargo, & Guillemin, 1977a; Bloom, Rossier, Battenberg, Vargo, Minick, Ling, & Guillemin, 1977b; Bloom et al., 1978a; Elde, Johansson, Terenius, & Stein, 1976; Hökfelt, Elde, Johansson, Terenius, & Stein, 1977a; Hökfelt, Ljundahl, Terenius, Elde & Nilsson, 1977b; Watson, 1978; Watson, Akil, Sullivan, & Barchas, 1977; Watson, Akil, Richards, & Barchas, 1978a; Watson, Richards, & Barchas, 1978b); and (3) the functional analysis of the most appropriate peptides on the correct target cells (Davies & Dray, 1978; Davies & Duggan, 1974; Duggan, Hall, & Headley, 1976b; Hill, Mitchell, & Pepper, 1976; Nicoll, Siggins, Ling, Bloom, & Guillemin, 1977; Segal, 1977) usually by microiontophoresis. The results of such studies lead to the conclusion that B-E-containing cells exist independently from enkephalins and from pituitary endorphins (Austen & Smyth, 1977; Bloom et al., 1977a,b; Bloom et al., 1978a,b; Teschemacher et al., 1976). Furthermore, comparison of the actions of B-E with those of enkephalin on central and peripheral receptors has led to the postulation that there may not simply be a single monolithic class of endorphin receptors used by all the peptides, but rather subclasses of opioid receptors that may be peptide-specific (Henriksen, Bloom, McCoy, Ling, & Guillemin, 1978a; Henriksen, McCoy, French, & Bloom, 1978b; Lord, Waterfield, Hughes, & Kosterlitz, 1977).

Much recent work attests to the view that M-e is, in fact, not derived from B-E, and that L-e may also have its own precursor peptide that is completely unrelated to the molecules of the β-LPH-derived endorphin series (Goldstein, Tachibana, Lowney, Hunkapiller, & Hold, 1979; Kangawa & Matsuo, 1979; Stern, Lewis, Kimura, Rossier, Gerberg, Brink, Stein, & Udenfriend, 1979). In addition, a very active opioid peptide, dynorphin, isolated from pituitary sources by Goldstein and associates, contains the L-e pentapeptide at its NH_2-terminus, with at least eight more amino acid residues extending from the Leucine in the C-terminal direction (Goldstein et al., 1979). Although the full sequence of dynorphin has not yet been reported, current evidence suggests that, because of its greater potency than L-e on the guinea pig ileum and mouse vas deferens assays, dynorphin is unlikely to be a simple precursor of L-e. Whether the reduced activity of L-e indicates instead that the short peptide is more properly regarded as a by-product of dynorphin remains to be determined. However, preliminary tests in our laboratory in which dynorphin is applied by iontophoresis to cells of the hippocampus indicate that the longer peptide lacks the potent excitatory action (Henriksen & Bloom, unpublished data) produced by enkephalins and by B-E (French & Siggins, 1980; Nicoll et al., 1977; Taylor, Hoffer, Zieglgänsberger, & Siggins, 1979). The latter observation suggests, from yet another direction, the potential diversity of central endorphin receptor functions of possible relevance to eventual studies on behaviors.

II. DISTRIBUTION OF ENKEPHALINS

Using antisera developed independently in several laboratories against NH_2-terminally conjugated enkephalins, a series of progressively refined reports have described the quantitative and cytological distribution of these pentapeptides in the nervous system and other tissues.

A. CNS Enkephalin Content

Distributional studies on the content of enkephalins in brain were reported (Hughes, 1975; Pasternak, Simantov, & Snyder, 1976; Simantov & Snyder, 1976a,b) before the availability of the specific radioimmunoassay (RIA). These data, obtained through the use of radioreceptor displacement assays on non-purified extracts, provided numbers that can now be considered of doubtful significance. The original methods, in retrospect, gave poor recovery of endogenous peptides due both to incomplete inactivation of degrading peptidases or to suboptimal extraction procedures (Rossier et al., 1977a).

Simantov and Snyder (1976a,b) were the first to report the results of a regional assessment of rat brain with an enkephalin radioimmunoassay in which the antisera showed at least 10% cross-reactivity between L-e and M-e immunogens. These values were tenfold less than reported by Yang, Hong, and Costa (1977) with results on rats killed with focused microwave irradiation. Both of these studies indicated that the amounts of M-e were some 5–10 times greater than L-e, that enkephalins were generally low in cortical regions and highest in diencephalic regions, and that they were highest of all in the corpus striatum (Hong, Yang, & Costa, 1977a). Similar results were also obtained by others (Gillin, Hong, Yang, & Costa, 1978; Gros, Pradelles, Rougeot, Bepoldin, Dray, Fournie-Zaluski, Roques, Pollard, Llorens-Cortes, & Schwartz, 1978b; Miller, Chang, Cooper, & Cuatrecasas, 1978) but not by Simantov and Snyder (1976a,b) who reported that, in cow brain, the ratio of M-e:L-e was 0.1. More recent studies have shown that the ratio of M-e:L-e in cows is about 5–10 (Lewis, Stein, Gerber, Rubinstein, & Udenfriend, 1978) as it is in the more common laboratory species such as rat and rabbit. Methods in which L-e and M-e are separated by various chromatographic methods and then independently assayed by radioimmunoassay (Bayon, Rossier, Mauss, Bloom, Iversen, Ling, & Guillemin, 1978; Hong, Yang, Fratta, & Costa, 1978), by radioreceptor assay (Lewis et al., 1978), or by biological assay (Hughes, Kosterlitz, & Smith, 1977) yield similar results. Alternatively, M-e activity can be destroyed by treatment with cyanogen bromide, leaving L-e values unaffected. Thus, by differential assay before and after treatment with cyanogen bromide, values for both M-e and L-e may be obtained (Hughes et al., 1977).

The estimates of enkephalin content by RIA are in general agreement with the recent quantitative regional and histological estimates of the distribution of the "opiate receptors" (Atweh & Kuhar, 1977a,b,c). Nevertheless, the distribution of the enkephalin and B-E systems does not always parallel the distribution of opiate receptors (see what follows).

B. CNS Enkephalin Immunohistochemistry

Although there have been further detailed studies on enkephalin-containing regions (Kobayashi et al., 1978), the best guide to the distribution of enkephalin content in brain has been derived from immunohistochemistry. All of the studies describing immunocytochemical distribution patterns largely agree on the location of nuclear groups exhibiting nerve terminals in untreated rats, and immunoreactive (ir) perikarya in rats pretreated with colchicine (Sar, Stumpf, Miller, Chang, & Cuatrecasas, 1978; Simantov, Kuhar, Uhl, & Snyder, 1977; Uhl, Goodman, Childers, & Snyder, 1978a; Wamsley, Young, & Kuhar, 1980). Such general agreement eliminates the necessity for dwelling on the possibility that slight differences in immunogens, in tissue fixation, antibody detection, or other covert causes of artifactual result have any significant bearing on the results. All of the results are to be considered as specific for enkephalins, although none per se is fully capable of absolute discrimination between M-e and L-e. The reasons underlying the reproducible finding that colchicine pretreatment facilitates the localization of enkephalin-containing perikarya remain unclear; the unproven assumption is that by disaggregating microtubules and depressing cellulofugal transport of stored peptide, immunoreactive materials accumulate within the perikaryon. However, we find that enkephalin immunoreactive perikarya are visible in avian brains by standard methods and without using colchicine (Bayon, Koda, Battenberg, Azad, & Bloom, 1980a).

If observations are restricted to colchicine pretreated rats, enkephalin immunoreactive cells are detected throughout the CNS (Hökfelt, Elde, Johansson, Terenius, & Stein, 1977a; Sar et al., 1978; Uhl et al., 1978a; Wamsley et al., 1980; and our own results with antisera donated by R. J. Miller, University of Chicago). Few specific pathways have linked immunoreactive perikarya with specific sets of nerve terminals, except for the studies that will be discussed separately.

In the telencephalon, enkephalin-reactive neurons are found in the nucleus accumbens, septum, the caudate nucleus and putamen (especially at the caudal pole), the interstitial nucleus of the stria terminalis, and the central nucleus of the amygdala. Nerve terminal-like patterns of immunoreactivity are densest in the globus pallidus and central nucleus of the amygdala, and show generally heavy innervation of the nucleus accumbens and the bed nucleus of the stria terminalis. Occasional patches of intense fiber staining around pyramidal cells in the posterior cingulate gyrus and in restricted patches within the hippocampus have also been described (Sar et al., 1978).

In the diencephalon, immunoreactive perikarya are found in the magnocellular supraoptic and parvocellular regions of the paraventricular nuclei in the anterior hypothalamus, in scattered small cells in the perifornical, dorsomedial, ventromedial, and arcuate nuclei of the midhypothalamus, and in the mamillary nuclei of the posterior hypothalamus. Fibers have been seen in almost all hypothalamic nuclei, but are especially dense in the periventricular area, in the external layer of the median eminence, and in the ventromedial nucleus. Enkephalin is less densely distributed in the thalamus than in the hypothalamus. Most of the thalamic enkephalin lies within fibers of the midline and intralaminar nuclei; however, both small cells and fibers are found in the paraventricular thalamic nuclei. In the epithalamus, both habenular nuclei contain enkephalin fibers.

In the brainstem, enkephalin immunoreactive perikarya have been reported in the interpeduncular nucleus, ventral lateral geniculate body, dorsal cochlear nucleus, medial vestibular nucleus, and the spinal vestibular nucleus, as well as in isolated small neurons throughout the pontine central grey and within the general confines of the reticular formation. Nerve fibers are found throughout the central grey as well, and specifically innervate those nuclei made popular by their monoamine content: the zona compacta dopamine cells of the substantia nigra, the noradrenergic neurons of the locus coeruleus, and the serotonergic neurons of the midline raphé nuclei. Reactive fibers penetrate into the fourth, seventh, tenth, and twelfth cranial nerve motor nuclei as well as the nucleus tractus solitarius, and the spinal portion of the trigeminal nucleus and the substantia gelatinosa. The brain region containing the lowest (ir) enkephalin levels is the cerebellum; however, a few cerebellar enkephalinergic Golgi cells have been reported by Sar et al. (1978). The distribution of these Golgi cells and the presence of enkephalinergic (ir) mossy fibers in the cerebella of different species have recently been characterized by Schulman, Finger, Brecha, and Karten (in preparation).

In the spinal cord, enkephalin reactive fibers are most pronounced in the dorsal Laminae I, II, and III where some immunoreactive perikarya can also be seen. The ventral horn and central gray regions are less densely innervated by enkephalin fibers.

C. Possible Enkephalin Circuits in CNS

Although the technology for tracing connectivity patterns of neuronal circuits is currently undergoing a renaissance due to the advent of new orthograde and retrograde markers, only five possible enkephalin-containing circuits have as yet been proposed from the detailed descriptions of cells and fibers that are immunoreactive. These proposed circuits have been based either on the results of large transecting lesions or retrograde tracing; none have been tested physiologically.

1. Hökfelt et al. (1977b) proposed on the basis of dorsal rhizotomy and total spinal transections that the enkephalin reactive cells responsible for the immunoreactive fibers of the dorsal horn are primarily spinal interneurons.

2. Cuello and Paxinos (1978) also used transections to propose that the enkephalinergic innervation of the globus pallidus may arise from the reactive neurons of the caudate nucleus and putamen, a conclusion that had been suggested earlier by radioimmunoassay results (Hong, Yang, Fratta, & Costa, 1977b).

3. Uhl, Kuhar, and Snyder (1978b) proposed, on the basis of transection studies, that the enkephalinergic innervation of the bed nucleus of the stria terminalis arises from the central nucleus of the amygdala and travels through the stria terminalis, a pathway first described by De Olmos (1972) without

chemical designation. However, after transections of the stria terminalis, Gros, Pradelles, Dray, LeGal La Salle, and Ben-Ari, (1978a) were unable to observe any decrease of (ir) M-e levels in the bed nucleus of the stria terminalis.

4. Immunocytochemical retracing of pathways already known to exist also underlies our proposal that the enkephalinergic neurons of the paraventricular and supraoptic nuclei provide the enkephalin reactive fibers to the neural lobe. Changes in these fibers and neural lobe enkephalin content have been shown to parallel functional or genetic changes in vasopressin content, leading to the proposal that the hypothalamic–pars nervosa enkephalinergic pathway may modulate neurohypophysial neurosecretion (Rossier, Battenberg, Pittman, Bayon, Koda, Miller, Guillemin, & Bloom, 1979).

5. Using a fluorescent retrograde tracer in combination with immunohistochemistry, Hökfelt, Terenius, Kuypers and Dann (1979) showed that the large paraolivary cells in the medulla that contain enkephalin project to the spinal cord.

D. Enkephalin Fibers and Opiate Receptors

These general immunocytochemical results can be compared with the distribution of the chemically detected or autoradiographically detected opiate binding sites. As acknowledged by Simantov et al. (1977), there are at least five major discrepancies that remain unexplained:

1. receptors are very dense in caudate and putamen, but not in globus pallidus where the heaviest fibers are seen;

2. the caudate and putamen are perplexingly slight in immunoreactive fibers although some perikarya are seen here after colchicine treatment;

3. cerebral cortex that has receptors if studied by binding and electrophysiological techniques (Zieglgänsberger & Tulloch, 1979) shows almost no immunoreactivity;

4. although nerve cells and fibers are exceedingly dense in the central nucleus of the amygdala, receptors are distributed more or less evenly throughout the entire amygdaloid complex; and

5. within the spinal cord, the fibers of the ventral gray regions are not represented by equivalent opiate receptor values.

Finally, we found that the colchicine injections required to demonstrate enkephalin reactive perikarya in rodents result in a redistribution of enkephalin content in only one area (Bayon et al., 1980b). Therefore, at least in the rodent brain, there is no reason to suspect enkephalin (ir) is an artifact of the disrupted microtubular transport system.

III. DISTRIBUTION OF β-ENDORPHIN

A. CNS β-Endorphin Content

Because B-E is the COOH-terminal 31 amino acid fragment of β-LPH, most antisera to B-E will also bind equimolar ratios of B-E, β-LPH, and the 31,000 molecular weight (MW) prohormone, pro-opiocortin (Mains, Eipper, & Ling, 1977; Roberts & Herbert, 1977; Rossier et al., 1977a). However, separation of the immunoreactive components can be achieved by gel filtration (Rossier et al., 1977a,b; Smyth, Snell, & Massey, 1978; Zakarian & Smyth, 1979). Consistently, two peaks of immunoreactivity have been resolved from extracts of rat brain. One peak coincides precisely with the location of synthetic B-E; the other peak elutes in a broad zone of larger molecular weight (10,000–30,000), which does not coincide closely with the elution pattern of either β-LPH or the 31,000 MW prohormone (also see Lewis, Stern, Kimura, Rossier, Stein, & Udenfriend, 1980).

When the same brain regions are compared for B-E and enkephalin content, the two classes of opioid peptides vary independently from region to region (Rossier et al., 1977b). Globus pallidus, caudate nucleus, and more caudal brainstem structures that contain enkephalin have virtually no B-E. Thus, these data strongly indicate that B-E and enkephalin are segregated within different neuronal systems in the brain.

A controversial aspect of the distribution of endorphins is associated with the relationship between pituitary endorphins and brain endorphins. Moldow and Yalow (1978) have suggested that brain ACTH arises, not from brain neurons, but diffuses retrogradely from the pituitary (also see Ogawa, Panerai, Lee, Forsbach, Haulicek, & Friesen, 1979). This view stems from the correlation between ACTH tissue content and distance from the basal hypothalamus and infundibulum, and from the finding that certain commercially prepared hypophysectomized animals still exhibit ACTH immunoreactivity in plasma and in tissue scrapings from the sella turcica. Presumably, this speculation on the nature of the brain ACTH source may also be leveled at the brain's B-E content if it could be firmly established that B-E and ACTH were made in (and released simultaneously from) the same nerve cells. We recently investigated this problem once again to determine the validity of our earlier observations that brain and pituitary endorphin content were unrelated (Rossier et al., 1977a,b). That conclusion was based on results obtained in rats that were maintained for several months after commercial hypophysectomy, and also on the basis of endocrine manipulations found to result in increases or depletions of adenohypophyseal endorphin and ACTH content. Devoting special attention to the issue of basal secretion in animals that showed no plasma B-E or ACTH, and no detectable increase after ether stress, we also found no detectable ACTH or B-E immunoreactivity in scrapings of the sella turcica. Yet, these animals showed a distribution of brain B-E and ACTH immunoreactivity that was indistinguishable from that of normal rats (Guillemin et al., unpublished observations). Therefore, the intrinsic neurons of the basal hypothalamus are the most logical source of brain endorphin, and are independent of the pituitary.

B. CNS β-Endorphin Immunohistochemistry

Using the well-characterized anti-B-E serum RB 100 of Guillemin et al. (1977), we examined in detail the distribution of B-E immunoreactivity in the rat and mouse CNS (Bloom et al., 1978a). Even in untreated rats, perikarya and neuronal processes exhibit immunoreactivity that is most pronounced within the diencephalon. The immunoreactive neurons exist exclusively within two adjacent clusters in the tuberal zone of the hypothalamus and extend from dorsolateral portions of the middle of the arcuate nucleus anterolaterally below the ventromedial nucleus, reaching almost to the lateral border of the hypothalamus (Fig. 2). No other immunoreactive β-endorphin cells or enkephalin immunoreactive neurons are observed, even after colchicine pretreatment. Extensive processes can be traced away from the B-E immunoreactive neurons, some extending laterally into the medial amygdala, but most extending into the anterior hypothalamic–preoptic area, where they turn dorsally at the level of the anterior commissure to project caudally along the dorsal midline of the thalamus into the brainstem. Fibers extend into the central grey within the dorsal longitudinal fasciculus moving laterally toward the locus coeruleus. Caudal to the locus coeruleus, the fibers become extremely difficult to detect. Specific nuclei in which there appear to be extensive arborization of B-E immunoreactive fibers are the periventricular, paraventricular, ventromedial, and dorsomedial nuclei of the hypothalamus, the paraventricular nucleus of the thalamus, the medial amygdala, septum, and the bed nucleus of the stria terminalis (BNST). The dorsal raphé and locus coeruleus nuclei are both innervated, but the substantia nigra receives only a few fibers in the most lateral zone.

Patterns of immunoreactivity with antisera to unconjugated human β-LPH (Watson et al., 1977) are similar, if not identical, to those observed by us with RB 100 (Bloom et al., 1978a) and with anti-ACTH serum (from D. Krieger, New York, R. DiAugustine, NIEHS, or R. Benoit, Salk Institute). These data may be most conservatively explained, as we and others have proposed (Bloom et al., 1978a; Watson et al., 1978): namely, that β-LPH, B-E, and ACTH (or α-MSH) may well be present within the same neurons of the rat CNS as they are known to coexist within the cells of the intermediate lobe and within

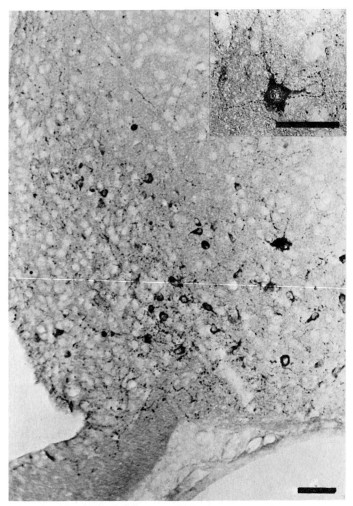

Fig. 2. β-Endorphin immunoreactive cells in the rat medial basal hypothalamus. Insert: high power view of β-endorphin cell. Bar = 50 μm.

the corticotropic cells of the adenohypophysis. The cells and fibers observed in the human hypothalamus with antisera to human β-LPH (Pelletier, Desy, Lissitszky, Labrie, & Li, 1978) are consistent with this idea as well. Further support is derived from studies of the 31,000 MW precursor, pro-opiocortin, which contains β-LPH, B-E, and ACTH (Mains et al., 1977; Nakanishi, Inoue, & Kita, 1979; Roberts & Herbert, 1977). Nevertheless, the answer must be regarded as incomplete pending more detailed studies of the precise patterns and ultrastructural details of the neuronal elements stained with each of these antisera (see McGinty & Bloom, 1980).

C. Morphological Distinctions between Endorphin and Enkephalin Neurons

The neurons containing β-endorphin and enkephalin have distinct morphological features that may be of functional significance. Table 1 lists the characteristics that distinguish the β-endorphin system from the enkephalin system in the CNS.

As already indicated, β-endorphin and enkephalin are synthesized from different precursor molecules. The 31KMW pro-opiocortin precursor peptide contains β-LPH, β-endorphin, ACTH, α-, β-, and γ-MSH (Nakanishi et al., 1979). All these peptides are found in corticotropic and intermediate lobe pituitary cells and in the same cluster of neurons found only in the medial basal hypothalamus (MBH) (Bloom et al., 1978a; Bloom, Battenberg, Shibasaki, Benoit, Ling, & Guillemin, 1980; Watson et al., 1978). Whereas enkephalin immunoreactive cells are widespread throughout the nervous system, opiocortin (ir) neurons comprise a relatively homogeneous cell type, fusiform in shape and approximately 10–20 μm in diameter, close to the third ventricle (Fig. 2—insert). There is enough peptide in the opiocortin perikarya to be visualized without an axoplasmic transport blocker like colchicine. In contrast, colchicine is needed in mammals to visualize enkephalin perikarya that are heterogeneous in shape and size (Fig. 3). Enkephalin cells are primarily small, local circuit neurons (Fig. 3C) with very fine axons, although the enkephalinergic paraolivary cells in the medulla are large (Fig. 3A) and project to the spinal cord (Hökfelt et al., 1979). The functionally heterogeneous neuronal systems containing enkephalin stand in marked contrast to the periventricular system that contains the opiocortin neurons and other neuroendocrine related peptidergic neurons.

TABLE 1

Morphological Distinctions between Opioid Systems

Endorphin	Enkephalin
Pro-opiocortin precursor	Multiple putative precursors
Single neuronal population	Multiple cell body populations
Homogeneous cell size and shape	Heterogeneous cell sizes and shapes
Perikarya seen without colchicine	Perikarya seen with colchicine
Large, varicose fibers	Fine, varicose fibers
Long projection system	Multiple short projecting or local circuit systems
Periventricular distribution	Heterogeneous distributions
Endocrine relationships	Multiple functional relationships

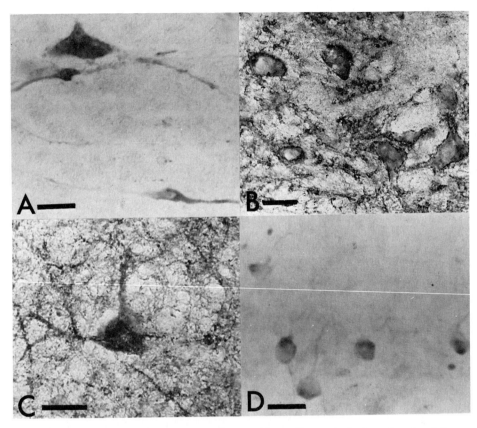

Fig. 3. Enkephalin immunoreactive cells in rat. (A) paraolivary reticular formation of medulla; (B) globus pallidus; (C) ventral medial nucleus of hypothalamus; (D) cerebellar cortex; (A and D courtesy of J. Schulman and H. Karten; B courtesy of E. Battenberg). Bar = 25 μm.

D. Endorphins and the Periventricular System

Standard neuroanatomical techniques have demonstrated that the periventricular system (PVS) is the major efferent pathway emanating from the medial hypothalamus (Krieg, 1932; Nauta & Haymaker, 1969). It is primarily composed of heterogeneous unmyelinated axons that course rostrally and turn dorsally along the midline at the anterior commissure to run through the paraventricular thalamic nucleus in a rostro–caudal direction. As the PVS enters the brainstem in the central gray, it is classically known as the dorsal longitudinal fasciculus of Schutz, distributing principally to nuclei around the ventricles like locus coeruleus and autonomic preganglionic centers like nucleus tractus solitarius.

Other fibers continue through the central grey into the spinal cord. The PVS also contains a major reciprocal ascending limb largely composed of short and long unmyelinated axons of the monoaminergic system (Lindvall & Bjorklund, 1974), distributing principally to the medial diencephalon and septum.

The axons of the opiocortin neurons primarily course through the classically defined PVS as far caudal as the locus coeruleus. Joining the opiocortin axons in the descending limb of the PVS are fibers containing the neuroendocrine peptides, luteotropic releasing hormone (LRH) (Sternberger & Hoffman, 1978), vasopressin (VP), and oxytocin (OX) (Buijs, Swaab, Dogterom, & van Leeuwen, 1978; Sofroniew & Weindl, 1979). The LRH, OX, and VP cell bodies reside primarily in the preoptico–hypothalamic area. In addition to their median eminence–pituitary terminations, they have extensive extrahypothalamic projections that distribute to most of the opiocortin terminal fields (Fig. 4). Among their common fields are the bed nucleus of the stria terminalis, septum, amygdala (the only nonperiventricular field), the paraventricular thalamic nucleus, the periaqueductal grey, and the organum vasculosum of the lamina terminalis (OVLT).

It will be important for future work to investigate the endocrine-related functions of each of these extrahypothalamic peptidergic systems with regard to possible synchronous links with pituitary hormone release. An example of such synchrony is illustrated by studies in which LRH infused into the periaqueductal grey elicited sexual receptivity (lordosis) in ovariectomized, estrogen primed female rats (Moss, 1979). In normal females approaching estrus, LRH neurons that project to the median eminence elicit LH release from the pituitary, causing ovulation; synchronously, they activate the central circuits regulating sexually receptive behavior.

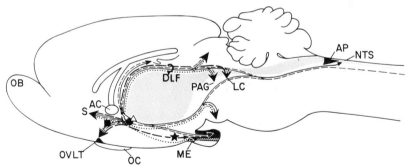

Fig. 4. The periventricular system of rat. Opiocortin (β-endorphin–ACTH–MSH) (★--★), luteotropic releasing hormone (■···■), oxytocin (●---●), and vasopressin (△—△) cell bodies originating primarily in the ventral diencephalon (symbols) project their fibers into the median eminence (ME) and dorsal longitudinal fasciculus (DLF). Abbreviations: AC, anterior commissure; NTS, nucleus tractus solitarius; OC, optic chiasm; AP, area postrema; PAG, periacqueductal grey; LC, locus coeruleus; OB, olfactory bulb; S, septum.

Two other endocrine peptides, somatostatin and thyrotropic releasing hormone, also have extensive intra- and extrahypothalamic distributions (see Hökfelt, Elde, Fuxe, & Johansson, 1978). However, their projection pathways are not yet well enough described to evaluate their possible contribution to the PVS. Moreover, their multiple extrahypothalmic cell body populations and many nonperiventricular terminal fields suggest that these latter peptides are contained in much more heterogeneous systems than those of endorphin, LRH, OX, and VP.

Enkephalin neurons are also distributed to many of the PVS terminal fields just described. However, their predominantly local circuit nature makes enkephalin a less likely candidate for a long-distance synchronizer of CNS–endocrine activity.

The dense peptidergic innervation of OVLT (Fig. 4) may be particularly significant. The OVLT is a circumventricular organ (CVO) of the same type as the median eminence–posterior pituitary and the area postrema. Circumventricular organs are characterized by their dense vascularity, fenestrated capillaries, and their lack of a glial sheath. Thus, molecules in the systemic circulation that cannot cross the blood–brain barrier may gain access to the brain and cerebrospinal fluid through CVOs. For example, ^{3}H-estradiol injected systemically is found in greatest concentration at periventricular sites, particularly in and around the circumventricular organs (Stumpf & Sar, 1975). At the OVLT where LRH terminals contact intracerebral blood vessels (G. Hoffmann, personal communication), ^{3}H-estradiol may be taken up into the LRH neurons or into unidentified neurons with reproductive significance. Endorphin terminals may also contact capillaries and specialized ependyma in the OVLT and median eminence (Fig. 5). If so, unidentified molecules circulating in the blood or CSF may be taken up into endorphin neurons. Conversely, β-endorphin and ACTH may also be released into the CSF through these contacts.

IV. EFFECTS OF OPIOID PEPTIDES ON NEURONAL ACTIVITY

A. Introduction

Electrophysiological research on opiates and opioid peptides has used extracellular, and more rarely, intracellular, single unit recordings directed in CNS areas with either a high density of opiate receptors or involvement in nociception. To document that neuronal responses to opioids involve stereospecific opiate binding sites, two tests have been relied upon: (1) blockade of effects

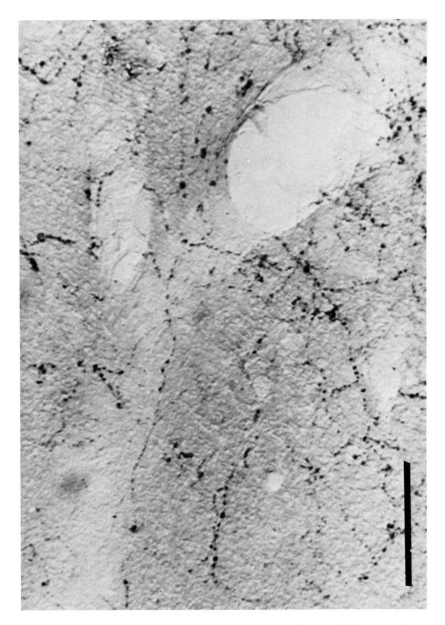

Fig. 5. β-Endorphin immunoreactive terminals contacting capillary in OVLT. Bar = 50 μm.

with an antagonist like naloxone; and (2) mimicry of actions by agonists like levorphanol but not by its inactive D+ enantiomer, dextrophan.

In general, the effects of opioid peptides are the same as the effects of opiate alkaloids (however, see Carette & Poulain, 1978; Duggan et al., 1976a; Segal, 1977). Most of these stereospecific, naloxone antagonized actions are inhibitions of single-unit discharge that are qualitatively similar throughout the mammalian central and peripheral nervous systems. However, some major exceptions exist: Naloxone reversible excitatory responses were seen with pyramidal cells in the hippocampus (Hill et al., 1976a; Nicoll et al., 1977), Renshaw cells in the spinal cord and some less well-identified cells in various parts of the CNS (Davies, 1976; Davies and Dray, 1978; Davies & Duggan, 1974; Duggan, Davies, & Hall, 1976a; Duggan et al., 1976b; Nicoll et al., 1977). The excitatory response of hippocampal neurons may be viewed as a primary inhibitory effect resulting in excitation by disinhibition (see following sections).

B. Cerebral Cortex

Only the frontal cortex contains notable concentrations of opiate binding sites (Pert, Kuhar, & Snyder, 1975) and relatively modest to negligible amounts of enkephalin immunoreactivity (Hökfelt et al., 1977a; Sar et al., 1978). Neurons in frontal or parietal areas are depressed by opiate agonists via stereospecific naloxone reversible opiate receptors (Satoh, Zieglgänsberger, Fries, & Herz, 1974; Satoh, Zieglgänsberger, & Herz, 1975; Satoh, Zieglgänsberger, & Herz, 1976). Like opiate alkaloids, M-e and L-e depress spontaneous and l-glutamate-induced discharge of most units tested in this part of the cortex (Zieglgänsberger, Fry, Herz, Moroder, & Wunsch, 1976).

C. Striatum

Despite the high concentration of opiate binding sites, the role of the striatum in the pharmacology or physiology of opiates and opioid peptides remains unclear. Striatal neurons are depressed by opioid and opiate agonists applied microiontophoretically (Frederickson & Norris, 1976; Gayton & Bradley, 1976; Nicoll et al., 1977; Zieglgänsberger & Fry, 1976) or systemically (Bigler & Eidelberg, 1976; Chan, Lee, & Wong, 1977). The speeding of some neurons in substantia nigra by systemic applications (Iwatsuto & Clouet, 1977) would be in accord with neurochemical evidence that opiates increase the synthesis and release of dopamine at striatal nerve terminals. Iontophoretic studies indicate that neuronal inhibition produced by opioids or by dopamine are mediated by separate receptors (Bradley & Gayton, 1976; Gayton & Bradley, 1976).

D. Thalamus

Noxious thermal and mechanical stimulation excites cells in the ventrobasal complex, the intralaminar nuclei, and the nucleus lateralis anterior of the thalamus (Emmers, 1976; Hill & Pepper, 1976). The responses are depressed by phoretically or systemically administered opiates or opioids (Frederickson, Norris, & Hewes, 1975; Hill et al., 1977a,b; Hill, Pepper, & Mitchell, 1976b,c; Hill & Pepper, 1976). Some of these effects are antagonized by systemic naloxone, whereas phoretically applied antagonists give strong spike blocking side effects (Duggan & Hall, 1977). However, the nociceptive responsiveness of intralaminar thalamic neurons is also inhibited by electrical stimulation of the ventrobasal complex in a non-naloxone sensitive fashion (Benabid, Henriksen, McGinty, & Bloom, 1980). Therefore, pain modulation at the thalamic level is regulated by both opioid sensitive and insensitive mechanisms.

E. Hippocampus

In contrast to neurons in most other regions of the rat's brain, hippocampal pyramidal neurons are predominantly excited by phoretically and systemically applied opiate agonists and opioid peptides (Hill et al., 1976a; Nicoll et al., 1977). This excitatory effect is in accord with the excitatory (Chou & Wang, 1976) and seizure-inducing properties of systemically, topically, and intraventricularly applied endorphins (Dunwiddie, Mueller, Palmer, Stewart & Hoffer, 1980; Henriksen, Bloom, Ling, & Guillemin, 1977; Henriksen et al., 1978a,b; Taylor et al., 1979). The excitation is clearly not due to an interaction with the muscarinic cholinergic input to pyramidal cells (French, Siggins, Henriksen, & Ling, 1977). This type of hippocampal unit is most frequently encountered when conventional multibarreled electrodes are employed (Nicoll et al., 1977). However, all hippocampus studies report both excitation and inhibition from application of opioids. Because some authors were unable to antagonize these excitations with phoretically applied naloxone, the specificity of the opiate-related effects, and other unspecific actions remained uncertain (Fry, Zieglgänsberger, & Herz, 1978; Segal, 1977). Interestingly, on some hippocampal neurons, inhibitory effects were stereospecific, and tolerance developed to the excitatory effect of opiate agonists in morphine tolerant–dependent animals (Fry et al., 1978).

Recent experiments employing single and simultaneous double unit recordings from hippocampal pyramidal and basket cells revealed that some of the excitatory responses were brought about by naloxone sensitive inhibitory actions of opiate agonists (morphine, M-e, D-Ala^2-enkephalin, β-endorphin) on nearby inhibitory interneurons leading to disinhibition of pyramidal cells (Zieglgänsberger, French, Siggins, & Bloom, 1979). These results indicate that the effects of a phoretically applied drug depend not only on the receptors involved but also on the circuitry in a given brain region.

F. Spinal Cord

1. Extracellular Recordings

Although most neurons located in the dorsal horn of the spinal cord are involved in somatosensory perception or in processing nociception, neurons in Laminae I and V are considered to play the major roles in nociceptive processes (see Yaksh & Rudy, 1977, 1978). Opiate binding sites and the enkephalin-containing small neurons are concentrated in Laminae II and III (Elde et al., 1976; Hökfelt et al., 1977a; Simantov et al., 1977; Watson et al., 1977). Some, but not all, inhibitory responses to phoretically and systemically applied opiates or opioid peptides were antagonized by naloxone, whereas most excitatory responses were obviously not replicated by stereospecific opiate agonists nor blocked by naloxone. Dorsal horn neurons are particularly sensitive to inhibition when opiate agonists are applied to Laminae II and III (corresponding to the substantia gelatinosa of Rolandi (Davies & Dray, 1978; Duggan et al., 1976a,b). The mechanism of these inhibitory effects is unknown. The data would support a primary distal dendritic action of opiates, although an excitatory action on inhibitory interneurons in this layer or an effect directly on primary afferent terminals cannot now be excluded. Renshaw cells in the ventral horn of the spinal cord (Curtis & Duggan, 1969; Duggan & Curtis, 1972; Felpel, Sinclair, & Yim, 1970) are probably not directly involved in nociception. Subsequent studies showed that excitations induced by opiate alkaloids (Davies & Duggan, 1974) are mimicked by opiate peptides and are stereospecific (Davies & Dray, 1978; Davies, 1976; Duggan et al., 1976). This excitatory effect seems to be a unique property of these specialized interneurons because naloxone also antagonizes the excitatory nicotinic actions of acetylcholine.

2. Intracellular Studies

Early studies of morphine agonists in the spinal cord showed that intravenous administration depressed polysynaptic excitatory postsynaptic potentials (EPSPs). This effect was reduced by opiate antagonists (Jurna 1966, 1973). More recent studies employing intracellular recording and simultaneous extracellular microiontophoretic application revealed that morphine and opioid peptides do not change membrane potentials or resting membrane resistance (Zieglgänsberger

& Bayerl, 1976; Zieglgänsberger & Fry, 1976). Nevertheless, opiates decrease the rate of rise of the EPSPs in motoneurons, interneurons, and neurons in the dorsal horn involved in somatosensory perception. These effects are antagonized by naloxone and are stereospecific.

In addition, opiates and opioids depress l-glutamate-induced depolarization (Zieglgänsberger & Bayerl, 1976). Microiontophoretically applied l-glutamate is thought to cause an increase in the permeability of the postsynaptic membrane to sodium ions (Zieglgänsberger & Puil, 1972). Zieglgänsberger proposed that opiates interfere with the chemically excitable sodium channel. Because these depolarizing responses are clearly antagonized by opiate agonists, the opiate receptors involved in this effect must be located on the postsynaptic membrane.

The *in vivo* antiglutamate actions of opiates and opioids have recently been confirmed with spinal neurons grown in tissue culture (Barker, Neale, Smith, & MacDonald, 1978a). An analysis of the kinetics of antiglutamate indicates that the inhibitory action of the opioid peptides are brought about by a "noncompetitive mechanism" on the postsynaptic sodium conductance mechanism. A similar interpretation was suggested by studying the effects of opiates upon the depolarizing response seen in neuroblastoma–glioma hybrid cells (Myers, Livengood, & Shain, 1975; Kumakura, Karoum, Guidotti, & Costa, 1980).

V. STRUCTURAL–FUNCTIONAL CORRELATIONS

When we speak of trying to determine the functional role of a particular group of neurons, several different sorts of classifying schemes may be used to make the group designation. The most frequently applied classifying schemes are based upon four definable properties: (1) presumptive function (e.g., a sensory, motor, or internuncial neuron); (2) cell shape or location (e.g., pyramidal, mitral, Purkinje, or locus coeruleus neuron); (3) cellular activity patterns (e.g., pacemakers, followers, bursters); or (4) neurotransmitter product (e.g., noradrenergic, cholinergic, or, in the present context, endorphinergic). With the emergence of immunocytochemical and physiological techniques, it is possible to classify neurons on the basis of more than one of the previous categories. When this is done for the endorphin-containing neurons, the multiparameter designations help to discriminate the various types of functional roles in which these neurons may participate. At this point, we shall consider two sorts of structural–functional correlations with existing information to contrast characteristics of presumptive enkephalinergic and endorphinergic neurons. We must recognize at the outset, however, that a crucial missing datum is the absolute identification of what these neurons actually secrete.

A. Hierarchical Position

Present concepts of the structural organization of the brain rely heavily on two principles of connectivity. In the primary sensory and motor pathways, the prevailing principle has been the classical concept of hierarchical relationships in which the transmission of information is highly sequential and specific. The key element of the hierarchical or through-put system is that destruction of one link incapacitates the chain. No complete set of transmitter molecules for any such sensory–motor through-put system has been identified; so, it is not yet possible to determine what specific chemicals are used to transmit between specific levels of such hierarchies. However, the anatomy of the central endorphin-containing neurons can be viewed as a pivotal link in a sensory–motor through-put system with medial basal hypothalamic neurons connected to a series of target neurons in various endocrine control fields of the periventricular system. Thus, it seems reasonable to include the opiocortin neuronal system as a component of the endocrine peptidergic network in the CNS. In this respect, the functional significance of β-endorphin in the CNS might best be investigated with an eye toward its synergistic relationship with the endocrine system as a whole.

The second widely applied principle of neuronal connectivity is that of the local circuit neuron (Rakic, 1975): typically a small, frequently unipolar, neuron whose efferent processes bear the morphology of dendrites. The basic feature of a local circuit neuron is that its connections are established exclusively within the local vicinity in which the cell body is found. Such small interneurons can exert significant control over information flowing through that locale, and may do so through "dendritic" release of their transmitter and without action potentials (Rakic, 1975). In some cases of inhibitory local circuits, the amino acids gamma-aminobutyric acid (GABA) or glycine (GLY) have been implicated as the transmitter (see Alger & Nicoll, 1978). The neuroanatomy and heterogeneous morphology of enkephalin neurons suggest that enkephalin may be a local circuit mediator of synaptic events of significance to many diverse areas and functions of the nervous system.

B. Molecular Mechanisms

Conceivably, enkephalin-containing and endorphin-containing neural circuits could have identical actions at the molecular level, and differ only in the overall functional control regulated by the neuronal subsystems in which these chemically identified links are components. One might even speculate that the longer distances connected by endorphin-containing neurons and the longer length of the endorphin peptide are somehow related, and similarly, the shorter enkephalin-linked distances are related to the shorter chemical length. The postsynaptic receptor for the appropriate locally released peptide may also respond to the other peptide if it were administered into the proper site. Regardless of these structure–activity properties of the two sets of opioid peptides, and the possibility of separate subclasses of endorphin-responsive receptors, there is only a modest amount of information characterizing the molecular mechanisms of opioid peptide receptor actions, and the ways in which these actions may differ from those executed either by other neuropeptides or by other neurotransmitters.

Molecular mechanisms attributable to transmitters of the classical variety such as amino acids or acetylcholine at the neuromuscular junction emphasize the altered ionic permeability produced in the responsive cell. Although endorphin peptides can produce changes in ion permeability in neurons of the autonomic ganglia (North, Katayama, & Williams, 1979) and locus coeruleus (Pepper & Henderson, 1980), these effects may be less universal than a new class of peptide action in which the peptide acts to prevent the action of another transmitter substance (Barker et al., 1978a; Barker, Smith, & Neale, 1978b; Barker & Smith, 1979; Zieglgänsberger & Tulloch, 1979).

To grasp the possible significance of the peptide action as a preventor of some other chemical signal, we can compare this effect with the actions of monoamine transmitters. In other theoretical overviews (Bloom, 1978, 1980), the idea was developed that, at those noradrenergic synapses in the mammalian brain where postsynaptic receptors are linked to activation of adenylate cyclase, norepinephrine "modulates" the responsiveness of target cells to other neurotransmitter input. More explicitly, norepinephrine may "enable" these target cells to respond more effectively to inputs from noncatecholaminergic neurons. Such enabled responses are a deducible quality of the transmembrane hyperpolarizations with increased membrane resistance at specific monoaminergic synapses. Similar enabling actions could also be an element of the response repertoire of some cells responding to neuropeptides. In fact, in some cases, it seems that a peptidergic neuron may release both an amine and a peptide, and that pairs of transmitters may create both long- and short-term complementary actions on the target cell (Hökfelt, Johansson, Ljungdahl, Lundberg, & Schultzberg, 1980). However, a new peptide action exemplified by certain tests with enkephalins (Barker et al., 1978a,b; Barker & Smith, 1979; Zieglgänsberger & Tulloch, 1979) may be viewed as "disenabling." In this case, the target cell responds to the peptide in such a way that simultaneous responses to glutamate or acetylcholine (acting individually at their own glutamatergic or cholinergic receptors as a classical excitant with depolarizing ionic conductance increases) are antagonized by the peptide. However, when examined for actions of the peptide alone, no obvious membrane property changes are seen. Because the endorphin and the classical transmitter each act through individual receptors, an interaction between the two separate receptor mechanisms must occur to diminish the glutamate response. In keeping with the "enabling" action of monoamines, this effect of peptides may therefore be disenabling; thus, the peptide receptor alters the membrane of the target cell in some way to prevent the membrane from executing responses that would have been elicited by other signaling molecules, such as amino acids or monoamines.

In Chapter 12 we shall indicate how we view the preceding cellular information with respect to possible explanations of behavioral phenomena and to consider schema that attempt to attribute behavioral phenomena to specific chemical transmitters, especially those that are usually termed "memory" or "learning."

REFERENCES

Alger, B. E., & Nicoll, R. A. (1978). GABA-mediated biphasic inhibiting responses in hippocampus. Nature, (London), 281, 315–317.

Atweh, S. F., & Kuhar, M. J. (1977a). Autoradiographic localization of opiate receptors in rat brain. I. Spinal cord and lower medulla. Brain Research, 124, 53–67.

Atweh, S. F., & Kuhar, M. J. (1977b). Autoradiographic localization of opiate receptors in rat brain. II. The brain stem. Brain Research, 129, 1–12.

Atweh, S. F., & Kuhar, M. J. (1977c). Autoradiographic localization of opiate receptors in rat brain. III. The telencephalon. Brain Research, 134, 393–405.

Austen, B. M., & Smyth, D. G. (1977). The NH^2-terminus of C-fragment is resistant to the action of aminopeptidases. Biochemical and Biophysical Research Communications, 77, 86–94.

Barker, J. L., Neale, J. H., Smith, T. G., Jr., & MacDonald, R. L. (1978a). Opiate peptide modulation of amino acid responses suggests novel form of neuronal communication. Science, 199, 1451–1453.

Barker, J. L., & Smith, T. G. (1979). Three modes of communication in the nervous system. Advances in Experimental Medicine & Biology, 116, 3–25.

Barker, J. L., Smith, T. G., Jr., & Neale, J. H. (1978b). Multiple membrane actions of enkephalin revealed using cultured spinal neurons. Brain Research, 154, 153–158.

Bayon, A., French, E., Henriksen, S. J., Siggins, G. R., Segal, D., Browne, R., Ling, N., & Guillemin, R. (1978). β-endorphin cellular localization, electrophysiological and behavioral effects. In E. Costa & M. Trabuchi (Eds.), The Endorphins: Advances in Biochemical Pharmacology, Vol. 18, pp. 89–109. New York: Raven.

Bayon, A., Koda, L., Battenberg, E., Azad, R., & Bloom, F. E. (1980a). Regional distribution of endorphin, met-enkephalin and leu-enkephalin in the pigeon brain. Neuroscience Letters, 16, 75–80.

Bayon, A., Koda, L., Battenberg, E., & Bloom, F. (1980b). Redistribution of endorphin and enkephalin immunoreactivity in the rat brain and pituitary after in vivo treatment with colchicine or cytochalasin 3. Brain Research, 183, 103–111.

Bayon, A., Rossier, J., Mauss, A., Bloom, F. E., Iversen, L. L., Ling, N., & Guillemin, R. (1978). In vitro release of [5-methionine] enkephalin and [5-leucine] enkephalin from the rat globus pallidus. Proceedings of the National Academy of Sciences, U.S.A., 75, 3503–3506.

Benabid, A., Henriksen, S., McGinty, J., & Bloom, F. (1980). Medial thalamic response to noxious inputs: Inhibitory interactions with lateral thalamic stimulation. Society for Neuroscience Abstracts, 6, 245.

Bigler, E. D., & Eidelberg, E. (1976). Nigrostriatal effects of morphine in two mouse strains. *Life Sciences*, **19**, 1399–1406.

Bloom, F. (1978). Is there a neurotransmitter code in the brain? *In Advances in Pharmacology Therapy*, Vol. 2, Neurotransmitters, pp. 205–216. Oxford: Pergamon.

Bloom, F. (1980) Preface in *Peptides: Integrators of Cell and Tissue Function*, New York: Raven Press.

Bloom, F., Battenberg, E., Rossier, J., Ling, N., & Guillemin, R. (1978a). Neurons containing β-endorphin in rat brain exist separately from those containing enkephalin: Immunocytochemical studies. *Proceedings of the National Academy of Sciences, U.S.A.*, **75**, 1591–1595.

Bloom, F., Battenberg, E., Shibasaki, T., Benoit, R., Ling, N., & Guillemin, R. (1980). Localization of γ-melanocyte stimulating hormone (γMSH): Immunoreactivity in rat brain and pituitary. *Regulatory Peptides* **1**, 202–205.

Bloom, F., Battenberg, E., Rossier, J., Ling, N., Leppaluoto, J., Vargo, T. M., & Guillemin, R. (1977a). Endorphins are located in the intermediate and anterior lobes of the pituitary gland, not in the neurohypophysis. *Life Sciences*, **20**, 43–48.

Bloom, F., Rossier, J., Battenberg, E., Vargo, T., Minick, S., Ling, N., & Guillemin, R. (1977b). Regional distribution of β-endorphin and enkephalin in rat brain: A biochemical and cytochemical study. *Society for Neuroscience Abstracts*, **3**, 286.

Bloom, F., Segal, D., Ling, N., & Guillemin, R. (1976). Endorphins: Profound behavioral effects in rats suggest new etiological factors in mental illness. *Science*, **194**, 630–632.

Bradbury, A. F., Feldberg, W. F., Smyth, D. G., & Snell, C. R. (1976a). Lipotropin C fragment: An endogenous peptide with potent analgesic activity. *In* H. W. Kosterlitz (Ed.), *Opiates and Endogenous Opioid Peptides*, pp. 9–17. Amsterdam: Elsevier.

Bradbury, A. F., Smyth, D. G., Snell, C. R., Birdsall, N. J. M. & Hulme, E. C. (1976b). C-fragment of lipotropin has a high affinity for brain opiate receptors. *Nature (London)*, **260**, 793–795.

Bradbury, A. F., Smyth, D. G., Snell, C. R., Deaken, J. F. W. & Nendlandt, S. (1977). Lipotropin: Precursor to two biologically active peptides. *Biochemical and Biophysical Research Communications*, **74**, 748–754.

Bradley, P. B., & Gayton, R. J. (1976). Actions and interactions of morphine and dopamine on single neurons in the rat caudate nucleus. *British Journal of Pharmacology*, **57**, 425–426.

Buijs, R., Swaab, D., Dogterom, J., & van Leeuwen, F. (1978). Intra- and extrahypothalamic vasopressin and oxytocin pathways in the rat. *Brain Research*, **186**, 423–433.

Carette, B., & Poulain, P. (1978). Inhibitory action of iontophoretically applied methionine-enkephalin and leucine-enkephalin on tuberal hypothalamic neurons. *Neuroscience Letters*, **7**, 137–140.

Chan, S. H., Lee, C. M., & Wong, P. C. L. (1977). Suppression of caudate neuron activities by morphine and the involvement of dopaminergic neurotransmission. *Federation Proceedings*, **36**, 668.

Childers, S. W., Schwarcz, R., Coyle, J. T., & Snyder, S. H. (1978). Radioimmunoassay of enkephalins: Levels of methionine- and leucine-enkephalins in morphine-dependent and kainic acid lesioned rat brains. *In* E. Costa & M. Trabuchi (Eds.), *The Endorphins: Advances in Biochemical Pharmacology*, Vol. 18, pp. 161–173. New York, Raven.

Chou, T., & Wang, S. C. (1976). Effects of morphine and benzodiazepines on limbic system. *Federation Proceedings*, **35**, 357.

Cox, B. M., Opheim, K. E., Teschemacher, H., & Goldstein, N. A. (1975). A peptide-like substance from pituitary that acts like morphine. *Life Sciences*, **16**, 1777–1782.

Cuello, A. C., & Paxinos, G. (1978). Evidence for a long leu-enkephalin striopallidal pathway in rat brain. *Nature (London)*, **271**, 178–180.

Curtis, D. R., & Duggan, A. W. (1969). The depression of spinal inhibition by morphine. *Agents and Actions*, **1**, 15–19.

224   FLOYD E. BLOOM AND JACQUELINE F. MCGINTY

Davies, J. (1976). Effects of morphine and naloxone on Renshaw cells and spinal interneurons in morphine dependent and non-dependent rats. *Brain Research*, **113**, 311–326.

Davies, J., & Dray, A. (1978). Pharmacological and electrophysiological studies of morphine and enkephalin on rat supraspinal neurones and cat spinal neurones. *British Journal of Pharmacology*, **63**, 87–96.

Davies, J., & Duggan, A. W. (1974). Opiate agonist-antagonist effects on Renshaw cells and spinal interneurones. *Nature (London)*, **250**, 70–71.

De Olmos, J. (1972). The amygdaloid projection field in the rat as studied with the cupric-silver method. In B. Eleftheriou (Ed.), *The Neurobiology of the Amygdala*, pp. 145–204. New York: Plenum.

Doneen, B., Chung, D., Yamashiro, D., Law, P., Loh, H. & Li, C. (1977) Beta-endorphin structure-activity relationships in guinea pig ileum and opiate receptor binding assays. *Biochemical and Biophysical Research Communications*, **74**, 656–662.

Dragon, N., Seidah, N. G., Lis, M., Routhier, R., and Chretien, M. (1977). Primary structure and morphine-like activity of human beta-endorphin. *Canadian Journal of Biochemistry*, **55**, 666–670.

Duggan, A. W., & Curtis, D. R. (1972). Morphine and the synaptic activation of Renshaw cells. *Neuropharmacology* **11**, 189–196.

Duggan, A. W., Davies, J., & Hall, J. G. (1976a). Effects of opiate agonists and antagonists on central neurons of the cat. *Journal of Pharmacology & Experimental Therapeutics*, **196**, 107–120.

Duggan, A. W., & Hall, J. G. (1977). Morphine, naloxone, and the responses of medial thalamic neurons in the cat. *Brain Research*, **122**, 49–57.

Duggan, A. W., Hall, J. G., & Headley, P. M. (1976b). Morphine, enkephalin, and the substantia gelatinosa. *Nature (London)*, **264**, 456–458.

Dunwiddie, T., Mueller, A., Palmer, M., Stewart, J., & Hoffer, B. (1980). Electrophysiological interactions of enkephalins with neuronal circuitry in rat hippocampus. I. Effects on pyramidal cell activity. *Brain Research*, **184**, 311–330.

Elde, R., Hökfelt, T., Johansson, O., & Terenius, L. (1976). Immunohistochemical studies using antibodies to leucine-enkephalin: Initial observations on the nervous system of the rat. *Neuroscience*, **1**, 349–351.

Emmers, R. (1976). Thalamic mechanisms that process a temporal pulse code for pain. *Brain Research*, **103**, 425–441.

Feldberg, W., & Smyth, D. G. (1977). Analgesia produced in cats by the C-fragment of lipotropin and by a synthetic pentapeptide. *Journal of Physiology (London)*, **265**, 25–27.

Felpel, L. P., Sinclair, J. G., & Yim, G. F. W. (1970). Effects of morphine on Renshaw cell activity. *Neuropharmacology*, **9**, 203–210.

Frederickson, R. C. A., & Norris, F. H. (1976). Enkephalin-induced depression of single neurons in brain areas with opiate receptors: Antagonism by naloxone. *Science*, **194**, 440–442.

Frederickson, R. C. A., Norris, F. H., & Hewes, C. R. (1975). Effects of naloxone and acetylcholine on medial thalamic and cortical units in naive and morphine dependent rats. *Life Sciences*, **17**, 81–82.

French, E. D., & Siggins, G. R. (1980). An iontophoretic survey of opioid peptide actions in rat limbic system. In search of opiate epileptogenic mechanisms. *Regulatory Peptides* **1**, 115–126.

French, E. D., Siggins, G. R., Henriksen, S. J., & Ling, N. (1977). Iontophoresis of opiate alkaloids and endorphins accelerates hippocampal unit firing by a non-cholinergic mechanism. *Society for Neuroscience Abstracts*, **3**, 291.

Fry, J., Herz, A., & Zieglgänsberger, W. (1978). Non-specific actions of opioids in the rat hippocampus. *Journal of Physiology (London)*, **280**, 15P.

Gayton, R. J., & Bradley, P. B. (1976). Comparison of the effects of morphine and related substances with those of dopamine on single neurons in the cat caudate nucleus. In H. W. Kosterlitz (Ed.) *Opiates and Endogenous Opioid Peptides*, pp. 213–219. Amsterdam: Elsevier.

Geisow, M. J., Deakin, J. F. W., Dostrovsky, J. O., & Smyth, D. G. (1977). Analgesic activity of lipotropin C-fragment depends on carboxylterminal tetrapeptide. *Nature (London)*, **269**, 167–168.

Gillin, J. C., Hong, J. S., Yang, H. Y. T., & Costa, E. (1978). [Met]enkephalin content in brain regions of rats treated with lithium. *Proceedings of the National Academy of Sciences, U.S.A.*, **75**, 2991–2993.

Gintzler, A. R., Gershon, M. D., & Spector, S. (1978). A nonpeptide morphine-like compound: Immunocytochemical localization in the mouse brain. *Science*, **199**, 477–478.

Goldstein, A. (1976). Opioid peptides (endorphins) in pituitary and brain. *Science*, **193**, 1081–1083.

Goldstein, A., Tachibana, S., Lowney, L. I., Hunkapiller, M., & Hold, L. (1979). Dynorphin-(1–13), an extraordinarily potent opioid peptide. *Proceedings of the National Academy of Sciences U.S.A.*, **76**, 6666–6670.

Graf, L., Ronai, A. Z., Bajusz, C., Cseh, G., & Szekely, J. I. (1976a). Opioid agonist activity of β-lipotropin fragments: A possible biological source of morphine-like substances in the pituitary. *FEBS Letters*, **64**, 181–185.

Graf, L., Szekely, J. I., Ronai, A. Z., Dunai-Kovacs, A., & Bajusz, S. (1976b). Comparative study on analgesic effect of met^5-enkephalin and related lipotropin fragments. *Nature (London)*, **263**, 240–242.

Gros, C., Pradelles, P., Humbert, J., Dray, F., Le Gal La Salle, G., & Ben-Ari, Y. (1978a). Regional distribution of met-enkephalin within the amygdaloid complex and bed nucleus of the stria terminalis. *Neuroscience Letters* **10**, 193–196.

Gros, C., Pradelles, P., Rougeot, C., Bepoldin, O., Dray, F., Fournie-Zaluski, M. C., Roques, B. P. Pollard, H., Llorens-Cortes, C., & Schwartz, J. C. (1978b). Radioimmunoassay of methionine- and leucine-enkephalin in regions of rat brain and comparison with endorphins estimated by a radioreceptor assay. *Journal of Neurochemistry*, **31**, 29–39.

Guillemin, R. (1978). Peptides in the brain: The new endocrinology of the neuron. *Science*, **202**, 390–401.

Guillemin, R., Ling, N., & Burgus, R. (1976). Endorphins, hypothalamic and neurohypophyseal peptides with morphinomimetic activity: Isolation and molecular structure of alpha-endorphin. *Comptes Rendus Academy Science of Paris, Series D*, **283**, 783–785.

Guillemin, R., Ling, N., & Vargo, T. M. (1977). Radioimmunoassays for alpha-endorphin and beta-endorphin. *Biochemical & Biophysical Research Communications*, **77**, 361–366.

Henriksen, S. J., Bloom, F. E., Ling, N., & Guillemin, R. (1977). Induction of limbic seizures by endorphins and opiate alkaloids: Electrophysiological and behavioral correlates. *Society for Neuroscience Abstracts*, **3**, 293.

Henriksen, S. J., Bloom, F. E., McCoy, F., Ling, N., & Guillemin, R. (1978a). β-endorphin induces non-convulsive limbic seizures. *Proceedings of the National Academy of Sciences, U.S.A.*, **75**, 5221–5225.

Henriksen, S. J., McCoy, F., French, E., & Bloom, F. E. (1978b). β-endorphin induced epileptiform activity: effects of lesions and specific opiate receptor agonists. *Society for Neuroscience Abstracts* **4**, 409.

Hill, R. G., Mitchell, J. F., & Pepper, C. M. (1976a). The excitation and depression of hippocampal neurons by iontophoretically applied enkephalins. *Journal of Physiology (London)*, **272**, 50–51.

Hill, R. G., & Pepper, C. M. (1976). The effects of morphine and met-enkephalin on nociceptive neurons in the rat thalamus. *British Journal of Pharmacology*, **58**, 459–460.

Hill, R. G., & Pepper, C. M. (1977a). A novel population of neurons in the rat thalamus which respond to noxious stimuli. *Journal of Physiology (London)*, **269**, 378.

Hill, R. G., & Pepper, C. M. (1977b). Studies on the pharmacology of nociceptive neurones in the rat thalamus. *In* R. W. Ryall and J. S. Kelly (Eds.), *Microiontophoresis and Transmitter Mechanisms in the Mammalian Central Nervous System*. Amsterdam: Elsevier.

226   FLOYD E. BLOOM AND JACQUELINE F. MCGINTY

Hill, R. G., Pepper, C. M., & Mitchell, J. F. (1976b). Depression of nociceptive and other neurones in the brain by iontophoretically applied met-enkephalin. *Nature (London)*, 262, 604–606.

Hill, R. G., Pepper, C. M., & Mitchell, J. F. (1976c). The depressant action of iontophoretically applied met-enkephalin on single neurons in rat brain. *In* H. W. Kosterlitz (Ed.), *Opiates and Endogenous Opioid Peptides*, pp. 225–230. Amsterdam: Elsevier.

Hökfelt, T., Elde, R., Fuxe, K., & Johansson, O. (1978). Aminergic and peptidergic pathways in the nervous system with special reference to the hypothalamus. *In The Hypothalamus*, pp. 69–136. New York: Raven.

Hökfelt, T., Elde, R., Johansson, O., Terenius, L., & Stein, L. (1977a). The distribution of enkephalin-immunoreactive cell bodies in rat central nervous system. *Neuroscience Letters*, 5, 25–31.

Hökfelt, T., Johansson, O., Ljungdahl, A., Lundberg, J., & Schultzberg, M. (1980). Peptidergic neurones. *Nature (London)*, 284, 515–521.

Hökfelt, T., Ljungdahl, A., Terenius, L., Elde, R., & Nilsson, G. (1977b). Immunohistochemical analysis of peptide pathways possibly related to pain and analgesia: Enkephalin and substance P. *Proceedings of the National Academy of Sciences, U.S.A.*, 74, 3081–3085.

Hökfelt, T., Terenius, L., Kuypers, H., & Dann, O. (1979). Evidence for enkephalin immunoreactive neurons in the medulla oblongata projecting to the spinal cord. *Neuroscience Letters*, 14, 55–60.

Hollt, V., & Herz, A. (1978). In vivo receptor occupation by opiates and correlation to the pharmacological effect. *Federation Proceedings*, 37, 158–161.

Hong, J. S., Yang, H. Y. T., & Costa, E. (1977a). On the location of methionine-enkephalin neurones in rat striatum. *Neuropharmacology*, 16, 451–453.

Hong, J. S., Yang, H. Y., Fratta, W., & Costa, E. (1977b). Determination of methionine enkephalin in discrete regions of rat brain. *Brain Research*, 134, 383–386.

Hong, J. S., Yang, H. Y. T., Fratta, W., & Costa, E. (1978). Elevation of methionine-enkephalin concentration in rat striatum induced by chronic administration of haloperidol, pimozide, chlorpromazine but not clozapine. *Journal of Pharmacology & Experimental Therapeutics*, 205, 141–147.

Hughes, J. (1975). Isolation of an endogenous compound from the brain with pharmacological properties similar to morphine. *Brain Research*, 88, 295–308.

Hughes, J., Kosterlitz, H. W., & Smith, T. W. (1977). The distribution of methionine-enkephalin and leucine-enkephalin in the brain and peripheral tissues. *Journal of Pharmacology* 61, 639–647.

Hughes, J., Smith, T. W., Kosterlitz, H. W., Fothergill, L. A., Morgan, B. A., & Morris, H. R. (1975). Identification of two related pentapeptides from the brain with potent opiate agonist activity. *Nature (London)*, 258, 577–580.

Iwatsubo, K., & Clouet, D. H. (1977). Effects of morphine and haloperidol on the electrical activity of rat nigrostriatal neurons. *Journal of Pharmacology & Experimental Therapeutics*, 202, 429.

Jurna, I. (1966). Inhibition of the effects of repetitive stimulation on spinal motoneurons of the cat by morphine and pethidine. *International Journal of Neuropharmacology*, 5, 117–123.

Jurna, I., Grossman, W., & Theres, C. (1973). Inhibition by morphine of repetitive activation of cat spinal motoneurons. *Neuropharmacology*, 12, 983–993.

Kangawa, K., & Matsuo, H. (1979). Alpha-Neo-endorphin: A "big" Leu-enkephalin with potent opiate activity from porcine hypothalami. *Biochemical & Biophysical Research Communications*, 86, 153–160.

Kobayashi, R. M., Palkovits, M., Miller, R. J., Chang, K. J., & Cuatrecasas, P. (1978). Brain enkephalin distribution is unaltered by hypophysectomy. *Life Sciences*, 22, 527–530.

Kosterlitz, H. W., & Hughes, J. (1978). Development of the concepts of opiate receptors and their ligands. *In* E. Costa and M. Trabuchi (Eds.), *The Endorphins: Advances in Biochemical Pharmacology*, Vol. 18, pp. 31–44. New York: Raven.

Krieg, W. S. (1932). The hypothalamus of the albino rat. *Journal of Comparative Neurology*, 55, 19–89.

Krieger, D. T., Liotta, A., Suda, T., Palkovits, M., & Brownstein, M. J. (1977). Presence of immunoassayable beta-lipotropin in bovine brain and spinal cord: Lack of concordance with ACTH concentrations. *Biochemical & Biophysical Research Communications*, 76, 930–936.

Kumakura, K., Karoum, F., Guidotti, A., & Costa, E. (1980). Modulation of nicotinic receptors by opiate agonists in cultured adrenal chromaffin cells. *Nature (London)*, 283, 489–492.

Lazarus, L. H., Ling, N., & Guillemin, R. (1976). Beta-lipotropin as a prohormone for the morphinomimetic peptides endorphins and enkephalins. *Proceedings of the National Academy of Sciences U.S.A.*, 73, 2156–2159.

Lewis, R. V., Stein, S., Gerber, L. D., Rubinstein, M., & Udenfriend, S. (1978). High molecular weight opioid containing proteins in striatum. *Proceedings of the National Academy of Sciences, U.S.A.*, 75, 4021–4023.

Lewis, R. V., Stern, A. S., Kimura, S., Rossier, J., Stein, S., & Udenfriend, S. (1980). An about 50,000 Dalton protein in adrenal medulla: A common precursor of [Met] and [Leu] enkephalin. *Science*, 208, 1459–1461.

Li, C. H. (1964). Lipotropin: A new active peptide from pituitary glands. *Nature (London)*, 201, 924.

Li, C. H., Barnafi, L., Chretien, M., & Chung, D. (1965). Isolation and amino acid sequence of beta-LPH from sheep pituitary glands. *Nature (London)*, 208, 1093–1094.

Li, C. H., & Chung, D. (1976). Isolation and structure of an untriakontapeptide with opiate activity from camel pituitary glands. *Proceedings of the National Academy of Sciences, U.S.A.*, 73, 1145–1148.

Li, C. H., Yamashiro, D., Tseng, L. F., & Loh, H. H. (1977). Synthesis and analgesic activity of human beta-endorphin. *Journal of Medicinal Chemistry*, 20, 325–328.

Lindvall, O., & Bjorklund, A. (1974). The organization of the ascending catecholamine neuron systems in the rat brain. *Acta Physiologica Scandinavica, Supplement*, 412.

Ling, N. (1977). Solid phase synthesis of porcine alpha-endorphin and gamma-endorphin. *Biochemical & Biophysical Research Communications*, 74, 248–256.

Ling, N., Burgus, R., & Guillemin, R. (1976). Isolation, primary structure, and synthesis of alpha-endorphin and gamma-endorphin. *Proceedings of the National Academy of Sciences, U.S.A.*, 73, 3942–3946.

Ling, N., & Guillemin, R. (1976). Morphinomimetic activity of synthetic fragments of beta-lipotropin and analogs. *Proceedings of the National Academy of Sciences, U.S.A.* 73, 3308–3310.

Loh, H. H., Tseng, L. F., Wei, E., & Li, C. H. (1976). Beta-endorphin is a potent analgesic agent. *Proceedings of the National Academy of Sciences, U.S.A.*, 73, 2895–2898.

Lord, J. A. H., Waterfield, A. A., Hughes, J., & Kosterlitz, H. W. (1977). Endogenous opioid peptides: Multiple agonists and receptors. *Nature (London)*, 267, 495–499.

Mains, R. E., Eipper, B. A., & Ling, N. (1977). Common precursor to corticotrophins and endorphins. *Proceedings of the National Academy of Sciences U.S.A.*, 74, 3014–3018.

McGinty, J. F., & Bloom, F. E. (1980). Double immunocytochemical labeling demonstrates distinctions among opioid peptidergic neurons. *Society for Neuroscience Abstracts*, 6, 354.

Miller, R. J., Chang, K. G., Cooper, B., & Cuatrecasas, P. (1978). Radioimmunoassay and characterization of enkephalins in rat tissues. *Journal of Biological Chemistry*, 253, 531–538.

Moldow, R. L., & Yalow, R. S. (1978). Extrahypothalamic distribution of corticotropin as a function of brain size. *Proceedings of the National Academy of Sciences, U.S.A.*, 75, 994–998.

Moss, R. (1979). Actions of hypothalamic-hypophysiotropic hormones on the brain. *Annual Review of Physiology*, 41, 617–631.

Myers, P. R., Livengood, D. R., & Shain, W. (1975). Effect of morphine on a depolarizing dopamine response. *Nature (London)*, **257**, 238–240.

Nakanishi, S., Inoue, A., Kita, T., Nakamura, M., Chang, A. C. Y., Cohen, S. N., & Numa, S. (1979). Nucleotide sequence of cloned cDNA for bovine corticotropin-β-lipotropin precursor. *Nature (London)*, **278**, 423–427.

Nauta, W.J.H., & Haymaker, W. (1969). Hypothalamic nuclei and fiber connections. In *The Hypothalamus*. Springfield, Illinois: Thomas.

Nicoll, R. A., Siggins, G. R., Ling, N., Bloom, F. E., & Guillemin, R. (1977). Neuronal actions of endorphins and enkephalins among brain regions: A comparative microiontophoretic study. *Proceedings of the National Academy of Sciences, U.S.A.*, **75**, 1591.

North, R. A., Katayama, Y., & Williams, J. T. (1979). On the mechanism and site of action of enkephalin on single myenteric neurons. *Brain Research*, **165**, 67–77.

Ogawa, N., Panerai, A., Lee, S., Forsbach, G., Haulicek, V., & Friesen, H. (1979). Beta-endorphin concentration in the brain of intact and hypophysectomized rats. *Life Sciences*, **25**, 317–326.

Pasternak, G. W., Simantov, R., & Snyder, S. H. (1976). Characterization of an endogenous morphine-like factor (enkephalin) in mammalian brain. *Molecular Pharmacology*, **12**, 504–513.

Pelletier, G., Desy, L., Lissitszky, J. C., Labrie, F., & Li, C. H. (1978). Immunohistochemical localization of β-LPH in the human hypothalamus. *Life Sciences*, **22**, 1799–1804.

Pepper, C. M., & Henderson, G. (1980). Opiates and opioid peptides hyperpolarize locus coeruleus neurons in vitro. *Science*, **209**, 394–395.

Pert, C. B., Kuhar, M. J., & Snyder, S. H. (1975). Autoradiographic localization of the opiate receptor in rat brain. *Life Sciences*, **16**, 1849–1853.

Rakic, P. (1975). Local circuit neurons. *Neuroscience Research Program Bulletin*, **13**.

Roberts, J. L., & Herbert, E. (1977). Characterization of a common precursor to corticotropin and beta-lipotropin: Identification of beta-lipotropin peptides and their arrangement relative to corticotropin. *Proceedings of the National Academy of Sciences, U.S.A.*, **74**, 5300–5304.

Ross, M., Dingledine, R., Cox, B. M., & Goldstein, A. (1977). Distribution of endorphins in pituitary. *Brain Research*, **124**, 523–532.

Rossier, J., Bayon, A., Vargo, T., Ling, N., Guillemin, R., & Bloom, F. (1977a). Radioimmunoassay of brain peptides: Evaluation of a methodology for the assay of β-endorphin and enkephalin. *Life Sciences*, **21**, 847–852.

Rossier, J., Battenberg, E., Pittman, Q., Bayon, A., Koda, L., Miller, R., Guillemin, R., & Bloom, F. (1979). Hypothalamic enkephalin neurones may regulate the neurohypophyses. *Nature (London)*, **277**, 653–655.

Rossier, J., Vargo, T. M., Minick, S., Ling, N., Bloom, F., & Guillemin, R. (1977b). Regional dissociation of beta-endorphin and enkephalin content in rat brain and pituitary. *Proceedings of the National Academy of Sciences, U.S.A.*, **74**, 5162–5165.

Sar, M., Stumpf, W. E., Miller, R. J., Chang, K. J., and Cuatrecasas, P. (1978). Immunohistochemical localization of enkephalin in rat brain and spinal cord. *Journal of Comparative Neurology*, **182**, 17–38.

Satoh, M., Zieglgänsberger, W., Fries, W., & Herz, A. (1974). Opiate agonist-antagonist interaction at cortical neurones of naive and tolerant/dependent rats. *Brain Research*, **82**, 378–382.

Satoh, M., Zieglgänsberger, W., & Herz, A. (1975). Interaction between morphine and putative excitatory neurotransmitters in cortical neurones in naive and tolerant rats. *Life Sciences*, **17**, 75–80.

Satoh, M., Zieglgänsberger, W., & Herz, A. (1976). Actions of opiates upon single unit activity in the cortex of naive and tolerant rats. *Brain Research*, **115**, 99–110.

Schally, A. (1978). Aspects of hypothalamic regulation of the pituitary gland. *Science*, **202**, 18–28.

Segal, M. (1977). Morphine and enkephalin interactions with putative neurotransmitters in rat hippocampus. *Neuropharmacology*, **16**, 587–592.

Simantov, R., Kuhar, M. J., Uhl, G. R., & Snyder, S. H. (1977). Opioid peptide enkephalin: Immunohistochemical mapping in rat central nervous system. *Proceedings of the National Academy of Sciences,U.S.A.*, **74**, 2167–2171.

Simantov, R., & Snyder, S. H. (1976a). Morphine-like peptides in mammalian brain: Isolation, structure elucidation, and interaction with the opiate receptor. *Proceedings of the National Academy of Sciences, U.S.A.*, **73**, 2515–2519.

Simantov, R., & Snyder, S. H. (1976b). Brain pituitary opiate mechanisms. In H. W. Kosterlitz (Ed.), *Opiates and Endogenous Peptides*, pp. 41–48. Amsterdam: Elsevier.

Simon, E., & Hiller, J. M. (1978). The opiate receptors. *Annual Review of Pharmacology & Toxicology*, **18**, 371–394.

Smyth, D. G., & Snell, C. R. (1977). Metabolism of the analgesic peptide lipotropin C-fragment in rat striatal slices. *FEBS Letters*, **78**, 225–228.

Smyth, D. G., Snell, C. R., & Massey, D. E. (1978). Isolation of the C-fragment and C′-fragment of lipotropin from pig pituitary and C-fragment from brain. *Biochemical Journal*, **175**, 261–270.

Snyder, S. H. (1978). The opiate receptor and morphine-like peptides in the brain. *American Journal of Psychiatry*, **135**, 645–652.

Sofroniew, M., & Weindl, A. (1979). Projections from the parvocellular vasopressin and neurophysin-containing neurons of the suprachiasmatic nucleus. *American Journal of Anatomy*, **153**, 391–430.

Stern, A. S., Lewis, R. V., Kimura, S., Rossier, J., Gerber, L. D., Brink, L., Stein, S., & Udenfriend, S. (1979). Isolation of the opioid heptapeptide Met-enkephalin [Arg^2,Phe^3] from bovine adrenal medullary granules and striatum. *Proceedings of the National Academy of Sciences, U.S.A.*, **76**, 6680–6683.

Sternberger, L., & Hoffman, J. (1978). Immunocytology of luteinizing hormone-releasing hormone. *Neuroendocrinology*, **25**, 111–128.

Stumpf, W., & Sar, M. (1975). Hormone architecture of the mouse brain with ^{3}H-estradiol. In W. Stumpf and L. Grant (Eds.), *Anatomical Neuroendocrinology*. Basel: Karger.

Taylor, D., Hoffer, B., Zieglgänsberger, W., Siggins, G., Ling, N., Seiger, Å., & Olson, L. (1979). Opioid peptides excite pyramidal neurons and evoke epileptiform activity in hippocampal transplants in oculo. *Brain Research*, **176**, 135–142.

Teschemacher, H., Opheim, K. E., Cox, B. M., & Goldstein, A. (1976). A peptide-like substance from pituitary that acts like morphine. I. Isolation. *Life Sciences*, **16**, 1771–1775.

Terenius, L. (1978a). Endogenous peptides and analgesia. *Annual Review of Pharmacology and Toxicology*, **18**, 189–204.

Terenius, L. (1978b). Significance of endorphins in endogenous antinociception. In E. Costa and M. Trabuchi, (Eds.), *The Endorphins: Advances in Biochemical Pharmacology*, Vol. 18, pp. 321–332. New York: Raven.

Uhl, G. R., Goodman, R. R., Childers, S. R., & Snyder, S. H. (1978a). Immunohistochemical mapping of enkephalin containing cell bodies, fibers and nerve terminals in the brain stem of the rat. *Brain Research*, **166**, 75–94.

Uhl, G., Kuhar, M., & Snyder, S. (1978b). Enkephalin-containing pathway: Amygdaloid efferents in the stria terminalis. *Brain Research*, **149**, 223–228.

Urca, G., Frenk, H., Liebeskind, J. C., & Taylor, A. N. (1977). Morphine and enkephalin: Analgesic and epileptic properties. *Science*, **197**, 83–86.

Volavka, J., Marya, A., Baig, S., & Perez-Cruet, J. (1977). Naloxone in chronic schizophrenia. *Science*, **196**, 1227–1228.

Wamsley, J. K., Young, W. S., & Kuhar, M. J. (1980). Immunohistochemical localization of enkephalin in rat forebrain. *Brain Research*, **190**, 153–174.

Watson, S. J., Akil, H., Richards, C. W., III, & Barchas, J. D. (1978a). Evidence for two separate opiate peptide neuronal systems. *Nature (London)*, **275**, 226–228.

Watson, S. J., Akil, H., Sullivan, S., & Barchas, J. D. (1977). Immunocytochemical localization of methionine enkephalin: Preliminary observations. *Life Sciences*, 21, 733–738.

Watson, S. J., Richards, Ш, C. W., & Barchas, J. D. (1978b). Adrenocorticotropin in rat brain: Immunocytochemical localization in cells and axons. *Science*, 200, 1180–1182.

Yaksh, T. L. & Rudy, T. A. (1977). Studies on the direct spinal action of narcotics in the production of analgesia in the rat. *Journal of Pharmacology & Experimental Therapeutics*, 202, 411–428.

Yaksh, T. L., & Rudy, T. A. (1978). Narcotic analgesics: CNS sites and mechanisms of action as revealed by intracerebral injection techniques. *Pain*, 4.

Yang, H. Y., Hong, J. S., & Costa, E. (1977). Regional distribution of leu- and met-enkephalin in rat brain. *Neuropharmacology*, 16, 303–307.

Zakarian, S., & Smyth, D. (1979). Distribution of active and inactive forms of endorphins in rat brain and pituitary. *Proceedings of the National Academy of Sciences, U.S.A.*, 76, 5972–5976.

Zieglgänsberger, W., & Bayerl, H. (1976). The mechanism of inhibition of neuronal activity by opiates in the spinal cord of cats. *Brain Research*, 115, 111–138.

Zieglgänsberger, W., French, E., Siggins, G. R., & Bloom, F. E. (1979). Opioid peptides may excite hippocampal pyramidal neurons by inhibiting adjacent inhibitory interneurons. *Science*, 205, 415–417.

Zieglgänsberger, W., & Fry, J. P. (1976). Actions of enkephalin on cortical and striatal neurones of naive and morphine tolerant/dependent rats. *In* H. W. Kosterlitz (Ed.), *Opiates and Endogenous Opioid Peptides*, pp. 231–238. Amsterdam: Elsevier.

Zieglgänsberger, W., Fry, J. P., Herz, A., Moroder, L., & Wunsch, E. (1976). Enkephalin-induced inhibition of cortical neurons and the lack of this effect in morphine tolerant/dependent rats. *Brain Research*, 115, 160–164.

Zieglgänsberger, W., & Puil, E. A. (1972). Actions of glutamic acid on spinal neurones. *Experimental Brain Research*, 17, 35–49.

Zieglgänsberger, W., Siggins, G., French, E., & Bloom, F. (1978). Effects of opioids on single unit activity. *In* J. van Ree and L. Terenius (Eds.), *Characteristics and Functions of Opiods*, pp. 75–86. Amsterdam: Elsevier.

Zieglgänsberger, W., & Tulloch, I. F. (1979). The effects of methionine- and leucine-enkephalin on spinal neurons of the cat. *Brain Research*, 167, 53–64.

CHAPTER 11

Endorphin Influences on Learning and Memory

Gábor L. Kovács and David de Wied

Endorphins and enkephalins, a newly discovered class of neuropeptides derived from β-lipotropin, have been shown to affect behavioral adaptation. Adaptation is a complex neurobiological and humoral response that involves learning and memory components with goal directed aspects of attention, motivation, and perception. The influence of endorphins on learning and memory processes has therefore been studied in detail. This chapter describes the effects of β-endorphin and its fragments, α-endorphin and γ-endorphin, on a fear-motivated passive avoidance response. This measure has been widely used to assess the influence of various classes of neuropeptides on learning and memory. β-Endorphin facilitates retention of a passive avoidance response in a dose- and time-dependent manner. This facilitation of passive avoidance behavior was not reversible by pretreatment with a specific opiate antagonist, naltrexone, suggesting that β-endorphin affected the fear-motivated passive avoidance behavior independently, either from opiate receptors in the brain or at least from naltrexone-sensitive endogenous binding sites. α-Endorphin and γ-endorphin, both possible endogenous products generated by the biotransformation of β-endorphin, exerted opposite effects on passive avoidance behavior. α-Endorphin caused a time-dependent facilitation, whereas γ-endorphin caused a time-dependent attenuation of information processing. These data suggest an opposite influence of α-type and γ-type of endorphins on learning and memory processes and suggest that one aspect of the global influence of endorphins on adaptive mechanisms may be changes in consolidation and retrieval of newly acquired information.

ENDOGENOUS PEPTIDES
AND LEARNING AND MEMORY PROCESSES

I. INFLUENCE OF ENDORPHINS ON BEHAVIORAL PROCESSES: INFORMATION PROCESSING AS AFFECTED BY BIOTRANSFORMATION OF β-ENDORPHIN

It is thought that β-lipotropin may be a source of biologically active peptides, such as β-endorphin (β-LPH$_{61-91}$), α-endorphin (β-LPH$_{61-76}$), γ-endorphin (β-LPH$_{61-77}$), Des-tyrosine^{61}-γ-endorphin (β-LPH$_{62-77}$), Met-enkephalin (β-LPH$_{61-65}$), and β-MSH (β-LPH$_{41-58}$) (Bradbury, Smyth, Snell, Birdsall, & Hulme, 1976; Burbach, Loeber, Verhoef, Wiegant, de Kloet, & de Wied, 1980; Chrétien, Benjannet, Dragon, Seidah, & Lis, 1976; Gráf, Rónai, Bajusz, Cseh, & Székely, 1976; Li & Chung, 1976). That pituitary neuropeptides related to β-LPH influence functions of the central nervous system (CNS) other than nociception has been shown earlier. ACTH and fragments of ACTH such as ACTH$_{4-10}$ influence motivational, attentional, and learning processes (Bohus, Hendrickx, van Kolfschoten, & Krediet, 1975; de Wied, Witter, & Greven, 1975; Kastin, Sandman, Stratton, Schally, & Miller, 1975). More recently, β-endorphin has also been implicated in the control of adaptive behavior (de Kloet & de Wied, 1980; de Wied, Bohus, van Ree, & Urban, 1978b). Small amounts of β-endorphin (0.1–0.3 μg, administered subcutaneously) increased resistance to extinction of a pole-jumping avoidance paradigm (Bohus & Gispen, 1978; de Wied et al. 1978b). In a similar manner to ACTH$_{4-10}$, β-endorphin also caused a shift in the dominant frequency of hippocampal theta activity toward higher frequencies during paradoxical sleep (de Wied et al. 1978b). Thus, it seems that β-endorphin increased the arousal state of the animals and, thereby, may have affected pain-motivated behavioral processes, such as active avoidance behavior.

There is also evidence that ACTH and ACTH-like peptides affect memory processes (de Wied, 1966; Rigter, Elbertse & van Riezen, 1975). This effect of ACTH is compatible with the hypothesis that ACTH-like peptides may affect motivational process (de Wied et al. 1975), as motivational effects operate in most of the test situations used for measurement of learning and memory. Because β-endorphin and some fragments of this neuropeptide profoundly affect the extinction of pole-jumping (de Wied et al., 1978b; de Wied, Kovács, Bohus, van Ree, & Greven, 1978c) and, as this test situation involves motivational, learning, and memory components, it has been predicted that neuropeptides related to β-endorphin might also influence memory processes (Kovács, Bohus, Greven, & de Wied, 1978a). Retention of passive avoidance behavior provides a sensitive criterion for analyzing the nature of neuropeptide action on memory processes (Bohus, Kovács, & de Wied, 1978a; Kovács, Vécsei, & Telegdy, 1978b). Therefore, the influence of β-endorphin, α-endorphin, and γ-endorphin on the retention of a one-trial passive avoidance learning response was studied.

The behavior was investigated as described in detail by Ader, Weijnen and Moleman (1972). Briefly, the apparatus consists of a dark compartment and an illuminated, elevated pedestal attached to the outside. Because rats prefer dark, the animals readily leave the pedestal and enter the dark compartment through a small opening. After one habituation trial to the dark compartment, three pretraining trials are given on the following day. At the end of the third trial, an unavoidable electric footshock (0.25 mA for 2 sec) is delivered through the grid floor of the dark compartment. This comprises the learning trial. Retention of the passive avoidance response is tested 24 hr after the learning trial by placing each animal on the pedestal and measuring the latency to reenter the dark compartment. Since consolidation of memory, the input stage of information processing, takes place within the first few hours after training, treatments administered shortly after the learning trial may affect memory consolidation (McGaugh, 1966). If, however, a treatment is administered shortly before the retention trial that affects retention performance, retrieval processes might have been altered. The term retrieval or read-out mechanism designates the process by which acquired information is made available during recall.

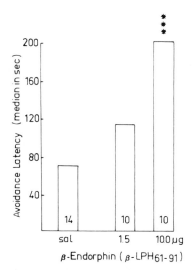

Fig. 1.   Effect of β-endorphin (β-LPH$_{61-91}$) on the retention of a one-trial learning, passive avoidance behavior, following subcutaneous administration of the peptide immediately after the learning trial. The numbers in bars indicate the number of animals used in each dose group. Probability level of tests of significance: ***$p < .01$; Mann-Whitney U-test.

Subcutaneous administration of human β-endorphin obtained from Organon, Oss, The Netherlands influenced retention of the passive avoidance response in a time- and dose-dependent manner. When injected immediately after the learning trial, 1.5 μg/rat of β-endorphin did not affect passive avoidance behavior, whereas a higher dose of 10.0 μg/rat significantly ($p < .01$) facilitated passive avoidance behavior as measured 24 hr after the peptide treatment (see Fig. 1).

In contrast, if β-endorphin was injected 1 hr before the retention trial, the lower dose of 1.5 μg caused a significant facilitation of the passive avoidance response (see Fig. 2). The higher dose of the peptide (10.0 μg), however, was ineffective.

These data indicate that β-endorphin exerts a dose-dependent influence on passive avoidance retention. The neuropeptide facilitates retention performance of a passive avoidance response when administered either immediately after the learning trial or shortly before the retention trial. However, different findings have been reported by Martinez and Rigter (1980). These authors administered β-endorphin i.p. to rats in a dose of 100.0 ng/kg and investigated the retention of a passive avoidance response. Given immediately after a single learning trial, β-endorphin caused a retention deficit. Interestingly enough, this effect was absent if the peptide treatment was postponed 90 min after the learning trial. It is important to note, however, that Martinez and Rigter (1980) used the i.p. route of peptide administration, whereas in our experiments, β-endorphin was administered subcutaneously. It is possible that different routes of administration resulted in different uptake or metabolism of β-endorphin. It may also be important that, in our experiments, the effect of the peptide was measured 24 hr after treatment, whereas Martinez and Rigter tested their animals 72 hr after training. Whether these or other factors such as differences in pretraining

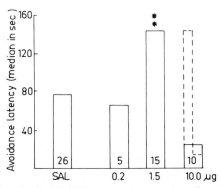

Fig. 2. Effect of β-endorphin (β-LPH$_{61-91}$) on the retention of passive avoidance behavior following subcutaneous administration of the peptide 1 hr prior to the retention trial. Probability level of tests of significance: ** $p < .02$.

procedures resulted in opposite findings remains to be answered. In any case, the conclusion of Martinez and Rigter that the direction of an endorphin effect on learning and memory processes is dependent upon the experimental parameters seems to be justified.

De Wied et al. (1978b) found that β-endorphin delayed the extinction of a pole-jumping avoidance response in rats. A similar effect was also observed when fragments of β-endorphin were administered. Either α-endorphin or Met-enkephalin also delayed extinction. Interestingly, α-endorphin was even more potent that β-endorphin. However, γ-endorphin produced opposite effects and facilitated extinction. These data prompted de Wied et al. (1978b) to conclude that the β-endorphin molecule may carry opposite information and the eventual effect on avoidance behavior may depend on biotransformation of this peptide. The present data obtained with passive avoidance behavior seem to confirm this hypothesis.

In our experiments, we found that lower doses of β-endorphin facilitated passive avoidance behavior if the peptide was administered shortly before retention testing. However, a higher dose (10.0 μg/rat) of β-endorphin was ineffective. Further analysis of the distribution of avoidance latencies revealed that the distribution of high and low avoidance latencies in the β-endorphin treated group was significantly different from that of control rats. Whereas in the saline-treated control group, the majority of the latencies were between 20 and 100 sec following β-endorphin treatment; the data clearly showed a bimodal distribution. Mostly high (more than 100 sec) or low (less than 20 sec) latencies were found in this group. A medium range of latencies characteristic of control rats was absent following this relatively high dose of β-endorphin. It is of interest to note that the same relatively high dose of α-endorphin caused a unidirectional shift toward high latencies whereas that of γ-endorphin caused a unidirectional shift toward lower latencies. Thus, it seems that the effects of relatively high doses of β-endorphin partially resemble that of α-endorphin and partially those of γ-endorphin (Table 1).

It might be hypothesized, that β-endorphin in low doses facilitates avoidance behavior because of an inherent behavioral potency of the peptide itself. In contrast to this unidirectional facilitation of passive avoidance behavior by low doses of β-endorphin, higher doses caused a bimodal distribution of avoidance latencies, presumably because, with high doses, the behavior was a result of the influence of β-endorphin and also α- and γ-type endorphins, generated as a result of biotransformation of β-endorphin in the brain.

This concept has recently received biochemical support. Austen, Smyth, and Snell (1977) showed that α-endorphin and γ-endorphin are generated as a result of the degradation of β-endorphin in an *in vitro* synaptosomal fraction of the rat brain. Burbach, Loeber, Verhoef, de Kloet, and de Wied (1979) found evidence for the presence of aminopeptidase and endopeptidase activity

TABLE 1

Effects of β-, α-, and γ-Endorphin on the Distribution of Passive Avoidance Latencies[a]

Treatment	N	Low (%)[b]	Medium (%)[c]	High (%)[d]	p
Saline	20	5	75	20	—
β-Endorphin	10	50	10	40	< .001 versus saline
α-Endorphin	7	0	29	71	< .05 versus saline
γ-Endorphin	9	78	22	0	< .001 versus saline

[a] Rats were treated s.c. with saline or 10.0 μg of the peptides 1 hr before the retention test of one-trial learning passive avoidance behavior. The data were statistically analyzed with a Chi square test.
[b] Low: latencies below 20 sec.
[c] Medium: latencies between 21–100 sec.
[d] High: latencies higher than 100 sec.

in synaptosomal plasma membrane preparations. In addition, Burbach et al. (1980) reported that α-endorphin and γ-endorphin together with the respective Des-tyrosine[61]-analogs are formed from β-endorphin. It appears that differences in pH in a narrow, near-physiological range determined whether the biotransformation to γ-endorphin and subsequently to Des-tyrosine[61]-γ-endorphin or to α-endorphin and then to Des-tyrosine[61]-α-endorphin was the determining factor in the enzymatic conversion of β-endorphin (Burbach et al. 1980).

The possibility that the effects of β-endorphin on passive avoidance behavior were mediated by an interaction with endogenous opiate binding sites (i.e., receptors), was tested in the following experiments. For this purpose, naltrexone was used as a specific opiate receptor antagonist (Martin, Jasinski, & Mansky, 1973). Naltrexone treatment alone (90 μg, s.c.) did not affect passive avoidance behavior except for a slight, nonsignificant, attenuation. This finding does not exclude the possibility, however, that endogenous opiate binding sites are involved in the control of adaptive behavior. In fact, Messing, Jensen, Martinez, Spiehler, Vasquez, Soumireu-Mourat, Liang, and McGaugh (1979) found that naloxone, another opiate receptor antagonist, facilitates retention in a one-trial inhibitory avoidance task when administered immediately and 30 min after training. These authors concluded that endogenous opioid systems are involved in memory consolidation processes. The two findings, however, are not directly comparable because of the different drugs, doses, and different times of administration relative to the learning trial. It might be possible that higher doses of naltrexone would have affected passive avoidance behavior, as has been shown for the extinction of active avoidance behavior (de Wied et al., 1978a). In this experiment, however, we were mainly interested in using a test dose that would not affect passive avoidance behavior. This dose was used for further experimentation.

As shown in Fig. 3, naltrexone pretreatment did not prevent the facilitation of passive avoidance behavior, which occurred after administration of a low dose of β-endorphin challenge. The same amount of naltrexone, however, effectively antagonized the antinociceptive action of higher doses of β-endorphin (Bohus, unpublished observation) and that of morphine (Kovács, Bohus, & de Wied, in preparation). Thus, these data support earlier findings by de Wied et al. (1978b) that endorphins influence adaptive behavior by a mode of action that is probably independent of opiate receptors. Similar conclusions were reached by Rigter (1978) and by Rigter, Greven, and van Riezen (1977). These authors administered Met-enkephalin or Leu-enkephalin subcutaneously in microgram doses to rats shortly before retention testing in a passive avoidance situation and observed that both peptides attenuated CO_2-induced amnesia. In contrast, analgesic activity of enkephalins is only present when administered into the cerebral ventricles. Naloxone treatment did not prevent the antiamnesic effect of the opioid peptides (Rigter et al., 1977), whereas the antinociceptive effect of these peptides was reversible with naloxone treatment. These findings are in agreement with those of Kastin, Scollan, King, Schally, and Coy (1976) who found that rats made fewer errors in a maze-learning paradigm after enkephalin treatment, even in the presence of naloxone. Naloxone in high doses (10 mg/kg), on the other hand, reverses the effects of Leu-enkephalin (Rigter, Hannan, Messing, Martinez, Vasquez, Jensen, Veliquette, & Mc-Gaugh, 1980) and Met-enkephalin (Martinez, Rigter, & van der Gugten, 1980) on active avoidance conditioning.

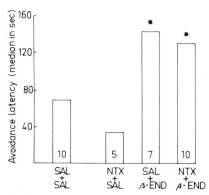

Fig. 3.   Effect of β-endorphin on the retention of passive avoidance behavior in normal and naltrexone-pretreated rats. Naltrexone hydrochloride (90.0 μg s.c.) was administered 120 min before the retention trial. β-Endorphin (1.5 μg s.c.) was injected 1 hr after naltrexone treatment. Probability level of tests of significance: * $p < .05$.

That β-endorphin influences CNS processes involved in learning by a mechanism of action independent from opiate receptors is also supported by findings from other types of experiments. Chronic intravenous administration of non-analgesic amounts of β-endorphin influenced operantly conditioned behavior. Although the peptide did not affect the number of trials to achieve the learning criterion, it significantly lengthened response latencies (Gorelick, Catlin, George, & Li, 1978).

In conclusion, these data support the notion that β-endorphin affects adaptive behavior via non-opioid mechanisms, partly by metabolic conversion of the opioid peptide to shorter neuropeptide sequences.

II. TIME-GRADIENT EFFECTS OF α-ENDORPHIN AND γ-ENDORPHIN ON PASSIVE AVOIDANCE BEHAVIOR

In the light of the preceding data, it is of interest to note that α-endorphin delayed extinction of pole-jumping avoidance behavior, whereas γ-endorphin had an opposite effect (de Wied et al. 1978a; de Wied, Kovács, Bohus, van Ree, & Greven, 1978c). Thus, two possible endogenous metabolites of β-endorphin influenced behavior in an opposite direction suggesting that metabolism of β-endorphin may ensure the homeostatic modulation of adaptive behavior. This concept has received experimental support from other laboratories as well. For example, γ-endorphin appeared to facilitate extinction of shuttlebox avoidance (Király, Borsy, Tapfer, & Gráf, 1979), and Le Moal, Koob, and Bloom (1979) have replicated the data of de Wied et al. (1978b) and confirmed that α-endorphin delayed, while γ-endorphin facilitated extinction of pole-jumping avoidance behavior.

As both learning and memory are dependent upon many CNS processes that they have in common, it is rather difficult to distinguish between them. Such a distinction, however, would be important for both conceptual and practical reasons. The term *learning* usually refers to those biological responses involved in the acquisition of information. This includes perception of sensory information and a variety of performance factors operating during any environmental situation leading to memory formation (Zornetzer, 1978). The few data available on the influence of endorphins on learning are rather conflicting. In addition, a major problem with testing the influence of opioid peptides on avoidance learning is that these substances, when injected before the learning trial, induce overall changes in an organism's response to its environment (Meyerson, 1979; Meyerson & Terenius, 1977), including effects on sensory inputs at the perceptual and attentional level. These overall effects can be clearly demonstrated with central administration of endorphin fragments.

Peripheral injection of relatively low doses of α- or γ-endorphin does not seem to affect learning in a shuttlebox (Bohus, in press). Higher doses of γ-endorphin (10.0–100.0 μg), however, seem to retard learning in a similar test situation (Király et al., 1979). In contrast, γ-endorphin caused a slight, but significant facilitation of problem-solving behavior in a food-rewarded multiple-maze situation (Bohus, in press). Because animals treated with γ-endorphin made fewer errors than the controls, Bohus (in press) concluded that this neuropeptide improved cognitive function. In contrast, rats receiving α-endorphin made more errors in approaching the goal compartment, suggesting that cognitive processes might be influenced by the two fragments of β-lipotropin in an opposite way.

Thus, an effect of endorphins on motivational processes cannot be excluded when peptide treatments are given before the learning trial. A positive influence of α-endorphin on some specific motivational substrate such as fear would explain the resistance to extinction in the active avoidance reaction (de Wied et al., 1978b) as well as improved performance in the passive avoidance situation. Conversely, a decrease in the motivational value of fearful environmental stimuli would be an explanation for the facilitation of active avoidance performance and the attenuation of passive avoidance behavior, which occurs after γ-endorphin treatment (de Wied et al., 1978b,c; Király et al., 1979; Le Moal et al., 1979). It is not likely, however, that α- and γ-endorphin are specifically effective in aversively motivated situations. Dorsa, van Ree, and de Wied (1979) showed that α-endorphin facilitates self-stimulation behavior of the ventral tegmental area in rats. Thus, this neuropeptide also affects behavior in a typical reward situation.

However, the data thus far indicate that the effect of α- and γ-endorphin might be different in aversively and appetitively motivated behavioral situations. Unlike the effects seen with aversively motivated behavior, both α-endorphin and γ-endorphin increased resistance to extinction of an appetitive runway task for a water reward (Le Moal et al., 1979). In a runway response for a food reward, however, the Des-tyrosine[61] analogs of both α-endorphin (β-LPH$_{62-76}$) and of γ-endorphin (β-LPH$_{62-77}$) facilitate extinction (Grossi & Bohus, personal communication).

In addition to these psychological explanations, the influence of endorphins on avoidance learning can also be analyzed from a psychopharmacological point of view. γ-Endorphin, and Des-tyrosine[61]-γ-endorphin both affect adaptive behavior in learning situations in a manner similar to the known neuroleptic compound, haloperidol (de Wied et al., 1978b; de Wied, Bohus, van Ree, Kovács, & Greven, 1978c; Kovács & de Wied, 1978a). Thus, a neuroleptic-like effect of endorphin fragments closely related to γ-endorphin might explain some influences of endorphins on avoidance behavior.

A further factor that might complicate our understanding of the influence of endorphins on learning is their possible action on locomotor or exploratory activity. This may be crucial, because changes in locomotor activity produced by endorphin treatment might result in false conclusions regarding learning and memory. In relatively high doses, the influence of endorphins on exploratory (open-field) behavior is controversial. Veith, Sandman, Walker, Coy, and Kastin (1978) reported that 100.0 μg β-endorphin consistently increased locomotor activity of rats in an open-field test. The same group (King, Kastin, Olson, & Coy, 1979), however, showed that 80.0 μg/kg of β-endorphin significantly depressed activity in the home cage of the animals. These authors concluded that a novel environment, such as the open field, might have modified the effect of β-endorphin on exploratory activity. However, data of Meyerson (1979; Meyerson & Terenius, 1977) indicate that the effect of β-endorphin on exploratory behavior depends on variables other than a novel environment. These authors studied the exploratory and sociosexual behavior of male rats and established that certain patterns of the behavioral repertoire, such as amicable behavior of males, were reduced by β-endorphin treatment in male–male encounters, but increased in male–female encounters. Thus, they concluded that β-endorphin affects the sensory information that the animal samples from its environment.

Although larger amounts of endorphins affect locomotor activity, such an effect is not likely to interfere with the changes in avoidance behavior because of the low doses of the peptides used. De Wied et al. (1978c) showed that α-endorphin, γ-endorphin, Des-tyrosine^{61}-γ-endorphin, and β-endorphin do not significantly affect the rate of ambulation, rearing, or grooming in doses of 10.0–50.0 μg, administered subcutaneously. The same peptides, however, are highly effective in altering adaptive behavior in doses of 0.03–1.5 μg (de Wied et al., 1978a,b,c).

In contrast to the term learning, which refers to the biological responses operating in the presence of an environmental stimulus leading to acquisition, the term *memory* refers to those neurobiological responses that take place after the learning situation has terminated. Therefore, an influence of neuropeptides on memory processes or on information processing can only be demonstrated if the peptide is administered after the learning trial and the animal is trained in a drug-free state. Memory processes are conventionally thought to involve consolidation, storage, and retrieval of acquired information. A substance that influences memory consolidation processes should be effective during the first few hours after the learning trial (critical period). Furthermore, it is also necessary to demonstrate that the same treatment is ineffective if given after this sensitive period has passed. Thus, a time-gradient effect of the substance should be demonstrated (McGaugh, 1966). Very few data are available on the influence of endorphins on memory processes.

Taking these criteria into account, the time-gradient effect of low doses of α-endorphin and γ-endorphin was investigated, using the one-trial passive avoidance task.

α-Endorphin was injected s.c. at various time intervals after the single learning trial, and passive avoidance retention was tested 24 hr after the learning trial. Given immediately after the learning trial α-endorphin facilitated passive avoidance behavior. A slight facilitation occurred in passive avoidance responding when the peptide treatment was postponed 3 hr after the learning trial, but the neuropeptide was ineffective when administered 6 hr after the training. α-Endorphin facilitated performance of a passive avoidance response when the treatment was given 1 hr before the retention test (Table 2).

Essentially, the opposite effect was observed in rats receiving γ-endorphin. When injected immediately or 3 hr after the learning trial, γ-endorphin attenuated the passive avoidance response. The same dose of the peptide, however, was ineffective if administered 6 hr after training. A highly significant attenuation of passive avoidance retention was again observed if the peptide was given 1 hr before the retention testing (Table 3).

The time-dependent nature of the effects of both α- and γ-endorphin indicates that these peptides may influence memory processes. Because consolidation of newly acquired information is generally believed to take place in the brain within the first 3 hr after learning (de Wied, van Wimersma Greidanus, Bohus, Urban, & Gispen, 1976; McGaugh, 1966); one can assume that α-endorphin facilitates, whereas γ-endorphin attenuates memory consolidation in a passive avoidance task. Neither α-endorphin, nor γ-endorphin influenced the avoidance response when drug administration was postponed until the labile phase of memory formation had terminated. This finding indicates that an anterograde drug effect of the peptides on retention performance is not likely to have occurred. Thus, the influence of α-endorphin and γ-endorphin on retention of passive avoidance behavior might, in fact, be related to memory consolidation

TABLE 2

Time-Gradient Effect of α-Endorphin (β-LPH$_{61-76}$) on a One-Trial Passive Avoidance Learning Task

Time of treatment after the learning trial	Passive avoidance behavior 24 hr after the learning trial		
	Saline[a]	α-Endorphin[a,b]	p
0 hr (immediately after the learning trial)	70 (14)	168 (9)	<.01
3 hr	80 (5)	121 (6)	n.s.
6 hr	66 (5)	72 (5)	n.s.
23 hr	68 (14)	183 (6)	<.02

[a] Median avoidance latency in seconds.

[b] α-Endorphin was administered s.c. in a dose of 1.5 μg/animal; numbers in parentheses indicate the number of animals in each treatment group.

TABLE 3

Time-Gradient Effect of γ-Endorphin (β-LPH$_{61-77}$) on a One-Trial Passive Avoidance Learning Task

Time of treatment after the learning trial	Passive avoidance behavior 24 hr after the learning trial		
	Saline[a]	γ-Endorphin[a,b]	p
0 hr (immediately after the learning trial)	70 (14)	15 (10)	<.05
3 hr	80 (5)	22 (10)	<.05
6 hr	66 (5)	52 (10)	n.s.
23 hr	68 (14)	6 (9)	<.01

[a] Median avoidance latency in seconds.
[b] γ-Endorphin was injected s.c. in a dose of 10.0 μg/rat.

processes. The second peak effectiveness that appears 1 hr before retention testing suggests that the influence of endorphins is not only related to the input stage of memory; α-endorphin also appears to facilitate, whereas γ-endorphin attenuates memory retrieval. Martinez and Rigter (1980) studied the effect of α- and γ-endorphin on retention of a one-trial learning passive avoidance task. They observed that γ-endorphin administered i.p. immediately after training and α-endorphin given i.p. either before or after training was without effect on retention of the response. Further experiments are needed to resolve the contradiction between these and our data.

Even though endorphins affect memory processes, they cannot be regarded as a unique class of neuropeptides capable of influencing information processing. It has been shown that neuropeptides of posterior pituitary origin, vasopressin and oxytocin, exert even more specific effects on memory processes (Bohus, Urban, van Wimersma Greidanus, & de Wied, 1978a; de Wied, 1971; de Wied et al., 1976; Kovács et al., 1978b; Schulz, Kovács, & Telegdy, 1974). Vasopressin induces long-term facilitation of avoidance responding (Bohus et al., 1978a,b; de Wied, 1971; de Wied et al., 1976; Telegdy & Kovács, 1979). Another neurohypophyseal hormone, oxytocin, facilitates extinction of bench-jumping (Schulz et al., 1974) and pole-jumping avoidance reactions (Bohus et al., 1978a), and attenuates step-down (Kovács et al., 1978b) and step-through (Bohus et al., 1978a) passive avoidance responding. Since oxytocin exerts time-dependent effects on passive avoidance responding (Bohus et al., 1978b), it has been concluded that oxytocin may function as an amnesic neuropeptide.

Despite the striking similarities that exist between the influence on memory processes of vasopressin and α-endorphin, on the one hand, and of oxytocin and γ-endorphin, on the other hand, the effects of endorphins are basically different from those of the neurohypophyseal peptides. The neurohypophyseal hormones have CNS effects that are probably more specifically related to memory formation. Endorphins, although clearly affecting information processing, might have more general CNS effects primarily by influencing behavioral adaptation. In support of this notion, vasopressin improves the performance of animals, not only in fear-motivated behavioral reactions, but also in sexually motivated approach behavior. Male rats, trained in a maze to run for receptive females, choose the correct arm of the maze more frequently following treatment with desglycinamide-lysine vasopressin (Bohus, 1977). β-Endorphin treated male rats show less sexual behavior than normal rats (Meyerson, 1979). Also, vasopressin facilitates cognitive functions, as measured in a multiple-maze paradigm in food deprived animals (Bohus et al., in preparation) or in nondeprived animals that run for palatable food (Kovács, Bohus, & de Wied, unpublished observations). α-Endorphin seems to improve information processing primarily when the task is closely related to fear motivation. As outlined previously, this neuropeptide interferes with cognitive functions in an appetitively motivated maze-learning situation. A further, basically different, aspect of their CNS effect is that, whereas vasopressin decreases brain-stimulation reward (Schwarzberg, Hartmann, Kovács & Telegdy, 1976), α-endorphin appears to facilitate it (Dorsa et al., 1979).

These findings indicate the mode of action of neuropeptides derived from β-lipotropin and those of posterior pituitary origin might be different, despite the similar appearance of their effects in fear motivated behavioral situations. Recent biochemical findings support this notion. Vasopressin facilitates catecholamine turnover, mainly in limbic structures (Tanaka, de Kloet, de Wied, & Versteeg, 1977), and the influence of this peptide on memory consolidation is most likely mediated by an interaction of the neuropeptide with limbic–midbrain terminals of the coeruleotelencephalic (dorsal noradrenergic) bundle (Kovács, Bohus, & Versteeg, 1979). It is not entirely clear, however, what biochemical mechanism is responsible for the effects of endorphins on adaptive behavior. Although endorphins may act as neuroregulators in the CNS by affecting the metabolism of catecholamines and indolamines (Izumi, Motomatsu, Chrétien, Butterworth, Lis, Seidah, & Barbeau, 1977; van Loon, & Kim, 1977), acetylcholine or GABA (Moroni, Cheney, & Costa, 1978), the doses needed to produce these biochemical effects were in the same range as required to induce analgesia. Naltrexone or naloxone pretreatment abolished the majority of these biochemical effects.

It might be more than a coincidence that dopamine turnover in certain brain areas was oppositely influenced by α- and γ-type endorphins. This finding is parallel to the changes induced by these peptides in avoidance behavior. Versteeg, de Kloet, and de Wied (1979) administered β-endorphin, α-endorphin, and Des-tyrosine[61]-γ-endorphin in rather low doses (100.0 ng into the cerebral ventricles) and studied the turnover of noradrenaline and dopamine in discrete brain regions. They found opposite effects of α-type and γ-type endorphins that were restricted to dopamine turnover in the zona incerta and in the hypothalamic magnocellular area (nucleus paraventricularis).

When considering the similar effects of neurohypophyseal hormones and endorphins on avoidance behavior, one must keep in mind that the opiate alkaloid morphine affects the release of various pituitary hormones (de Wied, van Ree, & de Jong, 1974; Kokka, Garcia, George, & Elliott, 1972; Martin et al., 1975). More recently, it has been proposed that pituitary enkephalin fibers play a role in vasopressin secretion (Rossier, Battenberg, Pittman, Bayon, Koda, Miller, Guillemin, & Bloom, 1979). Like the enkephalins, β-endorphin also inhibits stimulus evoked vasopressin release (Iversen, Iversen, & Bloom, 1980) and decreases vasopressin levels in rat blood plasma (van Wimersma Greidanus, Thody, Verspaget, de Rotte, Goedemans, Croiset, & van Ree, 1979). Interestingly, this latter effect of β-endorphin could not be blocked by naltrexone (van Wimersma Greidanus et al., 1979). Because of these interactions, more experimental work is needed to reveal similarities and dissimilarities between the influence of endorphin fragments and neurohypophyseal hormones on information processing to understand the possible interrelationships between these two classes of neuropeptides.

In conclusion, the data favor the hypothesis that peripheral administration of relatively low amounts of α-endorphin facilitates, whereas that of γ-endorphin attenuates memory processes. These CNS effects might appear to be one aspect of the wide-scale influence of endorphins on adaptive processes. Further experiments are required to determine the mode of action of α- and γ-endorphin on the brain.

REFERENCES

Ader, R., Weijnen, J.A.W.M., & Moleman, P. (1972). Retention of a passive avoidance response as a function of the intensity and duration of electric shock. *Psychonomic Science*, 26, 125–128.
Austen, B. M., Smyth, D. G., & Snell, C. R. (1977). γ-Endorphin, α-endorphin and Met-enkephalin are formed extracellularly from lipotropin C-fragment. *Nature*, 269, 619–621.
Bohus, B. (1977). Effect of desglycinamide-lysine vasopressin (DG-LVP) on sexually motivated T-maze behavior in the male rat. *Hormones and Behavior*, 8, 52–61.
Bohus, B. (in press). Endorphins and behavioral adaptation. *Advances in Biological Psychiatry*.
Bohus, B. & Gispen, W. H. (1978). The role of endorphins in behavior. In J. M. van Ree & L. Terenius (Eds.), *Characteristics and Function of Opioids*, pp. 367–376. Amsterdam: Elsevier/North Holland.
Bohus, B., Hendrickx, H. H. L., van Kolfschoten, A. A., & Krediet, T. G. (1975). Effect of ACTH$_{4-10}$ on copulatory and sexually motivated approach behavior in the male rat. In M. Sandler & G. L. Gessa (Eds.), *Sexual Behavior: Pharmacology and Biochemistry*, pp. 269–275. New York: Raven.

Bohus, B., Kovács, G. L., & de Wied, D. (1978a). Oxytocin, vasopressin and memory: Opposite effects on consolidation and retrieval processes. *Brain Research*, **157**, 414–417.

Bohus, B., Urban, I., van Wimersma Greidanus, Tj. B., & de Wied, D. (1978b). Opposite effects of oxytocin and vasopressin on avoidance behavior and hippocampal theta rhythm in the rat. *Neuropharmacology*, **17**, 239–247.

Bradbury, A. F., Smyth, S. G., Snell, C. R., Birdsall, N. J. M., & Hulme, E. C. (1976). C-Fragment of lipotropin has a high affinity for brain opiate receptors. *Nature*, **260**, 793–795.

Burbach, J. P. H., Loeber, J. G., Verhoef, J., de Kloet, E. R., & de Wied, D. (1979). Bio-transformation of endorphins by a synaptosomal plasma membrane preparation of rat brain and by human serum. *Biochemical and Biophysical Research Communications*, **86**, 1296–1303.

Burbach, J. P. H., Loeber, J. G., Verhoef, J., Wiegant, V. M., de Kloet, E. R., & de Wied, D. (1980). Selective conversion of β-endorphin into peptides related to γ- and α-endorphin. *Nature*, **283**, 96–97.

Chrétien, M., Benjannet, S., Dragon, N., Seidah, N. G., & Lis, M. (1976). Isolation of peptides with opiate activity from sheep and human pituitaries: Relationship to β-LPH. *Biochemical and Biophysical Research Communications*, **72**, 472–478.

de Kloet, E. R., & de Wied, D. (1980). The brain as target tissue for hormones of pituitary origin: Behavioral and biochemical studies. *In* L. Martini & W. H. Ganong (Eds.), *Frontiers in Neuroendocrinology*, pp. 157–201. New York: Raven.

de Wied, D. (1966). Inhibitory effect of ACTH and related peptides on extinction of conditioned avoidance behavior in rats. *Proceedings of the Society for Experimental Biology and Medicine*, **122**, 28–32.

de Wied, D. (1971). Long-term effect of vasopressin on the maintenance of a conditioned avoidance response in rats. *Nature*, **232**, 58–60.

de Wied, D., Bohus, B., van Ree, J. M., Kovács, G. L., & Greven, H. M. (1978a). Neuroleptic-like activity of (Des-Tyr¹)-γ-endorphin in rats. *Lancet*, **1**: 1046.

de Wied, D., Bohus, B., van Ree, J. M., & Urban, I. (1978b). Behavioral and electrophysiological effects of peptides related to lipotropin (β-LPH). *Journal of Pharmacology and Experimental Therapeutics*, **204**, 570–580.

de Wied, D., Kovács, G. L., Bohus, B., van Ree, J. M., & Greven, H. M. (1978c). Neuroleptic activity of the neuropeptide β-LPH 62–77 (Des-Tyr¹-γ-endorphin; DTγE). *European Journal of Pharmacology*, **49**, 427–436.

de Wied, D., van Ree, J. M., & de Jong, W. (1974). Narcotic analgesics and the neuroendocrine control of anterior pituitary function. *In* E. Zimmerman & R. George (Eds.), *Narcotics and the Hypothalamus*, pp. 251–266. New York: Raven.

de Wied, D., van Wimersma Greidanus, Tj.B., Bohus, B., Urban, I., & Gispen, W. H. (1976). Vasopressin and memory consolidation. *In* M. A. Corner & D. F. Swaab (Eds.), *Perspectives in Brain Research: Progress in Brain Research* (Vol. 45), pp. 181–194. Amsterdam: Elsevier.

de Wied, D., Witter, A., & Greven, H. M. (1975). Behaviorally active ACTH analogs. *Biochemical Pharmacology*, **24**, 1463–1468.

Dorsa, D. M., van Ree, J. M., & de Wied, D. Effects of (Des-Tyr¹)-γ-endorphin and α-endorphin on substantia nigra self-stimulation. *Pharmacology, Biochemistry and Behavior*. 1979, **10**, 899–905.

Gorelick, D. A., Catlin, D. H., George, R., & Li, C. H. (1978). Beta-endorphin is behaviorally active in rats after chronic intravenous administration. *Pharmacology, Biochemistry and Behavior*, **9**, 385–386.

Gráf, L., Rónai, A. Z., Bajusz, S., Cseh, G., & Székely, J. I. (1976). Opioid agonist activity of β-lipotropin fragments: A possible biological source of morphine-like substances in the pituitary. *FEBS Letters*, **64**, 181–184.

Iversen, L. L., Iversen, S. D., & Bloom, F. E. (1980). Opiate receptors influence vasopressin release from nerve terminals in rat neurohypophysis. *Nature*, **284**, 350–351.

Izumi, K., Motomatsu, T., Chrétien, M., Butterworth, R. F., Lis, M., Seidah, N., & Brabeau, A. (1977). β-Endorphin induced akinesia in rats: Effect of apomorphine and α-methyl-*p*-tyrosine and related modifications on dopamine turnover in the basal ganglia. *Life Sciences*, **20**, 1149–1156.

Kastin, A. J., Sandman, C. A., Stratton, L. O., Schally, A. V., & Miller, L. H. (1975). Behavioral and electrographic changes in rat and man after MSH. In W. H. Gispen, Tj. B. van Wimersma Greidanus, B. Bohus, & D. de Wied (Eds.), *Hormones, Homeostasis, and the Brain: Progress in Brain Research* (Vol. 42), pp. 143–150. Amsterdam: Elsevier.

Kastin, A. J., Scollan, E. L., King, M. G., Schally, A. V., & Coy, D. H. (1976). Enkephalin and a potent analog facilitates maze performance after intraperitoneal administration in rats. *Pharmacology, Biochemistry and Behavior*, **5**, 691–695.

King, M. G., Kastin, A. J., Olson, R. D., & Coy, D. H. (1978). Systemic administration of Met-enkephalin, (D-Ala²)-Met-enkephalin, β-endorphin, and (D-Ala²)-β-endorphin: Effects on eating, drinking and activity measures in rats. *Pharmacology, Biochemistry and Behavior*, **11**, 407–411.

Király, I., Borsy, J., Tapfer, M., & Gráf, L. (1979). Study on the neuroleptic activity of endorphins. *Third Congress of the Hungarian Pharmacological Society*, Budapest, August 22–25, Abstract Book, p. 46.

Kokka, N., Garcia, J. F., George, R., & Elliott, H. W. (1972). Growth hormone and ACTH secretion: Evidence for and inverse relationship in rats. *Endocrinology*, 1972, **90**, 735–743.

Kovács, G. L., & de Wied, D. (1978b). Effects of amphetamine and haloperidol on avoidance behavior and exploratory activity. *European Journal of Pharmacology*, **53**, 103–107.

Kovács, G. L., Bohus, B., Greven, H. M., & de Wied, D. (1978a) The effects of endorphins on memory processes in rats. *Proceedings of the Ninth International Congress of the Psycho-neuroendocrine Society*, **42**.

Kovács, G. L., Bohus, B., & Versteeg, D. H. G. (1979). Facilitation of memory consolidation by vasopressin: Mediation by terminals of the dorsal noradrenergic bundle? *Brain Research*, **172**, 73–85.

Kovács, G. L., Vécsei, L., & Telegdy, G. (1978c). Opposite action of oxytocin to vasopressin in passive avoidance behavior in rats. *Physiology and Behavior*, **20**, 801–802.

Le Moal, M., Koob, G. F., & Bloom, F. E. (1979). Endorphins and extinction: Differential actions on appetitive and aversive tasks. *Life Sciences*, **24**, 1631–1636.

Li, C. H., & Chung, D. (1976). Isolation and structure of an untriakontapeptide with opiate activity from the camel pituitary. *Proceedings of the National Academy of Sciences, U.S.A.*, **73**, 1145–1148.

Martin, W. R., Jasinski, D. R. & Mansky, P. A. (1973). Naltrexone, an antagonist for the treatment of heroin dependence. *Archives of General Psychiatry*, **28**, 784–791.

Martinez, Jr., J. L., & Rigter, H. (1980). Endorphins alter acquisition and consolidation of an inhibitory avoidance response in rats. *Neuroscience Letters*, **19**, 197–201.

Martinez, Jr., J. R., Rigter, H., & van der Gugten, J. (1980). Enkephalin effects on avoidance conditioning are dependent on the adrenal cortex. *Proceedings of the International Union of Physiological Sciences XIV*. Budapest, (Abstract No. 2352), 568.

McGaugh, J. L. (1966). Time-dependent processes in memory storage. *Science*, **153**, 1351–1358.

Messing, R. B., Jensen, R. A., Martinez, Jr., J. L., Spiehler, V. R., Vasquez, B. J., Soumireu-Mourat, B., Liang, K. C., & McGaugh, J. L. (1979). Naloxone enhancement of memory. *Behavioral and Neural Biology*, **27**, 266–275.

Meyerson, B. J. (1979). Hypothalamic hormones and behavior. *Medical Biology*, **57**, 69–83.

Meyerson, B. J., & Terenius, L. (1977). β-Endorphin and male sexual behavior. *European Journal of Pharmacology*, **42**, 191–192.

Moroni, F., Cheney, D. L., & Costa, E. (1978). Turnover rate of acetylcholine in brain nuclei of rats injected intraventricularly and intraseptally with alpha- and beta-endorphin, *Neuropharmacology*, **17**, 191–196.

Rigter, H. (1978). Attenuation of amnesia in rats by systemically administered enkephalins. *Science*, **200**, 83–85.

Rigter, H., Elbertse, R., & van Riezen, H. (1975). Time-dependent anti-amnestic effect of ACTH$_{4-10}$ and desglycinamide-lysine vasopressin. *In* W. H. Gispen, Tj. B. van Wimersma Greidanus, B. Bohus, & D. de Wied (Eds.), *Hormones Homeostasis and the Brain: Progress in Brain Research* (Vol. 42), pp. 163–171. Amsterdam: Elsevier.

Rigter, H., Greven, H. M., & van Riezen, H. (1977). Failure of naloxone to prevent reduction of amnesia by enkephalins. *Neuropharmacology*, **16**, 545–547.

Rigter, H., Hannan, T. J., Messing, R. B., Martinez, Jr., J. L., Vasquez, B. J., Jensen, R. A., Veliquette, J., & McGaugh, J. L. (1980). Enkephalins interfere with acquisition of an active avoidance response. *Life Sciences*, **26**, 337–345.

Rossier, J., Battenberg, E., Pittman, Q., Bayon, A., Koda, L., Miller, L., Guillemin, R., & Bloom, F. (1979). Hypothalamic enkephalin neurones may regulate the neurohypophysis, *Nature*, **277**, 653–655.

Schulz, H., Kovács, G. L., & Telegdy, G. (1974). Effect of doses of vasopressin and oxytocin on avoidance behavior in rats. *Acta Physiologica Academiae Scientiarum Hungaricae*, **45**, 211–215.

Schwarzberg, H., Hartmann, G., Kovács, G. L., & Telegdy, G. (1976). Effect of intraventricular oxytocin and vasopressin on self-stimulation in rats. *Acta Physiologica Academiae Scientiarum Hungaricae*, **47**, 127–131.

Tanaka, M., de Kloet, E. R., de Wied, D., & Versteeg, D.H.G. (1977). Arginine^{8}-vasopressin affects catecholamine metabolism in specific brain nuclei. *Life Sciences*, **20**, 1799–1808.

Telegdy, G., & Kovács, G. L. (1979). Role of monoamines mediating the action of hormones on learning and memory. *In* M.A.B. Brazier (Ed.), *Brain Mechanisms in Memory and Learning: From Single Neuron to Man. IBRO Monograph Series* (Vol. 4), pp. 249–268. New York: Raven.

van Loon, G. R., & Kim, C. (1977). Effect of β-endorphin on striatal dopamine metabolism. *Research Communications of Chemical and Pathological Pharmacology*, **18**, 171–174.

van Wimersma Greidanus, Tj. B., Thody, T. J., Verspaget, H., de Rotte, G. A. Goedemans, H.J.H., Croiset, G., & van Ree, J. M. (1979). Effects of morphine and β-endorphin on basal and elevated plasma levels of α-MSH and vasopressin. *Life Sciences*, **24**, 579–586.

Veith, J. L., Sandman, C. A., Walker, J. M., Coy, D. H., & Kastin, A. J. (1978). Systemic administration of the endorphins selectively alters open-field behavior of rats. *Physiology and Behavior*, **20**, 539–542.

Versteeg, D.H.G., de Kloet, E. R., & de Wied, D. (1979). Effects of α-endorphin, β-endorphin and (Des-tyr^1)-γ-endorphin on the α-MPT-induced catecholamine disappearance in discrete regions of the rat brain. *Brain Research*, **179**, 85–92.

Zornetzer, S. F. (1978). Neurotransmitter modulation and memory: A new neuropharmacological phrenology? *In* M. A. Lipton, A. Di Mascio, & K. F. Killam (Eds.), *Psychopharmacology. A Generation of Progress*, pp. 637–649. New York: Raven.

CHAPTER 12

Enkephalin and Endorphin Influences on Appetitive and Aversive Conditioning

George F. Koob, Michel Le Moal, and Floyd E. Bloom

In experiments using aversively motivated learning and memory tasks, opiate receptor antagonists such as naloxone tend to improve general performance, whereas opiate alkaloids tend to impair performance, even when administered after training in tests of memory. The effects of opioid peptides on these tasks are not as clear-cut, but in general, endorphins and enkephalins tend to attenuate amnesia or delay extinction of aversively motivated learning. In experiments designed to explore further the nature of peptide effects on behavior, we found that α-endorphin inhibits the extinction of a pole-jump avoidance task but that γ-endorphin produces more rapid extinction. However, in a passive avoidance task, α-endorphin increased retention latencies whereas γ-endorphin decreased them. Therefore, γ-endorphin appears to produce behavioral effects that are opposite to those of other endorphins or enkephalins.

Few studies of peptide effects on behavior have been performed using appetitively motivated tasks. In a continuous reinforcement lever-press situation for food reward, α-endorphin delayed, whereas γ-endorphin slightly facilitated extinction. In a runway task with water reward, both α- and γ-endorphin delayed extinction, an effect that was not blocked by naloxone. In general, with the exception of γ-endorphin, the endorphins tend to improve performance and prolong extinction in appetitively motivated tasks. However, the different effect of γ-endorphin on food- and water-motivated tasks emphasizes the need for further studies of peptide effects with different motivational status.

Gaining an understanding of the neurobiological basis of the different behavioral effects of these peptides will require research into their molecular sites and mechanisms of action, their cellular interactions, and their behavioral interactions. Furthermore, it is unclear whether systemically

249

ENDOGENOUS PEPTIDES
AND LEARNING AND MEMORY PROCESSES

administered peptides exert their behavioral effects centrally, peripherally, or both. It is likely that centrally derived endorphins may act as classical opiate receptors, where they produce morphine-like actions dependent on a high concentration of circulating agonist. In contrast, low concentrations of peripherally derived endorphins may act as nonclassical opioid receptors to produce hormone-like actions that are not reversible by naloxone.

I. INTRODUCTION

Two major approaches are being used to study the role of opiates and opioid peptides in learning. The first employs central or peripheral administration of the opiate agonist, either an opiate, a synthetic opioid, or alkaloid peptide, to act directly on receptors whose location is unknown. The second approach is even more indirect. It infers that an endogenous opioid peptide has a role in learning because the behavior is susceptible to blockade with naloxone, the opiate antagonist.

Both of these approaches require the same major assumption: The natural agonists and synthetic antagonists both act directly and specifically at the same opioid receptors. However, studies have already suggested that different populations of opiate receptors probably exist with different affinities for agonists and antagonists. Martin, Eades, Thompson, Huffler, and Gilbert (1976) and Gilbert and Martin (1976) have identified three such receptors, mu, gamma, and kappa. The mu receptor is considered the classical opiate receptor in that morphine-like drugs appear to interact with this receptor, whereas certain benzomorphan compounds such as ketocyclazocine appear to act at the kappa receptor. Others (Lord, Waterfield, Hughes, & Kosterlitz, 1977) have identified a fourth receptor, called delta and differentiated from the mu receptor by a weak naloxone antagonism. Examination of the pharmacological responses of two different *in vitro* physiological preparations provided evidence that there are indeed heterogeneous receptor populations. In the guinea pig ileum, for example, the endorphin peptides appear to interact mainly with the mu receptor, whereas in the mouse vas deferens, the endorphins probably act more through the putative delta receptor (Lord et al., 1977). The guinea pig ileum also appears to have more kappa receptors than the mouse vas deferens.

In addition, the definition of opioid activity by the criterion of naloxone antagonism has been questioned by recent findings. First, there is the problem of pharmacological specificity of the antagonist itself as, in very high concentrations, naloxone also antagonizes γ-aminobutyric acid (GABA) (Dingledine, Iversen, & Breuker, 1978). Second, recent behavioral work with peripheral injections of peptides has shown that some of these effects are not reversible by naloxone (de Wied, Bohus, van Ree, & Urban, 1978b; Rigter, Greven, & van Riezen, 1977a). The possibility of multiple opioid receptors with different affinities for particular agonists and antagonists has important implications for the study of the role of opioids in learning.

In the present chapter, we shall first review the effects of opiates and naloxone on appetitive and aversive conditioning. Second, more recent results using opioid peptides and, in particular, some of our work on the effects of peripheral injections of endorphins will be discussed. Finally, we will attempt to integrate these findings into a hypothesis regarding the role of opioid peptides in learning.

II. OPIATE ALKALOIDS AND NALOXONE: AVERSIVELY MOTIVATED TESTS

The role of opiates in pain perception has been known for centuries and, therefore, it is not surprising to observe that opiates can also influence the learning or retention of pain motivated tasks (Banerjee, 1971; Castellano, 1975; Cook & Weidley, 1957; Domino, Karoly, & Walker, 1963; Gallagher & Kapp, 1978). For example, even in studies where the opioid is administered after the presentation of the unconditioned stimulus, its action could conceivably prolong or shorten the motivational state (i.e., fear, produced by the unconditioned stimulus; Kapp & Gallagher, 1979). In early work, morphine was indeed found to disrupt learning and performance of conditioned avoidance responses (Banerjee, 1971; Cook & Weidley, 1957; Domino et al., 1963; Verhave, Owen, & Robbins, 1959). This disruption is least severe in highly overtrained animals (Domino et al., 1963), and most severe when the drug is injected prior to each training session (Banerjee, 1971). Presumably the disruption of this avoidance behavior is related to the ability of morphine to reduce the alarm response associated with a conditioned situation motivated by fear or anxiety. Two groups (Cook & Weidley, 1957; Verhave et al., 1959) reported such effects using doses of morphine that clearly produced analgesia. However morphine, even at subanalgesic doses (0.25 mg/kg) caused a disruption of the acquisition of a conditioned (pole-jump) avoidance response (Banerjee, 1971). Withdrawal from morphine improved avoidance responding with a recovery of the previously suppressed conditioned emotional responses (i.e., piloerection, trembling, shaking, micturition etc.; Banerjee, 1971). This suggests that the suppression of the alarm or fear response by morphine may be independent of its analgesic effects.

Consistent with this more cognitive–emotional effect of morphine are the results of more recent work involving posttraining (consolidation) injections of morphine. Using a discriminative avoidance test, a water maze, Castellano (1975) reported that mice injected with morphine or heroin immediately after daily conditioning sessions showed retrograde amnesia or a deficit in retention performance. These effects were dependent on the training–drug treatment interval; injection 2 hr after training had no effect. This retention deficit was blocked by naloxone. A recent replication of these results used an inhibitory avoidance test with rats, again showed a significant retrograde amnesia after injection of morphine (Jensen, Martinez, Messing, Speihler, Vasquez, Sou-mireu-Mourat, Liang, & McGaugh, 1978; also see Chapter 20, Messing et al., this volume). Morphine injected intracerebroventricularly (i.c.v.), at a low dose (3.0 μg), immediately after the training (shock) trial also produced a decreased latency for reentry into the shock compartment (i.e., disrupted retention; however, a higher dose (40 μg) facilitated retention; Jensen et al., 1980). A similar facilitation of retention was observed by Stein and Belluzzi (1978) also using similarly high doses of morphine (20 μg/rat) or methionine enkephalin (200 μg/rat). Interestingly, naloxone has been reported to produce the opposite result or an increase in latency to reenter the shock compartment in an inhibitory avoidance task and a facilitation of retention in an active avoidance task (Messing, Jensen, Martinez, Spiehler, Vasquez, Soumireu-Mourat, Liang, & McGaugh, 1979). These results are consistent with either an increase in the motivational value of the aversive stimulus and/or a facilitation of associative learning.

These observations with morphine and naloxone also appear to generalize to other learning situations using shock (Izquierdo, 1979) and attest to the robustness of the phenomenon. Izquierdo found that naloxone enhances and morphine depresses retention of an avoidance test with two different shock contingencies during either the training or test situation (i.e., four different training–test combinations). In the Pavlovian (P) contingency, shock always followed tone regardless of the response of the rat; in the avoidance (A) contingency, shock was omitted when the rat made a correct avoidance response. Animals trained in the P paradigm were tested either in the P or A contingency and vice versa, providing the four training–testing situations. In all cases, immediate posttraining injections of naloxone (0.4 mg/kg/i.p.) enhanced performance during the test 7 days later. Morphine (1.0 mg/kg) decreased performance and morphine plus naloxone produced no effect. This pharmacologic interaction is predictable based on the logic that naloxone and opiates appear to interact in dynamic competition at the level of the receptor (Stinus, Koob, Ling, Bloom, & Le Moal, 1980). In this case, morphine, itself, presumably displaces naloxone from the receptor to reverse any naloxone effects.

In fact, there are only two inconsistencies in the literature with regard to the effects of peripheral injections of morphine and naloxone on the acquisition and retention of aversively motivated tasks. Castellano (1980) recently reported that low doses of heroin injected pre- and posttrial actually improved performance in a five-choice pattern discrimination task motivated by shock; Messing et al. (1979) reported that naloxone given 30 min instead of immediately after the pairing of shock and the conditioned stimulus produced amnesia. Curiously, little or no work has been reported on the effects of opiates or naloxone on acquisition or retention of appetitively motivated tasks.

Recent work with intracerebral injections of opiates and opiate antagonists has suggested that a specific endorphinergic neural system may be involved in these opiate effects. Posttraining injections of the opiate agonist, levorphanol, into the amygdala of rats produced a time-dependent and dose-dependent decrease in the latency to reenter the shock compartment in an inhibitory avoidance test (Gallagher & Kapp, 1978). These effects appeared to be stereospecific, as dextrophan had no effect. In addition, posttraining administration of naloxone into the amygdala significantly increased the latency to reenter the shock compartment and the combination of levorphanol and naloxone produced no effect.

Although these results collectively argue for an action of the opiates on the neural processes responsible for the formation of associations, or memory, other explanations cannot entirely be ruled out. For example, as discussed by Kapp and Gallagher (1979), it is possible that opiate administration attenuates the emotional state associated with aversive situations, particularly the establishment of fear that presumably outlasts the actual duration of the shock in an inhibitory avoidance task. A theoretical model for this latter explanation of the effects of postconditioning challenge has been proposed (Spevack & Suboski, 1969) and may explain why so many different drugs appear to interfere with the formation of the memory substrate.

III. ENDORPHINS AND ENDORPHIN ANALOGS

A. Aversively Motivated Tests

The effects of opiate-like peptides on aversive conditioning are not nearly as clear as the results that have developed regarding the opiate alkaloids and naloxone. In fact, to a large extent, the results appear to be exactly opposite (see Table 1). In a series of experiments by Rigter and colleagues, endorphins were found to attenuate amnesia in rats after systemic injection (Rigter, 1978; Rigter, Greven, & van Riezen, 1977a; Rigter, Shuster, & Thody, 1977b). Here, β-endorphin at a low dose (10.0 μg/rat) injected subcutaneously (s.c.) 1 hr before the retention test reversed the disruption of a one-trial inhibitory avoidance task caused by posttraining exposure to carbon dioxide. Like $ACTH_{4-10}$, β-endorphin only acted when injected 1 hr pretest and not when injected 1 hr preacquisition. However, extraordinarily low doses of Met- and Leu-enkephalin (0.0003–30.0 μg/rat) produced identical results to β-endorphin when injected subcutaneously 1 hr before either the acquisition or the retention test. This attenuation of CO_2-induced amnesia by enkephalins was not reversible

TABLE 1

Summary of the Effects of Opiates and Endorphins on Avoidance Conditioning

	Active avoidance		Inhibitory avoidance	Amnesia reversal	References
	Acquisition	Extinction			
Opiate alkaloids	↓b	—	↓a		Cook and Weidley (1957)
					Verhave et al. (1959)
					Banerjee (1971)
					Jensen et al. (1978)
Naloxone	↑a	—	↑a		Messing et al. (1979)
					Izquierdo (1979)
Endorphins					
β	—	↑c	↑d ↓a	↑d	Rigter et al. (1977a)
α	—	↑c	↑d	↑d	Rigter et al. (1977b)
					de Wied et al. (1978a)
					de Wied et al. (1978c)
γ	↑b	↓c	↑b ↓d	—	de Wied et al. (1980)
					Le Moal et al. (1979)
					Martinez and Rigter (1980)
Enkephalins	↓b	↑c	↑d	↑d	Rigter et al. (1980)
					Rigter (1978)

[a] injection immediately posttraining.
[b] injection <1 hour pretraining.
[c] injection immediately after first 10 extinction trials.
[d] injection 1 hr pretest.

↑ improvement in performance (i.e., delayed extinction in active avoidance; increased latency to reenter in inhibitory avoidance).
↓ disruption of performance (i.e., facilitated extinction in active avoidance; decreased latency to reenter in inhibitory avoidance).

by naloxone (Rigter et al., 1977). Although these results are consistent with an interpretation of an antiamnesia action for endorphins, broader hypotheses regarding changes in arousal, fear motivation, or response to stress were not explored. Similar results and a similarly difficult interpretation can be found in the results of recent studies by de Wied and his colleagues (de Wied, et al., 1978b; de Wied, Kovács, Bohus, van Ree, & Greven, 1978c; de Wied, van Ree, & Greven, 1980). Here peripheral injections of Met-enkephalin, β-endorphin, and α-endorphin (β-LPH$_{61-76}$) all delayed the extinction of a pole-jump avoidance task at doses of less than 3.0 μg/rat when injected peripherally and at doses less than 0.1 μg/rat when injected intraventricularly (i.v.) (de Wied et al., 1978a). Curiously, naltrexone produced the opposite effect, facilitating extinction of the pole-jump avoidance task, but failing to block the delay in extinction produced by ACTH$_{4-10}$ and α-endorphin. This appears to be in direct contrast to the memory-enhancing effects of naloxone referred to earlier.

Even more curiously, γ-endorphin (β-LPH$_{61-77}$) facilitated the extinction of pole-jump avoidance (de Wied et al., 1978c, 1980). This facilitation of extinction and the fact that Des-Tyr^1-γ-endorphin appeared to have "cataleptogenic" properties similar to haloperidol (de Wied et al., 1978c) prompted de Wied to speculate that Des-Tyr^1-γ-endorphin could be an endogenous neuroleptic (de Wied et al., 1978a, 1980). The differential effects of α- and γ-endorphin on extinction of pole-jump avoidance after peripheral injection have been replicated (Le Moal, Koob, & Bloom, 1979), but to date no one has observed cataleptogenic action of Des-Tyr^1-γ-endorphin (Stinus et al., 1980; Pert, A., personal communication; Weinberger, Arnsten, & Segal, 1979), nor has Des-Tyr^1-γ-endorphin been isolated from rat brain in significant quantities.

In contrast to what might be expected from these studies with peripheral injections of endorphins 1 hr pretraining or pretest, Met^5-enkaphalin and Leu^5-enkephalin and a D-ala^2-analog of Leu^5-enkephalin, D-Ala^2-D-Leu^5-enkephalin (DADLE) actually impaired acquisition of an active avoidance task when injected immediately before a series of eight trials, but produced no effect when injected 15 min prior to the training (Rigter, et al., 1980). The effective doses were 40.0–400.0 μg/kg for Leu^5- and Met^5-enkephalin and 4.0 μg/kg for the DADLE. γ-Endorphin injected 30 min prior to training in a one-trial inhibitory avoidance test, actually enhanced retention (Martinez & Rigter, 1980). In the same study, β-endorphin injected immediately after training produced a retention deficit (Martinez & Rigter, 1980). Although difficult to interpret in light of these findings and Rigter's previous work, the results suggest that the time course of the training testing protocol may be a very important factor in determining the mechanism of action of peripherally injected endorphins.

Several other studies have also attempted to examine the effects of endorphins on retention of inhibitory avoidance and, in general, the results appear to be consistent with the findings using carbon dioxide induced amnesia and the extinction of active avoidance. Posttraining or preretention treatment with α-endorphin significantly increased reentry latencies during the retention test of a one-trial inhibitory avoidance task, whereas a similar treatment regime with Des-Tyr[1]-γ-endorphin decreased latencies (de Wied et al., 1978c). In a series of experiments in our laboratory, we have replicated the effects of some of the endorphins on the extinction of aversively motivated tasks, and later extended those observations to extinction of appetitively motivated situations (see Section III-B).

Endorphins were tested in rats during extinction of a pole-jump avoidance test (de Wied, Witter, & Greven, 1975). In this task, male Wistar rats (100–150 g, Charles River) were trained to jump onto a pole to avoid shock (0.2–0.4 mA AC 60 Hz) using a light that was on for a duration of 5 sec as a conditioned stimulus (CS), exactly as described by de Wied et al. (1978a). The experiments with α-endorphin were conducted with 30 training trials, and 40 training trials were used for the experiment with γ-endorphin. However, to reach the performance requirements outlined as acceptable in the de Wied experiments (de Wied et al., 1975), a special shaping procedure was employed. During the first set of 10 trials, each rat was allowed to experience the CS and the shock on the first trial. On the second trial, if the rat failed to jump spontaneously onto the pole after the shock onset (and very few did), the animal was gently placed on the pole by the experimenter. Spontaneous jumping to the shock or CS generally occurred after this trial without further experimenter aid. However, throughout the training procedure, rats that failed to climb down the pole after the 30 sec intertrial interval, were also gently turned around on the pole and faced downward, at which point they always descended. These somewhat rigorous training procedures were necessary for us to attain the performance criterion of 7–10 successful avoidances during the first extinction session as reported by de Wied et al. (1975).

The results are shown in Fig. 1. Injection of 1.0 μg of α-endorphin (dissolved in saline and injected s.c.) per rat, significantly inhibited extinction in rats that had received 30 acquisition trials, whereas, in rats that received 40 acquisition trials (to delay extinction in the control group), 1.0 μg of γ-endorphin produced a faster extinction.

In the passive avoidance experiments, a similar restrictive and rigorous training procedure was employed to produce similar latencies as reported by de Wied et al. (1978c). The testing apparatus consisted of a two-chambered cage, each side measuring 12 cm × 28 × 20 cm and separated by a black Plexiglas sliding door. On Day 1 of a 3-day procedure, the subjects were placed in the side constructed of clear Plexiglas over which a 25 watt bulb was suspended at a distance of 40 cm, and then were allowed to enter the dark compartment for a 2 min familiarization period. On Day 2, the same procedure was repeated 3 times followed by shock (1.2 mA, scrambled, constant current, biphasic

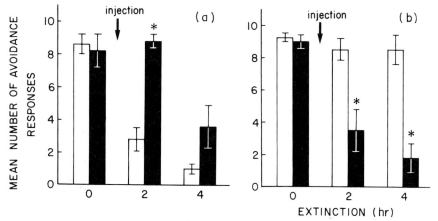

Fig. 1.   (a) Effects of α-endorphin and (b) effects of γ-endorphin on the extinction of pole-jump avoidance behavior. Peptides (1.0 µg/rat) were injected s.c. immediately after the first extinction session. Black bars indicate animals receiving (a) α-endorphin and (b) γ-endorphin. White bars indicate animals receiving saline injections. (Student's *t*-test). In (a) $n = 5$, (b) $n = 4$. Probability levels of test of significance: *$p < .05$. (Figs. 1–4 reprinted by permission from the chapter entitled "The role of endorphins in neurobiology behavior and psychiatric disorders." *In* G. B. Nemeroff and A. J. Dunn (Eds.), *Behavioral Neuroendocrinology.* Copyright © 1981 by Spectrum Publications, Inc., New York.)

square wave produced by a SGS-003 stimulator (from BRS/LVE Division of Technical Services, Inc.) on the third and last trial. The intertrial interval was approximately 5 min. Only those animals that entered the dark side in less than 10 sec on the third training trial were injected with peptide (s.c., dissolved in saline) 23 hr later, or 1 hr prior to the retention test (retrieval). This restrictive criterion was used to reduce variability for between-group comparisons and was based on the observation that, regardless of treatment, a rat showing a long initial latency during training showed a long latency for reentry during the test.

The results are shown in Fig. 2. Here, α-endorphin (10.0 µg/rat, s.c.) significantly increased the latency to reenter the dark box, whereas γ-endorphin (10.0 µg/rat) decreased reentry latency, Met⁵-enkephalin (30.0 µg/rat) had effects similar to α-endorphin, but Met⁵-enkephalin conjugated to horseradish peroxidase (by Dr. J. Rossier) had no effect.

Results of these experiments and the literature review of Sections I, II, and III are summarized in Table 1. With few exceptions, morphine injected either before or immediately after avoidance learning disrupts performance and retention, whereas naloxone facilitates it. In contrast, endorphins injected during extinction of active avoidance prolong extinction, and when injected 1 hr before retention of an inhibitory avoidance response, reverse CO_2-induced amnesia or facilitate retention.

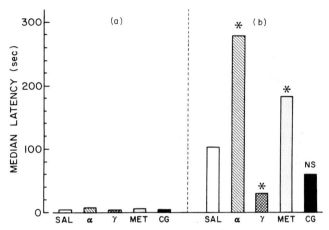

Fig. 2. Effects of endorphins on retention of an inhibitory avoidance task. Values represent median latency (a) to enter a dark box during the last training trial, and (b) during the subsequent test trial 24 hr later. Peptides were injected s.c. 1 hr before the test trial; α- and γ-endorphin: 10.0 μg/rat, Met-enkephalin, 30.0 μg/rat. $*p < .05$, significantly different from saline, (Mann-Whitney U-Test); $n = 8$.

B. Appetitively Motivated Learning

In what appears to be the first reported study of the effects of endorphins on any learning task, and the only study using appetitively motivated learning, Met[5]-enkephalin and a potent analog [D-Ala[2]]-Met[5]-enkephalin-NH$_2$, facilitated performance of hungry rats in a complex 12-choice maze (Kastin, Scollan, King, Schally, & Coy, 1976a). Here, the peptide was injected at a dose of 80.0 μg/kg intraperitoneally (i.p.) 15 min before the test each day. Interestingly, a Met-enkephalin analog with virtually no opiate activity [D-Phe[2]]-Met[5]-enkephalin also facilitated maze performance, whereas morphine produced the opposite results. This contrary effect of endorphin peptides and morphine is consistent with the observations of opioid effects on aversive conditioning described in previous sections.

The purpose of these experiments was to extend our observations with avoidance learning to appetitively motivated tasks to test the generality of the conclusions. If the actions of the endorphins were similar for both the aversive and appetitive tasks, a primary learning or memory substrate might be proposed as a mechanism of action. Alternatively, if the effects of the peptides are task specific, other mechanisms of action such as motivational variables may be more important.

In our laboratory, instead of studying learning acquisition, we attempted to alter the extinction of learning motivated by positive reward; our aim was a direct comparison of experimental results obtained with negative reward as previously demonstrated by de Wied and associates (1978a,b,c) and by ourselves (Le Moal et al., 1979). The results we obtained in these studies provide an intriguing insight into the mechanism of action of the peripherally injected endorphin peptides. First, in a continuous reinforcement lever press task for food reward, α-endorphin again delayed and γ-endorphin again slightly facilitated extinction (see Fig. 3). The rats were trained for 10 days as described by Mason and Iversen (1977) and, on the eleventh day, a series of extinction tests were performed every 2 hr. As with the active avoidance task, peptide or saline (10.0 μg/rat) was injected s.c. immediately after the first extinction trial. The peptide treated rats were significantly different from control injected rats only at the first 2 hr extinction trial.

A potentially more interesting result, but also more difficult to understand, was obtained in a test using water as a reward. Here, rats deprived of water for 23 hr/day were trained for 7 days (one trial per day) to run down an alley constructed with a 45 degree angle to lick a water tube for 30 sec. On the eighth day, a series of extinction trials were conducted every 2 hr and saline or a peptide was injected (10.0 μg/rat s.c.) after the first extinction trial. Surprisingly, both α- and γ-endorphin delayed the extinction of a runway task for water (Le Moal et al., 1979). The rats receiving α- and γ-endorphin continued to run to the dry tube even at the fourth extinction trial (see Fig. 4a). In a replication of this water task experiment with γ-endorphin (Le Moal et al., 1979), naloxone (5.0 mg/kg) not only failed to block the delay in extinction, it, in fact, produced similar effects by itself. The naloxone plus γ-endorphin group actually ran to the dry tube faster than the γ-endorphin plus saline group

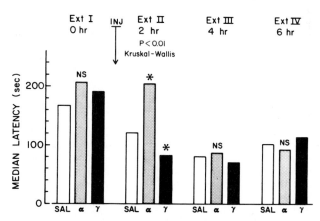

Fig. 3. Effects of endorphins on extinction of a continuously reinforced lever press response. Values represent the median latency to reach an extinction criteria of 2 min without lever pressing (see Mason & Iversen, 1977). Peptides (10.0 μg/rat) were injected s.c. immediately after the first extinction session. Three more extinction sessions were performed at 2 hr intervals. Significantly different from saline, $^*p < .05$, (Mann-Whitney U-test); $n = 8$.

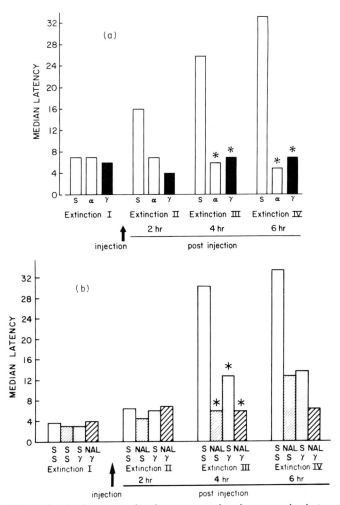

Fig. 4. Effects of endorphins on median latency to touch a dry water tube during successive extinction sessions. Rats were injected s.c. with peptides (10.0 μg/rat) immediately after the first extinction session. Significantly different from saline group, *$p < .05$ (Mann-Whitney U-test); $n = 7$. (b) Effect of γ-endorphin with and without naloxone pretreatment on the median latency to touch a dry water tube during successive extinction sessions. Rats were injected i.p. immediately after the first extinction session with saline or naloxone (5.0 mg/kg) followed by injection s.c. of γ-endorphin (10.0 μg/rat). Significantly different from saline group, *$p < .05$ (Mann-Whitney U-test); $n = 8$.

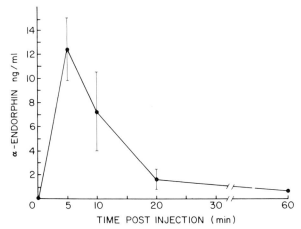

Fig. 5. Levels of α-endorphin in the blood following injection of 10.0 μg/rat, s.c. Values represent the mean ± SEM for separate groups of injected rats sacrificed at different times postinjection.

(see Fig. 4b). In both series of experiments, peptide actions on extinction were detectable for 4–6 hr after injection.

Thus, in shock motivated tests, α-endorphin delays extinction and γ-endorphin facilitates extinction. Similarly, opposite effects of smaller magnitude were observed in the extinction of a lever press response for food. However, in a water motivated test, *both* α- and γ-endorphin delayed extinction. This effect was not blocked by naloxone and, in fact, naloxone appeared to produce results similar to the endorphins. This effect of naloxone is consistent with the earlier effects observed with naloxone in aversively motivated tasks (Section II).

To examine the prolonged time course of apparent peptide action relative to the disappearance of the endorphin following peripheral administration, endorphins were measured in blood, hypothalamus, and brain following s.c. injection. Four groups of rats were injected s.c. with α-endorphin at a dose of 10.0 μg/rat and each group was sacrificed by decapitation at 5, 10, 20, or 60 min postinjection. α-Endorphin was measured in the samples by radioimmunoassay (Rossier, Bayon, Vargo, Ling, Guillemin, & Bloom, 1977b). As can be seen in Fig. 5, α-endorphin levels in the blood peak 5 min after injection and virtually disappear by 20 min postinjection. No detectable levels of α-endorphin were ever seen in hypothalamus or in the rest of the brain.

IV. CONCLUSIONS

In summary, opiate alkaloids appear to impair general performance of aversively motivated learning even when injected posttrial in memory tests, whereas naloxone appears to improve performance in similar situations. In dramatic contrast, however, endorphins and enkephalins appear to reverse amnesia or delay extinction of aversively motivated learning. An exception to these findings is that γ-endorphin produces effects opposite to other endorphins or enkephalins on extinction of either active or passive avoidance. Also when enkephalin analogs are injected immediately before the test session (Rigter et al., 1980) rather than 1 hr or more before the test (de Wied et al., 1978b,c; Rigter et al., 1977a,b; Rigter, 1978; and ourselves) the peptides also impair performance.

With appetitively motivated behavior, the general tendency seen in a smaller number of studies, is also for endorphins to improve performance or to prolong extinction. The results showing differences in the effects of γ-endorphin on food motivated versus water motivated learning emphasize the need for further studies examining motivational states as independent variables.

The finding of rapid disappearance of α-endorphin in the blood after peripheral injection is in accord with previous work showing a rapid clearance of peptide from the blood. The half-time disappearance of radioactivity after the injection of labeled Met^5-enkephalin is less than 1 min (Kastin, Nissen, Schally, & Coy, 1976b). Thus, the behavioral effects of the endorphins appear not to require persistence of the peptide in the blood. In fact, there is evidence that behavioral effects immediately after injection (Rigter et al., 1980) may be different from those produced by behavioral tests 1 hr postinjection (Rigter et al., 1977a,b; Rigter, 1978). Determining the actual active compound or mechanism by which these peptides produce this delayed effect is thus of major importance for future work.

At least one group argues that radiolabeled Met-enkephalin (^{3}H-Tyrosine[1]) crosses the blood–brain barrier (Kastin et al., 1976b). Using the Oldendorf (1970) procedure, they found a brain uptake index (BUI) of 15 for tritiated Met-enkephalin even though the subsequent radioactivity in brain represented only 0.84% of the radioactivity injected into the carotid (Kastin et al., 1976b; Kastin, Coy, Olson, Panksepp, Schally, & Sandman, 1979). These results would suggest that peripherally derived peptides could actually act directly on brain receptors to produce their behavioral effects, although the amounts reaching brain sites would be incredibly small.

The seemingly divergent behavioral results between opiate alkaloids and endorphins, are difficult to integrate into a single hypothesis explaining the functional significance of natural endorphins. The role of central endorphins in the relief of pain and in producing behavioral activation is relatively clear, particularly as these effects appear to be naloxone reversible and as they can be attributed to specific central nervous system (CNS) pathways (Kelley, Stinus, & Iversen, 1980; Stinus et al., 1980). In addition, for analgesia and behavioral activation, all the endorphin analogs appear to act in the same qualitative direction after either central or peripheral injection, to produce analgesia or hyperactivity with varying potencies.

In part, solution of the functional riddle of endorphins may require better understanding of possible peripheral receptors to blood-borne versus neuronally released peptides. Endorphins released from the pituitary may have a similar but more subtle role in producing central activation with appropriately integrated peripheral actions during stressful situations. β-Endorphin is released from the anterior pituitary during stress (Rossier, French, Rivier, Ling, Guillemin, & Bloom, 1977a) and β-LPH-like material can reverse some of the behavioral deficits observed after hypophysectomy (de Wied, 1977). Moreover, in normal animals, the endorphins appear to be most potent in altering aversively motivated learning, although they also clearly influence appetitively motivated learning. Furthermore, after peripheral injections, different analogs may be of particular importance in determining the direction of the behavioral effect (de Wied et al., 1978a,b; Le Moal et al., 1979). In this regard, it is interesting to note that *in vivo* β-endorphin stimulates the release of vasopressin (Weitzman, Fister, Minnick, Ling, & Guillemin, 1977), but it does not do so when tested directly on the magnocellular neurons or their axons (Iversen, Iversen, & Bloom, 1980; Pittman, Hallow, & Bloom, 1979). It remains to be determined whether these subtle behavioral effects arising from peripheral administration of minute amounts of α- or γ-endorphin actually indicate a unique role for the metabolic products of naturally released β-endorphin.

A critical set of questions that ultimately require confrontation center on three groups of unknowns:

1. Molecular interactions: the essential sites and mechanisms of action of the various endorphins and enkephalins in the CNS (see Chapter 10, Bloom & McGinty, this volume).

2. Cellular interactions: the result of such molecular interactions on the activity of receptive neurones.

3. Behavioral interactions: the classes of central cellular systems that operate in the behavioral operations underlying memory tasks.

Until each of these three levels of information is more solidly constrained by experimental results, the innumerable possible interpretations will preclude a compelling demonstration that any endorphin or enkephalin is either "involved in" or "mediates" such a behavior. Nevertheless, approaches to this fundamental series of demonstrations would seem to be essential for attempts to extrapolate from animal behavior, experiments susceptible to pharmacological manipulation, to human behavioral syndromes. These now appear, on superficial grounds, to be related to the endorphins, such as mental illnesses or substance abuse syndromes. Our current tentative position is that the behavioral action

TABLE 2

Behavioral Effects of β-Lipotropin Derivatives

Action	"Classical–traditional"	Non-"classical–traditional"
Behavior	Opiate-like analgesia psychomotor stimulation	Nonopiate-like, ACTH-like learning–memory? motivation? emotion?
Administration		
Type	Central i.c.v. or local i.c. injection	Peripheral or central injection
Site	For analgesia effects: CG, Raphé For stimulant effects: VTA, SN(DA)	Unknown
Endorphin analogs	Nonspecificity	Specificity
Naloxone	Blockade at low doses	No blockade
Latency and duration of action	Latency is shortly after injection lasting approx. 90 min, depending on dose	Unknown
Dose range for effect	Greater than 0.2 μg/rat by intracerebral injection	Less than 10 μg/rat by subcutaneous injection

CG: central gray
VTA: ventral tegmental area
SN: substantia nigra
DA: dopamine

of an opioid peptide and its "opiateness" depends crucially on the cellular source of the peptide, the receptive cell and its receptor class, and on the manner of transmittal between the source and the response site.

Consideration of a two-level (central versus peripheral) functional mode for endorphins may be pertinent to the hypothesis of multiple opiate receptors (Chang & Cuatrecasas, 1979). This concept is outlined in Table 2. Here, centrally derived endorphins may act at classical opiate receptors where they produce morphine-like actions that are dependent on a given (high) level of circulating agonist. For example, high doses of enkephalins can mimic the action of morphine when injected i.c.v. (Stein & Belluzzi, 1978). In contrast, peripherally derived endorphins may act at nonclassical opiate-like receptors (Chang & Cuatrecasas, 1979) to produce hormone-like actions that are not naloxone reversible, and that require only low levels of circulating agonist (see Table 2). Identification and characterization of the site of action of the effects of peripherally injected endorphins will go far toward clarifying these putative distinctions.

ACKNOWLEDGMENTS

Supported by DA 01785 and AA 03504. We thank Nancy Callahan and Yvonne Byszewski for manuscript assistance.

REFERENCES

Banerjee, U. (1971). Acquisition of conditioned avoidance response in rats under the influence of addicting drugs. *Psychopharmacologia*, 22, 133–134.

Castellano, C. (1975). Effects of morphine and heroin on discrimination learning and consolidation in mice. *Psychopharmacologia*,42, 235–242.

Castellano, C. (1980). Dose dependent effects of heroin on memory in two inbred strains of mice. *Psychopharmacology*, 67, 235–239.

Chang, K. J., & Cuatrecasas, P. (1979). Multiple opiate receptors. *Journal of Biological Chemistry*, 254, 2610–2618.

Cook, L., & Weidley, E. (1957). Behavioral effects of some psychopharmacological agents. *Annals of the New York Academy of Sciences*, 66, 740–752.

de Wied, D. (1977). Peptides and behavior. *Life Sciences*, 20, 195–204.

de Wied, D., Bohus, B., van Ree, J. M., Kovács, G. L., & Greven, H. M. (1978a). Neuroleptic-like activity of [des-tyr]-γ-endorphin in rats. *Lancet*, 1, 1046.

de Wied, D., Bohus, B., van Ree, J. M., & Urban, I. (1978b). Behavioral and electrophysiological effects of peptides related to lipotropin (β-LPH). *Journal of Pharmacology and Experimental Therapeutics*, 204, 570–580.

de Wied, D., Kovács, G. L., Bohus, B., van Ree, J. M., & Greven, H. M. (1978c). Neuroleptic activity of the neuropeptide β-LPH. *European Journal of Pharmacology*, 49, 427–436.

de Wied, D., van Ree, J. M., & Greven, H. M. (1980). Neuroleptic-like activity of peptides related to [Des-tyr]-γ-endorphin: Structure activity studies. *Life Sciences*, 26, 1575–1579.

de Wied, D., Witter, A., & Greven, H. M. (1975). Behaviorally active ACTH analogues. *Biochemical Pharmacology*,24, 1463–1468.

Dingledine, R., Iversen, L. L., & Breuker, E. (1978). Naloxone as a GABA antagonist: Evidence from iontophoretic, receptor binding, and convulsant studies. *European Journal of Pharmacology*, 47, 19–27.

Domino, E. F., Karoly, A. J., & Walker, F. L. (1963). Effects of various drugs on a controlled avoidance response in dogs resistant to extinction. *Journal of Pharmacology and Experimental Therapeutics*, 41, 92–99.

Gallagher, M., & Kapp, B. S. (1978). Opiate administration into the amygdala: Effects on memory processes. *Life Sciences*,23, 1973–1978.

Gilbert, P. E., & Martin, W. R. (1976). The effects of morphine and nalorphine-like drugs in the nondependent, morphine dependent and cyclazocine-dependent chronic spinal dog. *Journal of Pharmacology and Experimental Therapeutics*, 198, 66–82.

Iversen, L. L., Iversen, S. D., & Bloom, F. E. (1980). Opiate receptors influence vasopressin release from nerve terminals in rat neurohypophysis. *Nature (London)*, 284, 350–351.

Izquierdo, I. (1979). Effect of naloxone and morphine on various forms of memory in the rat: Possible role of endogenous opiate mechanisms in memory consolidation. *Psychopharmacology*, 66, 199–204.

Jensen, R. A., Martinez, Jr., J. L., Messing, R. B., Speihler, V., Vasquez, B. J., Soumireu-Mourat, B., Liang, K. C., & McGaugh, J. L. (1978). Morphine and naloxone alter memory in the rat. *Society for Neuroscience Abstracts*, 4, 260.

Kapp, B. S., & Gallagher, M. (1979). Opiates and memory. *Trends in Neuroscience*, 177–180.

Kastin, A. J., Scollan, E. L., King, M. G., Schally, A. V., & Coy, D. H. (1976a). Enkephalin and a potent analog facilitate maze performance after intraperitoneal administration in rats. *Pharmacology, Biochemistry and Behavior*, 5, 691–695.

Kastin, A. J., Nissen, C., Schally, A. V., & Coy, D. H. (1976b). Blood brain barrier, half-time disappearance, and brain distribution for labeled enkephalin and a potent analog. *Brain Research Bulletin*, 1, 583–589.

Kastin, A. J., Coy, D. H., Olson, R. D., Panksepp, J., Schally, A. V., & Sandman, C. A. (1979). Behavioral effects of the brain opiates enkephalin and endorphin. In R. Collu (Ed.), *Central Nervous System Effects of Hypothalamic and Other Peptides*, pp. 273–281. New York: Raven.

Kelley, A. E., Stinus, L., & Iversen, S. D. (1980). Interactions between d-ala-met-enkephalin, and A10 dopaminergic neurons and spontaneous behavior in the rat. *Behavioral Brain Research*, 1, 3–24.

Le Moal, M., Koob, G. F., & Bloom, F. E. (1979). Endorphins and extinction: Differential actions on appetitive and aversive tasks. *Life Sciences*, 24, 1631–1636.

Lord, J.A.H., Waterfield, A. A., Hughes, J. & Kosterlitz, H. W. (1977). Endogeneous opioid peptides: Multiple agonists and receptors. *Nature (London)*, 267, 495–499.

Martin, W. R., Eades, C. G., Thompson, J. A., Huppler, R. E., & Gilbert, P. E. (1976). The effects of morphine- and nalorphine-like drugs in the nondependent and morphine-dependent chronic spinal dog. *Journal of Pharmacology and Experimental Therapeutics*, 197, 517–532.

Martinez, J. L., Jr., & Rigter, H. (1980). Endorphins alter acquisition and consolidation of an inhibitory avoidance response in rats. Submitted. *Neuroscience Letters*, 19, 197–201.

Mason, S. J., & Iversen, S. D. (1977). Effects of selective forebrain noradrenaline loss on behavioral inhibition in the rat. *Journal of Comparative and Physiological Psychology*, 91, 165–173.

Messing, R. B., Jensen, R. A., Martinez, Jr., J. L., Spiehler, V. R., Vasquez, B. J., Soumireu-Mourat, B., Liang, K. C., & McGaugh, J. L. (1979). Naloxone enhancement of memory. *Behavioral and Neural Biology*, 27, 266–275.

Olendorf, W. H. (1970). Measurement of brain uptake of radiolabeled substances using a tritiated water in external standard. *Brain Research*, 24, 372–376.

Pittman, Q. J., Hatton, J. D., & Bloom, F. E. (1979). Opiates selectively depress spontaneous activity in the hypothalamic slice preparation. *Society for Neuroscience Abstracts*, 5, 233.

Rigter, H. (1978). Attenuation of amnesia in rats by systemically administered enkephalin. *Science*, 200, 83–85.

Rigter, H., Greven, H., & van Riezen, H. (1977a). Failure of naloxone to prevent reduction of amnesia by enkephalins. *Neuropharmacology*, 16, 545–547.

Rigter, H., Shuster, S., & Thody, A. J. (1977b). ACTH, α-MSH and β-LPH: Pituitary hormones with similar activity in an amnesia test in rats. *Journal of Pharmacy and Pharmacology*, 29, 110–111.

Rigter, H., Hannan, T. J., Messing, R. B., Martinez, J. L., Jr., Vasquez, B. J., Jensen, R. A., Veliquette, J., & McGaugh, J. L. (1980). Enkephalins interfere with acquisition of an active avoidance response. *Life Sciences*, 26, 337–346.

Rossier, J., French, E. D., Rivier, C., Ling, N., Guillemin, R., & Bloom, F. E. (1977a). Foot-shock induced stress increases beta-endorphin levels in blood but not in brain. *Nature (London)*, 270, 618–620.

Rossier, J., Bayon, A., Vargo, T. J., Ling, N., Guillemin, R., & Bloom, F. E. (1977b). Radioimmunoassay of brain peptides: Evaluation of a methodology for the assay of beta-endorphin and enkephalin. *Life Sciences*, 21, 847–852.

Spevack, A. A., & Suboski, M. D. (1969). Retrograde effects of electroconvulsive shock on learned responses. *Psychological Bulletin*, 72, 66–76.

Stein, L., & Belluzzi, J. D. (1978). Brain endorphins and the sense of well-being: A psychobiological hypothesis. *In* E. Costa & M. Trabucchi (Eds.), *The Endorphins: Advances in Biochemical Psychopharmacology*, Vol. 18. New York: Raven.

Stinus, L., Koob, G. F., Ling, N., Bloom, F. E., & Le Moal, M. (1980). Locomotor activation induced by infusion of endorphins into the ventral tegmental area: Evidence for opiate-dopamine interactions. *Proceedings of the National Academy of Sciences, U.S.A.*, 77, 2323–2327.

Verhave, T., Owen, J. E., & Robbins, E. B. (1959). The effect of morphine sulfate on avoidance and escape behavior. *Journal of Pharmacology and Experimental Therapeutics*, 125, 248–251.

Weitzman, R., Fisher, D., Minnick, S., Ling, N., & Guillemin, R. (1977). Beta-endorphin stimulates secretion of arginine vasopressin *in vivo*. *Endocrinology*, 101, 1643–1646.

Weinberger, S. B., Arnstein, A., & Segal, D. C. (1979). Des-tryosine-γ-endorphin and haloperidol: Behavioral and biochemical differentiation. *Life Sciences*, 24, 1637–1644.

CHAPTER 13

Endogenous Opioids, Memory Modulation, and State Dependency

Iván Izquierdo, Marcos L. Perry, Renato D. Dias,
Diogo O. Souza, Elaine Elisabetsky, María A. Carrasco,
Otto A. Orsingher, and Carlos A. Netto

Memory for various tasks, in the rat, is enhanced by the posttraining intraperitoneal administration of naloxone (0.2–5.0 mg/kg), and depressed by that of morphine (1.0–3.0 mg/kg), β-endorphin (1.0–10.0 μg/kg), Met-, Leu-, and Des-Tyr-Met-enkephalin (0.32–10.0 μg/kg). β-Endorphin and Met-enkephalin (5.0–25.0 ng/rat) also cause retrograde amnesia when given by the intracerebroventricular route. The effect of morphine and of the opioid peptides is antagonized by naloxone. The antagonism between naloxone and the opioid peptides is competitive. Administration of β-endorphin (2.0 or 20.0 μg/kg) prior to training also causes amnesia, which is reversible by posttraining naloxone (0.2 mg/kg) injection. β-Endorphin is released from the rat brain at a rate of 30–50 ng/brain/25 min during various forms of training, including shuttle avoidance and habituation training, pseudoconditioning, and the repeated presentation of footshocks alone. The amount released is compatible with doses of the drug that, when injected, cause amnesia. Electroconvulsive shock treatment releases β-endorphin from the brain in somewhat larger amounts, but not from the hypophysis. These findings support the concept of an endogenous amnesic mechanism mediated

269

ENDOGENOUS PEPTIDES
AND LEARNING AND MEMORY PROCESSES

by β-endorphin and perhaps by other opioid peptides as well. In addition, it seems possible that this mechanism also acts proactively during multitrial learning sessions such as shuttle avoidance and habituation. Its role during acquisition or relearning would be to make the animal forget adventitious learnings that interfere with the specific task for which they are being trained, so that the latter may partially survive. Too little or too much β-endorphin during acquisition may lead to a deficit in acquisition.

I. INTRODUCTION

Harlow, McGaugh, and Thompson (1971) suggested that forgetting may be the most salient phenomenon of memory, and that one important cause of forgetting could be interference with consolidation of long-term memory. This interference may be brought about by a variety of treatments, such as electro-convulsive shock, CO_2 intoxication, cranial trauma, or a number of drugs. When these treatments are applied shortly after a learning experience, the effect is called retrograde amnesia. Most, if not all, of these treatments may act by interfering with the development of internal neurohumoral states related to arousal that modulate the consolidation process (Gold & McGaugh, 1975). Too little or too much arousal and, consequently, too small or too large a neurochemical change leads to a deficit in memory consolidation. Conversely, an appropriate degree of stimulation of the neurochemical systems that modulate consolidation causes a selective enhancement of the build-up of the diverse memory "channels." A channel is operationally defined as a form of memory more susceptible to some modulatory influences than to others (Elisabetsky, Vendite, & Izquierdo, 1979; Izquierdo, 1979; Izquierdo & Elisabetsky, 1978). Thus, if it is assumed that consolidation depends on selective modulation of internal neurohumoral systems, it may be concluded that the term *retrograde amnesia* actually designates one major form of forgetting, and does not nec-essarily have any pathological connotation.

Among the neurohumoral systems that play important facilitatory roles in the modulation of the early stages of memory, central dopaminergic and no-radrenergic mechanisms are implicated by a vast amount of evidence (see Zornetzer, 1978). Treatments (amphetamine, nicotine, etc.) that release one or the other catecholamine, or that mimic their effects (apomorphine, etc.), usually cause retrograde memory facilitation (Evangelista & Izquierdo, 1971, 1972; Grecksch & Matthies, 1981; Krivanek & McGaugh, 1969). Interference with the biosynthesis, release, or action of dopamine and/or norepinephrine usually results in a disruption of consolidation (Dismukes & Rake, 1972; Ful-giniti, Molina, & Orsingher, 1976; Grecksch & Matthies, 1981; Izquierdo, Beamish, & Anisman, 1979; Lauzi Gozzani, & Izquierdo, 1976; McGaugh, Gold, Handwerker, Jensen, Martinez, Meligeni, & Vasquez, 1979; Meligeni, Ledergerber, & McGaugh, 1978). There is substantial evidence in favor of the concept of a critical amount of release, and of a thereby critical brain level of norepinephrine, that would ensure that an adequate degree of consolidation will take place (Gold, 1981; Gold & van Buskirk, 1978).

Endogenous opioid peptides have recently been shown to inhibit both dopamine and norepinephrine release at a number of central and peripheral synapses (Christina Jones, personal communication; see Izquierdo, Dias, Souza, Carrasco, Elisabetsky, & Paiva, 1980b, and Snyder & Childers, 1979, for references). This finding led us to become interested in the possibility that endogenous opioid peptides may be involved in the modulation of memory consolidation.

II. EARLY EVIDENCE THAT THE PRIMARY ROLE OF ENDOGENOUS OPIOIDS MIGHT BE IN THE REGULATION OF BEHAVIOR

The discovery of the endogenous opioid peptides, enkephalins (Hughes, Smith, Kosterlitz, Fothergill, Morgan, & Morris, 1975) and endorphins (Bradbury, Smyth, & Snell, 1976; Chrétien, Benjannet, Dragon, Seidah, & Lis, 1976; Cox, Goldstein, & Li, 1976; Guillemin, Ling, & Burgus, 1976) was followed by a search for a physiological role for these substances. Most initial investigations were biased by the notion that they should exert morphine-like activities such as analgesia and euphoria in normal subjects. This notion is not necessarily valid, and would make it difficult to explain why addicts go out of their way to obtain morphine or heroin when their brains are producing morphine analogs all the time. The enkephalins and endorphins share many properties with opiate compounds, including effects upon smooth muscle, binding to opiate binding sites (Kosterlitz & Hughes, 1978; Morley, 1980), the induction of analgesia, and self-administration into the brain ventricles (Stein & Belluzzi, 1978). However, exceedingly large doses of the peptides are needed to produce analgesia (Belluzzi, Grant, Garsky, Sarantakes, Wise, & Stein, 1976; Kastin, Jemison, & Coy, 1979a) or to induce self-administration (Stein & Belluzzi, 1978). Since, in the case of the enkephalins, these doses are 10^2 to 10^3 times higher than the *total* brain content of opioid peptides (Bloom, Rossier, Battenberg, Bayon, French, Henriksen, Siggins, Segal, Browne, Ling, & Guillemin, 1978; DiGiulio, Majane, & Yang, 1979; Izquierdo, Souza, Carrasco, Dias, Perry, Eisinger, Elisabetsky, & Vendite, 1980d; Yang, Hong, Fratta, & Costa, 1978), a doubt was raised as to the physiological significance of those results (Izquierdo et al., 1980b,d; Izquierdo, Paiva, & Elisabetsky, 1980c). Pharmacological effects obtained with very large doses of an endogenous compound do not necessarily reflect its physiological function.

Several experiments raised the suspicion that the primary role of the endogenous opioids might be to regulate behavior, rather than to mimic the effect of opium alkaloids. Kastin, Scollan, King, Schally, and Coy (1976) obtained facilitation of maze learning in rats with 80 μg/kg of enkephalins; a dose 10^3 times higher did not produce analgesia. Veith, Sandman, Walker, Coy, and Kastin (1978) observed changes in the open-field behavior of rats injected i.p. with either 1.0 mg/kg of α-, β-, or γ-endorphin or their ^{2}D-Ala analogs, given i.p. Katz, Carroll, and Baldrighi (1978) observed that the i.c.v. administration of 6.0–50.0 μg of ^{2}D-Ala-Leu- or Met-enkephalin caused an increase in the general activity of mice; the effect was similar to that of a very large dose of morphine (50 mg/kg, i.p.), and opposite to that of an equally large dose of naloxone (8.0 mg/kg, i.p.).

These experiments (for a review see Kastin, Olson, Schally, & Coy, 1979b) may, of course, be criticized on the same grounds as those on analgesia and self-administration mentioned earlier; namely, that the doses used were disproportionally high. However, a series of papers from various laboratories that appeared in 1978 substantially contributed to the concept that endogenous opioids may be primarily involved in the regulation of behavior. De Wied, Bohus, van Ree, and Urban (1978a) reported that the s.c. injection of 0.3–3.0 μg/rat of Met-enkephalin, α- or β-endorphin retarded the extinction of a pole-climbing avoidance response, whereas the opiate antagonist, naltraxone, had an opposite effect. Subsequently, de Wied, Kovacs, Bohus, van Ree, and Greven (1978b) found that, in this and other tests, similarly low doses of γ- or Des-Tyr-γ-endorphin also had an opposite effect to that of α- or β-endorphin. Their data were confirmed by Le Moal, Koob, and Bloom (1979), who also reported that the differential effect of α- and γ-endorphin (10.0–40.0 μg/kg) on extinction depended on the task. Rigter (1978) found that very low doses (0.03–3.0 μg/kg) of Met- or Leu-enkephalin, given i.p. to rats, counteracted the amnesic effect of CO_2 in a step-through inhibitory avoidance paradigm when the drugs were given prior to the retention session; the effect of Met-enkephalin was also observed when the drug was given prior to training. Jensen, Martinez, Messing, Spiehler, Vasquez, Soumireu-Mourat, Liang, & McGaugh (1978) reported that the posttraining i.p. administration of 1.0 mg/kg naloxone caused memory facilitation of the step-through task in rats, and Gallagher and Kapp (1978a,b) obtained a similar effect using microinjections of nanogram doses of the drug into the amygdaloid nucleus.

The findings of de Wied et al. (1978a,b), or Rigter (1978) were difficult to interpret at that time, in terms of learning or memory, because they gave the drugs prior to training and/or testing the animals (see Izquierdo et al., 1979; Zornetzer, 1978). However, the findings of Jensen et al. (1978) and Gallagher and Kapp (1978a,b) clearly suggested a role of the endogenous opioids in the modulation of memory consolidation; particularly as, at that time, agonist effects of naloxone (see, for example, Jean-Baptiste & Rizack, 1980; Woolf, 1980) were virtually unknown (Terenius, 1978; Snyder & Childers, 1979).

The present chapter reviews a series of experiments carried out in this laboratory since January 1979 to test the hypothesis that opioid peptides may be involved in memory modulation. As will be seen, some of the results help to interpret the findings of de Wied et al. (1978a,b) and Rigter (1978).

III. EFFECT OF NALOXONE ON MEMORY CONSOLIDATION

The modulation of memory consolidation may be considered as a complex physiological function, consisting of facilitatory and inhibitory influences.

To establish a role of endogenous opioids in any given physiological function, it must be shown that: (1) the function is modified by the opiate antagonists, naloxone or naltrexone; (2) opiates and endogenous opioids have an effect opposite to that of, and antagonized by, naloxone or naltrexone (unless, of course, the latter are agonists on that particular system); (3) at least one opioid peptide must be released from some part of the brain or from other storage sites during the operation of the physiological variable under study; (4) the amount released must be compatible with a dose of the peptide(s) that, when injected, has a definite effect (Izquierdo et al., 1980b,d).

As mentioned earlier, the first studies to show that posttraining naloxone administration causes memory facilitation in rats were those of Jensen et al. (1978) and Gallagher and Kapp (1978a,b) using the step-through inhibitory avoidance paradigm. The former used i.p. injections, and the latter, microinjections into the amygdaloid nucleus. Jensen et al. found that the drug has an inverted U-shaped dose– response curve. Such curves, with or without opposite effects at one extreme of the curve, are typical of substances that affect learning and memory (Gold & McGaugh, 1975; Izquierdo et al., 1980b). Jensen et al. obtained memory facilitation with 1.0, but not with 0.1 or with 10.0 mg/kg of naloxone. The effect was antagonized by morphine, which, on its own, at a dose of 1.0 or 3.0 mg/kg, was amnesic. These authors also obtained an inverted U-shaped dose–response curve for morphine using i.c.v. injections; a dose of 0.3 μg had no effect, with 3.0 μg there was retrograde amnesia, and with 40.0 μg there was memory facilitation. This agrees with the literature on the effect of posttraining i.p. injections of morphine. Low doses (1.0 or 3.0 mg/kg) produce amnesia (Izquierdo, 1979; see also Castellano, 1975); high doses (30 mg/kg, Stäubli & Huston, 1980a; 40.0 or 100.0 mg/kg, Mondadori & Waser, 1979) cause facilitation. The latter may be attributable (Stäubli & Huston, 1980a) either to the well-known aversive properties of high doses of this drug (Le Blanc & Cappell, 1975), or to some effect upon other aspects of reinforcement (Mondadori & Waser, 1979). In the experiment of Gallagher and Kapp (1978a,b), posttraining intra-amygdaloid injection of 0.1–2.5 nmol of naloxone caused memory facilitation; the opiate agonist, levorphanol (1.0–5.0 nmol) caused amnesia; the two drugs, when given together, canceled each other's effects; and the inactive enantiomer, dextrorphan (5.0 nmol) had no effect. In their full paper published one year later, Messing, Jensen, Martinez, Spiehler, Vasquez, Soumireu-Mourat, Liang, and McGaugh (1979) reported, in addition to the effect on the step-through task, a memory-facilitating effect of naloxone in a one-way active avoidance paradigm for which, however, larger doses of the drug and a combination of pre- and posttrial treatments were needed.

Our first experiments (Izquierdo, 1979) were aimed at testing the generality of the findings of Jensen et al. (1978) and Gallagher and Kapp (1978a,b) which had, so far, been obtained using only one task. For this purpose, we selected a variety of multitrial paradigms: two forms of shuttle avoidance (one with, and the other without, CS–US pairing), classical conditioning in the shuttlebox, and habituation of a rearing response to a tone. In this, and in all other experiments from this laboratory commented on here, the subjects were female Wistar rats from our own breeding stock, 60–80 days of age and weighing 100–180 g. Aversive training or test sessions consisted of 50 trials at randomly variable intervals of 10–50 sec and lasted for 25 min. The CS was a 5-sec, 1 kHz, 70 dB tone; and the US was a 2-sec, 60 Hz, 1.0 mA footshock. In the standard shuttle avoidance paradigm, and in the classical conditioning procedure, the tone was immediately followed by the footshock (CS–US pairing); in the former, the footshock was omitted from those trials in which there was a shuttle response to the tone (avoidance contingency); in the classical paradigm, the shock was given on all trials regardless of whether responses were made to the tone (Izquierdo & Elisabetsky, 1978). In the avoidance conditioning paradigm without CS–US pairing, the CS–US interval was randomly varied between 5 and 35 sec, and shuttle responses to the tone resulted in cancelation of the next scheduled shock. Habituation consisted of 20 presentations of the tone, at intervals of 5–35 sec, over 4–6 min. Except when stated otherwise, the training–test interval was 7 days in the aversive tasks, and 24 hr in the habituation paradigm. In all cases, the drugs were dissolved in saline to an injection volume of 1.0 or 2.0 ml/kg. Control animals received saline injections. For a further description of the behavioral procedures, see Izquierdo (1979); Izquierdo et al. (1980c).

Immediate posttraining i.p. administration of naloxone facilitated retention of all tasks. Retention of the avoidance paradigm without CS–US pairing was facilitated by 0.2 mg/kg of the drug; the habituation task was sensitive only to 0.8 mg/kg; and classical conditioning, the standard shuttle avoidance procedure, and memory "transfers" between classical and avoidance conditioning, were enhanced either by 0.4 or by 0.8 mg/kg of naloxone. Antagonism with morphine was studied in the standard shuttle avoidance task, and in the habituation paradigm. In both cases, immediate posttraining i.p. injection of 1.0 mg/kg morphine caused retrograde amnesia. When morphine and naloxone were given together, they canceled each other's effects (Izquierdo, 1979).

Thus, the generality of the effect of naloxone on memory consolidation was substantiated; and it was clearly shown to be independent of the response requirements of the tasks (activity, response inhibition), and of the presence of pain during the training session because the habituation task involves no pain (Izquierdo, 1979).

IV. POSSIBLE MECHANISM OF THE EFFECT OF NALOXONE ON CONSOLIDATION

This was investigated both in the habituation task, and in the standard shuttle avoidance paradigm. In the habituation task (Izquierdo & Graudenz, 1980), the facilitatory effect of a posttraining i.p. injection of 0.8 mg/kg naloxone was blocked by the simultaneous administration of haloperidol (0.5 mg/kg) or propranolol HCl (0.5 mg/kg), but not by phenoxybenzamine HCl (2.0 mg/kg) or tolazoline HCl (2.0 mg/kg). This suggested that dopaminergic and beta, but not alpha, adrenergic systems may be involved in the effect of naloxone on memory consolidation. A simple explanation could be that naloxone releases dopaminergic and noradrenergic neurons from an inhibitory influence of opiods. This interpretation was supported by the following experiment. Contrary to findings in many other tasks (Evangelista & Izquierdo, 1971, 1972; Zornetzer, 1978), posttraining i.p. administration of 0.5 or 2.5 mg/kg d-amphetamine sulfate, or of 0.2 or 0.5 mg/kg nicotine, did not cause memory facilitation of the habituation task (Izquierdo & Graudenz, 1980). This was probably due to the fact that learning of this task presumably takes place with very little endogenous catecholamine mobilization (Izquierdo, Dias, Perry, Souza, Elisabetsky, & Carrasco, 1980a). However, when amphetamine or nicotine were given together with a normally ineffective dose of naloxone (0.2 mg/kg), a very pronounced facilitatory effect upon retention was obtained (Izquierdo & Graudenz, 1980).

Results obtained in the avoidance task, and shown in Table 1, were consonant with this interpretation of the mode of action of naloxone on the consolidation process. Animals received, 3 hr before training, an i.p. injection of either saline, or the catecholamine synthesis inhibitor, alpha-methyl-p-tyrosine (60.0 mg/kg). Immediately after training, they received an i.p. injection of either

TABLE 1

Performance of Shuttle Avoidance Responses[a]

Pretraining treatment[a]	Posttraining treatment[a]	n	Number of avoidance responses in	
			Training sessions[d]	Test sessions[d]
Saline	Saline	11	16.8 ± 1.7	21.8 ± 1.1^b
α-m-p-tyrosine	Saline	10	14.9 ± 1.6	22.5 ± 2.0^b
Saline	Naloxone	11	18.1 ± 1.9	28.0 ± 1.9^c
α-m-p-tyrosine	Naloxone	10	16.9 ± 1.6	23.1 ± 1.8^b

[a] Pretraining treatment consisted of saline or alpha-methyl-p-tyrosine (60.0 mg/kg, i.p.) 3 hr prior to training; posttraining treatment consisted of saline or naloxone (0.4 mg/kg, i.p.) immediately after training. The training–test interval was 7 days.

[b] Different from training session performance at 1 or 5% level in Duncan Multiple Range Test.

[c] Different from test session performance of all other groups at 1 or 5% level; difference in training session scores among groups, not significant.

[d] Numbers before and after ± refer to means and SEM, respectively.

naloxone (0.4 mg/kg) or saline. Animals were tested for retention 7 days later. Pretreatment with alpha-methyl-p-tyrosine canceled the memory-facilitating effect of naloxone.

These findings suggest that naloxone facilitates memory consolidation through release of catecholaminergic systems from an opioid-mediated inhibition (Izquierdo et al., 1980a,b; Izquierdo & Graudenz, 1980). This interpretation is in agreement with the known inhibitory effect of opioids on catecholamine release (see preceding section), and fits with all the observed drug interactions. If this interpretation is correct, this effect of opioid peptides should be more important for the modulation of memory consolidation than others, such as the recently suggested depression of inhibitory inputs to A10 dopaminergic neurons (Kelley, Stinus, & Iversen, 1980) and tonic stimulation of vasopressin release (Lightman & Forsling, 1980). Dopaminergic neuron stimulation (Grecksch & Matthies, 1981) and vasopressin release (Kovács, Bohus, & Versteeg, 1979) would be expected to enhance, rather than to depress, memory consolidation.

V. EFFECT OF OPIOID PEPTIDES ON CONSOLIDATION

The effect of opioid peptides on consolidation was studied using the habituation and shuttle avoidance tasks. These two procedures were selected because the former involves no pain, and the two involve opposite response requirements (Izquierdo et al., 1980c). From the start, it was obvious to us that if we were to obtain data that reflected a physiological role of the opioid peptides they had to be given at doses as low as those used, e.g., by de Wied et al. (1978a,b), or Rigter (1978). In all experiments mentioned in this section the training–test interval was 24 hr.

The immediate posttraining i.p. injection of 10.0 μg/kg of Leu-enkephalin or of β-endorphin caused complete retrograde amnesia for the two tasks (Izquierdo et al., 1980c). Subsequent experiments showed that much smaller doses also were effective and that a similar amnesic effect could be obtained with other opioid peptides. In the avoidance task, the amnesic ED_{50} of β-endorphin was found to be 1.0 μg/kg ≅ 140 ng/rat (Izquierdo et al., 1980d). The amnesic ED_{50} of Met-, Leu-, and Des-Tyr-Met-enkephalin in the same task was found to be lower (≅ 0.4 μg/kg; no significant differences among the three enkephalins, Izquierdo & Dias, 1981). The effect of the four opioid substances was antagonized by the posttraining administration of a low dose of naloxone that had no effect on its own (0.2 mg/kg; Izquierdo, 1980b; Izquierdo & Dias, 1981). The antagonism between naloxone and the three enkephalins was competitive, as ascertained by Lineweaver-Burk plots (Izquierdo & Dias, 1981). The antagonism between naloxone and β-endorphin appears also to be competitive (Izquierdo, 1981). β-Endorphin and Met-enkephalin also cause retrograde amnesia for the shuttle avoidance task when given i.c.v. at doses between 5.0 and 25.0 ng/rat (Izquierdo, 1981).

There is an inverted U-shaped dose–response curve for the effect of posttraining i.p. injections of opioid peptides on memory consolidation. In the shuttle avoidance task, β-endorphin and Met-, Leu-, and Des-Tyr-Met-enkephalin cause retrograde amnesia at doses up to 10.0 μg/kg (Izquierdo, 1980b, 1981; Izquierdo et al., 1980a,c,d); however, a dose of 20 μg/kg is ineffective (Izquierdo, 1980b, 1981). In the step-through inhibitory avoidance paradigm, Martinez and Rigter (1980) obtained retrograde amnesia with 0.1, but not with 1.0 or more μg/kg of β-endorphin. The displacement to the left of the curve obtained by Martinez and Rigter relative to ours could be due to differences in sensitivity either of the task or of the animals used by both groups (Izquierdo, 1980b; see the following section). The inverted U-shaped dose-response curve for the effect of opioids on memory is remindful of the one obtained with morphine (Izquierdo et al., 1980a,b,d; Jensen et al., 1978). It seems likely that at very high doses opioid peptides may cause memory facilitation, as occurs with morphine (Stäubli & Huston, 1980a). Stein and Belluzzi (1978) described a memory facilitating effect of a very large dose (200 μg, i.c.v.) of Met-enkephalin in a step-down inhibitory avoidance task in rats. This dose is equal to about 10^3 times the total brain content of the substance and thus the effect is unlikely to reflect a physiological phenomenon (Izquierdo et al., 1980a,d). There are several possible explanations for the eventual memory facilitatory effect of very high doses of opiates or opioids; among them, aversive properties of the drugs (Stäubli & Huston, 1980a), toxic effects (Izquierdo et al., 1980a,d), or hippocampal and/or neocortical spreading depression (Sprick, Oitzl, Ornstein & Huston, 1980). Contrary to some earlier reports, we found recently that posttraining hippocampal spreading depression facilitates several types of memory in the rat (Elisabetsky et al., 1979).

Thus, when doses within the physiological range are considered, the second criterion needed to establish the role of endogenous opioids as inhibitory memory modulators is fulfilled. The opioids are amnesic at low doses, and their effect is antagonized competitively by naloxone (Izquierdo, 1981; Izquierdo & Dias, 1981).

These findings provide no clue as to the site of action of the opioid peptides and, therefore, on the brain structures involved in the operation of the proposed amnesic mechanism. The experiments of Gallagher and Kapp (1978a,b) and some recent biochemical findings with naloxone (Dias, Carrasco, Souza, & Izquierdo, 1979) suggest that the amygdala may be an important site of action. However, this structure, while rich in Met-enkephalin, contains relatively little β-endorphin (Bloom et al., 1978; Dupont, Barden, Cusan, Mérand, Labrie, & Vaudry, 1980); and β-endorphin has a powerful amnesic effect. It is possible that there is more than one site of action and therefore more than one brain structure involved in the amnesic effect of opioids. Indeed, the fact that β-endorphin, Met-, Leu-, and Des-Tyr-Met-enkephalin were all effective at similar dose levels, and that their effect was reversed by naloxone, is in itself a suggestion that there may be several sites and receptors involved. Clearly, the fact that Des-Tyr-Met-enkephalin had a potency similar to that of Met- or Leu-enkephalin (Izquierdo & Dias, 1981) suggests that at least some of the receptors involved in the amnesic effect are different from traditional opiate receptors, inasmuch as Des-Tyr-Met-enkephalin is usually inactive at smooth muscle opiate binding sites (Morley, 1980).

VI. TRAINING RELEASES β-ENDORPHIN FROM THE RAT BRAIN

β-Endorphin immunoreactivity was measured in the hypothalamus and the rest of the brain (minus olfactory bulb, midbrain, and cerebellum), hypophysis, and plasma, in rats submitted to 25-min training or test sessions of shuttle avoidance (50 paired tone–footshock presentations, shock omitted from those trials in which there was a shuttle response to the tone), pseudoconditioning in the shuttlebox (50 tones and 25 footshocks at random), habituation (50 tones alone), or 50 footshocks. Training–test interval was 7 days. The data for the hypothalamus and the rest of the brain have been published (Izquierdo et al., 1980d), and those for the hypophysis and plasma are shown in Table 2.

TABLE 2

β-Endorphin Immunoreactivity in the Hypophysis (ng/gland) and Plasma (ng/ml) of Rats[a]

	β-Endorphin immunoreactivity in	
Training procedure	Hypophysis (ng/gland)	Plasma (ng/ml)
Control (untrained)	221.9 ± 23.2 (20)	11.0 ± 1.0 (7)
Shuttle avoidance	243.1 ± 42.0 (14)	9.6 ± 1.3 (8)
Tones alone	231.0 ± 36.2 (16)	10.2 ± 2.3 (8)
Footshocks alone	115.7 ± 12.0 (16)	18.0 ± 3.8[b] (8)

[a] Rats submitted to a 25-min training session of shuttle avoidance, tones alone, or footshocks alone. Means ± SEM; number of animals per group in parentheses.

[b] Different from all other groups at 5% level in Duncan Multiple Range Test. (Values not corrected for the presence of β-lipotropin in the samples.)

A decrease of β-endorphin immunoreactivity, equivalent to 20 to 40 ng of β-endorphin per brain, was observed in the rest of the brain after training with any of the four procedures; no changes were detected in the hypothalamus, or after test sessions (Izquierdo et al., 1980a,d).

An increase in plasma β-endorphin levels and a concomitant decrease of pituitary gland β-endorphin immunoreactivity were observed after footshock training, but not after training with tones alone or in the avoidance paradigm (Table 2).

The increase of plasma β-endorphin levels by footshocks confirms a previous report by Rossier, French, Rivier, Ling, Guillemin, and Bloom (1977), who observed that the levels increase concomitantly with ACTH levels. Indeed, various forms of stress release both substances simultaneously from the hypophysis into the blood stream (Rossier, French, Gros, Minick, Guillemin, & Bloom, 1979), and this is part of the evidence that both originate from a common precursor molecule. It seems clear that footshock training was different from the other forms of training, in that the latter did not release β-endorphin from the pituitary gland into the plasma.

Thus, the effect of the four forms of training on brain β-endorphin immunoreactivity cannot be attributed to a release into the general circulation (Izquierdo et al., 1980a); nor can it be attributed to an inhibition of the synthesis of this substance because this is very slow and would not bring about a reduction of β-endorphin levels even if it were completely arrested for 25 min (see

Izquierdo et al., 1980a,b,d for references). Therefore, the decrease of β-endorphin immunoreactivity in the rest of the brain after training can only be explained either by metabolism without release, or by release followed by metabolism. The former possibility is unlikely, since it would make it a unique case among neurotransmitters and/or modulators, by which β-endorphin-containing terminals would be inhibited by intracellular destruction of the substance. Therefore, the most likely explanation for the decrease of β-endorphin immunoreactivity caused by training, is that the substance is released and subsequently metabolized.

This release is probably not a consequence of learning, but rather of non-associative concomitants of learning (or effects of stimulation), because it occurred in all groups (Izquierdo et al., 1980a,d).

It should be noted that the amount released during training sessions is compatible with amnesic doses of this substance (5.0–25.0 ng/rat, i.c.v.; or 140 ng/rat i.p. of which only 20% or less reaches the brain within 120 min from injection; see preceding and Izquierdo, 1981; Izquierdo et al., 1980d; Rapoport, Klee, Pettigrew, & Ohno, 1980).

Thus, the last two criteria needed to establish a role for opioid peptides in memory modulation (release from the brain during training in amounts compatible with an effective dose) were fulfilled.

VII. ELECTROCONVULSIVE SHOCK RELEASES β-ENDORPHIN FROM BRAIN BUT NOT PITUITARY GLAND

Electroconvulsive shock is one of the most widely used posttraining amnesic treatments. Recently (Dias, Perry, Carrasco, & Izquierdo, 1981), it was found that 10 or 30 min after electroconvulsive shock there is a very large decrease of β-endorphin immunoreactivity in the hypothalamus and in the rest of the brain of rats. The decrease may be estimated to be of approximatley 60–80 ng per brain (hypothalamus + the rest). No significant changes were detected in the pituitary gland or plasma. These data suggest that the amnesic effect of electroconvulsive shock could be due to a massive release of β-endorphin from the brain but not from the pituitary gland.

VIII. POSSIBLE ROLE OF A PHYSIOLOGICAL AMNESIC MECHANISM

The opioid mediated amnesic mechanism could, in principle, operate before, during, or after consolidation. In the first case, its role would be to prevent the access of traces to consolidation. In the others, it could act either by interfering with this process, or by erasing freshly (and partly) fixed traces, respectively. The fact that naloxone and the opioids act when given after multitrial learning sessions suggests that these substances act a posteriori, because there must be some consolidation during such training sessions (Izquierdo et al., 1980a,b,c,d; Messing et al., 1979). This could explain the rapid decay of memory that takes place in the first few seconds or minutes after an experience (Ebbinghaus, 1885; Harlow et al., 1971), without resorting to hypothetical forms of memory of different duration (see Gold & McGaugh, 1975, for references) or to hypothetical repressor mechanisms.

Most of the neurohumoral systems that have been shown to have a facilitatory influence on consolidation are triggered into action by nonassociative events (arousal, etc.) that occur during and/or shortly after training (Gold & McGaugh, 1975; McGaugh et al., 1979). The inhibitory mechanism mediated by opioid peptides also appears to be triggered by nonassociative events, and it would operate precisely by interfering with catecholaminergic facilitatory systems (Izquierdo & Graudenz, 1980).

Because β-endorphin is released during training and it is amnesic, the question arises of what could be its role during multitrial learning in which there must be some consolidation from trial to trial.

The effect of opioid substances given prior to training or test sessions has been investigated by several groups, with apparently conflicting results. As mentioned earlier, Kastin et al. (1976) observed facilitation of maze learning in rats with 80.0 μg/kg of enkephalins given i.p.; and de Wied et al. (1978a,b) obtained retarded extinction of a one-way active task using low doses of some opioid peptides, and accelerated extinction using other opioids or naltrexone, given prior to the test. Rigter (1978) described an "antiamnesic" effect of low doses of Met- and Leu-enkephalin. However, Rigter et al. (1980) obtained a depression of one-way active avoidance learning using 40.0 μg/kg of Met-enkephalin or 400.0 μg/kg of Leu-enkephalin in American Fisher F344 rats, and just 1.0 or 10.0 μg/kg of either compound in the apparently much more sensitive Dutch Wistar rats (Martinez, Rigter, & van der Gugten, 1980). In the Fisher rats, the effect of Leu-enkephalin was blocked by correspondingly high doses of naloxone; whereas in the Dutch rats, naloxone, at low doses, was able to antagonize the effect of Met- but not that of Leu-enkephalin. Thus, it would seem as if the influence of opioids upon learning and relearning could depend on the task, on the strain, and/or on whether the drugs are given prior to training or to retest sessions.

In 1964, Belleville described state-dependent learning using morphine. It occurred to us that, at least part of these conflicting results might be explained by assuming that learning of the various tasks is dependent on the endogenous release of β-endorphin and, perhaps, of other opioid peptides as well. This is tantamount to postulating that learning depends on a particular state induced by endogenous substances. From a neurochemical point of view, this may be considered to be a rather complex state, inasmuch as opioids are known to influence the release of various other transmitters, modulators, and hormones, all of which have been shown to play a role in memory modulation (see preceding sections, and Izquierdo et al., 1980a,b; Snyder & Childers, 1979; Terenius, 1978, for references). As was shown in the previous section, β-endorphin is not released in a test session when it is carried out with the same procedures used for training. It follows that, if learning were normally state-dependent upon β-endorphin, all experiments using training–test session designs would actually be experiments in dissociated learning or, experiments in which the state prevalent during training would not be the same that prevails during testing (Izquierdo et al., 1980a).

The dependency of each task on the opioids could be due to a proactive amnesic influence of these substances exerted preferentially upon the various adventitious learning that occurs during, and interferes with, any form of training (Izquierdo, 1980a,b; Izquierdo et al., 1980a,b,d). Thus, according to this hypothesis, the main task for which the animals are trained would be freed from the interfering influence of the adventitious learning as a consequence of the amnesic effect of the opioids. There are many reasons why the main task would survive this effect. For example, its relative overlearning due to the particular design of the training procedures, or its peculiar biological significance or "meaningfulness" (Izquierdo et al., 1980a).

This hypothesis leads to the prediction that there should be an optimum dose for each opioid peptide, at which the specific or main task would be preserved and the adventitious memories would be erased. The dose would depend on the drug, the task, the peculiar sensitivity of each strain or stock, and the amount of opioid substances, if any, released endogenously. At a dose below the optimum, adventitious learning would tend to interfere with acquisition of the main task; but retention of the latter would not necesarily be impaired; in fact, it could even be enhanced, in relative terms, by the absence of the amnesic opioids. At a high dose of the opioids, their proactive amnesic effect would "spread" over all behaviors, and both acquisition and retention would be impaired. In test sessions, in which β-endorphin is not released, the administration of β-endorphin or a similar compound, at an appropriate dose, should enhance performance of a previously learned task. This would result from a combination of two factors: (1) the positive bias on performance caused by retention of that task; (2) the proactive amnesic effect of the substance that is now unable to block the much strengthened "main" task, would trim it still further by erasing all other memories that might compete with it (Izquierdo et al., 1980a).

In view of the apparent difference in sensitivity of the diverse tasks and/or strains to the opioids discussed, this hypothesis, which implies a special case of state dependency, must be tested using low and high doses of at least one opioid peptide in the same task, and using the same type of animal throughout. This was done in our laboratory, using our own breed of Wistar rats, with two doses of β-endorphin (2.0 and 20.0 μg/kg, i.p.), and two tasks: the standard shuttle avoidance paradigm, and habituation of a rearing response to a tone. The results have been published elsewhere (Izquierdo, 1980a,b) and will be briefly summarized here.

Naloxone (0.8 and 1.6 mg/kg, i.p.), given prior to training, impaired acquisition of the two tasks. This was manifested as decreased responding in the avoidance task, and as increased responding in the habituation paradigm, and cannot be attributed to proactive effects upon the response requirements of the tasks, or upon pain mechanisms. The effect of naloxone might be attributed to an interference with the action of the β-endorphin that is normally released during these two types of training. Animals trained under naloxone, however, were left with comparatively good retention of whatever they were able to learn because they showed large training–test performance differences when tested on the next day. Naloxone given prior to the test session had no effect on performance in that session. This was to be expected, as there was no endogenous β-endorphin release during that session in either of the two tasks.*

Given prior to training at a dose of 2.0 μg/kg, β-endorphin had no apparent effect on acquisition in any of the two tasks; the animals, however, were left amnesic; that is, their test session performance showed no improvement over training session levels. This finding suggests that the amount of this substance that is endogenously released during training is sufficient for a normal rate of acquisition; but that a slight excess, while leaving acquisition relatively unscathed (at least as far as can be determined by evaluating training session scores), is already sufficient to bring about amnesia later on. At a dose of 20.0 μg/kg, β-endorphin depressed acquisition of the two tasks, and also left the animals amnesic. The depressant effect on acquisition, which is coincident with the one described by Rigter et al. (1980) or Martinez et al. (1980) for the enkephalins in a one-way avoidance task, was antagonized by naloxone (1.6 mg/kg). This was despite the fact that each of the two treatments when given separately had a depressant influence of its own. The amnesic effect of β-endorphin (2.0 or 20.0 μg/kg) given prior to training, was reversed by the posttraining administration of a low dose of naloxone (0.2 mg/kg), which had no effect on retention on its own. This suggests that the amnesic effect of pretraining β-endorphin injection resulted mainly from effects upon the posttraining consolidation period (Izquierdo, 1980b).

---

* When the procedure used for training is different from the one used for testing, as is the case when animals are tested for extinction, there is some release of β-endorphin during the test session (Izquierdo et al., 1980a). This might explain the facilitatory effect of naltrexone on extinction of a pole-climbing avoidance response reported by de Wied et al. (1978a).

Given before the test session, β-endorphin (2.0 μg/kg) caused a marked improvement in relearning–retention performance in the two tasks. This agrees with the previous finding of de Wied et al. (1978a) on the effect of this drug on retrieval of a one-way active response. In animals that had been rendered amnesic by treatment with this same substance during training, the facilitatory effect of 2.0 μg/kg β-endorphin on test session performance was also present (Izquierdo, 1980a). This finding agrees with the antiamnesic effect of the enkephalins described by Rigter (1978) using the step-through inhibitory avoidance paradigm.

Therefore, these findings fully agree with all the predictions derived from the hypothesis stated above on the role of the opioid mediated amnesic mechanism in acquisition; and explain most of the apparently conflicting data on the proactive effect of opioids on learning and relearning. It must be pointed out, however, that some discrepancies still remain. For example, Martinez and Rigter (1980) were unable to find an amnesic effect with 0.1–10.0 μg/kg of β-endorphin given i.p. 30 min prior to step-through inhibitory avoidance training, whereas the lower dose was amnesic when given immediately posttrial. It is possible that a considerable amount of the substance was metabolized during the 30 min interval used by Martinez and Rigter, so that very little was available in the posttraining period at any dose. In several papers from this laboratory (Elisabetsky et al., 1979; Izquierdo et al., 1979; Izquierdo & Elisabetsky, 1978), we presented arguments in favor of the need for caution when comparing data obtained in different tasks; and, particularly, direct extrapolations from findings on retention of the step-through inhibitory avoidance task to other forms of memory, which are consolidated and stored through different channels, involving different neurohumoral modulatory processes.

IX. FINAL COMMENT

The amnesic mechanism mediated by β-endorphin, and perhaps by other opioid peptides as well, explains several hitherto obscure facts about memory, such as the early rapid decay of memories, why memory is not perfect, and perhaps some aspects of acquisition as well.

The possible role of that mechanism in acquisition suggests a special case of state dependency which might, at the same time, be very general. In particular, all learning, or at least several different forms of learning, may normally be dissociated. This raises the question of what the neurohumoral state may be in each of the various types of learning, or in each particular training–test design. Endogenous opioids inhibit the release of catecholamines (Izquierdo et al., 1980b; Snyder & Childers, 1979), and of several other neurotransmitters as well, some of which, such as Substance P (Stäubli & Huston, 1980b), may also be involved in memory or learning (see Zornetzer, 1978). In addition, opioids may also modulate transmitter effects at a postsynaptic level (Izquierdo et al., 1980b), and have a tonic stimulant effect on vasopressin release (Lightman & Forsling, 1980). Although the latter effect is not likely to be involved in the amnesic action of opioids discussed in the present chapter, inasmuch as vasopressin is generally thought to facilitate memory consolidation (Kovács et al., 1979), it might play a role in other forms of learning, different from those studied here.

Kety (1976) suggested that each form of learning is accompanied by a peculiar constellation of neurohumoral changes. Izquierdo and Elisabetsky (1978) proposed that consolidation along each memory channel, the form of memory that originates from each type of learning, is subserved by different modulatory processes. As was discussed earlier, several of these modulatory processes are initiated precisely by those nonassociative events that accompany each type of training (Gold & McGaugh, 1975; Izquierdo et al., 1980a,b,d). Peripheral factors should also be included as a major influence mediating these neurohumoral states. Agents that modify peripheral catecholamine effects or metabolism strongly affect various forms of learning and memory (Izquierdo & Thaddeu, 1975; Lauzi Gozzani & Izquierdo, 1976; McGaugh, Martinez, Jensen, Hannan, Vasquez, Messing, Liang, Brewton & Spiehler, 1981; Rachid, de Souza, & Izquierdo, 1977). Also, the presence of an intact adrenal medullary function seems important for the effect of enkephalins upon acquisition (Martinez et al., 1980), either because the latter act upon the adrenal medulla, or because they need the presence of circulating epinephrine to exert their central action.

Perhaps a good problem for further study is the eventual matching or unmatching of the internal neurohumoral states that accompany training and/or consolidation, with those that occur during testing. This strategy would lead biochemical studies of memory away from the more traditional approach of a search for a specific molecule or a specific change typical of each form of behavior. This might not be a great loss, as that approach has been remarkably unsuccessful so far, due to two insurmountable difficulties. First, memory modulation is such a powerful influence that it may be impossible to study anything else in experiments designed to measure biochemical correlates of consolidation (Dunn, 1980). Second, it also seems impossible to design a control group in which there is no learning, but in which the nonassociative concomitants of learning are still present (Dunn, 1980; Souza, Elisabetsky, & Izquierdo, 1980).

ACKNOWLEDGMENTS

Work supported by FAPERGS, PROPESP-UFRGS, PROPLAN-UFRGS, and CNPq, Brasil. Elaine Elisabetsky is a fellow of FAPESP, Brasil. Otto A. Orsingher is a visiting professor from

Departamento de Farmacología, Facultad de Ciencias Químicas, Universidad de Córdoba, Argentina.

REFERENCES

Belleville, R. E. (1964). Control of behavior by drug-produced internal stimuli. *Psychopharmacologia*, **5**, 95–105.

Belluzzi, J. D., Grant, N., Garsky, V., Sarantakes, D., Wise, C. D., & Stein, L. (1976). Analgesia induced in vivo by central administration of enkephalin in rat. *Nature (London)*, **260**, 625–626.

Bloom, F. E., Rossier, J., Battenberg, E. L. F., Bayon, A., French, E., Henriksen, S. J., Siggins, G. R., Segal, D., Browne, R., Ling, N., & Guillemin, R. (1978). β-Endorphin: Cellular localization, electrophysiological and behavioral effects. *In* E. Costa & M. Trabucchi (Eds.), *The Endorphins*, pp. 89–110. New York: Raven.

Bradbury, A. F., Smyth, D. G., & Snell, C. R. (1976). Lipotropin: Precursor to two biologically active peptides. *Biochemical and Biophysical Research Communications*, **69**, 950–956.

Castellano, C. (1975). Effects of morphine and heroin on discrimination learning and consolidation in mice. *Psychopharmacologia*, **42**, 235–242.

Chrétien, M., Benjannet, S., Dragon, N., Seidah, N. G., & Lis, M. (1976). Isolation of peptides with opiate activity from sheep and human pituitaries: Relation to β-lipotropin. *Biochemical and Biophysical Research Communications*, **72**, 472–478.

Cox, B. M., Goldstein, A., & Li, C. H. (1976). Opioid activity of a peptide, β-lipotropin-(61–91), derived from β-lipotropin. *Proceedings of the National Academy of Sciences, U.S.A.* **73**, 1821–1823.

de Wied, D., Bohus, B., van Ree, J., & Urban, I. (1978a). Behavioral and electrophysiological effects of peptides related to lipotropin (β-LPH). *Journal of Pharmacology and Experimental Therapeutics*, **204**, 570–580.

de Wied, D., Kovács, G. L., Bohus, B., van Ree, J. M., & Greven, H. M. (1978b). Neuroleptic activity of the neuropeptide β-LPH$_{62-77}$ ([Des-Tyr^1] γ-endorphin; DTγE). *European Journal of Pharmacology*, **49**, 427–436.

Dias, R. D., Carrasco, M. A., Souza, D. O., & Izquierdo, I. (1979). Effect of naloxone, haloperidol and propranolol on cyclic 3′:5′ adenosine monophosphate content of rat amygdala. *European Journal of Pharmacology*, **60**, 345–347.

Dias, R. D., Perry, M. L. S., Carrasco, M. A., & Izquierdo, I. (1981). Effect of electroconvulsive shock on β-endorphin immunoreactivity of rat brain, pituitary gland, and plasma. *Behavioral and Neural Biology*, **32**, 265–268.

Dupont, A., Barden, N., Cusan, L., Mérand, Y., Labrie, F., & Vaudry, H. (1980). β-Endorphin and Met-enkephalins: Their distribution, modulation by estrogens and haloperidol, and role in neuroendocrine control. *Federation Proceedings*, **39**, 2544–2550.

Di Giulio, A. M., Majane, E. M., & Yang, H-Y. T. (1979). On the distribution of [Met^5]- and [Leu^5]-enkephalins in the brain of the rat, guinea pig, and calf. *British Journal of Pharmacology*, **66**, 297–301.

Dismukes, R. K., & Rake, A. V. (1972). Involvement of biogenic amines in memory formation. *Psychopharmacologia*, **23**, 17–25.

Dunn, A. J. (1980). Neurochemistry of learning and memory: An evaluation of recent data. *Annual Review of Psychology*, **31**, 343–390.

Ebbinghaus, H. (1885). *Über das Gedachtniss*. Leipzig: Drucker & Humblat.

Elisabetsky, E., Vendite, D. A., & Izquierdo, I. (1979). Memory channels in the rat: Effect of post-training application of potassium chloride on the hippocampus. *Behavioral and Neural Biology*, **27**, 354–361.

Evangelista, A. M., & Izquierdo, I. (1971). The effect of pre- and post-trial amphetamine injections on avoidance responses of rats. *Psychopharmacologia*, 20, 42–47.

Evangelista, A. M., & Izquierdo, I. (1972). Effects of atropine on avoidance conditioning: Interaction with nicotine and comparison with N-methyl-atropine. *Psychopharmacologia*, 27, 241–248.

Fulginiti, S., Molina, V. A., & Orsingher, O. A. (1976). Inhibition of catecholamine biosynthesis and memory processes. *Psychopharmacology*, 51, 65–69.

Gallagher, M., & Kapp, B. S. (1978a). Opiate administration into the amygdala: Effects on memory processes. *Life Sciences*, 23, 1973–1978.

Gallagher, M., & Kapp, B. S. (1978b). Opiate administration into the amygdala: Effects on memory processes. *Society for Neuroscience Abstracts*, 4, 258.

Gold, P. E. (1981). Catecholamine modulation of memory formation. In H. Matthies (Ed.), *Sixth International Neurobiological Symposium on Learning and Memory*, New York: Raven.

Gold, P. E., & McGaugh, J. L. (1975). A single-trace, two-process view of memory storage processes. In D. & J. A. Deutsch (Eds.), *Short-term Memory*, pp. 355–378. New York: Academic Press.

Gold, P. E., & van Buskirk, R. (1978). Effects of α- and β-adrenergic receptor antagonists on post-trial epinephrine modulation of memory: Relationship to post-training brain norepinephrine concentrations. *Behavioral Biology*, 24, 168–184.

Grecksch, G., & Matthies, H. (1981). Involvement of hippocampal dopaminergic receptors in memory consolidation in rats. In H. Matthies (Ed.), *Sixth International Neurobiological Symposium on Learning and Memory*. New York: Raven.

Guillemin, R., Ling, N., & Burgus, R. (1976). Endorphines, peptides d'origine hypothalamique et neurohypophysaire à l'activité morphinomimétique. Isolement et structure moléculaire de l' α-endorphine. *Comptes Rendues de l'Academie des Sciences (Paris)*, 282, 783–785.

Harlow, H. F., McGaugh, J. L., & Thompson, R. F. (1971). *Psychology*. San Francisco: Albion Press.

Hughes, J., Smith, T. W., Kosterlitz, H. W., Fothergill, L., Morgan, B. A., & Morris, H. R. (1975). Identification of two related pentapeptides from the brain with potent opiate agonist activity. *Nature (London)*, 258, 577–579.

Izquierdo, I. (1979). Effect of naloxone and morphine on various forms of memory in the rat: Possible role of endogenous opiate mechanisms in memory consolidation. *Psychopharmacology*, 66, 199–203.

Izquierdo, I. (1980a). Effect of beta-endorphin and naloxone on acquisition, memory and retrieval of shuttle avoidance and habituation learning in rats. *Psychopharmacology*, 69, 111–115.

Izquierdo, I. (1980b). Effect of a low and a high dose of β-endorphin on acquisition and retention in the rat. *Behavioral and Neural Biology*, 30, 460–464.

Izquierdo, I. (1981). Effect of opioid peptides on learning and memory: Single or dual effect? In S. Saito & J. L. McGaugh (Eds.), *Pharmacology of Learning and Memory*, in preparation. London: Pergamon.

Izquierdo, I., Beamish, D. G., & Anisman, H. (1979). Effect of an inhibitor of dopamine-beta-hydroxylase on the acquisition and retention of four different avoidance tasks in mice. *Psychopharmacology*, 63, 173–178.

Izquierdo, I., & Dias, R. D. (1981). Retrograde amnesia caused by Met-, Leu-, and des-Tyr-Met-enkephalin in the rat and its reversal by naloxone. *Neuroscience Letters*, 22, 189–193.

Izquierdo, I., Dias, R. D., Perry, M. L., Souza, D. O., Elisabetsky, E., & Carrasco, M. A. (1980a). A physiological amnesic mechanism mediated by endogenous opioid peptides and its possible role in learning. In H. Matthies (Ed.), *Sixth International Neurobiological Symposium on Learning and Memory*, in press. New York: Raven.

Izquierdo, I., Dias, R. D., Souza, D. O., Carrasco, M. A., Elisabetsky, E., & Perry, M. L. (1980b). The role of opioid peptides in memory and learning. *Behavioral Brain Research*, 1, 451–468.

Izquierdo, I., & Elisabetsky, E. (1978). Four memory channels in the rat brain. *Psychopharmacology*, 57, 215–222.

Izquierdo, I., & Graudenz, M. (1980). Memory facilitation by naloxone is due to release of dopaminergic and beta-adrenergic systems from tonic inhibition. *Psychopharmacology*, 67, 265–268.

Izquierdo, I., Paiva, A. C. M., & Elisabetsky, E. (1980c). Posttraining intraperitoneal administration of Leu-enkephalin and beta-endorphin causes retrograde amnesia for two different tasks in rats. *Behavioral and Neural Biology*, 28, 246–250.

Izquierdo, I., Souza, D. O., Carrasco, M. A., Dias, R. D., Perry, M. L., Eisinger, S., Elisabetsky, E., & Vendite D. A. (1980d). Beta-endorphin causes retrograde amnesia and is released from the rat brain by various forms of training and stimulation. *Psychopharmacology*, 70, 173–177.

Izquierdo, I., & Thaddeu, R. C. (1975). The effect of adrenaline, tyramine, and guanethidine on two-way avoidance conditioning and on pseudoconditioning. *Psychopharmacologia*, 43, 85–87.

Jean-Baptiste, E., & Rizack, M. A. (1980). In vitro cyclic AMP-mediated lipolytic activity of endorphins, enkephalins, and naloxone. *Life Sciences*, 27, 135–141.

Jensen, R. A., Martinez, J. L., Jr., Messing, R. B., Spiehler, V., Vasquez, B. J., Soumireu-Mourat, B., Liang, K. C., & McGaugh, J. L. (1978). Morphine and naloxone alter memory in the rat. *Society for Neuroscience Abstracts*, 4, 260.

Kastin, A., Jemison, M. T., & Coy, D. H. (1979a). Analgesia after peripheral administration of enkephalin and endorphin analogues. *Pharmacology, Biochemistry and Behavior*, 11, 713–716.

Kastin, A. J., Olson, R. D., Schally, A. V., & Coy, D. H. (1979b). CNS effects of peripherally administered brain peptides. *Life Sciences*, 25, 401–414.

Kastin, A. J., Scollan, E. L., King, M. G., Schally, A. V., & Coy, D. H. (1976). Enkephalin and a potent analog facilitate maze performance after intraperitoneal administration in rats. *Pharmacology, Biochemistry and Behavior*, 5, 691–695.

Katz, R. J., Carroll, B. J., & Baldrighi, G. (1978). Behavioral activation by enkephalins in mice. *Pharmacology, Biochemistry and Behavior*, 8, 493–496.

Kelley, A. E., Stinus, L., & Iversen, S. D. (1980). Interactions between D-Ala-Met-enkephalin, A10 dopaminergic neurones, and spontaneous behavior in the rat. *Behavioral Brain Research*, 1, 3–24.

Kety, S. S. (1976). Biological concomitants of affective states and their possible role in memory processes. In M. R. Rosenzweig & E. L. Bennett (Eds.), *Neural Mechanisms in Learning and Memory*, pp. 321–326. Cambridge, Massachusetts: MIT Press.

Kosterlitz, H. W., & Hughes, J. (1978). Development of the concepts of opiate receptors and their ligands. In E. Costa & M. Trabucchi (Eds.), *The Endorphins*, pp. 31–44. New York: Raven.

Kovács, G. L., Bohus, B., & Versteeg, D. H. (1979). The effects of vasopressin on memory processes. The role of noradrenergic neurotransmission. *Neuroscience*, 4, 1529–1537.

Krivanek, J. A., & McGaugh, J. L. (1969). Facilitating effects of pre- and posttrial amphetamine administration on discrimination learning in mice. *Agents & Actions*, 1, 36–42.

Lauzi Gozzani, J., & Izquierdo, I. (1976). Possible peripheral adrenergic and central dopaminergic influences in memory consolidation. *Psychopharmacology*, 49, 109–111.

Le Blanc, A. E., & Cappel, H. (1975). Antagonism of morphine-induced aversive conditioning by naloxone. *Pharmacology, Biochemistry and Behavior*, 3, 185–188.

Le Moal, M., Koob, G. F., & Bloom, F. E. (1979). Endorphins and extinction: Differential actions on appetitive and aversive tasks. *Life Sciences*, 24, 1631–1636.

Lightman, S. L., & Forsling, M. L. (1980). Evidence for endogenous opioid control of vasopressin release in man. *Journal of Clinical Endocrinology and Metabolism*, 50, 569–571.

McGaugh, J. L., Martinez, Jr., J. L., Jensen, R. A., Hannan, T. J., Vasquez, B. J., Messing, R. B., Liang, K. C., Brewton, C. B., & Spiehler, V. R. (1981). Modulation of memory storage by treatments affecting peripheral catecholamines. In H. Matthies (Ed.), Sixth International Neurobiological Symposium on Learning and Memory, New York: Raven.

McGaugh, J. L., Gold, P. E., Handwerker, M. J., Jensen, R. A., Martinez, J. L., Jr., Meligeni, J. A., & Vasquez, B. J. (1979). Altering memory by electrical and chemical stimulation of the brain. In M. A. B. Brazier (Ed.), Brain Mechanisms in Memory and Learning: From the Single Neuron to Man, pp. 151–164. New York: Raven.

Martinez, Jr., J. L., & Rigter, H. (1980). Endorphins alter acquisition and consolidation of an inhibitory avoidance response in rats. Neuroscience Letters, 19, 197–201.

Martinez, Jr., J. L., Rigter, H., & van der Gugten, J. (1980). Enkephalin effects on avoidance conditioning are dependent on the adrenal glands. Proceedings of the International Union of Physiological Sciences, 15, 568.

Meligeni, J. A., Ledergerber, S., & McGaugh, J. L. (1978). Norephinephrine attenuation of amnesia produced by diethyldithiocarbamate. Brain Research, 149, 155–164.

Messing, R. B., Jensen, R. A., Martinez, J. L., Jr., Spiehler, V. R., Vasquez, B. J., Soumireu-Mourat, B., Liang, K. C., & McGaugh, J. L. (1979). Naloxone enhancement of memory. Behavioral and Neural Biology, 27, 266–275.

Mondadori, C., & Waser, P. G. (1979). Facilitation of memory processing by posttrial morphine: Possible involvement of reinforcement mechanisms? Psychopharmacology, 63, 297–300.

Morley, J. S. (1980). Structure–activity relationships of enkephalin-like peptides. Annual Review of Pharmacology and Toxicology, 20, 81–110.

Rachid, C., de Souza, A. S., & Izquierdo, I. (1977). Effect of pre- and posttrial tyramine and guanethidine injections on an appetitive task in rats. Behavioral Biology, 21, 294–299.

Rapoport, S. I., Klee, W. A., Pettigrew, K. D., & Ohno, K. (1980). Entry of opioid peptides into the central nervous system. Science, 207, 84–86.

Rigter, H. (1978). Attenuation of amnesia in rats by systemically administered enkephalins. Science, 200, 83–85.

Rigter, H., Hannan, T. J., Messing, R. B., Martinez, Jr., J. L., Vasquez, B. J., Jensen, R. A., Veliquette, J., & McGaugh, J. L. (1980). Enkephalins interfere with acquisition of an active avoidance response. Life Sciences, 26, 337–345.

Rossier, J., French, E. D., Rivier, C., Ling, N., Guillemin, R., & Bloom, F. (1977). Footshock induced stress increases β-endorphin levels in blood but not brain. Nature, 270, 618–620.

Snyder, S. H., & Childers, S. R. (1979). Opiate receptors and opioid peptides. Annual Review of Neuroscience, 2, 35–64.

Souza, D. O., Elisabetsky, E., & Izquierdo, I. (1980). Effect of various forms of training and stimulation on the incorporation of ^{32}P into nuclear phosphoproteins of the rat brain. Pharmacology, Biochemistry and Behavior, 12, 481–486.

Sprick, U., Oitzl, M-S., Ornstein, K., & Huston, J. P. (1980). Spreading depression induced by microinjection of enkephalins into the hippocampus and neocortex. Brain Research.

Stäubli, U., & Huston, J. P. (1980a). Avoidance learning enhanced by posttrial morphine injection. Behavioral & Neural Biology, 28, 487–490.

Stäubli, U., & Huston, J. P. (1980b). Facilitation of learning by post-trial injection of Substance P into the medial septal nucleus. Behavioral Brain Research, 1, 245–255.

Stein, L., & Belluzzi, J. D. (1978). Brain endorphins and the sense of well-being: A psychobiological hypothesis. In E. Costa & M. Trabucchi (Eds.), The Endorphins, pp. 299–311. New York: Raven.

Terenius, L. (1978). Endogenous peptides and analgesia. Annual Review of Pharmacology and Toxicology, 18, 189–204.

Veith, J. L., Sandman, C. A., Walker, J. M., Coy, D. H., & Kastin, A. J. (1978). Systemic administration of endorphins selectively alters open field behavior of rats. Physiology and Behavior, 20, 539–542.

Woolf, C. J. (1980). Analgesia and hyperalgesia produced in the rat by intrathecal naloxone. *Brain Research*, **189**, 593–597.

Yang, H-Y. T., Hong, J. S., Fratta, W., Costa, E. (1978). Rat brain enkephalins: Distribution and biosynthesis. *In* E. Costa & M. Trabucchi (Eds.), *The Endorphins*, pp. 149–159. New York: Raven.

Zornetzer, S. F. (1978). Neurotransmitter modulation and memory: A new neuropharmacological phrenology? *In* M. A. Lipton, A. DiMascio, & K. F. Killam (Eds.), *Psychopharmacology: A Generation of Progress*, pp. 637–649. New York: Raven.

CHAPTER 14

Facilitation of Long-Term Memory by Brain Endorphins

James D. Belluzzi and Larry Stein

The memory-enhancing activity of opiate agonists was studied in a one-trial step-down passive avoidance task. Rats were shocked after stepping down from a shelf and were then given intraventricular (i.v.) injections of morphine (20 μg), Met-enkephalin (100–200 μg), Leu-enkaphalin (100–200 μg), or intraperitoneal (i.p.) injections of a potent enkephalin analog, N(Me)Tyr-D-Ser-Gly-N(Me)Phe-D-Ser-NH$_2$ (Wy-42,896, 5–10 mg/kg). Three days later, the animals that had received morphine, Met-enkephalin, or Wy-42,896 showed facilitated retention of the passive avoidance response. Leu-enkephalin, surprisingly, failed to produce a significant enhancement of memory, despite the fact that it is a more potent reinforcer than Met-enkephalin in self-administration tests. Naloxone (10 μg) blocked the memory-enhancing effects of morphine, but, at the dose used, had no effect on Met-enkephalin mediated memory facilitation. Carry-over or punishing effects of the opiate treatments were ruled out by appropriate control experiments, supporting the conclusion that posttraining administration of opiates in high doses enhances memory consolidation. With similar results from other laboratories, the present findings support the hypothesis that the activation of brain endorphin systems closely following training enhances long-term memory formation. In apparent contradiction to this hypothesis, opiates are reported to produce amnesia when administered in low doses immediately after training. However, the amnesic and memory-enhancing effects of opiates could result from separate, dose-related actions at presynaptic and postsynaptic sites. Low doses could actually reduce postsynaptic opiate receptor activity by presynaptic inhibition of endogenous endorphin release, whereas high doses could increase postsynaptic opiate receptor activity by direct agonist action.

ENDOGENOUS PEPTIDES
AND LEARNING AND MEMORY PROCESSES

I. ENDORPHINS AND BEHAVIOR

It has long been known that lasting behavioral changes (learning) follow the reinforcement of specific responses. Such long-term modification of behavior by reinforcement would seem to be based on a reorganization of long-term memory. Hence, it is likely that reinforcement and memory mechanisms are intimately related, and it is possible that highly similar or identical neurochemical processes are involved in both. Some evidence already exists that certain noradrenaline systems in the brain participate both in behavioral reinforcement and long-term memory formation (Stein, Belluzzi, & Wise, 1975; Stein, 1978). Here we suggest that central endorphin systems also may exert dual regulation of reinforcement and memory processes.

A role for endorphins in the mediation of behavioral reinforcement is suggested by several lines of evidence: (1) injections of Met- or Leu-enkephalin or a degradation resistant analog may serve as reinforcement for self-administration behavior (Belluzzi & Stein, 1977a; Mello & Mendelson, 1978); (2) electrical stimulation of many enkephalin-rich regions (e.g., midbrain central gray and globus pallidus) supports high rates of self-stimulation (Belluzzi & Stein, 1977a; Stein & Belluzzi, 1978); and (3) such brain stimulation reinforcement may be antagonized in a dose-related fashion by naloxone (Belluzzi & Stein, 1977a; Childress, 1979; Kelsey, Belluzzi, & Stein, 1979; Stapelton, Merriman, Coogle, Gelbard, & Reid, 1979; Stein & Belluzzi, 1979). Thus, results from self-administration testing (which demonstrates reinforcement from exogenously adminstered endorphins) and from self-stimulation testing (which demonstrates reinforcement from endogenously released endorphins) both are consistent with the hypothesis that brain endorphins may regulate or mediate behavioral reinforcement.

The self-administration and self-stimulation tests, while providing convenient analyses of reinforcement mechanisms, have certain limitations. Administration of pure peptides in the self-administration experiments strictly controls the chemical nature of the reinforcing injection, but, because the distribution of the injected peptide to active sites cannot exactly duplicate that of naturally released endorphins, the ensuing pattern of receptor activation could be artifactual or misleading. However, although electrical activation of peptide pathways during self-stimulation presumably releases the chemical messenger in a relatively natural distribution at appropriate postsynaptic sites, it must also cause the simultaneous release of many different transmitters and neurohormones, including some that are still unknown. Because both self-stimulation and self-administration typically allow the animal subjects to respond freely on schedules of continuous or densely spaced reinforcement, direct carry-over effects of drug injections and electrical brain stimulation on closely following responses pose serious problems. These drug–behavior and stimulation–behavior interactions may be avoided by use of the single-trial learning test described in the following sections.

II. OPIATE RECEPTOR ACTIVATION AND MEMORY

We have already reported that enkephalins and morphine enhance retention of a learned response when given immediately after training in a one-trial passive avoidance test (Belluzzi & Stein, 1977b). Here we fully describe this and further experimental evidence (Belluzzi & Stein, in preparation) suggesting that opiate receptor activation may indeed facilitate memory formation. Other workers also have investigated the role of endorphins in memory formation, but no consensus has been reached, because opiates have been found both to facilitate and to inhibit retention. We review this literature here and attempt to solve the apparent contradictions.

The subjects were male albino Charles River rats (Sprague-Dawley derived), weighing 300–400 g and were individually housed with food and water available *ad libitum*. Some rats had permanently indwelling 27-gauge stainless steel cannulas stereotaxically implanted in the lateral ventricle to permit administration of drugs directly into the brain. All rats were gentled by frequent handling for 7–10 days prior to avoidance training. Animals were trained in a Plexiglas chamber with a grid floor for delivery of electric shock from a constant current shock source; an electrically insulated aluminum shelf extended along the back wall of the apparatus 9.4 cm above the grid floor. On the training day, each rat was placed on the shelf and the time to step completely off the shelf (step-down latency) was recorded automatically within 0.1 sec by the operation of a microswitch under the shelf (Stein et al., 1975). Trial 1 was a habituation trial; no shock was given after the step-down response and the rat was permitted to explore the box for 3 min. Trial 2 was the learning trial; the rat received a mildly painful footshock immediately after stepping off the shelf. Mild intensities of footshock were used to produce moderate levels of avoidance learning and thus provide a sensitive baseline for measurement of memory enhancement by drugs. The rat was removed from the apparatus 10 sec after the shock, injected either intraventricularly or subcutaneously (s.c.) with test compound or vehicle, and returned to its home cage. Long-term memory was evaluated in a retention test 3 days later by placing the rat on the shelf, exactly as before, and recording the step-down latency. If no response occurred within 180 sec, the trial was terminated and a score of 180 sec was recorded.

In the first experiment, different groups of rats received intraventricular injections of morphine, enkephalins, or Ringer's solution either immediately or 15 min after the shock, as indicated in Table 1. The Ringer's controls exhibited moderate learning and stayed on the shelf in the retention test for an average

TABLE 1

Effects of Posttraining Injections of Morphine and Enkephalin on Step-Down Passive Avoidance

Treatment	Dose (µg)	Number of rats	Step-down time (sec) Means ± SEM
Ringer's	—	17	36.2 ± 9.4
Morphine	20	19	78.2 ± 12.1*
Leu-enkephalin	100	10	49.2 ± 22.0
	200	8	47.0 ± 19.7
Met-enkephalin	100	5	8.8 ± 1.8
	200	10	114.2 ± 27.7*
Morphine (shock withheld)	20	8	5.8 ± 0.9**
Morphine (15 min delay)	20	7	39.5 ± 23.7

* $p < .02$ versus Ringer's.
** $p < .001$ versus Morphine.

of 36.2 sec as compared to their preshock latency of 8.2 sec (Fig. 1). Significantly increased step-down latencies were observed in the group treated immediately after the shock with morphine. This morphine-induced increase in retention latency could be due to an enhanced memory of the shock, but other explanations are possible. First, direct inhibitory effects of morphine on performance in the retention test must be ruled out. Because the retention test occurred 3 days after the drug injection, such carry-over effects seem unlikely, and, indeed, are ruled out by the results of the "morphine delayed" group. These rats received

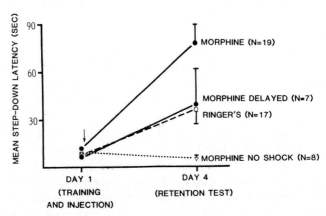

Fig. 1.   Intraventricular administration of morphine (20 µg) immediately after step-down passive avoidance training (arrow) facilitates ($p < .02$) retention measured three days later. Morphine had no effect on retention scores if the injections were delayed for 15 min after training or if the shock was omitted during training. Bars indicate standard errors.

the same training as the morphine group, but the injection was delayed for 15 min after the shock. In this case, the same carry-over effects should occur, but the retention latencies did not differ from those of the Ringer's group (Fig. 1).

Direct effects on retention performance seem to be an unlikely explanation of the increased retention latencies, but other direct effects of the morphine injections could possibly have produced these results. For example, the morphine injection might have produced a rewarding or euphoric state in the animal. However, if morphine reinforced the step-down response, one would expect animals to step off the shelf more quickly in the retention test. Evidence also exists that a first injection of morphine can be punishing (Jacquet, 1973). In such a case, the step-down latency in the retention test would be increased and could erroneously be interpreted as memory facilitation. To test for such direct punishing effects of morphine, an additional group of rats was trained and injected with morphine immediately after the step-down response, but these rats were not shocked. If the morphine injection was punishing, these animals should also exhibit long retention latencies. However, if morphine acts to strengthen the memory of the shock, then animals in this group should step down quickly in the retention test, as there was no shock to remember. Figure 1 shows such quick step-down latencies in the morphine no-shock group, suggesting that the morphine injection had little or no punishing effects under the conditions of these experiments, and that the long latencies in the morphine group may be attributed to memory enhancement.

Because temporal contiguity between the response and the reinforcement is crucial for memory formation, another group of rats was trained and received the shock in a normal manner, but the morphine injection was delayed for 15 min after the shock. Delaying the morphine injection beyond the usual consolidation period should demonstrate whether time-dependent memory processes are involved (McGaugh, 1966). This delayed administration completely abolished the memory-enhancing effects of morphine (Fig. 1, Morphine Delayed), suggesting that the opiate effects are time-dependent and must occur soon after the learning experience. These results are consistent with the idea that the opiate effects influence brain processes involved in the posttraining consolidation of the memory and that such consolidation mechanisms are facilitated by opiate receptor activation.

III. PEPTIDE EFFECTS ON MEMORY

In the same experiment, the possible memory-enhancing effects of Met- and Leu-enkephalin were examined. In different groups of rats, these peptides were injected in the lateral ventricle at doses of 100 or 200 μg per injection. The

200 μg dose of Met-enkephalin given immediately after training significantly facilitated retention of the learned response as compared to the Ringer's control (Fig. 2, Table 1). No significant effect on retention was observed following either dose of Leu-enkephalin (Table 1), but it is extremely interesting that the 100 μg dose of Met-enkephalin produced an apparent amnesia that failed to achieve statistical significance.

The memory-enhancement induced by Met-enkephalin is consistent with the results produced by morphine. The failure of Leu-enkephalin to share this effect is unexpected, especially because Leu-enkephalin is a more potent reinforcer than Met-enkephalin in the self-administration test (Belluzzi & Stein, 1977a). The present result may simply reflect differences in the dose–response curves for the two pentapeptides; it is possible, however, that functional differences exist between Met- and Leu-enkephalin systems, and that the present findings reflect the possible separation of the reinforcing and memory-facilitating effects.

The second experiment examined the effects of naloxone on memory facilitation induced by morphine and by Met-enkephalin. Demonstration of naloxone blockade would suggest involvement of opiate receptors in these effects. Animals received avoidance training and administration of morphine (20 μg) or Met-enkephalin (200 μg) immediately after training as before. Two additional groups were given morphine or Met-enkephalin in a solution also containing

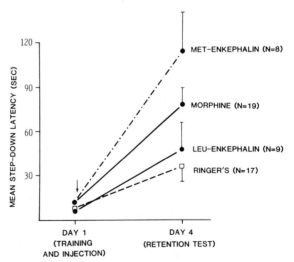

Fig. 2. Intraventricular administration of Methionine-enkephalin (200 μg) immediately after step-down passive avoidance training (arrow) facilitates (*p* < .02) retention measured 3 days later. Identical administration of Leucine-enkephalin had no significant effects. Bars indicate standard errors.

naloxone (10 μg). Naloxone was delivered in the same solution as the opiates so that all groups would be injected only once following training. Naloxone completely blocked the memory-enhancing effects of morphine ($p < .05$), but had no significant effect on the facilitation induced by Met-enkephalin (Table 2). We do not know whether a higher dose of naloxone would antagonize the action of Met-enkephalin or if the effect of this peptide is nonreversible by naloxone.

In a final experiment, the memory-enhancing effects of a peripherally active enkephalin analog (Wy-42,896; N(Me)Tyr-D-Ser-Gly-N(Me)Phe-D-Ser-NH_2) were examined. Again, experimental procedures were identical to the previous tests except that the peptide was adminstered s.c. Potent analgesic activity by this route has been demonstrated for this compound (Belluzzi & Stein, unpublished data). Different groups of rats received injections of either the enkephalin analog (Wy-42,896) or the vehicle. Control rats, injected s.c. with saline immediately after the shock, exhibited moderate learning and stayed on the shelf in the retention test, on the average, for 49.8 sec. Significant facilitation ($p < .02$) of the learned response was observed in the group given immediate postshock treatments (5–10 mg/kg) of Wy-42,896 (Table 3; Fig. 3). If the injection was delayed for 15 min after the shock trial (15' Delay group), or if the shock was omitted (No-Shock group) the memory-enhancing effect of Wy-42,896 was abolished. These results replicate our findings following central injections of morphine. Similar subcutaneous treatments with equianalgesic doses of morphine (2.5–10 mg/kg) or a related analog (Wy-42,186; Tyr-D-Ala-Gly-Phe-D-Pro-NH_2, 5–10 mg/kg) failed to alter retention performance. A 2.5-mg/kg dose of Wy-42,896 also failed to enhance memory formation, although this dose produces analgesia.

TABLE 2

Effects of Naloxone on Morphine- and Enkephalin-Induced Memory Formation

Treatment	Dose (μg)	Number of rats	Step-down latency (sec) Mean ± SEM
Ringer's	—	18	7.1 ± 2.2
Morphine	20	14	38.3 ± 9.0**
Morphine + naloxone	20 10	8	5.5 ± 2.3
Met-enkephalin	200	18	31.0 ± 13.1*
Met-enkephalin + naloxone	200 10	6	28.6 ± 12.9*

* Significantly different from Ringer's, $p < .05$.
** $p < .02$.

TABLE 3

Effects of Posttraining Injections of Enkephalin Analogs on Step-Down Passive Avoidance

Treatment	Dose (mg/kg, s.c.)	Number of rats	Step-down latency (sec) Mean ± SEM Training (preshock)	Retention (72 hr)
Vehicle	—	25	4.4 ± 0.8	49.8 ± 11.0
Wy-42,896[a]	2.5	6	10.6 ± 2.0	63.2 ± 31.6
	5.0	12	9.7 ± 3.5	100.8 ± 16.8*
	10.0	5	13.8 ± 6.6**	139.5 ± 19.2**
Wy-42,896 (no-shock)	5.0	8	8.1 ± 3.0	5.6 ± 1.7
Wy-42,896 (15′delay)	5.0	5	5.1 ± 1.8	57.1 ± 34.8
Morphine	2.5	10	9.6 ± 3.0	43.6 ± 22.9
	5.0	13	9.3 ± 3.7	42.9 ± 17.6
	10.0	11	13.0 ± 4.5*	46.7 ± 20.8
Wy-42,186[b]	5.0	11	3.4 ± 0.6	63.0 ± 23.1
	10.0	11	7.8 ± 3.4	25.1 ± 6.8

[a] N(Me)Tyr-D-Ser-Gly-N(Me)Phe-D-Ser-NH$_2$

[b] Tyr-D-Ala-Gly-Phe-D-Pro-NH$_2$

* Significantly different from vehicle, $p < .02$.

** $p < .01$.

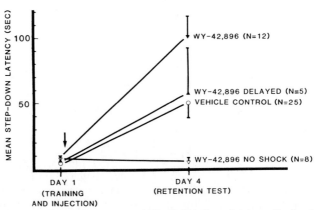

Fig. 3. Intraperitoneal administration of Wy-42,896 (5 mg/kg) immediately after step-down passive avoidance training (arrow) facilitates ($p < .02$) retention measured three days later. Wy-42,896 had no effect on retention scores if the injections were delayed for 15 min after training or if the shock was omitted during training. Bars indicate standard errors.

IV. MEMORY MODULATION BY OPIOID SYSTEMS

The present observations that morphine, Met-enkephalin, and a potent analog of enkephalin facilitate retention in one-trial avoidance learning supports the idea that opiate receptor activation is involved in memory formation. A number of other laboratories also have studied the effects of opiates and opioid peptides in memory and learning tests, but no consensus has emerged regarding their role in memory formation; in fact, contradictory conclusions have been reported. However, it may be possible to reconcile these findings by giving consideration to differences in training methods and drug administration procedures, as discussed below.

As we report here (and previously, Belluzzi & Stein, 1977b), studies in which morphine or enkephalins are administered in high doses immediately after a single training trial consistently find enhancement of learning. Thus, Jensen, Martinez, Messing, Spiehler, Vasquez, Soumireu-Mourat, Liang, and Mc-Gaugh (1978) found that intraventricular administration of morphine (40 μg) in rats immediately after training in one-trial passive avoidance facilitates retention performance. Mondadori and Waser (1979) similarly report facilitated retention of a passive avoidance response by posttraining administration of morphine, but in this experiment 40–100 mg/kg of the drug was injected intraperitoneally, and the subjects were mice. In the same experiment lithium chloride (LiCl), in a dose known to induce a conditioned taste aversion, did not affect retention of the step-down response. Because LiCl is a stronger punisher than morphine (Gorman, Obladia, Scott, & Reid, 1978), Mondadori and Waser (1979) suggested that morphine's effect on retention cannot be attributed to a punishing property of the injection and thus could reflect memory facilitation. We came to a similar conclusion using a different control for the possible punishing effects of morphine (see Fig. 1 and text). Stäubli and Huston (1980) also reported facilitated retention following posttraining morphine injections (10–30 mg/kg, i.p.) in the rat one-trial step-down test. These authors did not control for possible aversive properties of morphine, but such controls in the present study and in that by Mondadori and Waser argue against this possibility.

Especially noteworthy is a study by White, Major, and Siegel (1978) demonstrating memory facilitation by morphine in appetitive learning. Morphine (100 μg) administered intraventricularly immediately following a single training trial facilitated retention of a water-finding response. In this case, punishing and memory-enhancing effects of morphine cannot be confused, as drug induced punishment would increase retention latencies, whereas drug-induced memory enhancement would decrease retention latencies. However, problems of interpretation would arise if the morphine treatment had rewarding effects, because reinforcement of the water-finding response by morphine would also decrease retention latencies. This possibility seems to be ruled out by parallel experiments in rats who learned that water was not present in the familiar water tubes. In these rats, the morphine treatment produced increased retention latencies. This result is consistent with morphine-induced memory enhancement but not with morphine-induced reward; rewarding effects should have caused the rats to approach the empty water tube.

In the experiments just described, very high doses of opiates were used to facilitate learning. Deficits in retention are reported when low doses of opiates are administered immediately after training in a one-trial inhibitory avoidance task (Jensen et al. 1978; Martinez & Rigter, 1980). Jensen et al. (1978) found retention deficits after intraventricular administration of 3 μg of morphine, although 40 μg facilitated retention performance. Using peripheral administration, the same authors found that 1–3 mg/kg of morphine induced amnesia, whereas 10–30 mg/kg had no inhibitory or facilitatory effect. Martinez and Rigter similarly reported that 0.1 μg/kg of β-endorphin administered immediately after training produces a retention deficit, whereas 100 μg/kg of β-endorphin did not alter retention. It would be interesting to determine whether higher doses of β-endorphin (comparable to the effective doses of morphine and enkephalin in the positive studies described above) might also facilitate memory.

When opiates are administered in moderate doses after training in multiple-trial tests, both retention-enhancing and retention-inhibiting effects are observed. Castellano (1980) reported that heroin injections (0.5 mg/kg, i.p.) given immediately after each of five daily training sessions facilitated acquisition of pattern discrimination in a complex maze. Similar injections administered 2 hr after each training session had no effect. Izquierdo (1979) reported that morphine (1 mg/kg, i.p.) depressed retention of shuttlebox avoidance when given immediately after a 50-trial training session, and Izquierdo, Paiva, and Elisabetsky (1980) found that posttraining intraperitoneal administration of Leu-enkephalin (10 μg/kg, i.p.) or β-Endorphin (10 μg/kg, i.p.) produced retention deficits in both shuttlebox avoidance and habituation of a rearing response.

Reports have also appeared in which opiate agonists have been administered prior to training. Administration of drugs prior to training confounds learning and performance variables and introduces the possibility of state-dependence effects. Nevertheless, it may be of interest that the administration of opiates prior to training in appetitive tests apparently improves performance (Castellano, 1980; Kastin, Scollan, King, Schally, & Coy, 1976; Olson, Olson, Kastin, Green, Roig-Smith, Hill, & Coy, 1979), whereas such administration prior to training in avoidance tests inhibits performance (Castellano, 1975; Rigter, Hannan, Messing, Martinez, Vasquez, Jensen, Veliquette, & McGaugh, 1980; Rigter, Jensen, Martinez, Messing, Vasquez, Liang, & McGaugh, 1980).

The studies reviewed above present a complicated picture of the role of endogenous opiate systems in memory formation. Even if we consider only those experiments best designed to permit inferences on memory processes (that is, immediate posttraining opiate administration in one-trial learning tests), we still find apparently conflicting results. On the one hand, five laboratories report that high doses of enkephalin or morphine cause memory facilitation (Belluzzi & Stein, 1977b; Jensen et al, 1978; Mondadori & Waser, 1979; Stäubli & Huston, 1980; White et al., 1978); on theoretical grounds, this result is pleasing, as it is consistent with the well-established facilitatory relationship between natural rewards and learning. On the other hand, a number of studies demonstrate memory inhibition following administration of low doses of opiates (Jensen et al., 1978; Martinez & Rigter, 1980; see also Chapter 20). Dose-dependent opposite actions are also observed with apomorphine, clonidine, and other drugs (Langer, Briley, & Raisman, 1980), and are found with opiates in other behavioral tests (e.g., Belluzzi & Stein, 1978; Holtzman, 1975; Lorens & Mitchell, 1973). Opposite actions of high and low doses of classical postsynaptic catecholamine agonists have been explained by considering their effects at presynaptic receptors (Starke, 1979). The opposite effects on memory of high and low opiate doses may similarly be explained in terms of the relative importance of their presynaptic and postsynaptic actions, if it is assumed that low opiate doses act selectively at presynaptic receptors. Thus, low doses would produce amnesia by actual reduction of postsynaptic opiate receptor activity (through presynaptic inhibition of endogenous endorphin release), whereas high doses would enhance memory by an increase in postsynaptic opiate receptor activity (by direct agonist action).

ACKNOWLEDGMENTS

This work was performed, in part, at Wyeth Laboratories, Radnor, PA. Additional support was provided by NIDA Grant DA 02725. We thank Herman Morris for expert technical assistance and Dr. Victor Garsky for synthesis of the enkephalin analogs.

REFERENCES

Belluzzi, J. D., & Stein, L. (1977a). Enkephalin may mediate euphoria and drive-reduction reward. *Nature (London),* 266, 556–558.
Belluzzi, J. D., & Stein, L. (1977b). Enkephalin- and morphine-induced facilitation of long-term memory. *Society for Neuroscience Abstracts,* 3, 230.
Belluzzi, J. D., & Stein, L. (1978). Do enkephalin systems mediate drive reduction? *Society for Neuroscience Abstracts,* 4, 405.

Castellano, C. (1975). Effects of morphine and heroin on discrimination learning and consolidation in mice. *Psychopharmacologia*, **42**, 235–242.

Castellano, C. (1980). Dose-dependent effects of heroin on memory in two inbred strains of mice. *Psychopharmacology*, **67**, 235–239.

Childress, A. R. (1979). Naloxone suppression of brain self-stimulation: Evidence for endorphin-mediated reward. Unpublished doctoral dissertation, Bryn Mawr College, Bryn Mawr, Pennsylvania.

Gorman, J. E., Obladia, R. N., Scott, R. C., & Reid, L. D. (1978). Morphine injections in the taste aversion paradigm: Extent of aversions and readiness to consume sweetened morphine solutions. *Physiological Psychology*, **6**, 101–109.

Holtzman, S. G. (1975). Effects of narcotic antagonists on fluid intake in the rat. *Life Sciences*, **16**, 1465–1470.

Izquierdo, I. (1979). Effect of naloxone and morphine on various forms of memory in the rat: Possible role of endogenous opiate mechanisms in memory consolidation. *Psychopharmacology*, **66**, 199–203.

Izquierdo, I., Paiva, A. C. M., & Elisabetsky, E. (1980). Post-training intraperitoneal administration of Leu-enkephalin and β-endorphin causes retrograde amnesia for two different tasks in rats. *Behavioral and Neural Biology*, **28**, 246–250.

Jacquet, Y. F. (1973). Conditioned aversion during morphine maintenance in mice and rats. *Physiology and Behavior*, **11**, 527–541.

Jensen, R. A., Martinez, Jr., J. L., Messing, R. B., Spiehler, V., Vasquez, B. J., Soumireu-Mourat, B., Liang, K. C., & McGaugh, J. L. (1978). Morphine and naloxone alter memory in the rat. *Society for Neuroscience Abstracts*, **4**, 260.

Kastin, A. J., Scollan, E. L., King, M. G., Schally, A. V., & Coy, D. H. (1976). Enkephalin and a potent analog facilitate maze performance after intraperitoneal administration in rats. *Pharmacology, Biochemistry and Behavior*, **5**, 691–695.

Kelsey, J. E., Belluzzi, J. D., & Stein, L. (1979). Does naloxone suppress self-stimulation by decreasing reward or increasing aversion? *Society for Neuroscience Abstracts*, **5**, 530.

Langer, S. Z., Briley, M. S., & Raisman, R. (1980). Regulation of neurotransmission through presynaptic receptors and other mechanisms: Possible clinical relevance and therapeutic potential. In G. Pepeu, M. J. Kuhar, & S. J. Enna (Eds.), *Receptors for neurotransmitters and peptide hormones: Advances in biochemical psychopharmacology*, Vol. 21, pp. 203–212. New York: Raven.

Lorens, S. A., & Mitchell, C. L. (1973). Influence of morphine on lateral hypothalamic self-stimulation in the rat. *Psychopharmacologia*, **32**, 271–277.

Martinez, J. L., & Rigter, H. (1980). Endorphins alter acquisition and consolidation of an inhibitory avoidance response in rats. *Neuroscience Letters*, **19**, 197–201.

McGaugh, J. L. (1966). Time dependent processes in memory storage, *Science*, **134**, 328–342.

Mello, N. K., & Mendelson, J. H. (1978). Self-administration of an enkephalin analog by rhesus monkey, *Pharmacology, Biochemistry and Behavior*, **9**, 579–586.

Mondadori, C., & Waser, P. G. (1979). Facilitation of memory processing by posttrial morphine: Possible involvement of reinforcement mechanisms? *Psychopharmacology*, **63**, 297–300.

Olson, G. A., Olson, R. D., Kastin, A. J., Green, M. T., Roig-Smith, R., Hill, C. W., & Coy, D. H. (1979). Effects of an enkephalin analog on complex learning in the rhesus monkey. *Pharmacology, Biochemistry and Behavior*, **11**, 341–345.

Rigter, H., Hannan, T. J., Messing, R. B., Martinez, Jr., J. L., Vasquez, B. J., Jensen, R. A., Veliquette, J., & McGaugh, J. L. (1980). Enkephalins interfere with acquisition of an active avoidance response. *Life Sciences*, **26**, 337–345.

Rigter, H., Jensen, R. A., Martinez, Jr., J. L., Messing, R. B., Vasquez, B. J., Liang, K. C., & McGaugh, J. L. (1980). Enkephalin and fear-motivated behavior. *Proceedings of the National Academy of Sciences, U.S.A.*, **77**, 3729–3732.

Stapelton, J. M., Merriman, V. J., Coogle, C. L., Gelbard, S. D., & Reid, L. D. (1979). Naloxone reduces pressing for intracranial stimulation of sites in the periaqueductal gray area, accumbens nucleus, substantia nigra, and lateral hypothalamus. *Physiological Psychology, 7,* 427–436.

Starke, K. (1979). Presynaptic regulation of release in the central nervous system. In D. M. Paton (Ed.), *The Release of Catecholamines from Adrenergic Neurons*, pp. 143–184. Oxford: Pergamon.

Stäubli, U., & Huston, J. P. (1980). Avoidance learning enhanced by post-trial morphine injection. *Behavioral and Neural Biology, 28,* 487–490.

Stein, L. (1978). Reward transmitters: Catecholamines and opioid peptides. In M. A. Lipton, A. DiMascio, & K. F. Killam (Eds.), *Psychopharmacology: A Generation of Progress*, pp. 569–581. New York: Raven.

Stein, L. & Belluzzi, J. D. (1978). Brain endorphins and the sense of well-being: A psychobiological hypothesis. In E. Costa, and M. Trabucchi (Eds.), *The Endorphins: Advances in Biochemical Psychopharmacology*, Vol. 18, pp. 299–311. New York: Raven.

Stein, L., and Belluzzi, J. D. (1979). Brain endorphins: Possible mediators of pleasurable states. *In* E. Usdin, W. E. Bunney, Jr., & N. S. Kline (Eds.), *Endorphins in Mental Health Research*, pp. 375–389. London: Macmillan.

Stein, L., Belluzzi, J. D., & Wise, C. D. (1975). Memory enhancement by central administration of norepinephrine. *Brain Research, 84,* 329–335.

White, N., Major, R., & Siegel, J. (1978). Effects of morphine on one-trial appetitive learning. *Life Sciences, 23,* 1967–1972.

CHAPTER 15

Endorphin and Enkephalin Effects on Avoidance Conditioning: The Other Side of the Pituitary–Adrenal Axis

Joe L. Martinez, Jr., Henk Rigter, Robert A. Jensen,
Rita B. Messing, Beatriz J. Vasquez,
and James L. McGaugh

The present series of investigations examined the effects of several endorphins and Met- and Leu-enkephalin on avoidance conditioning. Peripherally administered β-endorphin (1.0 μg/kg) enhanced acquisition of an inhibitory avoidance response if given before training and β-endorphin (0.1 μg/kg) impaired consolidation of an inhibitory avoidance response if given after training. Both of these effects were time-dependent suggesting that the peptides acted to alter some aspect of learning and memory processes.

Also, microgram amounts of systemically administered Met- or Leu-enkephalin impaired acquisition of an active avoidance response. The effective dose (10.0–400.0 μg/kg) was dependent on the strain of the animal. As in the inhibitory avoidance task, the enkephalin effect is time-dependent. It was suggested that the enkephalins affect some aspect of acquisition of the avoidance response rather than performance of the response. We found that a stable metabolic analog of Leu-enkephalin, D-Ala-D-Leu-enkephalin (4.0 μg/kg) had no effect in rats that had successfully performed an avoidance response, but impaired acquisition in yoked control animals that received equivalent amounts of shock.

Naloxone (10.0 mg/kg), which by itself had no effect on avoidance conditioning, antagonized the impairing action of Met-enkephalin (10.0 μg/kg) but not Leu-enkephalin (10.0 μg/kg) on acquisition of the avoidance response. These findings are consistent with the interpretation that Leu-enkephalin is primarily a Type 2 opiate receptor ligand and that Type 2 receptors are naloxone resistant.

305

ENDOGENOUS PEPTIDES
AND LEARNING AND MEMORY PROCESSES

Adrenal medullectomy, which removes an endogenous store of enkephalins, abolished the effect on avoidance conditioning of exogenously administered enkephalin (10.0 μg/kg). Increasing the dose of Leu- but not Met-enkephalin to 100.0 or 1000.0 μg/kg stored behavioral activity. This finding, along with that of differences in naloxone reversibility, suggests that the impairing actions of the enkephalins on acquisition of the avoidance response are produced by different mechanisms.

Taken together, the results support the view that endorphins and enkephalins are stress-related hormones and are involved in the regulation of complex behaviors such as learning and memory. Finally, the adrenal medulla appears to be an important locus of enkephalin effects on fear-motivated behavior.

I. INTRODUCTION

The findings of numerous studies have shown that learning and memory are affected by adrenocorticotropic hormone (ACTH), peptide fragments of ACTH, and corticosteroids (see Beckwith & Sandman, 1978; de Wied, 1974, 1977; Leshner, 1978). These findings support the general hypothesis that hormones of the pituitary–adrenal axis released by the stress of a learning experience play an important role in modulating the strength of retention of the learned response. The catecholamines epinephrine and norepinephrine are also released from the adrenal medulla during stress in response to splanchnic nerve stimulation. These hormones have also been found to modulate learning and memory (Gold & McGaugh, 1977). An understanding of how the adrenocortical and adrenal medullary hormonal systems act to influence learning and memory processes has been complicated by recent findings that in rats β-endorphin is released from the pituitary along with ACTH in response to stress and that this release is blocked by the synthetic corticosteroid, dexamethasone (DEX) (Guillemin, Vargo, Rossier, Minick, Ling, Rivier, Vale, & Bloom, 1977; cf. Kalin, Risch, Cohen, Insel, & Murphy, 1980). Also, enkephalin-like materials are found in the adrenal medulla (Di Giulio, Yang, Lutold, Fratta, Hong, & Costa, 1978; Schultzberg, Lundberg, Hökfelt, Terenius, Brandt, Elde, & Goldstein, 1978; Viveros, Diliberto, Hazum, & Chang, 1980a) and are probably released concomitantly with catecholamines (Viveros et al., 1980a; Viveros, Wilson, Diliberto, Hazum, & Chang, 1980b). Thus, there may be another side of the pituitary–adrenal axis in which β-endorphin released from the pituitary gland modulates release of catecholamines and enkephalins from the adrenal medulla (see Fig. 1). The exact mechanism by which this may occur is not presently known. Kumakura, Karoum, Guidotti, and Costa (1980) have shown that opioid peptides inhibit the release of catecholamines elicited by stimulation of cultured adrenal cells and speculate that opioid peptides released from the pituitary may normally function in this manner.

The purpose of the present series of investigations was to examine the effects of several opioid peptides on avoidance conditioning, and to relate some of

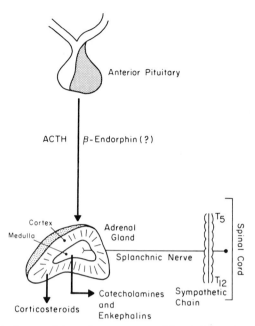

Fig. 1. Schematic diagram of the major components of the pituitary–adrenal axis. The adrenal gland is composed of two major parts, the adrenal cortex and the adrenal medulla. The adrenal cortex releases corticosteroids in response to adrenocorticotropic hormone (ACTH) from the pituitary gland. The adrenal medulla releases the catecholamines norepinephrine and epinephrine and possibly the opioids Met- and Leu-enkephalin in response to stimulation from the splanchnic nerve, which arises from the spinal cord. There may be another side of the pituitary–adrenal axis in which β-endorphin, released from the anterior pituitary gland, modulates release of catecholamines and enkephalins from the adrenal medulla.

these effects to adrenal medullary function. The findings summarized in this chapter demonstrate that microgram and, in one case, nanogram amounts of peripherally administered opioid peptides affect avoidance conditioning. Furthermore, some of these effects are altered by adrenal medullectomy. This finding suggests that the adrenal medulla may influence opioid effects on fear-motivated behavior. Finally, some of the effects we observed are reversed by high doses of naloxone.

The behavioral procedures used in these studies have been described in detail elsewhere (Martinez, McGaugh, Hanes, & Lacob, 1977; Martinez, Rigter, & van der Gugten, 1980a; Rigter, Jensen, Martinez, Messing, Vasquez, Liang, & McGaugh, 1980a). Two tasks were used, an inhibitory and a one-way active avoidance. Briefly, in the inhibitory avoidance task, a rat was allowed to step into a darkened shock compartment where it received a mild electric footshock. Endorphins were administered either before or after training and a retention test was given 1–3 days following training. The latency of the animal to reenter the shock compartment was taken as a measure of learning. In the active avoidance task, all animals were injected before training. A rat had 10 sec to avoid a footshock by either stepping through into a safe white compartment in one version of the task (Rigter et al., 1980a), or by stepping up onto a small platform, in the other version (Martinez et al., 1980a). Eight to 10 trials were run, and the number of avoidance responses the animal made was taken as the measure of acquisition performance.

II. ENDORPHIN EFFECTS ON INHIBITORY AVOIDANCE CONDITIONING

In this study, either β-endorphin (β-LPH$_{61-91}$), α-endorphin (β-LPH$_{61-76}$), γ-endorphin (β-LPH$_{61-77}$), or Met-enkephalin (β-LPH$_{61-65}$) was administered intraperitoneally (i.p.) to Wistar rats either 30 min before training or immediately after training (Martinez & Rigter, 1980a). It can be seen from Fig. 2a that γ-endorphin, at a dose of 1.0 μg/kg, significantly enhanced acquisition of the response if it was given before training. The effect was time-dependent; giving the peptide 90 min before training produced no effect (Fig. 2b). In contrast to the results obtained with pretrial administration of γ-endorphin, only β-endorphin affected retention of the response when given after training. Figure 3a shows that a dose of 0.1 μg/kg produced a significant impairment of retention performance. Also, as with γ-endorphin, the β-endorphin effect was time-dependent; no effect of β-endorphin was found when the injection was delayed for 90 min (Fig. 3b).

The results of this study indicate that endorphins modulate retention of an inhibitory avoidance response. Importantly, the modulatory actions of both γ-endorphin on acquisition (pretrial administration) and β-endorphin on consolidation (posttrial administration) are time-dependent, indicating that they acted on some aspect of the learning and memory processes (McGaugh, 1966). However, when γ-endorphin was given before training, it could not be determined whether the peptide affected variables other than memory, such as sensory, motivational, or emotional processes. This problem of interpretation does not occur when the peptide is given after training as the animals were trained in a normal state.

In a general way, these results are consistent with the research of Izquierdo, Perry, Dias, Souza, Elisabetsky, Carrasco, Orsingher, and Netto (Chapter 13) who reported that microgram amounts of peripherally administered β-endorphin given following training impaired retention of either a habituation or a shuttle avoidance task. We also found that posttrial administration of β-endorphin produced a memory deficit. However, our results do not agree with those of

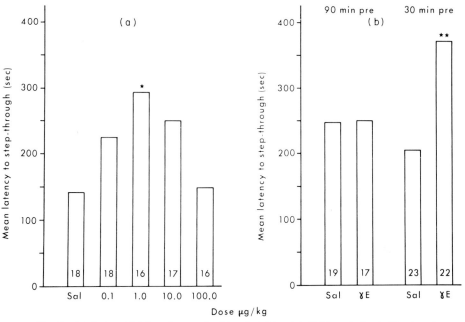

Fig. 2. (a) A dose of 1.0 μg/kg γ-endorphin administered to Wistar rats (i.p.), 30 min prior to training in a one-trial inhibitory task, significantly ($p < .05$) enhanced retention performance measured 72 hr later. The animals were trained using a 300 μA, 3 sec footshock. (b) A dose of 1.0 μg/kg γ-endorphin enhances retention performance when given 30 ($p < .01$), but not 90 min prior to training, indicating that the effect is time-dependent. The numbers within the bars are the numbers of animals in each dose condition. (From Martinez & Rigter, 1980a.)

de Wied, Kovács, Bohus, van Ree, and Greven (1978b) and Kovács and de Wied (Chapter 11) who reported that posttrial administration of β-endorphin enhanced memory consolidation of an inhibitory avoidance response. These authors also reported that α-endorphin enhanced retention, whereas we found that a wide dose range of α-endorphin had no effect. There were, however, differences between the experimental procedures used by the de Wied group and those we used in our studies. The most notable difference was the route of drug administration.

In the research just described (Martinez & Rigter, 1980a), there was no effect of a wide dose range of Met-enkephalin on retention of the inhibitory avoidance response in Wistar rats if the drug was administered either before or after training. Yet, in other unpublished experiments, we found that enkephalins administered after training affected inhibitory avoidance responding. In these experiments, Fischer 344 (F344) rats were used and the enkephalins were administered i.p. in divided doses both immediately and 30 min following training. The training–retention testing interval was 3 days. It can be seen from

310   JOE L. MARTINEZ, JR., ET AL.

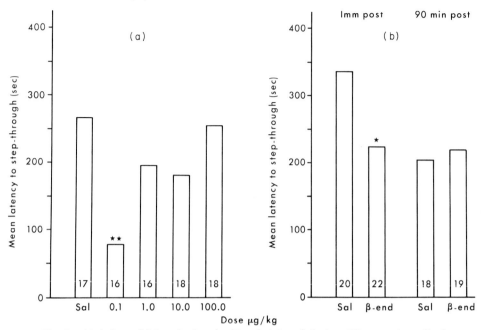

Fig. 3. (a) A dose of 0.1 μg/kg β-endorphin administered (i.p.) to Wistar rats immediately after training in a one-trial inhibitory avoidance task, significantly ($p < .02$) impaired retention performance measured 72 hr later. The animals were trained using a 300 μA, 3 sec footshock. (b) A dose of 0.1 μg/kg β-endorphin, impairs performance when given immediately ($p < .05$, one-tailed test), but not 90 min following training, indicating that the effect is time-dependent. The numbers within the bars are the number of animals in each dose condition. (From Martinez & Rigter, 1980a.)

Fig. 4a that 800.0 μg/kg of Leu-enkephalin enhanced retention performance. However, Fig. 4b shows that Met-enkephalin produced a biphasic dose–response function with a 3.13 μg/kg dose impairing retention performance, whereas a greater dose of 50.0 μg/kg tended to enhance retention. Although these data are preliminary, they indicate that the enkephalins may modulate memory storage processes. In this experiment, the animals were trained in a normal state and received the drug following training. These results agree with other published work in which we reported that systemically administered D-Ala-D-Leu-enkephalin (DADLE) given before training to F344 rats enhanced acquisition of an inhibitory avoidance response in which either footshock or handling was used as the aversive stimulus (Rigter et al., 1980a). These studies point out that factors such as strain, or perhaps the injection schedule, reveal differences in enkephalin responsiveness.

Taken together, the results of these studies demonstrate that microgram amounts of peripherally administered γ-endorphin influence the acquisition of an inhibitory avoidance response. On the other hand, β-endorphin, and perhaps the enkephalins, affect consolidation of an inhibitory avoidance response. These results are consistent with the notion that endorphins and enkephalins are stress-related hormones and one of their functions may be to regulate behavioral processes. Because all of the endorphins and enkephalins used in these studies are opiate receptor ligands (Cox, Goldstein, & Li, 1976; Hughes, Smith, Kosterlitz, Fothergill, Morgan, & Morris, 1975; Lazarus, Ling, & Guillemin, 1976), it is likely that their behavioral effects are due to direct action of these peptides on opioid receptors. However, it has not yet been determined whether these endorphin effects on inhibitory avoidance conditioning are reversible by opiate antagonists, although Izquierdo et al. (Chapter 13) report that memory deficits produced by posttrial administration of β-endorphin in both a habituation and shuttle avoidance task are naloxone reversible.

In general, the use of opiate antagonists to determine opiate receptor action of purported opioid agonists is of limited use in learning and memory experiments because opiate antagonists such as naloxone and naltrexone themselves affect memory processes (de Wied et al., 1978a; Izquierdo, 1979; Messing, Jensen, Martinez, Spiehler, Vasquez, Soumireu-Mourat, Liang, & McGaugh, 1979).

Finally, it is not known where the primary sites of action of these systemically administered endorphins may be. However, the low doses needed to induce a behavioral effect suggest that one site of action may be in the periphery (cf. Rapoport, Klee, Pettigrew, & Ohno, 1980). For example, the total dose of systemically administered β-endorphin needed to produce retrograde amnesia in the inhibitory avoidance task was approximatley 30 ng (Martinez & Rigter, 1980a).

III. ENKEPHALIN EFFECTS ON ACTIVE AVOIDANCE CONDITIONING

In this series of studies, we found that microgram amounts of systemically administered enkephalins impair acquisition of an active avoidance response. In the first study we found that either Leu- or Met-enkephalin administered i.p. immediately before training at a dose of 400.0 μg/kg impaired acquisition in F344 rats (Rigter, Hannan, Messing, Martinez, Vasquez, Jensen, Veliquette, & McGaugh, 1980b); and the metabolically more stable analog of Leu-enkephalin, DADLE, impaired acquisition of the avoidance response at a much lower dose of 4.0 μg/kg.

The actions of DADLE (4.0 μg/kg, i.p.) on avoidance conditioning are time-dependent. Fig. 5 (a) shows that DADLE administered immediately or 5 min before training impairs acquisition of the response. However, if the injection

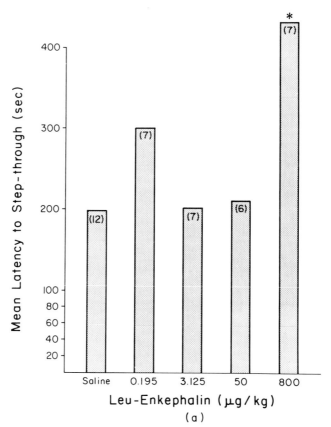

Fig. 4. (a) A dose of 800.0 μg/kg Leu-enkephalin administered (i.p.) to F344 rats in divided doses both immediately and 30 min following training, in a one-trial inhibitory avoidance task, significantly ($p < .05$) enhanced retention performance measured 72 hr later. The animals were trained using a 500 μA, 0.5 sec footshock. The numbers within the bars are the number of animals in each dose condition. (b) Met-enkephalin administered (i.p.) to F344 rats in divided doses both immediately and 30 min following training in a one-trial inhibitory avoidance task produced a biphasic dose–response function. A low dose of 3.125 μg/kg significantly ($p < .001$) impaired retention performance, whereas a higher dose of 50.0 μg/kg tended to enhance ($p < .05$, one-tailed test) retention performance measured 72 hr later. The animals were trained using a 500 μA, 0.5 sec footshock. The numbers within the bars are the number of animals in each dose condition.

is given 15 min prior to training, there is no effect. As the peptide is administered before training, it cannot be known whether the enkephalins impaired acquisition of the active avoidance response or whether they simply altered the performance of the response. We attempted to address this question by using a yoked control procedure. One group of rats was trained until each animal made its first active avoidance response. The rats were then returned to their

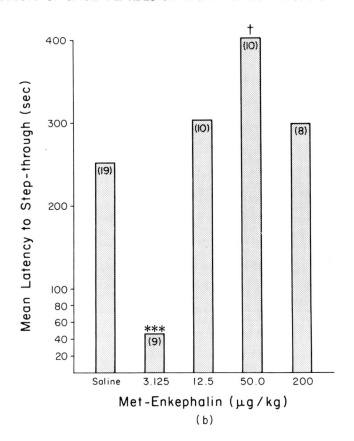

(b)

home cages and retrained 24 hr later in a regular training session. Administration of DADLE (4.0 µg/kg) or saline occurred 4 min before the retraining session. It can be seen from Fig. 5 (b) that the peptide did not depress avoidance responding if an animal had previously made at least one avoidance response. In this same experiment, other groups of rats served as yoked controls. For each rat that was allowed to acquire one correct avoidance response, another paired rat was handled identically and received the same duration of shock. However, these yoked controls were prevented from performing escape or avoidance responses by a closed door. These animals were also trained in a regular session after treatment with either DADLE or saline 24 hr later. Figure 5 (b) shows that DADLE impaired acquisition in the yoked-control animals. The results of this study indicate that enkephalins affect acquisition of the avoidance response rather than performance of the response, because DADLE had no effect in rats that had successfully performed an avoidance response, but it impaired acquisition in the yoked-control animals.

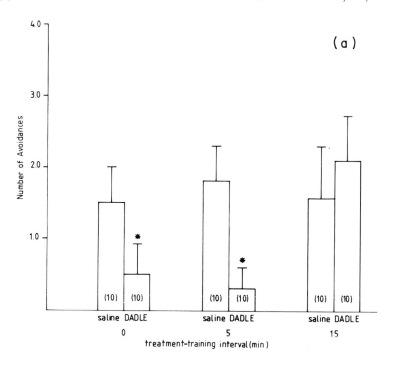

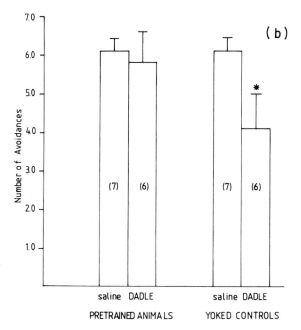

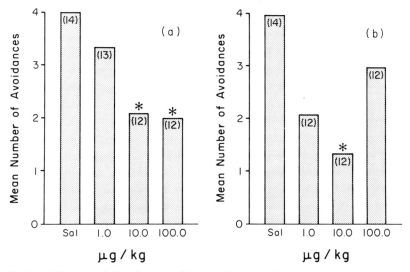

Fig. 6.   (a) Leu-enkephalin administered (i.p.) to Wistar rats 5 min before training significantly impairs acquisition of an active avoidance response at a dose of either 10.0 μg/kg ($p < .02$) or 100.0 μg/kg ($p < .02$). (b) Met-enkephalin similarly impairs acquisition of the response at a dose of 10 μg/kg ($p < .02$). The number of animals used in each dose condition is shown within the bars.

The basic finding that peripherally administered enkephalins impair acquisition of an active avoidance response was replicated in Wistar rats using a step-up response (Martinez & Rigter, 1980b; Martinez et al. 1980a). In these studies, either Met- or Leu-enkephalin was administered to the animals i.p. 5 min before training. It can be seen from Fig. 6b that 10.0 μg/kg Met- and 10.0 or 100.0 μg/kg Leu-enkephalin (Fig. 6a) impaired acquisition. The effective dose of Leu-enkephalin in Wistar rats was several orders of magnitude lower than the effective dose in F344 rats (400.0 μg/kg) indicating that important strain differences exist in enkephalin sensitivity.

Fig. 5.   (a) The impairing effect of D-Ala-D-Leu-enkephalin (DADLE) on acquisition of an active avoidance response is time-dependent. A dose of 4.0 μg/kg DADLE administered to F344 rats either immediately ($p < .05$) or 5 min before training ($p < .05$) impairs acquisition. However, if the injection is given 15 min before training, there is no effect. The values given are the mean number of avoidances ± SEM. The numbers within the bars are the number of animals in each dose condition. (From Rigter et al., 1980a.) (b) D-Ala-D-Leu-enkephalin (DADLE) in a dose of 4.0 μg/kg administered (i.p.) to F344 rats 5 min before training does not impair acquisition of the active avoidance response if the animals were allowed to make one avoidance response (pre-trained animals). However, in yoked-control rats which received equivalent amounts of shock but were prevented from making an avoidance response, the DADLE did impair acquisition ($p < .05$). The values given are the mean number of avoidances ± SEM. The numbers within the bars are the number of animals in each dose condition. (From Rigter et al., 1980b.)

IV. NALOXONE REVERSIBILITY

As noted previously, the enkephalins are endogenous ligands for opiate receptors (Hughes et al., 1975). The purpose of the following experiments was
to determine whether or not the enkephalin actions on avoidance conditioning
are reversible by the opiate antagonist naloxone. A preliminary study determined
that 1.0–100.0 mg/kg of naloxone administered before training did not affect
acquisition of the avoidance task. Based on the dose–response curves of enkephalin effectiveness in Wistar rats (Fig. 6), a dose of 10.0 μg/kg was chosen
to investigate naloxone reversibility. The peptides and naloxone (1.0 or 10.0
mg/kg) were administered as a cocktail 5 min before training (Martinez &
Rigter, 1980b; Martinez et al., 1980a). It can be seen from Fig. 7a that neither
1.0 nor 10.0 mg/kg of naloxone blocked the impairing action of Leu-enkephalin,
as the Leu-enkephalin plus naloxone groups were significantly different from
their respective saline control groups. In contrast, 10.0 mg/kg of naloxone did
antagonize the impairing actions of Met-enkephalin (Fig. 7b). In this case, the
group that received Met-enkephalin and 10.0 mg/kg of naloxone was significantly different from the group that received only 10.0 μg/kg Met-enkephalin
and was not significantly different from its saline control group.

The finding that Met- but not Leu-enkephalin actions on avoidance conditioning are antagonized by naloxone is difficult to interpret. Others have

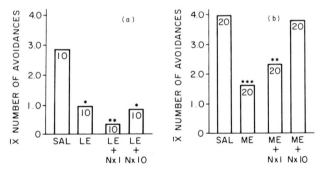

Fig. 7. (a) The impairing effect of 10.0 μg/kg Leu-enkephalin on acquisition of an active
avoidance response administered (i.p.) to Wistar rats 5 min before training as a cocktail with the
Leu-enkephalin was observed at a dose of either 1.0 mg/kg ($p < .01$) or 10.0 mg/kg ($p < .05$)
of naloxone. Also, the group treated only with Leu-enkephalin was different from its saline control
group ($p < .02$). (b) In contrast, 10.0 mg/kg of naloxone antagonized the impairing actions of
10.0 μg/kg of Met-enkephalin. The group that received both Met-enkephalin and 10.0 mg/kg of
naloxone was significantly different from the group that received only 10.0 μg/kg Met-enkephalin
($p < .05$) and was not significantly different from its saline control group. The group treated only
with Met-enkephalin was different from its saline control group ($p < .01$), as was the group that
received Met-enkephalin and 1.0 mg/kg naloxone ($p < .05$). The number of animals used in each
dose condition is shown in the bars. (From Martinez & Rigter, 1980b.)

reported that neither naloxone nor naltrexone block enkephalin effects on other conditioned behaviors (Rigter, Greven, & van Riezen, 1977; de Wied et al., 1980a). Furthermore, since Messing et al. (1979) showed that naloxone facilitates retention of a similar active avoidance task, it cannot be known whether the attenuating actions of naloxone in this task were specific to an antagonism of enkephalin actions at an opiate receptor. Nevertheless, in a preliminary study, we found that a wide dose range of naloxone did not affect acquisition of the response, thus, the results are consistent with the interpretation that naloxone specifically antagonized the actions of Met-enkephalin. Also, the finding that Leu-enkephalin induced impairment was not antagonized by naloxone is consistent with the results of Pert, Taylor, Pert, Harkenham, and Kent (1980) who suggest that Leu-enkephalin is primarily a Type 2 opiate receptor ligand and that Type 2 receptors are naloxone resistant. Finally, the differential action of naloxone in antagonizing the impairing actions of the enkephalins suggests that their effects on avoidance conditioning are mediated by distinct mechanisms. A further example of distinct behavioral effects of the two peptides is provided in the next series of studies in which the role of the adrenal medulla in mediating enkephalin actions on avoidance conditioning was examined.

V. ADRENAL MEDULLARY INVOLVEMENT

As noted earlier, enkephalin-like materials are found in the adrenal medulla (Di Giulio et al., 1978; Schultzberg et al., 1978; Viveros et al., 1980a,b). Thus, it seemed possible that enkephalin effects on avoidance conditioning are related to adrenal medullary function. To investigate this question, adrenal medullectomized rats were given doses of enkephalin that are known to impair acquisition of the avoidance response in normal animals. In these studies, the adrenal demedullations were performed on Wistar rats 2 wk prior to training (Martinez & Rigter, 1980b; Martinez et al., 1980a). As before, 10 μg/kg of either Met- or Leu-enkephalin was administered i.p. 5 min before training. Figure 8 shows that adrenal demedullation abolished the impairing action of both enkephalins. Ten μg/kg of the enkephalins impaired acquisition of the response in sham-operated (SHAM) animals but was ineffective in adrenal-medullectomized (ADXM) rats.

In the preceding experiment adrenal demedullation abolished the effect of both Leu- and Met-enkephalin; a final experiment determined whether increasing the dose of enkephalins to 100.0 or 1000.0 μg/kg in ADXM rats would restore their behavioral activity. Figure 9 shows that increasing the dose of Leu-enkephalin in ADXM animals restored its activity. Both a 100.0 and 1000.0 μg/kg dose were effective in impairing acquisition in both SHAM and ADXM

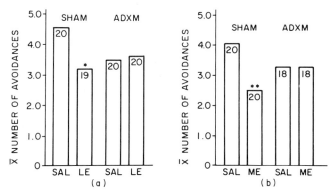

Fig. 8. (a) Leu-enkephalin at a dose of 10.0 μg/kg administered (i.p.) to Wistar rats 5 min before training impairs ($p < .05$) acquisition of an active avoidance response in sham-operated (SHAM) rats, whereas adrenal medullectomy (ADXM) abolishes the effect. (b) Met-enkephalin at a dose of 10.0 μg/kg also impairs acquisition of the response in SHAM controls ($p < .01$) but was ineffective in ADXM rats. The number of animals used in each dose condition is shown within the bars. (From Martinez & Rigter, 1980b.)

animals. In contrast, neither a 100.0, nor a 1000.0 μg/kg dose of Met-enkephalin produced any significant effect on acquisition of the response in either SHAM or ADXM rats.

Thus, adrenal medullectomy, which removes an endogenous store of enkephalins, abolished the effect on avoidance conditioning of a low dose of exogenously administered enkephalin. Importantly, increasing the dose of Leu-

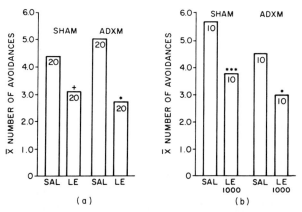

Fig. 9. (a) Leu-enkephalin at a dose of 100.0 μg/kg administered (i.p.) to Wistar rats 5 min before training significantly ($p < .05$, one-tailed test) impairs acquisition of an active avoidance response in sham-operated (SHAM) rats, and rats that have been adrenal medullectomized (ADXM) ($p < .05$). (b) A 1000.0 μg/kg dose of Leu-enkephalin was effective in both SHAM ($p < .02$) and ADXM animals ($p < .05$). The number of animals used in each dose condition is shown within the bars.

enkephalin restored behavioral activity, suggesting that its behavioral effect was produced by the combination of endogenous Leu-enkephalin released from the adrenal medulla and exogenous Leu-enkephalin administered through injection. However, increasing the dose of Met-enkephalin following adrenal medullectomy did not restore its behavioral effects. Therefore, unless we did not examine the proper dose of Met-enkephalin in ADXM rats, this finding suggests, as do the earlier findings with naloxone reversibility, that the impairing actions of Met-enkephalin are not produced in the same way as those of Leu-enkephalin.

An alternative explanation for the finding that adrenal demedullation abolished the actions of the enkephalins on avoidance conditioning is that adrenal demedullation may generally make rats less responsive to drugs that influence learning and memory. This interpretation is consistent with the findings reported here and elsewhere that adrenal demedullation abolished the effects of ACTH$_{4-10}$ (Martinez & Rigter, 1980a) on active avoidance conditioning and that of amphetamine and 4-OH amphetamine on inhibitory avoidance conditioning (Martinez, Vasquez, Rigter, Messing, Jensen, Liang, & McGaugh, 1980b). Nevertheless, these data do suggest that the enkephalin effects we observed on avoidance training are mediated, at least in part, through peripheral mechanisms, and an important locus of these enkephalin-induced actions on fear-motivated behavior appears to be the adrenal medulla.

VI. CONCLUSIONS

Taken together, the results of the studies reported in this chapter support the view that endorphins and enkephalins are stress-related hormones (Clement-Jones, Lowry, Rees, & Besser, 1980; Guillemin et al., 1977; Viveros et al., 1980a) and are involved in the regulation of complex behaviors such as learning and memory. Furthermore, enkephalin actions on avoidance conditioning may be related to adrenal medullary function suggesting that enkephalin regulation in general may be related to adrenal–sympathetic mechanisms (Viveros et al., 1980a,b) and perhaps to hormones of the other side of the pituitary–adrenal axis such as β-endorphin (see Fig. 1; Kumakura et al., 1980).

An important question that arises from these studies is whether the initial action of these systemically administered hormones is in the periphery or in the brain. Although the data reported in this chapter do not directly address this issue, they are consistent with a peripheral site of action. For example, we found that a dose of 3.125 μg/kg Met-enkephalin administered i.p. following training in an inhibitory avoidance task produced retrograde amnesia. In contrast, Belluzzi and Stein (Chapter 14) reported that a much greater dose of 200 μg/rat Met-enkephalin administered intracerebroventricularly (i.c.v.) was needed to impair retention of a similar inhibitory avoidance response. Thus, the low dose needed to induce a behavioral effect following systemic administration suggests that their primary site of action may be in the periphery.

Because all of the endorphins and enkephalins used in these studies are endogenous ligands for opiate receptors (Cox et al., 1976; Hughes et al., 1975; Lazarus et al., 1976), it is reasonable to ascribe their behavioral effects to a direct action on opiate receptors. A common way to assess opiate agonist specificity is by blocking the action of an agonist with a specific antagonist such as naloxone (Jaffe & Martin, 1980). However, as noted earlier, naloxone and naltrexone, by themselves, alter learning and memory processes (de Wied et al., 1978a; Izquierdo, 1979; Messing et al., 1979). As naloxone may have pharmacological actions unrelated to opiate receptor blockade (see Sawynok, Pinsky, & La Bella, 1979), the demonstration that naloxone antagonizes an endorphin effect on avoidance conditioning must be viewed with caution. Nevertheless, in one experiment, we did observe that a relatively high dose of naloxone, which by itself, had no effect on avoidance responding, antagonized the impairing action of Met- but not of Leu-enkephalin. These results are consistent with a current receptor typology proposed by Pert et al. (1980) in which Leu-enkephalin is primarily a Type 2 opiate receptor ligand and Type 2 receptors may be naloxone resistant. Thus, there is evidence that the enkephalin effects on avoidance conditioning that we observed may be related to a direct action of these substances on opiate receptors.

Another important issue is whether the enkephalins impaired acquisition of the avoidance response or simply altered performance of the response, as the peptides were administered before training. In a yoked-control study we showed that enkephalin did not depress avoidance responding if an animal had previously made at least one avoidance response, suggesting that the peptide does not impair performance but affects some other process related to acquisition of the response. It is also unlikely that the enkephalin-induced impairment is related to alterations in pain perception or pain reactivity because we found that Leu-enkephalin administered i.p. in a dose range of 4.0–1200.0 μg/kg does not alter flinch or jump thresholds (Rigter et al., 1980a).

The actions of enkephalin given after training on retention of an inhibitory avoidance response are consistent with the suggestion of Rigter et al. (1980a) that enkephalins may affect learning and memory through peptide-induced increases in fear or arousal. However, Rigter et al. (1980b) also suggest that enkephalins may affect learning and memory processes by strengthening the tendency of rats to inhibit behavior in the presence of cues previously associated with aversive stimulation. Whereas this interpretation would explain an impairment of acquisition of an active avoidance response because animals would tend not to escape the footshock, it does not easily explain the effects of enkephalin administered following training (Figs. 4a, b) because the animal is trained in a normal state. It is therefore possible that enkephalins exert multiple effects on behavior.

It is likely that Met- and Leu-enkephalin impair acquisition of an active avoidance response through distinct mechanisms. This conclusion is supported by results showing that naloxone antagonizes the impairing action of Met- but not of Leu-enkephalin. Also, following adrenal medullectomy, the impairing actions of Leu- but not of Met-enkephalin are restored by an increased dose. As noted earlier, these findings agree with other data demonstrating that the two peptides have different behavioral actions. Rigter (1978) reported that both enkephalins attenuated amnesia, induced by CO_2, for an inhibitory avoidance response, if the enkephalins were given shortly before the retention test. However, only Met-enkephalin was effective if given shortly before the training experience. It seems likely, with regard to behavior, that the two enkephalins may produce their behavioral actions in much the same way as do epinephrine and norepinephrine. For example, both catecholamines and both enkephalins are found in the adrenal medulla. All of these substances are probably released when an animal is stressed (Guillemin et al., 1977; Guyton, 1975; Viveros et al., 1980a,b). All of these substances, at the appropriate dosages, enhance retention of an inhibitory avoidance response (Gold & McGaugh, 1977; Rigter et al., 1980a). Both catecholamines are adrenergic receptor agonists with differing affinities for different types of adrenergic receptors (Weiner, 1980), and both enkephalins are opioid receptor agonists with differing affinities for different types of opiate receptors (Pert et al., 1980). Yet, as already discussed, when tested under the appropriate experimental conditions, the enkephalins have different behavioral actions as do epinephrine and norepinephrine (Gold & McGaugh, 1977).

The studies reported in this chapter raise many questions that will have to be answered by future studies. For example, it is not known by what mechanism(s) microgram amounts of these peripherally administered endorphins influence the acquisition and storage of information in the brain. The fact that catecholamines and enkephalins are found in the same cells and probably in the same storage granules in the chromaffin cells of the adrenal medulla (Viveros et al., 1980a,b) suggests that there may be a more widespread and intricate relationship between adrenergic and opioid systems in the regulation of behavior. For example, enkephalin-like immunoreactive material is also found in the sympathetic chain whose postganglionic efferent nerves are adrenergic (Hughes, Kosterlitz, & Smith, 1977). Finally, it is not known whether there are endorphin–endorphin relationships in peripheral systems that are important for behavior.

ACKNOWLEDGMENTS

This research was supported by Research Grants MH 12526 (to JLMcG, JLMJr, RAJ, RBM, and BJV) and AG 00538 (JLMcG, JLMJr, RAJ, and RBM) from the U.S. Public Health Service, BNS 76-17370 (JLMcG) from the National Science Foundation, and a postdoctoral research fellowship (JLMJr) from Organon International, Oss, The Netherlands. Some of the enkephalins used in these studies were synthesized by Dr. H. Greven and Dr. J. van Nispen, Organon.

REFERENCES

Beckwith, B. E., & Sandman, C. A. (1978). Behavioral influences of the neuropeptides ACTH and MSH: A methodological review. *Neuroscience and Biobehavioral Reviews*, **2**, 311–338.

Clement-Jones, U., Lowry, P. J., Rees, L. H., & Besser, G. M. (1980). Met-enkephalin circulates in human plasma. *Nature (London)*, **283**, 295–297.

Cox, B. M., Goldstein, A., and Li, C. H. (1976). Opioid activity of a peptide, β-lipotropin-(61–91), derived from β-lipotropin. *Proceedings of the National Academy Sciences, U.S.A.*, **73**, 1821–1823.

de Wied, D. (1974). Pituitary–adrenal system hormones and behavior. In F. O. Schmitt & F. G. Worden (Eds.), *The Neurosciences Third Study Program*, pp. 653–666. Cambridge, Massachusetts: MIT Press.

de Wied, D. (1977). Peptides and behavior. *Life Sciences*, **20**, 195–204.

de Wied, D., Bohus, B., van Ree, J. M., & Urban, I. (1978a). Behavioral and electrophysiological effects of peptides related to lipotropin (β-LPH). *Journal of Pharmacology and Experimental Therapeutics*, **204**, 570–580.

de Wied, D., Kovacs, G. L., Bohus, B., van Ree, J. M., & Greven, H. M. (1978b). Neuroleptic activity of the neuropeptide β-LPH $_{62-77}$ ([Des-Tyr^1] γ-endorphin; DTγE). *European Journal of Pharmacology*, **49**, 427–436.

Di Giulio, A. M., Yang, H. Y. T., Lutold, B., Fratta, W., Hong, J., & Costa, E. (1978). Characterization of enkephalin-like material extracted from sympathetic ganglia. *Neuropharmacology*, **17**, 989–992.

Gold, P. E., & McGaugh, J. L. Hormones and memory (1977). In L. H. Miller, C. A. Sandman, & A. J. Kastin (Eds.), *Neuropeptide Influences on Brain and Behavior*, pp. 127–143. New York: Raven.

Guillemin, R., Vargo, T., Rossier, J., Minick, S., Ling, N., Rivier, C., Vale, W., & Bloom, F. (1977). β-Endorphin and adrenocorticotropin are secreted concomitantly by the pituitary gland. *Science*, **197**, 1367–1369.

Guyton, A. C. (1975). *Textbook of Medical Physiology*. Philadelphia, Pennsylvania: Saunders.

Hughes, J., Kosterlitz, H. W., & Smith, T. W. (1977). The distribution of methionine-enkephalin and leucine-enkephalin in the brain and peripheral tissues. *British Journal of Pharmacology*, **61**, 639–647.

Hughes, J., Smith, T. W., Kosterlitz, H. W., Fothergill, L. A., Morgan, B. A., & Morris, H. R. (1975). Identification of two related pentapeptides from the brain with potent opiate agonist activity. *Nature (London)*, **258**, 577–580.

Izquierdo, I. (1979). Effect of naloxone and morphine on various forms of memory in the rat: Possible role of endogenous opiate mechanisms in memory consolidation. *Psychopharmacology*, **66**, 199–203.

Jaffe, J. H., & Martin, W. R. (1980). Opioid analgesics and antagonists. In A. G. Gilman, L. S. Goodman, & A. Gilman (Eds.), *Goodman and Gilman's The Pharmacological Basis of Therapeutics*, pp. 494–534. New York: Macmillan.

Kalin, N. H., Risch, S. C., Cohen, R. M., Insel, T., & Murphy, D. L. (1980). Dexamethasone fails to suppress β-endorphin plasma concentrations in humans and rhesus monkeys. *Science*, **209**, 827–828.

Kumakura, K., Karoum, F., Guidotti, A., & Costa, E. (1980). Modulation of nicotinic receptors by opiate receptor agonists in cultured adrenal chromaffin cells. *Nature (London)*, **283**, 489–492.

Lazarus, L. H., Ling, N., & Guillemin, R. (1976). β-lipotropin as a prohormone for the morphinomimetic peptides endorphins and enkephalins. *Proceedings of the National Academy of Sciences, U.S.A.*, **73**, 2156–2159.

Leshner, A. I. (1978). *An Introduction to Behavioral Endocrinology.* London and New York: Oxford University Press.

Martinez, Jr., J. L., McGaugh, J. L., Hanes, C. L., & Lacob, J. S. (1977). Modulation of memory processes induced by stimulation of the entorhinal cortex. *Physiology and Behavior*, **19**, 139–144.

Martinez, Jr., J. L., & Rigter, H. (1980a). Endorphins alter acquisition and consolidation of an inhibitory avoidance response in rats. *Neuroscience Letters*, **18**, 197–201.

Martinez, Jr., J. L. & Rigter, H. (1980b). Effects of enkephalin and ACTH$_{4-10}$ on avoidance conditioning. In H. Matthies (Ed.), *Neurobiology of Learning and Memory.* New York: Raven, in press.

Martinez, Jr., J. L., Rigter, H., & van der Gugten, J. (1980a). Enkephalin effects on avoidance conditioning are dependent on the adrenal glands. In E. Stark, G. B. Makara, Zs. Ács, & E. Endröczi (Eds.), *Endocrinology, Neuroendocrinology, Neuropeptides*, (Vol. 13), pp. 273–277. Budapest: Pergamon/Akadémiai Kiadó.

Martinez, Jr., J. L., Vasquez, B. J., Rigter, H., Messing, R. B., Jensen, R. A., Liang, K. C., and McGaugh, J. L. (1980b). Attenuation of amphetamine-induced enhancement of learning by adrenal demedullation. *Brain Research*, **199**, 433–443.

McGaugh, J. L. (1966). Time-dependent processes in memory storage. *Science*, **153**, 1351–1358.

Messing, R. B., Jensen, R. A., Martinez, Jr., J. L., Spiehler, V. R., Vasquez, B. J., Soumireu-Mourat, B., Liang, K. C., & McGaugh, J. L. (1979). Naloxone enhancement of memory. *Behavioral and Neural Biology*, **27**, 266–275.

Pert, C. B., Taylor, D. P., Pert, A., Herkenham, M. A., & Kent, J. L. (1980). Biochemical and autoradiographical evidence for type 1 and type 2 opiate receptors. In E. Costa & M. Trabucchi (Eds.), *Neural Peptides and Neuronal Communication*, pp. 581–589. New York: Raven.

Rapoport, S. I., Klee, W. A., Pettigrew, K. D., & Ohno, K. (1980). Entry of opioid peptides into the central nervous system. *Science*, **207**, 84–86.

Rigter, H. (1978). Attenuation of amnesia in rats by systemically administered enkephalins. *Science*, **200**, 83–85.

Rigter, H., Greven, H., & van Riezen, H. (1977). Failure of naloxone to prevent reduction of amnesia by enkephalins. *Neuropharmacology*, **16**, 545–547.

Rigter, H., Hannan, T. J., Messing, R. B., Martinez, Jr., J. L., Vasquez, B. J., Jensen, R. A., Veliquette, J., & McGaugh, J. L. (1980b). Enkephalins interfere with acquisition of an active avoidance response. *Life Sciences*, **26**, 337–345.

Rigter, H., Jensen, R. A., Martinez, Jr., J. L., Messing, R. B., Vasquez, B. J., Liang, K. C., & McGaugh, J. L. (1980a). Enkephalin and fear-motivated behavior. *Proceedings of the National Academy of Sciences U.S.A.*, **77**, 3729–3732.

Sawynok, J., Pinsky, C., & La Bella, F. S. (1979). On the specificity of naloxone as an opiate antagonist. *Life Sciences*, **25**, 1621–1632.

Schultzberg, M., Lundberg, J., Hökfelt, T., Terenius, L., Brandt, J., Elde, R. P., & Goldstein, M. (1978). Enkephalin-like immunoreactivity in gland cells and nerve terminals of the adrenal medulla. *Neuroscience*, **3**, 1169–1186.

Viveros, O. H., Diliberto, Jr., E. J., Hazum, E., & Chang, K.-J. (1980a). Enkephalins as possible adrenomedullary hormones: Storage, secretion and regulation of synthesis. In E. Costa & M. Trabucchi (Eds.), *Neural Peptides and Neuronal Communication,* pp. 191–204. New York: Raven.

Viveros, O. H., Wilson, S. P., Diliberto, Jr., E. J., Hazum, E., & Chang, K.-J. (1980b). Enkephalins in adrenomedullary chromaffin cells and sympathetic nerves. *In* E. Stark, G. B. Makara, B. Halász, & Gy. Rappay, *Endocrinology, Neuroendocrinology, Neuropeptides* (Vol. 14), pp. 349–353. Budapest: Pergamon/Akadémiai Kiadó.

Weiner, N. (1980). Norepinephrine, epinephrine and the sympathomimetic amines. *In* A. G. Gilman, L. S. Goodman & A. Gilman (Eds.), *Goodman and Gilman's The Pharmacological Basis of Therapeutics*, pp. 138–175. New York: Macmillan.

PART III

Vasopressin and Oxytocin

CHAPTER 16

Central Nervous System Distribution of Vasopressin, Oxytocin, and Neurophysin

Michael V. Sofroniew and Adolf Weindl

The distribution of vasopressin, oxytocin, and neurophysin in the central nervous system (CNS) has been investigated using specific antisera and the unlabeled antibody peroxidase antiperoxidase

327

ENDOGENOUS PEPTIDES
AND LEARNING AND MEMORY PROCESSES

immunohistochemical technique. The rat and guinea pig CNS have been extensively examined, and many of the findings have been confirmed in other mammals including humans. Perikarya producing these peptides were found only within the hypothalamus, or in groups extending a short distance beyond it. In addition to the magnocellular neurons producing vasopressin, oxytocin, and neurophysin, a portion of the parvocellular neurons of the suprachiasmatic nucleus (SCN) were found to produce vasopressin and neurophysin in various mammals including humans. The magnocellular neurons project not only to the neurohypophysis, but also, through extensive projections to neural target areas at various levels of the CNS, including: amygdala, mesencephalic central grey, nucleus of the solitary tract, dorsal motor nucleus of the vagus, lateral reticular nucleus, and portions of the spinal cord. The parvocellular vasopressin neurons of the SCN do not project to the neurohypophysis. Their projections to neural targets include: lateral septum, medial dorsal thalamus, lateral habenula, mesencephalic central grey, medial amygdala, and ventral hippocampus. In the various target areas, both vasopressin and oxytocin terminals contact neural perikarya and dendrites, suggesting that these projections are involved in modulating the activity of target neurons. The possible participation of these projections in various activities is discussed on the basis of neuroanatomical relationships, including: influence on behavior, cardiovascular regulation, neuroendocrine regulation, and pain.

I. INTRODUCTION

A. Vasopressin, Oxytocin, and Associated Neurophysins

Vasopressin and oxytocin were originally isolated and characterized as the hormones of the neurohypophysis responsible for the pressor, antidiuretic, oxytocic, and milk-ejection effects of neurohypophyseal extracts. Characterization and subsequent synthesis of vasopressin and oxytocin (Du Vigneaud, 1954) showed both hormones to be cyclic octapeptides that differ in only two amino acids. From comparative studies, it appears that the structure of oxytocin is the same in all mammals, whereas vasopressin is present in two variants. Most mammalian species produce vasopressin with arginine in the 8 position (arginine vasopressin); members of the swine family produce vasopressin with lysine in the 8 position (lysine vasopressin) (Acher, 1974; Sawyer, 1977). The neurophysins are low molecular weight proteins also present in neural lobe extracts (Van Dyke, Chow, Greep, & Rothen, 1942). Recent studies of the synthesis of vasopressin and oxytocin suggest that vasopressin and its associated neurophysin, and oxytocin and its associated neurophysin, are synthesized as portions of larger precursor molecules, from which they are subsequently cleaved (Gainer, Sarne, & Brownstein, 1977; Pickering, 1976). At present, no specific intrinsic biological activities have been clearly ascribed to the neurophysins (Breslow, 1979).

B. Neurosecretory Neurons

Although various lines of research have previously indicated that the neurohypophysis was the source of vasopressor, antidiuretic, and oxytocic hormones, the question of whether these hormones are produced locally in the gland by pituicytes, or are produced by hypothalamic neurons whose processes project into the gland, remained controversial until Bargmann (1949) demonstrated the secretory nature of magnocellular hypothalamic neurons. Many parallel and subsequent studies in various disciplines confirmed the concept that the hormones of the neurohypophysis are produced by magnocellular neurons of the hypothalamus, transported to the neurohypophysis in the axons of these neurons, and released into the general circulation from axon terminals (see Bargmann & Scharrer, 1951). Indeed, the secretory nature of the hypothalamic magnocellular neurons dominated the concepts concerning their function, so that various early reports of projections from these neurosecretory neurons into the brain, and apparent contacts with other neurons did not receive wide attention (Ananthanrazanan, 1955; Barry, 1956, 1961; Hild, 1951; Legait & Legait, 1957; Scharrer, 1951; Sterba, 1974). However, as a result of recent immunohistochemical investigations, it is now becoming clear that there is an extensive network of vasopressinergic and oxytocinergic projections from the hypothalamus to various levels of the CNS.

C. Methods of Localizing Vasopressin, Oxytocin, and Neurophysin in the Brain and Spinal Cord

1. Immunohistochemistry

The descriptions of vasopressin, oxytocin, and neurophysin distribution in the CNS presented in this report are based on our findings from the use of the unlabeled antibody–enzyme immunoperoxidase method of Sternberger (1974). Detailed descriptions of our modifications of this procedure, of the tests conducted to verify specificity of staining and of the production of the antisera used, are presented elsewhere (Sofroniew, Madler, Müller, & Scriba, 1978; Sofroniew, Weindl, Schinko, & Wetzstein, 1979a). In principle, the endogenous peptides are chemically fixed to the CNS tissue, and serial paraffin sections are collected. The sections are exposed to specific antisera directed against the peptides and subsequently to a series of other antisera coupling the enzyme peroxidase to the initial antibody–peptide complex. The peroxidase catalyzes a reaction making the complex visible. Detailed descriptions of the principles of immunohistochemistry are presented by Sternberger (1974). Our findings presented in this report, are based primarily on results obtained from study of the rat and guinea pig CNS. However, many of the findings have been confirmed in various other mammalian species including the human, rhesus monkey, tree shrew, rabbit, and mouse.

2. Radioimmunoassay

Radioimmunoassay provides a means of quantitative analysis of vasopressin and oxytocin in homogenates of brain tissue. Dogterom, Snijdewint, and Buijs (1978) and Hawthorn, Ang, and Jenkins (1980) investigated the vasopressin and oxytocin content of various brain areas by radioimmunoassay and have reported results that correlate fairly well with results obtained by immunohistochemistry, although there are some differences. The major disadvantage of detection by radioimmunoassay is the inability to ascertain whether the peptide found in a given area is present in perikarya, in fibers of passage, or in terminals. For this reason, the description of the CNS distribution of vasopressin, oxytocin, and neurophysin presented here is based in large part on immunohistochemical findings.

II. NEURONS PRODUCING VASOPRESSIN, OXYTOCIN, AND NEUROPHYSIN

A. Hypothalamic Magnocellular Neurons

Magnocellular neurons of the hypothalamus have been shown by various light (Fig. 1) and electron microscopic immunohistochemical studies to contain either vasopressin or oxytocin and their associated neurophysins (Aspeslagh, Vandesande, & Dierickx, 1976; Sofroniew et al., 1979a; Vandesande, Dierickx, & De Mey, 1975a; Zimmerman, 1976). Broadwell, Oliver, and Brightman (1979) successfully demonstrated at the ultrastructural level, that immunoreactive neurophysin in magnocellular hypothalamic neurons is present within cellular organelles associated with protein synthesis and packaging.

Some controversy has developed over whether or not there are magnocellular neurons capable of synthesizing both vasopressin and oxytocin. Most of the evidence appears to favor the one neuron–one hormone principle (Aspeslagh et al., 1976; Sofroniew et al., 1979a; Vandesande & Dierickx, 1975, 1979; van Leeuwen & Swaab, 1977). The report that certain magnocellular neurons may produce both vasopressin and oxytocin (Sokol, Zimmerman, Sawyer, & Robinson, 1976) has yet to be confirmed in other laboratories.

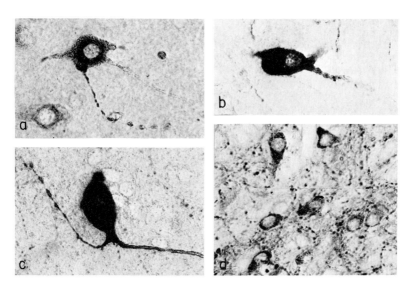

Fig. 1.   *Rat.* Hypothalamic magnocellular neurons stained for (a) oxytocin, (b) vasopressin, and (c) neurophysin. (d) Parvocellular neurons of the suprachiasmatic nucleus stained for vasopressin. × 480.

Recently, Sofroniew and Glasmann (1981) have analyzed the morphology of specific magnocellular neurons containing vasopressin, oxytocin, or neurophysin using Golgi-like immunoperoxidase staining. Their findings indicate that the axons of certain neurons have peptide-containing collateral branches, that certain neurons may give rise to two axons, and that peptide is present in non-axonal processes. In general, these neurons exhibit a more complex and heterogeneous morphology than previously appreciated, possibly reflecting their involvement in various different functions.

1. Supraoptic and Paraventricular Nuclei

The proportion of vasopressin to oxytocin neurons present in the supraoptic nuclei (SON) appears to vary greatly from species to species in various mammals. The two types of neurons are present in approximately equal numbers in the rat SON (Swaab, Nijveldt, & Pool, 1975; Vandesande & Dierickx, 1975). In the various other mammals investigated, vasopressin neurons clearly predominate in the SON, although there are always substantial numbers of oxytocin neurons in the SON; these mammals include: the human (Dierickx & Vandesande, 1977; Sofroniew, Weindl, Schrell, & Wetzstein, 1981; Zimmerman, 1976), the rhesus monkey (Sofroniew et al., 1981; Zimmerman & Antunes,

1976), the tree shrew (Sofroniew et al., 1981), the cow (Vandesande et al., 1975a), the cat (Reaves & Hayward, 1979), and the guinea pig (Sofroniew et al., 1979a).

In the paraventricular nuclei (PVN), there are not only species differences regarding the ratio of vasopressin to oxytocin neurons, but also regarding the topographic distribution of the two types of neurons in the PVN. In general, there appear to be about equal numbers of both types of neurons in the PVN in most mammals. The following authors report on the species indicated: human (Dierickx & Vandesande, 1977), cow (Vandesande et al., 1975a), cat (Reaves & Hayward, 1979), guinea pig (Sofroniew et al., 1979a), and rat (Swaab et al., 1975; Sofroniew & Glasmann, 1981).

2. Accessory Groups of Magnocellular Neurons

In addition to neurons in the SON and PVN there are a large number of magnocellular neurosecretory perikarya in various accessory groups that either appear as possible extensions of these two nuclei, or as separate clusters of neurons distinct from these two nuclei (Palkovits, Záborszky, & Ambach, 1974; Peterson, 1966). Some of the groups extend outside of the hypothalamus, for example, a group of oxytocin neurons extending from the PVN into the triangular nucleus of the septum in the guinea pig (Sofroniew et al., 1979a), or a group extending dorsally from the rostral PVN into the thalamus toward the stria terminalis in various species. The extent to which neurons in these accessory groups contribute to the extrahypothalamic projections described later in this chapter is not yet clear, although some findings suggest this (Sofroniew & Schrell, 1980). Reaves and Hayward (1979) report that both vasopressin and oxytocin neurons are present in the various accessory groups in the cat. Our findings in the rat are basically similar, although the anterior commissural nucleus of Peterson (1966) consists only of oxytocin neurons (Sofroniew & Glasmann, 1981).

B. Parvocellular Neurons of the Suprachiasmatic Nucleus

Immunohistochemical studies have revealed that a portion of the parvocellular neurons of the suprachiasmatic nucleus (SCN) produce vasopressin (Fig. 1d) and neurophysin, but not oxytocin. These neurons were first described in the rat (Vandesande, De Mey, & Dierickx, 1974; Vandesande, Dierickx, & De Mey, 1975b). A comparative investigation has shown them to be present in 13 mammalian species, including the human (Sofroniew, Weindl, & Wetzstein, 1979b; Sofroniew & Weindl, 1980). These neurons were present in all mammals examined. Only a portion of the SCN neurons produce vasopressin and neurophysin, with averages of about 17% in the rat and 31% in the human (Sofroniew & Weindl, 1980). At the ultrastructural level, the vasopressin present in SCN neurons has been shown to be contained within granules (van Leeuwen, Swaab, & de Raay, 1978).

III. VASCULAR PROJECTIONS OF VASOPRESSIN, OXYTOCIN, AND NEUROPHYSIN NEURONS

A. Neurohypophysis

On the basis of embryological, morphological, and functional criteria, the neurohypophysis is defined as consisting of both the neural lobe and median eminence (infundibulum) (Haymaker, 1969).

1. Neural Lobe

Numerous studies using neurosecretory staining procedures have firmly established that magnocellular neurons of the SON, PVN and various accessory nuclei project to the neural lobe (see Bargmann & Scharrer, 1951). These findings have been confirmed by the horseradish peroxidase tracing technique (Sherlock, Field, & Raisman, 1975), as well as by immunohistochemical staining for vasopressin, oxytocin, and neurophysin (Parry & Livett, 1973; Sofroniew et al., 1979a; Zimmerman, 1976). In a study combining the immunohistochemical detection of horseradish peroxidase transported from the neurohypophysis, with the immunohistochemical detection of vasopressin, oxytocin, and neurophysin in neighboring sections (Sofroniew, Schrell, Glasmann, Weindl, & Wetzstein, 1980), we found that virtually all magnocellular vasopressin and oxytocin neurons in the supraoptic nucleus, and a large number of those in the paraventricular nucleus and various accessory nuclei, project to the neurohypophysis. This projection to the neural lobe clearly constitutes the quantitatively most predominant projection of the magnocellular vasopressin, oxytocin, and neurophysin neurons. No projection from the parvocellular vasopressin neurons of the SCN to either the neural lobe or median eminence has been observed by tracing of fibers (Sofroniew & Weindl, 1978a), or by horseradish peroxidase tracing.

2. Median Eminence

In the median eminence, the majority of vasopressin and oxytocin fibers pass through the internal zone en route to the neural lobe. However, a number of vasopressin and neurophysin fibers, and a few oxytocin fibers pass from the internal zone into the external zone to contact portal capillaries (Dierickx, Vandesande, & De Mey, 1976; Parry & Livett, 1973; Silverman & Zimmerman, 1975; Sofroniew et al., 1979a). These external zone fibers may represent branches of fibers passing on to the neural lobe (Pittman, Blume, & Renaud, 1978; Sofroniew et al., 1979a). The amount of vasopressin in fibers contacting portal capillaries increases greatly following adrenalectomy (Dierickx et al., 1976), and this increase can be prevented by substitution with glucocorticoid but not with mineralocorticoid hormones (Sofroniew, Weindl, & Wetzstein, 1977; Stillman, Recht, Rosario, Seif, Robinson, & Zimmerman, 1977). Vasopressin has also been found in high concentrations in hypophyseal portal blood (Zimmerman, Carmel, Husain, Ferin, Tannenbaum, Frantz, & Robinson, 1973). The vasopressin fibers contacting portal capillaries in the external zone have been shown to arise from the PVN (Antunes, Carmel, & Zimmerman, 1977; Vandesande, Dierickx, & De Mey, 1977). Dornhorst, Carlson, Seif, Robinson, Zimmerman, and Gann (1978) have shown that electrical stimulation of the PVN, but not the SON, will increase peripheral levels of circulating ACTH. These various findings lend further support to the concept that vasopressin may be one of the factors involved in hypothalamic regulation of corticotropin release (Yates & Maran, 1974).

B. Other Circumventricular Organs

A small number of fibers, or an occasional single fiber, are present in various circumventricular organs (CVO) in some species. In the organum vasculosum of the lamina terminalis (OVLT) of the rat, a few neurophysin, oxytocin, and vasopressin fibers have often been observed (Buijs, Swaab, Dogterom, & van Leeuwen, 1978; Weindl & Sofroniew, 1978). We have been unable to find such fibers in the OVLT of the guinea pig and various other species, although they are present in the rhesus monkey. In the subfornical organ, pineal (Weindl & Sofroniew, 1978), and area postrema (unpublished), we have observed single neurophysin fibers in some animals. For these CVOs there were never more than one or two fibers in the CVO itself, although many fibers pass close by. When compared with the density of the fiber projections to capillaries in the median eminence or neural lobe, the functional significance of these single fibers should not be overestimated. We have never found any fibers in the subcommissural organ, although many vasopressin fibers pass close by (Weindl & Sofroniew, 1978). It seems likely that the vasotocin detected in the subcommissural organ by radioimmunoassay of dissected tissue (Rosenbloom & Fisher, 1975) is the result of cross-reaction with the vasopressin in these fibers.

In contrast to the reports that the pineal gland produces vasotocin (Pavel, 1971), and neurophysin (Reinharz, Czernichow, & Vallotton, 1974), we have never observed a pinealocyte stained positively for vasotocin or neurophysin.

Dogterom, Snijdewint, Pévet, and Swaab (1980) also have been unable to find vasotocin in the pineal, using a sensitive and specific radioimmunoassay.

C. Choroid Plexus

Although it is not a vascular target in the sense that it is an outlet to the general circulation with permeable capillaries, it is convenient to consider the choroid plexus at this point. Brownfield and Kozlowski (1977) have reported the presence of a "hypothalamochoroidal tract" in the rat, consisting of neurophysin fibers. In a previous study (Sofroniew & Weindl, 1978b) we were unable to confirm this finding. In more recent studies we have again examined this question. In several rats and guinea pigs where all serial sections were stained for neurophysin, we found in one rat a single fiber that actually entered the choroid fissure. In the guinea pig and other species we have to date not observed this. Our findings indicate that the large number of fibers passing dorsally out of the hypothalamus in this region move past the choroid fissure to enter the stria terminalis and pass to the amygdala in the rat, guinea pig, and several primates (Sofroniew & Weindl, 1978b; Sofroniew et al., 1981).

IV. NEURAL PROJECTIONS OF VASOPRESSIN, OXYTOCIN, AND NEUROPHYSIN NEURONS

We have organized the following description of the CNS distribution of vasopressin, oxytocin, and neurophysin fibers according to projections arising from magnocellular neurons, and projections arising from parvocellular vasopressin–neurophysin neurons of the SCN. The descriptions presented are from results obtained in our laboratory and are based on tracing immunohistochemically stained pathways in consecutive serial sections. The data derive primarily from examination of rat and guinea pig brains and spinal cords. However, we have been able to confirm many of the findings in various other species, including the human. Where relevant, species differences or similarities have been noted. The agreement as well as disagreement of our results with those of others are discussed in each section. It should be mentioned that, although the terminal fields seem fairly clear, it could not always be determined from which vasopressin or oxytocin neurons within the hypothalamus the projections to certain areas arise, and anatomical tracing studies will be required to establish this in the future. All projections appear to arise from vasopressin, oxytocin, and neurophysin neurons within the hypothalamus, or from neurons representing extensions of hypothalamic cell groups into immediately adjoining areas.

Figure 2 provides diagrammatic summaries of the projections arising from magnocellular neurons (represented by the PVN), or from the SCN. Figure 3a–n shows the distribution of vasopressin, oxytocin, and neurophysin fibers

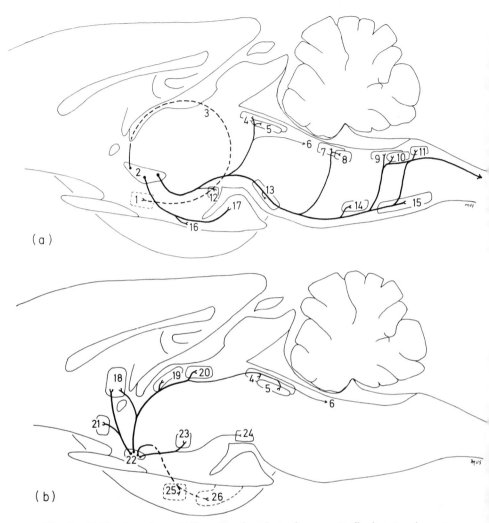

Fig. 2. (a) Drawing of a sagittally sectioned rat brain diagrammatically depicting the major fiber pathways arising from hypothalamic magnocellular vasopressin neurons or oxytocin (represented by the paraventricular nucleus), as determined by tracing immunohistochemically stained fibers in serial sections. (b) Drawing of a sagittally sectioned rat brain diagrammatically depicting the major fiber pathways arising from the parvocellular vasopressin–neurophysin neurons of the suprachiasmatic nucleus, as determined by tracing immunohistochemically stained fibers in serial sections.

The numbers refer to the following: 1-ca; 2-pvn; 3-ST; 4-mcg; 5-rd; 6-rcg; 7-pbd; 8-lc; 9-nts; 10-dX; 11-nco; 12-sum; 13-snc; 14-rm; 15-lr; 16-ez; 17-nl; 18-ls; 19-mdt; 20-lh; 21-td; 22-scn; 23-pph; 24-ip; 25-ma; 26-vh. See Table 1 for meaning of abbreviations.

TABLE 1

Abbreviations of Morphological Structures[a]

Abbreviation	Meaning	Abbreviation	Meaning
AC	Anterior commissure	nco	Nucleus commissuralis
acn	Anterior commissural nucleus of Peterson (1966)	nl	Neural lobe
bv	Blood vessel	nts	Nucleus of the solitary tract
C	Central canal	OCh	Optic chiasm
ca	Central amygdala	OT	Optic tract
CC	Corpus callosum	P	Tractus corticospinalis
coa	Cortical amygdala	pbd	Dorsal parabrachial nucleus
cu	Nucleus cuneiformis	pph	Periventricular posterior dorsal hypothalamus
DF	Dorsal funiculus	pvn	Paraventricular nucleus
dX	Dorsal nucleus of the vagus nerve	rcg	Rhombencephalic central grey
ez	External zone	rd	Nucleus raphé dorsalis
F	Fornix	rm	Nucleus raphé magnus
hgl	Hippocampal granular cell layer	ro	Nucleus raphé obscurus
hha	Hippocampal hilar area	rp	Nucleus raphé pontis
hml	Hippocampal molecular cell layer	scn	Suprachiasmatic nucleus
iml	Intermediolateral nucleus of the spinal cord	sgV	Substantia gelatinosa trigemini
ip	Interpeduncular nucleus	SM	Stria medullaris
lc	Locus coeruleus	snc	Substantia nigra pars compacta
lh	Lateral habenula	ST	Stria terminalis
LM	Lemniscus medialis	sum	Nucleus supramammillaris
lr	Lateral reticular nucleus	td	Nucleus tractus diagonalis Broca
ls	Lateral septum	TSV	Tractus spinalis nervi trigemini
ma	Medial amygdala	V	Ventricle
mcg	Mesencephalic central grey	vh	Ventral hippocampus
mdt	Mediodorsal thalamus	VF	Ventral funiculus

[a] Morphological structures in the rat central nervous system were identified and named according to König and Klippel (1963), Jacobowitz and Palkovits (1974), and Palkovits and Jacobowitz (1974). The same nomenclature was applied to corresponding structures in other species. In the human, the descriptions of Nieuwenhuys, Voogd, and Van Huijzen (1979) were used.

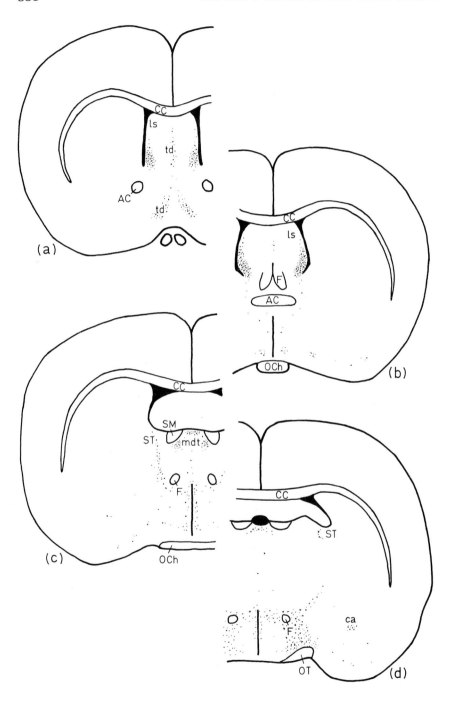

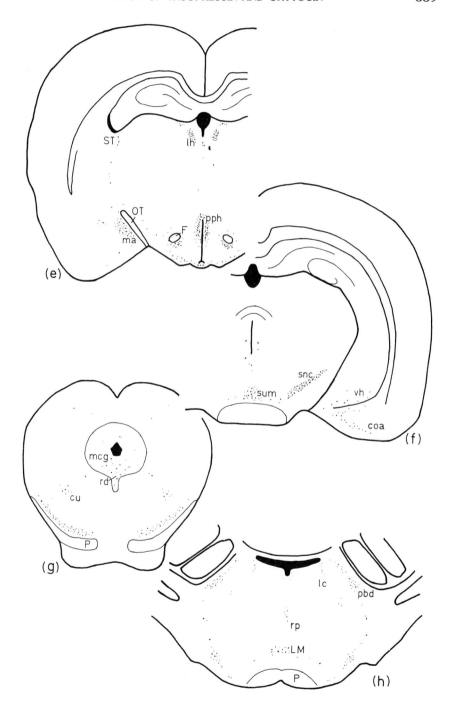

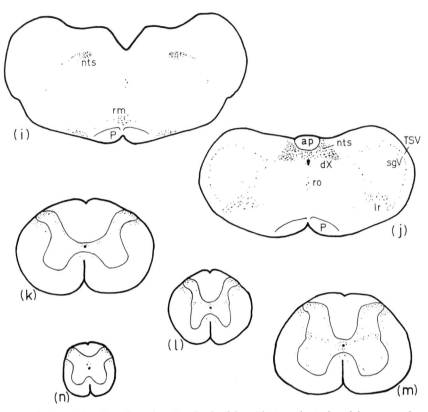

Fig. 3. Frontal sections through various levels of the rat brain and spinal cord drawn according to König and Klippel (1963), Jacobowitz and Palkovits (1974), and Palkovits and Jacobowitz (1974), showing the distribution of vasopressin, oxytocin, and neurophysin fibers in various regions. Refer to Table 2 for the ratios of vasopressin to oxytocin fibers in the areas shown. Levels of the spinal cord shown are representative of: k = midcervical; l = midthoracic; m = upper lumbar; n = sacral.

at various levels of the rodent CNS, and Table 2 gives an approximation of the density of fibers found in the areas shown, as well as an approximation of the ratio of vasopressin to oxytocin fibers in the various areas.

A. Projections from Magnocellular Neurons

1. The Amygdalar Nuclei

Several of the amygdalar nuclei receive vasopressin, oxytocin, and neurophysin projections that appear to originate from hypothalamic neurons via two pathways. The more heavily represented pathway consists of fibers probably originating in magnocellular neurons of the PVN and from a dorsal accessory

TABLE 2

Distribution of Vasopressin (VP) and Oxytocin (OT) Projections in the Rat

Target area	VP from SCN[a]	VP from PVN[b]	OT from PVN[c]
I. Telencephalon			
N. tractus diagonalis Broca[e]	++[f]		
Lateral septum	++++[f]		
Central, anterior, basal, and lateral amygdala		+	+
Medial amygdala	+++	·	·
Cortical amygdala	+[e]		
Ventral hipocampus	+		
II. Diencephalon			
Periventricular posterior dorsal hypothalamus	+++		
Mediodorsal thalamus	+++		
Lateral habenula	++++		
N. supramammillaris	·[d]		+
III. Mesencephalon			
Substantia nigra pars compacta		++	++++
Central grey	+	·	+
N. raphé dorsalis	+		+
Interpeduncular n.	·		
N. cuneiformis		·	+
IV. Pons			
Dorsal parabrachial n.[g]		+	++
Ventral parabrachial n.[g]		·	·
Locus coeruleus[g]	·	·	+
N. reticularis parvocellularis		·	
N. raphé pontis		·	
Central grey[g]	·	·	·
V. Medulla			
N. of the solitary tract[g]	·	++	++++
Dorsal n. of the vagus nerve[g]	·	++	++++
N. commissuralis[g]	·	++	++++
Lateral reticular n.		+	+++
N. ambiguus		·	·
Central grey[g]	·	·	
N. raphé magnus		+	++
N. raphé obscurus		·	
Substantia gelatinosa trigemini		·	+
VI. Spinal cord			
Laminae I–III		·	+
Lamina X		+	++
Intermediolateral n. of the spinal cord		·	+

[a] Distribution of VP projections from the supra chiasmatic nucleus (SCN).

[b] Distribution of VP projections from magnocellular neurons (represented by the paraventricular nucleus, PVN).

[c] Distributions of OT projections from the PVN.

[d] (·) Indicates regular presence of a few fibers and possibly terminals.

[e] (+) Indicates regular presence of a number of fibers and/or terminals.

[f] ++ − ++++ Indicate increasing density of fibers and/or terminals.

[g] In these areas the proportion of VP fibers from either the PVN or SCN is not certain.

[h] N. = nucleus.

magnocellular group. These fibers enter the stria terminalis to pass to the amygdala (Fig. 2a). One of the more prominent targets, receiving both vasopressin and oxytocin fibers appears to be the central amygdala in the rat (Sofroniew & Weindl, 1978b; Sofroniew, 1980), tree shrew, rhesus monkey (Sofroniew et al., 1981), and particularly in the guinea pig (Fig. 4) (Weindl & Sofroniew, 1976). In the rat and guinea pig, we have also found scattered vasopressin, oxytocin, and neurophysin fibers in the anterior, basal, medial, cortical, and lateral amygdalar nuclei (Fig. 3b–f), in agreement with the findings of Buijs (1978) and Swanson (1977). As was also described by these authors, a second, but quantitatively smaller, pathway consists of a number of scattered fibers from magnocellular neurons that reach the amygdala by a ventral directly piercing pathway. Using autoradiographic anterograde tracing in the rat brain, Conrad and Pfaff (1976) have also observed projections from the PVN to the amygdala, via the stria terminalis and a directly piercing ventral pathway. There is also a very prominent vasopressin projection to the medial amygdala which our current findings suggest arises from the SCN (see Section IV.B.4).

2. Scattered Single Fibers

Various single neurophysin fibers that appear to arise from hypothalamic magnocellular neurons are found scattered in the rostral septum, passing through the nucleus accumbens, or passing rostrally adjacent to or in the median forebrain bundle. On sagittal sections, a few fibers are seen entering the olfactory bulb. Other single fibers have been found in the thalamus, in the superior colliculus, and in various other areas. We have thus far observed such fibers in the normal rat, Brattleboro rat, and guinea pig. A detailed study of these fibers has not yet been conducted. These are all single fibers, and no prominent terminal fields have been observed in association with them, so that it is premature to speak of "projections" to any of these areas.

3. Mesencephalon, Brainstem, and Spinal Cord

The quantitatively most prominent extraneurohypophyseal projection of magnocellular neurons is that passing caudally from the hypothalamus. This projection consists primarily of oxytocin, but also contains vasopressin fibers. It appears to arise largely from the PVN, and possibly from accessory magnocellular neurons in the lateral posterior hypothalamus. Fibers can be traced on serial sagittal sections caudally from the PVN to pass laterally and caudally over the mammillary bodies and enter the pars compacta of the substantia nigra. The major pathway of fibers passes immediately dorsal to, as well as through the pars compacta of the substantia nigra (Figs. 3f and 5b), where

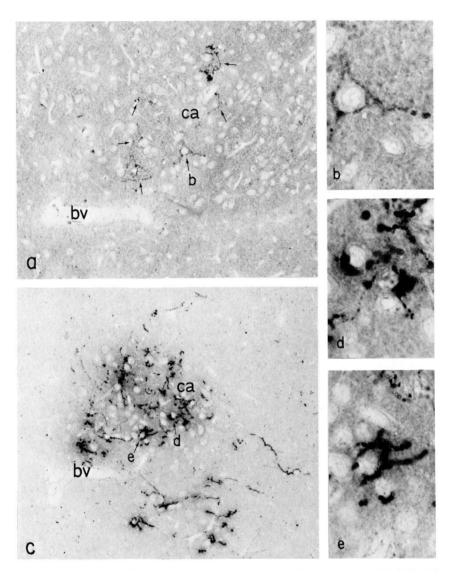

Fig. 4.   (a) Oxytocin fibers (arrows) in the central amygdala of guinea pig. × 155. (b) Detail of (a); oxytocin terminals contacting the cell soma and processes of a neuron. × 625. (c) Section neighboring that shown in (a); vasopressin fibers in the central amygdala. × 155. (d and e) Details of (c); vasopressin terminals contacting the somata of neurons. × 625.

some terminals may also be present, and continues ventrally through the brainstem to enter the lateral reticular nucleus. At various levels of the mesencephalon and brainstem, fibers can be seen to pass from this major ventral pathway to various target areas (Fig. 2a). At the level of the mammillary bodies, a number of fibers enter the supramammillary nucleus (Fig. 3f), and another small group of fibers leaves the main pathway dorsally and caudally to pass to the mesencephalic central grey. Here, fibers are distributed in the dorsal raphé nucleus (Figs. 3 and 5a) and in ventral portions of the rostral mesencephalic central grey (Fig. 3g). A number of vasopressin fibers from the SCN also project to the mesencephalic central grey via the thalamus and a dorsal pathway (see Section IV-B-2.). It is therefore not clear if vasopressin fibers found more caudally in the central grey arise from the SCN or from magnocellular neurons. Fibers are present along the entire length of the central grey in the midbrain and brainstem.

At the level of the exit of the fifth cranial nerve, a number of fibers pass caudally from the main pathway to the dorsal parabrachial nucleus (Fig. 3h). At this level, fibers are also present in the nucleus raphé pontis (Fig. 3h), in

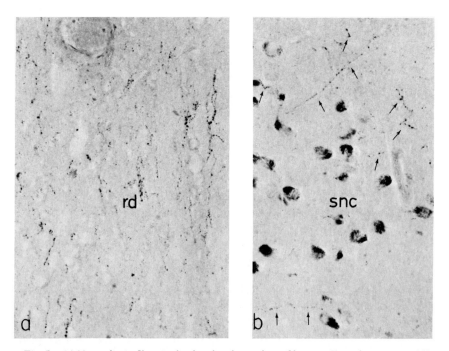

Fig. 5. (a) Neurophysin fibers in the dorsal raphé nucleus of human. Frontal section. × 160. (b) Neurophysin fibers (arrows) among the pigmented neurons of the substantia nigra of human. Frontal section. × 160.

the area of the medial lemniscus (Fig. 3h). We have now also confirmed the presence of fibers in the locus coeruleus (Fig. 3h), in contrast to our previous negative report (Sofroniew & Weindl, 1978b) and in agreement with the initial report of Swanson (1977). Somewhat more caudally, a fairly prominent projection to the nucleus raphé magnus is found (Fig. 3i).

The quantitatively most prominent projections to the brainstem are clearly those to the nucleus of the solitary tract (Figs. 3i, 3j, and 12b) dorsal motor nucleus of the vagus (Figs. 3j and 12b), and nucleus commissuralis. The fibers to these nuclei appear to pass dorsally from the main ventral pathway which, at these levels, passes through the lateral reticular nucleus (Fig. 3j). The majority of the fibers here are oxytocin fibers (Fig. 10c), but vasopressin fibers are definitely present (Fig. 10d) (Sofroniew, 1980). Fibers also pass laterally to, and are present throughout, the entire length of the marginal zone of the substantia gelatinosa of the spinal nucleus of the trigeminal nerve (Figs. 3j, and 6e). These fibers are continuous with fibers present in laminae I–III (defined according to Rexed, 1952) in the dorsal horn of the spinal cord (Figs. 3k–n, and 6f).

At the transition of the brainstem to the cervical spinal cord, some of the fibers continuing caudally pass to the lateral funiculus just ventral to the grey matter of the dorsal horn. This pathway is also described by Swanson and McKellar (1979). A few fibers also descend medially in the ventral funiculus close to the central grey (Figs. 6a and b), and pass caudally through the central grey dorsal to the central canal (Sofroniew & Weindl, 1978b; Sofroniew, 1980). In the spinal cord, fibers are found in laminae I–III (Figs. 3k–n and 6f) throughout the entire dorsal horn from cervical to cocygeal levels, and some fibers also appear to descend through these laminae. At thoracic and lumbar levels, fibers are also found in the intermediolateral nucleus (Figs. 3l,m, and 6c). This projection appears to be most prominent at upper lumbar levels (Sofroniew, 1980; Swanson & McKellar, 1979). The central grey or lamina X appears heavily innervated (Figs. 6a,b) at various levels, particularly at the lower thoracic and lumbar levels, and fibers are present passing between lamina X and the intermediolateral nucleus, as well as in the ventral horns at some levels (Figs. 6a,b). In the spinal cord, as in the brainstem, oxytocin fibers clearly predominate; however, vasopressin fibers are definitely present (Fig. 6d). In this our findings agree with Buijs (1978), but not with Swanson and McKellar (1979) who report only oxytocin fibers in the spinal cord. Direct projections from hypothalamic neurons in the PVN to the brainstem or spinal cord have also been demonstrated using various anatomical tracing techniques (Armstrong, Warach, Hatton, & McNeill, 1980; Cedarbaum & Aghajanian, 1978; Conrad & Pfaff, 1976; Hosoya & Matsushita, 1979; Ono, Nishino, Sasaka, Muramoto, Yano, & Simpson, 1978; Saper, Loewy, Swanson, & Cowan, 1976; Schober, 1978; Sofroniew & Schrell, 1980; Zemlan, Kow, Morrell, & Pfaff, 1979), and

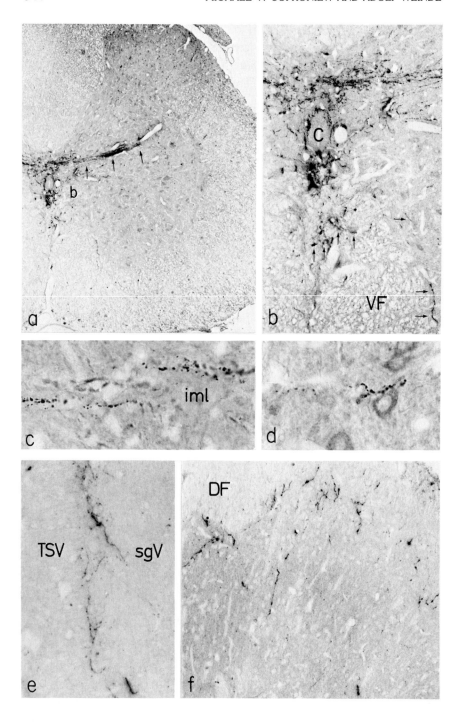

recently Sofroniew and Schrell (1981) have shown a direct projection specifically from oxytocin and vasopressin neurons in the hypothalamic paraventricular nucleus to the medulla oblongata by simultaneous immunohistochemical detection of both the tracer protein transported and peptide produced by the same neurons.

4. Findings in the Human

The previous description of oxytocin and vasopressin projections to the brainstem and spinal cord is based primarily on findings obtained in the rat. Swanson and McKellar (1979) have reported comparable findings in the spinal cord of the monkey. In human brainstems and cervical spinal cords obtained at autopsy we have observed oxytocin, vasopressin, and neurophysin terminals in the nucleus of the solitary tract (Figs. 10a,d), dorsal motor nucleus of the vagus (Figs. 10b,c), lateral reticular nucleus, and marginal zone of the substantia gelatinosa of the spinal nucleus of the trigeminal nerve, and in the substantia gelatinosa and nucleus proprius of the cervical spinal cord (Sofroniew, 1980; Sofroniew et al., 1981). More recently, we have also observed fibers in the human substantia nigra (Fig. 5b), supramammillary region, mesencephalic central grey, and dorsal raphé nucleus (Fig. 5a), and in the intermediolateral nucleus of the upper lumbar spinal cord. These preliminary findings indicate that oxytocin, vasopressin, and neurophysin fibers are present in the human midbrain, brainstem, and spinal cord in a distribution similar to that observed in the rat.

B. Projections from Parvocellular Vasopressin Neurons of the SCN

1. Lateral Septum, Nucleus of the Diagonal Tract and Rostral Fibers

Fibers leaving the SCN directly rostrally pass through the lamina terminalis to reach the ventral portion of the nucleus of the diagonal tract of Broca (Figs.

Fig. 6.   (a) Neurophysin fibers in the spinal cord (upper lumbar level) of rat. Survey showing fibers in lamina X, and a prominent bundle of fibers (arrows) passing between lamina X and the intermediolateral nucleus. × 48. (b) Detail of (a); neurophysin fibers in lamina X around the central canal, in the ventral horn (horizontal arrows) and in the ventral funiculus (vertical arrows). × 120. (c) Oxytocin fibers in the intermediolateral nucleus of the spinal cord (upper lumbar level) of rat. × 480. (d) Vasopressin fibers in the spinal cord (upper lumbar level) of rat, in the region between lamina X and the intermediolateral nucleus. × 480. (e) Neurophysin fibers in the marginal zone of the substantia gelatinosa of the spinal nucleus of the trigeminal nerve of rat. Frontal section. × 190. (f) Neurophysin fibers in laminae I–III of the dorsal horn of the spinal cord (upper cervical level) of rat. × 145.

2b and 3a) (Sofroniew & Weindl, 1978a). In the lamina terminalis of the rat, a few fibers appear to enter the OVLT as reported by Buijs et al. (1978). A large number of fibers pass rostrally and dorsally to reach the lateral septum (Sofroniew & Weindl, 1978a). The terminals in the lateral septum are largely confined to a distinct cluster in the ventral, caudal portion of the nucleus (Figs. 3a,b). The location of this cluster is most clearly seen in horizontal sections, and is quite prominent in the mouse (Fig. 7a), rat (Fig. 7b), tree shrew (Sofroniew et al., 1981), and guinea pig.

A number of vasopressin fibers from the SCN are present in the midline between the medial septal nuclei and probably represent terminals in the dorsal portion of the nucleus of the diagonal tract found here (Fig. 3a). Buijs (1978) has reported that these fibers pass dorsally to enter the dorsal hippocampus, but we are unable to confirm this. We find vasopressin fibers only in the ventral hippocampus (see Section VI.B.4.).

2. Dorsal Thalamus, Lateral Habenula, and Mesencephalic Central Grey

A large number of vasopressin–neurophysin fibers leave the SCN dorsally and pass along the third ventricle medial and rostral to the PVN (Fig. 8). The majority of the fibers can be clearly traced past magnocellular neurons in this area and continue dorsally to enter the medial dorsal thalamus (Fig. 8). In the dorsal thalamus a number of fibers appear to form terminals on neurons in the paratenial and periventricular thalamic nuclei (Sofroniew & Weindl, 1978a), but the majority of fibers pass caudally to enter and terminate in the medial portion of the lateral habenula (Fig. 3e; Sofroniew & Weindl, 1978a,b). This pathway has also been confirmed by Buijs (1978). A number of fibers continue past the lateral habenula to enter the mesencephalic central grey and dorsal raphé nucleus. They are joined here by the pathway of fibers from magnocellular neurons (see Section IV.A.3.). Thus it is uncertain whether vasopressin fibers found in more caudal portions of the central grey, and possibly in other brainstem areas, arise from parvocellular SCN or from magnocellular neurons. Recently, Kucera and Favrod (1979) have confirmed by several tracing techniques, a projection from the SCN to the mesencephalic central grey. Sakai, Touret, Salvert, Leger, and Jouvet (1977) have also reported that SCN neurons are labeled following injection of horseradish peroxidase into the region of the locus coeruleus in the cat. We have found a single neuron labeled in the SCN of one animal after injection of horseradish peroxidase into the medulla oblongata (Sofroniew & Schrell, 1980), while many SCN neurons were regularly labeled after injection into the mediodorsal thalamus (Schrell & Sofroniew, unpublished observations).

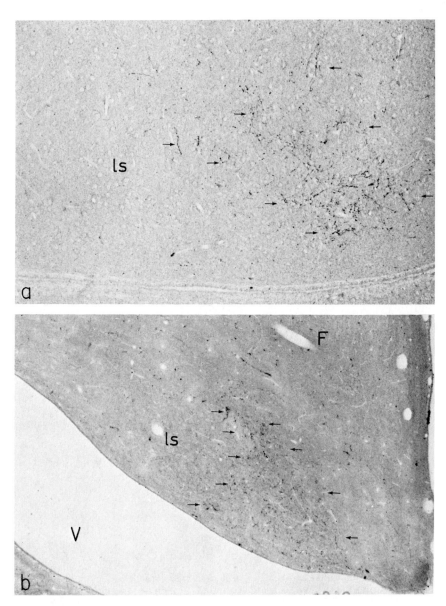

Fig. 7.   (a) Horizontal section showing vasopressin fibers in a distinct caudal portion (delineated by arrows) of the lateral septum of mouse. × 160. (b) Horizontal section showing neurophysin fibers in a distinct caudal portion (delineated by arrows) of the lateral septum of rat. × 64.

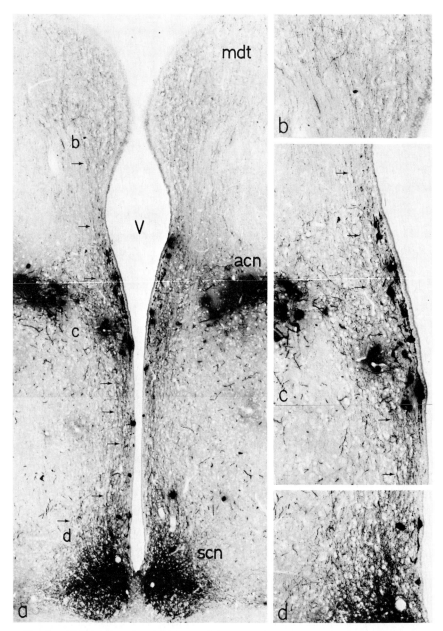

Fig. 8. (a) Frontal section darkly stained for neurophysin to allow tracing of fibers of rat. Neurophysin fibers can be traced (arrows) from the suprachiasmatic nucleus dorsally along the border of the ventricle to the medial dorsal thalamus. × 56. (b, c, and d) Details of (a); in (c), the neurophysin fibers (arrows) from the suprachiasmatic nucleus can be traced past the magnocellular neurons in this area. × 100.

3. Posterior Hypothalamus, Supramammillary Region, and Interpeduncular Nucleus

A number of vasopressin fibers leave the SCN dorsally and caudally to innervate the periventricular dorsal posterior hypothalamus (Sofroniew & Weindl, 1978a; Fig. 3e). Some of these fibers appear to continue caudally and possibly to terminate in the supramammillary region, and a few fibers appear to continue on to the interpeduncular nucleus.

4. Medial Amygdala and Ventral Hippocampus

Our previous description of the major projections of the vasopressin neurons of the SCN did not include a pathway to the medial amygdala and ventral hippocampus (Sofroniew & Weindl, 1978a). Although we had observed fibers passing laterally from the SCN through the lateral hypothalamus, we were unsure of their destination. Buijs et al. (1978) also observed such fibers. Recent extensive analysis of serial sections in the frontal, horizontal, and sagittal planes, stained for neurophysin, showed that these fibers from the SCN move laterally through the hypothalamus and pass over the optic tract caudally to the medial amygdala (Figs. 2b and 3e). The medial amygdala is heavily innervated and a number of fibers pass caudally and ventrally through the medial amygdala to enter the ventral hippocampus. Fibers are found only in the ventral portion of the hippocampus, distributed in the molecular layer and hilar regions (Figs. 3 and 9). Some fibers pass caudally and ventrally through the medial amygdala to the cortical amygdala. This pathway is quite prominent in the guinea pig (Sofroniew & Weindl, in preparation) and tree shrew (Sofroniew et al., 1981), and is also present (but not as prominent) in the rat. The medial amygdala thus appears to receive a large number of vasopressin fibers from the SCN, as well as a few oxytocin fibers from magnocellular neurons (see Section IV.A.1.). We have, to date, not observed any oxytocin fibers in the hippocampus.

C. Terminals in Neural Target Areas

Neural target areas are defined as areas where fibers appear to branch extensively and terminate. At the light microscopic level we have observed axosomatic, as well as axodendritic contacts stained for vasopressin (Figs. 4d,e and 10d), oxytocin (Figs. 4b and 10c), and neurophysin (Figs. 10a,b). Such terminals were observed not only in rodents (Figs. 4b,d,e), but also in humans (Fig. 10), and tree shrews (Sofroniew, 1980; Sofroniew et al., 1981). At the ultrastructural level, Buijs and Swaab (1979) successfully demonstrated synapse-like structures containing granules stained positively for vasopressin and oxytocin by the immunoperoxidase procedure (Fig. 11), contacting dendrites in various CNS areas where terminals have been seen at the light microscopic level.

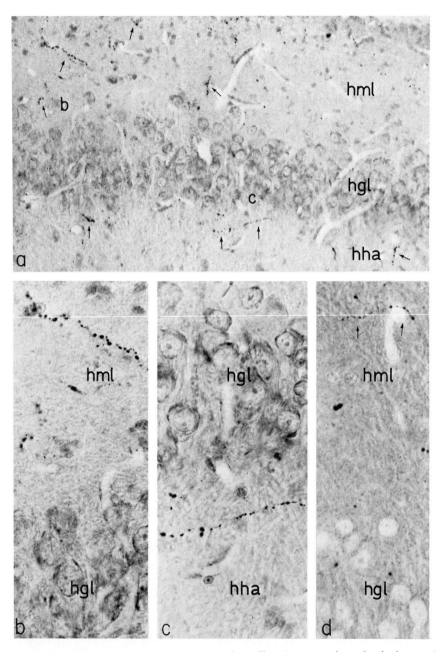

Fig. 9. (a) Horizontal section showing neurophysin fibers (arrows) in the molecular layer and hilar area of the ventral hippocampus. Hippocampal neurons are visible due to toluidine blue counterstain. × 210. (b and c) Details of (a). × 510. (d) Vasopressin fiber in the molecular layer of the ventral hippocampus, area similar to that shown in (b) and (c). × 510.

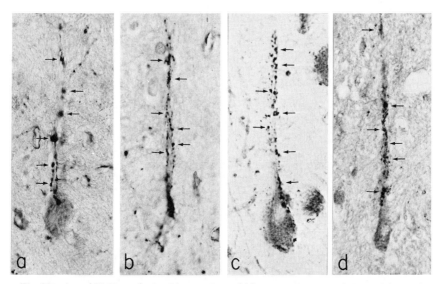

Fig. 10. (a and b) Neurophysin, (c) oxytocin, and (d) vasopressin terminals (arrows) lining the dendrites of toluidine blue counterstained neurons in the nucleus of the (a and d) solitary tract and (b and c) dorsal nucleus of the vagus of human. × 400.

V. VASOPRESSIN, OXYTOCIN, AND NEUROPHYSIN
IN THE CEREBROSPINAL FLUID (CSF)

Several reports have appeared on the presence of radioimmunoassayable vasopressin, oxytocin, and neurophysin in the CSF (Dogterom, van Wimersma Greidanus, & de Wied, 1978; Jenkins, Mather, & Ang, 1980; Robinson & Zimmerman, 1973). At present, the source of CSF levels of these peptides remains unknown. They may be specifically released into the CSF by hypothalamic neurons for unknown functions. Another possibility is that vasopressin, oxytocin, and neurophysin released in some of the terminal areas near ventricular surfaces may simply diffuse from these areas into the CSF. Many of the densely innervated areas either border directly on a ventricular surface or are close to such a surface. This possibility is supported by the finding of increased CSF levels of β-endorphin following periventricular stimulation of the posterior medial thalamus (Akil, Richardson, Barchas, & Li, 1978), an area containing β-endorphin immunoreactive terminals (Bloom, Battenberg, Rossier, Ling, & Guillemin, 1978; Sofroniew, 1979). Such diffusion of peptides and peptide fragments into the CSF may represent one means by which these peptides are removed or cleared from the areas in which they are active.

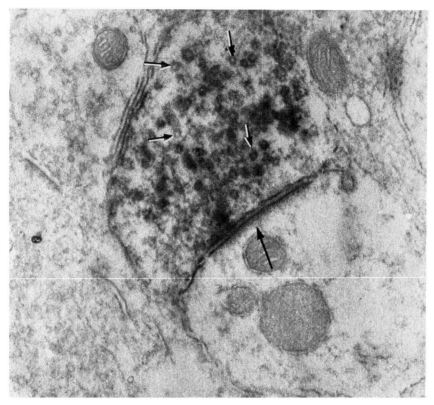

Fig. 11. Vasopressin positive terminal forming a synapse (black arrow) with an unlabeled dendrite in the lateral habenular nucleus of rat. Note the black diaminobenzidine deposit around clear vesicle-like structures (arrows). × 60,000. (Reprinted with permission from Buijs and Swaab, *Cell and Tissue Research* (1980). Copyright Springer: Berlin, Heidelberg, New York.)

VI. RELATION OF CNS VASOPRESSIN
AND OXYTOCIN TO PITUITARY STORES

Evidence has recently been provided that certain circulating peptides can enter the CNS (Rapoport, Klee, Pettigrew, & Ohno, 1980). These findings lend support to the various reports of the behavioral activities of peripherally applied peptides (de Wied, 1976; Roche & Leshner, 1979; van Wimersma Greidanus & Versteeg, 1980). However, it is still unknown if circulating vasopressin or oxytocin contribute to the CNS effects of these peptides under normal physiological conditions. It is clear from histological studies, as described in this report, that the vasopressin and oxytocin found in various specific CNS areas are transported to these areas in neural processes, and thus do not derive from pituitary stores. Whether or not depletion of pituitary stores affects the levels of these peptides in the CNS, or vice versa, awaits future investigation.

VII. FUNCTIONAL CONSIDERATIONS BASED ON NEUROANATOMICAL RELATIONSHIPS

The magnocellular vasopressin and oxytocin neurons of the hypothalamus have long been considered to be endocrine neurosecretory cells, which produce hormones to be released into the bloodstream. Nevertheless, numerous electrophysiological studies have shown that these neurons also have electrical characteristics similar to other neurons, and that release of hormone appears to be coupled to the electrical activity of the neurons (see Hayward, 1977). Thus, these peptide-producing neurosecretory cells can be considered functionally as neurons, and it seems likely that their projections to the various CNS target areas described in this chapter are involved in modifying the activity of other neurons in these areas. This is further supported by the observation that, in these target areas, terminals are not formed on capillaries, but rather on neural perikarya or on dendrites, as described in Section IV.C. The manner in which target neurons are affected is unknown, and may encompass changes in their electrical activity, or possibly in their biochemical activity. Barker (1976) has reported that vasopressin can alter the firing rates of certain neurons. Morris, Salt, Sofroniew, and Hill (1980) have found that oxytocin predominantly depresses the firing rate of neurons in the caudal medulla. At present, there are few reports on the electrophysiological investigation of projections from hypothalamic magnocellular neurosecretory neurons to extrahypophyseal CNS areas. Pittman, Blume, and Renaud (1978b) report antidromic activation of neurons in the paraventricular nucleus following stimulation of the amygdala or midbrain. Hayward (1974) has provided combined electrophysiological and morphological evidence that magnocellular neurosecretory neurons of the goldfish have branching axons.

The reasons for the separation of vasopressin and oxytocin neurons into a number of distinct nuclear groups are not understood, but may be related to evidence suggesting that different groups may project to different sites. The available evidence currently suggests:

1. neurons in the supraoptic nucleus project to the neurohypophysis but not to extrahypothalamic neural targets;

2. neurons in the paraventricular nucleus project to extrahypothalamic neural targets as well as to the neurohypophysis;

3. separate populations of neurons in the paraventricular nucleus project to the neurohypophysis or brainstem;

4. neurons in the paraventricular nucleus project to hypophyseal portal capillaries in the median eminence;

5. neurons in many accessory nuclei project to the neurohypophysis;

6. some accessory neurons may project to extrahypothalamic neural targets;

7. parvocellular SCN neurons project to various neural targets but not to the neurohypophysis (Antunes et al., 1977; Armstrong et al., 1980; Ono et al., 1978; Sofroniew, 1980; Sofroniew & Schrell, 1980, 1981; Sofroniew et al., 1980; Sofroniew & Weindl, 1978a; Swanson & Kuypers, 1980; Vandesande et al., 1977).

A. Possible Interaction of Vasopressin and Oxytocin with Neurotransmitters and Other Neuropeptides

It is not within the scope of this chapter to review the large number of recent reports on the CNS distribution of the various amines and neuropeptides, and studies on their possible interactions. However, it is worthwhile mentioning that many areas receiving vasopressin and oxytocin terminals either contain aminergic or other peptidergic neurons, or also receive aminergic or peptidergic terminals. A few examples are cited. In the midbrain and brainstem, several areas receiving oxytocin and vasopressin fibers appear to correspond to the location of the A1, A2, A5, A6, A8, and A9 catecholamine neurons (Dahlström & Fuxe (1964); Palkovits & Jacobowitz, 1974). It is of interest to note that Tanaka, de Kloet, de Wied, and Versteeg (1977) and Versteeg, de Kloet, van Wimersma Greidanus, and de Wied (1979) have shown that intraventricular administration of vasopressin or vasopressin antiserum can significantly alter catecholamine turnover rates in specific brain areas. These areas include the nucleus of the solitary tract, the A1 and A8 regions, the locus coeruleus (A6), and the dorsal raphé nucleus, areas in which we have found oxytocin and vasopressin fibers.

Oxytocin and vasopressin terminals are also present in various areas containing serotonin neurons (Steinbusch, 1981) or areas containing peptidergic neurons, such as the Substance P neurons found in the nucleus raphé magnus (Ljungdahl, Hökfelt, & Nilsson, 1978). Overlap of oxytocin and vasopressin terminals with other peptidergic terminals occurs in various areas, including: β-endorphin–ACTH terminals in the ventral, rostral mesencephalic central grey (Bloom et al., 1978; Sofroniew, 1979; Watson, Richard, & Barchas, 1978); Substance P terminals in the substantia gelatinosa of the spinal nucleus of the trigeminal nerve (SGV) and dorsal horn (Cuello & Kanazawa, 1978); somatostatin terminals in the parabrachial nucleus, SGV, and dorsal horn (Elde & Hökfelt, 1978), and enkephalin terminals in the dorsal horn (Elde & Hökfelt, 1978). Although no conclusive evidence for interaction of these peptides is available, the morphological overlap is quite conspicuous and provides an interesting basis for future study.

B. Vasopressin, Oxytocin, and Behavior

The exogenous application or depletion of vasopressin and oxytocin has effects on such behavior as memory processing (de Wied, 1976), activity, and grooming (Delanoy, Dunn, & Tinter, 1978). These effects have been reviewed by van Wimersma Greidanus and Versteeg (1980). Here, we would like to compare some of the previously conducted behavioral studies with the morphological data we have presented, and mention a few of the behavioral functions thought to be associated with some of the areas containing vasopresin and oxytocin terminals.

A number of studies have found that vasopressin enhances the retention of learned responses (de Wied, 1976; Flexner, Flexner, Hoffman, & Walter, 1977; Pfeifer & Bookin, 1978). Vasopressin also appears to enhance attention and memory in humans (Legros & Gilot, 1979). In line with these findings, vasopressin fibers and terminals are present in many of the areas thought to be involved in passive avoidance memory processes (Thompson, 1978), including: ventral hippocampus, caudal septum, amygdala, and mediodorsal thalamus. Van Wimersma Greidanus, Croiset, Bakker, and Bouman (1979) have shown that lesions of the amygdala can block the effect of vasopressin on the retention of learned responses. The septum and the amygdala are also involved in the regulation of aggressive and passive behavior (Kling, 1972; Kluever & Bucy, 1939; Miczek & Grossmann, 1972; Miley & Baenninger, 1972), and Roche and Leshner (1979) have shown that exogenously administered vasopressin enhances the learned submissiveness of nondominant male mice. Electrical stimulation of the dorsal raphé nucleus has been shown to disrupt memory (Fibiger, Lepiane, & Phillips, 1978), and both oxytocin and vasopressin terminals are found in this nucleus. Perhaps related to this are the findings of Bohus, Kovács, and de Wied (1978) that intraventricularly administered oxytocin has an amnesic effect. Current evidence also suggests that the hippocampal–septal axis is involved in the process of attention (Oades, 1979). Vasopressin terminals are present in both the septum and ventral hippocampus.

C. Vasopressin, Oxytocin, and Cardiovascular Regulation

When considering the various areas in the brainstem and spinal cord that receive oxytocin and vasopressin terminals, it is interesting to note that a number of areas currently thought to be involved in cardiovascular regulation (Loewy

& McKellar, 1980; Palkovits & Záborszky, 1977; Palkovits, Mezey, & Záborszky, 1979) receive innervation, including: the nucleus of the solitary tract, the commissural nucleus, the dorsal motor nucleus of the vagus, the lateral reticular nucleus, the nucleus ambiguus, the nucleus raphé magnus, the nucleus raphé obscurus, locus coeruleus, the A5 catecholamine cell group, and the intermediolateral nucleus of the spinal cord. In some of these areas we have confirmed the presence of oxytocin and vasopressin terminals in the human (Section IV-A-4.). These findings are of even greater interest when viewed in the context of various studies suggesting that circulating levels of vasopressin may be involved in different forms of pathological hypertension (Crofton, Share, Shade, Lee-Kwon, Manning, & Sawyer, 1979; Möhring, Arbogast, Düsing, Glänzer, Kintz, Liard, Maciel, Montani, & Schoun, 1980). Based on their own, and other experimental results, Möhring et al. (1980) have recently put forth the hypothesis that, in animals with intact cardiovascular reflex systems, the pressor effect of circulating vasopressin is specifically buffered so that a net increase in blood pressure does not occur. In situations where these reflex systems have been disrupted, either experimentally or pathologically, the pressor effects of circulating vasopressin are increased markedly. It appears that one major response of the reflex system in answer to increased levels of circulating vasopressin, is to reduce cardiac output in the effort to maintain constant blood pressure (Möhring et al., 1980). At present, the precise morphology of their proposed reflex system in unknown. That central vasopressin and oxytocin projections might be involved in such a reflex is an intriguing possibility. Electrical stimulation of the hypothalamic paraventricular nucleus produces brachycardia and hypotension (Folkow, Johansson, & Oberg, 1959; van Bogaert, Wellens, van Bogaert, Martin, & de Wilde, 1976). Recently, Sofroniew and Schrell (1981) have shown a direct projection from oxytocin and vasopressin neurons in the hypothalamic paraventricular nucleus to the medulla oblongata, by simultaneous detection of tracer protein and peptide in the same neurons. In the medulla oblongata, oxytocin, vasopressin, and neurophysin terminals are present in areas containing preganglionic vagal neurons (Fig. 12) (Sofroniew & Schrell, 1981), which in some cases appear to include the location of vagal cardioinhibitory neurons as identified by horseradish peroxidase tracing from cardiac branches of the vagus nerve (Nosaka, Yamamoto, & Yasunaga, 1979). Thus, oxytocin, vasopressin, and neurophysin terminals originating from hypothalamic neurons might represent a link in the reflex system hypothesized by Möhring et al. (1980). However, much future study will be required to test this possibility. In this context it may be mentioned that stimulation of the vagus nerve, carotid sinus, or atrial stretch receptors alters the electrical activity of SON and PVN neurons (Barker, Crayton, & Nicoll, 1971; Koizumi & Yamashita, 1978; Yamashita & Koizumi, 1979).

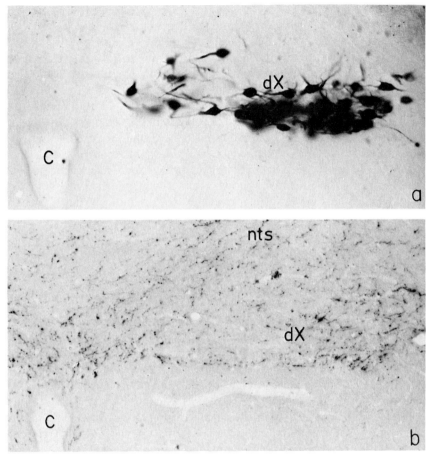

Fig. 12.   (a) Frontal 100 μm frozen section showing neurons in the dorsal nucleus of the vagus, labeled following the uptake of horseradish peroxidase from the cervical vagus nerve of rat. × 145. (From the unpublished work of U. Schrell and M. V. Sofroniew.) (b) Frontal 7 μm paraffin section showing neurophysin fibers and terminals in the dorsal nucleus of the vagus and nucleus of the solitary tract of rat. Level of section and area shown correspond to that of (a). × 160.

D. Oxytocin, Vasopressin, and Pain Mechanisms

Morphological findings strongly suggest that oxytocin and vasopressin may be involved in pain mechanisms. Oxytocin and vasopressin fibers are present in Laminae I–III throughout the entire length of the spinal cord from cervical to coccygeal levels and are continuous with fibers in the marginal zone of the substantia gelatinosa of the spinal nucleus of the trigeminal nerve. Fibers are also present in the nucleus raphé magnus and periaqueductal grey. These areas in the midbrain, brainstem, and spinal cord are known to be involved in pain transmission (see Fields & Basbaum, 1978). Oxytocin and vasopressin fibers present in the ventral rostral mesencephalic central grey overlap portions of areas densely innervated by β-endorphin–ACTH fibers (Bloom et al., 1978; Sofroniew, 1979; Watson et al., 1978), and several regions in the periventricular medial dorsal thalamus that receive β-endorphin–ACTH fibers also receive vasopressin fibers (personal observation). Electrical stimulation of the periventricular medial thalamus can produce analgesia in the human (Akil et al., 1978). Vasopressin and oxytocin terminals in these areas may or may not be directly involved in pain mechanisms. Since noxious stimuli can elevate circulating levels of vasopressin independently of changes in serum osmolarity (Kendler, Weitzman, & Fisher, 1978), and intraventricularly administrated vasopressin produces analgesia in rats (Berntson & Berson, 1980), this possibility seem worthy of further investigation.

E. Central Regulation of Neuroendocrine Secretion

There is little direct evidence that CNS projections of vasopressin and oxytocin neurons are involved in central mechanisms regulating neuroendocrine secretion, but the morphological evidence indicates that this is a possibility. The amygdalar nuclei are involved in regulating the secretion of vasopressin, such that electrical stimulation of the amygdala can greatly and directly increase circulating levels of vasopressin (Hayward, Murgas, Pavasuthipaisit, Perez-Lopez, & Sofroniew, 1977). Vasopressin and oxytocin terminals in the amygdala could be involved in amygdalar neuroendocrine regulation. In this context, Renaud (1976) has provided electrophysiological evidence that certain tubero-infundibular neurons send axon collaterals to both the median eminence and the amygdala. PVN projections may be involved in the regulation of ACTH secretion via the vasopressin projection to the hypophyseal portal capillaries as described in Section III-A-2. The extensive oxytocin projections to the brainstem may interact with the afferent pathway of the milk-ejection reflex (Tindal, Knaggs, & Turvey, 1967, 1968).

F. SCN and Biological Rhythms

Numerous studies have now shown that the SCN receives a direct input from the retina (Hendrickson, Wagoner, & Cowan, 1972; Moore, 1973; Moore & Lenn, 1972; Nishino, Koizumi, & Brooks, 1976), and that lesioning the SCN disrupts the regulation of various biological rhythms (Moore & Eichler,

1972; Rusak & Zucker, 1979; Stephan & Zucker, 1972; van den Pol & Powley, 1979). Parvocellular vasopressin neurons make up only the smaller portion of the population of SCN neurons, constituting about 17% of SCN neurons in the rat and about 31% in the human. The vasopressin neurons are located primarily in the rostral, dorsal, and medial portion of the rat SCN (Sofroniew & Weindl, 1978a), while the retinal input into the SCN appears to be located in the more caudal ventral and lateral portions of the SCN (Hendrickson et al., 1973; Moore & Lenn, 1972). Thus, we suggested that the vasopressin neurons of the SCN might not be involved in SCN regulation of biological rhythms. This indeed seems to be the case, as there seems to be no deficit in the circadian rhythms of Brattleboro rats which are genetically unable to produce vasopressin in magnocellular neurons or in parvocellular neurons of the SCN (Peterson, Watkins, & Moore, 1980; Stephan & Zucker, 1974).

VIII. FINAL REMARKS

In conclusion, results from our own and other laboratories, indicate that there is an extensive network of vasopressin and oxytocin projections from the hypothalamus to various parts of the CNS. These projections derive not only from magnocellular vasopressin and oxytocin neurons, but also from parvocellular vasopressin neurons in the SCN. In target areas, vasopressin and oxytocin terminals contact neural perikarya and dendrites, suggesting that these projections are involved in modulating the activity of target neurons. Based on current knowledge of neuroanatomical–functional interrelationships, these projections may be involved in several processes, including: influence on behavior, cardiovascular regulation, neuroendocrine regulation, and pain.

ACKNOWLEDGMENTS

The authors gratefully acknowledge the technical assistance of H. Asam, R. Köpp-Eckmann, A. Nekic, and I. Wild; and the editorial assistance of P. Campbell. This work was supported by the Deutsche Forschungsgemeinschaft (We 608/5).

REFERENCES

Acher, R. (1974). Chemistry of the neurohypophysial hormones: An example of molecular evolution. In R. O. Greep & E. B. Astwood (Eds.), Handbook of Physiology, Section 7: Endocrinology IV, pp. 119–130. Washington, D.C.: American Physiological Society.
Akil, H., Richardson, D. E., Barchas, J. D., & Li, C. H. (1978). Appearance of β-endorphin-

like immunoreactivity in human ventricular cerebrospinal fluid upon analgesic electrical stimulation. *Proceedings of the National Academy of Sciences, U.S.A.*, **75**, 5170–5172.

Ananthanrazanan, V. (1955). Nature and distribution of neurosecretory cells of the reptilian brain. *Zeitschrift für Zellforschung und mikroskopische Anatomie*, **43**, 8–16.

Antunes, J. L., Carmel, P. W., & Zimmerman, E. A. (1977). Projections from the paraventricular nucleus to the zona externa of the median eminence of the rhesus monkey: An immunohistochemical study. *Brain Research*, **137**, 1–10.

Armstrong, W. E., Warach, S., Hatton, G. I., & McNeill, T. H. (1980). Subnuclei in the rat hypothalamic paraventricular nucleus: A cytoarchitectural, horseradish peroxidase and immunocytochemical analysis. *Neuroscience*, **5**, 1931–1958.

Aspeslagh, M.-R., Vandesande, F., & Dierickx, K. (1976). Electron microscopic immuno-cytochemical demonstration of separate neurophysin–vasopressinergic and neurophysin–oxytocinergic nerve fibres in the neural lobe of the rat hypophysis. *Cell and Tissue Research*, **171**, 31–37.

Bargmann, W. (1949). Über die neurosekretorische Verknüpfung von Hypothalamus und Neurohypophyse. *Zeitschrift für Zellforschung und mikroskopische Anatomie*, **34**, 610–634.

Bargmann, W., & Scharrer, E. (1951). The origin of the posterior pituitary hormones. *American Scientist*, **39**, 255–259.

Barker, J. L. (1976). Peptides: Roles in neuronal excitability. *Physiological Reviews*, **56**, 435–452.

Barker, J. L., Crayton, J. W., & Nicoll, R. A. (1971). Antidromic and orthodromic responses of paraventricular and supraoptic neurosecretory cells. *Brain Research*, **33**, 353–366.

Barry, J. (1956). Les voies extra-hypophysaires de la neurosécrétion diencéphalique. *Bulletin de la Association des Anatomistes*, **89**, 264–276.

Barry, J. (1961). Recherches morphologiques et expérimentales sur la glande diencéphalique et l'appareil hypothalamo–hypophysaire. *Annales Scientifiques de l'Université Besancon, Zoologie, Physiologie*, **2**, 3–133.

Berntson, G. G., & Berson, B. S. (1980). Antinociceptive effects of intraventricular or systemic administration of vasopressin in the rat. *Life Sciences*, **26**, 455–459.

Bloom F., Battenberg, E., Rossier, J., Ling, N., & Guillemin, R. (1978). Neurons containing β-endorphin in rat brain exist separately from those containing enkephalin: Immunocytochemical studies. *Proceedings of the National Academy of Sciences, U.S.A.*, **75**, 1591–1595.

Bohus, B., Kovács, G. L., & de Wied, D. (1978). Oxytocin, vasopressin and memory: Opposite effects on consolidation and retrieval processes. *Brain Research*, **157**, 414–417.

Breslow, E. (1979). Chemistry and biology of the neurophysins. *Annual Review of Biochemistry*, **48**, 251–274.

Broadwell, R. D., Oliver, C., & Brightman, M. W. (1979). Localization of neurophysin within organelles associated with protein synthesis and packaging in the hypothalamo-neurohypophysial system: An immunocytochemical study. *Proceedings of the National Academy of Sciences, U.S.A.*, **76**, 5999–6003.

Brownfield, M. S., & Kozlowski, G. P. (1977). The hypothalamochoroidal tract. I. Immunohistochemical demonstration of neurophysin pathways to telencephalic choroid plexuses and cerebrospinal fluid. *Cell and Tissue Research*, **178**, 111–127.

Buijs, R. M. (1978). Intra- and extrahypothalamic vasopressin and oxytocin pathways in the rat. Pathways to the limbic system, medulla oblongata, and spinal cord. *Cell and Tissue Research*, **192**, 423–435.

Buijs, R. M., & Swaab, D. F. (1980). Immuno-electron microscopical demonstration of vasopressin and oxytocin synapses in the limbic system of the rat. *Cell and Tissue Research*, **204**, 355–365.

Buijs, R. M., Swaab, D. F., Dogterom, J., & van Leeuwen, F. W. (1978). Intra- and extrahypothalamic vasopressin and oxytocin pathways in the rat. *Cell and Tissue Research*, **186**, 423–433.

Cedarbaum, J. M., & Aghajanian, G. K. (1978). Afferent projections to the rat locus coeruleus as determined by a retrograde tracing technique. *Journal of Comparative Neurology*, **178**, 1–16.

Conrad, L.C.A., & Pfaff, D. W. (1976). Efferents from medial basal forebrain and hypothalamus in the rat. II. An autoradiographic study of the anterior hypothalamus. *Journal of Comparative Neurology*, 169, 221–262.

Crofton, J. T., Share, L., Shade, R. E., Lee-Kwon, W. J., Manning, M., & Sawyer, W. H. (1979). The importance of vasopressin in the development and maintenance of DOC–salt hypertension in the rat. *Hypertension*, 1, 31–38.

Cuello, A. C., & Kanazawa, I. (1978). The distribution of substance P immunoreactive fibers in the rat central nervous system. *Journal of Comparative Neurology*, 178, 129–156.

Dahlström, A., & Fuxe, K. (1964). Evidence for the existence of monoamine-containing neurons in the central nervous system. I. Demonstration of monoamines in the cell bodies of brain stem neurons. *Acta Physiologica Scandinavica*, 62, Supplementum 232, 1–55.

Delanoy, R. L., Dunn, A. J., & Tinter, R. (1978). Behavioral responses to intracerebroventricularly administered neurohypophyseal peptides in mice. *Hormones and Behavior*, 11, 348–362.

de Wied, D. (1976). Behavioral effects of intraventricularly administered vasopressin and vasopressin fragments. *Life Sciences*, 19, 685–690.

Dierickx, K., & Vandesande, F. (1977). Immunocytochemical localization of the vasopressinergic and the oxytocinergic neurons in the human hypothalamus. *Cell and Tissue Research*, 184, 15–27.

Dierickx, K., Vandesande, F., & De Mey, J. (1976). Identification, in the external region of the rat median eminence, of separate neurophysin–vasopressin and neurophysin–oxytocin-containing nerve fibres. *Cell and Tissue Research*, 168, 141–151.

Dogterom, J., Snijdewint, F.G.M., & Buijs, R. M. (1978). The distribution of vasopressin and oxytocin in the rat brain. *Neuroscience Letters*, 9, 341–346.

Dogterom, J., Snijdewint, F.G.M., Pévet, P., & Swaab, D. F. (1980). Studies on the presence of vasopressin, oxytocin and vasotocin in the pineal gland, subcommissural organ and fetal pituitary gland: Failure to demonstrate vasotocin in mammals. *Journal of Endocrinology*, 84, 115–123.

Dogterom, J., van Wimersma Greidanus, Tj. B., & de Wied, D. (1978). Vasopressin in cerebrospinal fluid and plasma of man, dog and rat. *American Journal of Physiology*, 234, E463–E467.

Dornhorst, A., Carlson, D. E., Seif, S. M., Robinson, A. G., Zimmerman, E. A., & Gann, D. S. (1978). Control of release of ACTH and vasopressin by supraoptic and paraventricular nuclei. *Society for Neuroscience Abstracts*, 4, 333.

Du Vigneaud, V. (1954). Hormones of the posterior pituitary gland: Oxytocin and vasopressin. *Harvey Lectures*, 50, 1–26.

Elde, R., & Hökfelt, T. (1978). Distribution of hypothalamic hormones and other peptides in the brain. In W. F. Ganong & L. Martini (Eds.), *Frontiers in Neuroendocrinology* (Vol. 5), pp. 1–33. New York: Raven.

Fibiger, H. C., Lepiane, F. G., & Phillips, A. G. (1978). Disruption of memory produced by stimulation of the dorsal raphé nucleus: Mediation by serotonin. *Brain Research*, 155, 380–386.

Fields, H. L. & Basbaum, A. I. (1978). Brainstem control of spinal pain-transmission neurons. *Annual Review of Physiology*, 40, 217–248.

Flexner, J. B., Flexner, L. B., Hoffman, P. L., & Walter, R. (1977). Dose–response relationships in attenuation of puromycin-induced amnesia by neurohypophyseal peptides. *Brain Research*, 134, 139–144.

Folkow, B., Johansson, B., & Oberg, B. (1959). A hypothalamic structure with a marked inhibitory effect on tonic sympathetic activity. *Acta Physiologica Scandinavica*, 47, 262–275.

Gainer, H., Sarne, Y., & Brownstein, M. J. (1977). Biosynthesis and axonal transport of rat neurohypophysial proteins and peptides. *Journal of Cell Biology*, 73, 366–381.

Hawthorn, J., Ang, V.T.Y., & Jenkins, J. S. (1980). Localization of vasopressin in the rat brain. *Brain Research*, 197, 75–81.

Haymaker, W. (1969). Hypothalamo–pituitary neural pathways and the circulatory system of the pituitary. *In* W. Haymaker, E. Anderson, & W.J.H. Nauta (Eds.), *The Hypothalamus*, pp. 219–250. Springfield, Ill.: Charles C Thomas.

Hayward, J. N. (1974). Physiological and morphological identification of hypothalamic magnocellular neuroendocrine cells in goldfish preoptic nucleus. *Journal of Physiology*, **239**, 103–124.

Hayward, J. N. (1977). Functional and morphological aspects of hypothalamic neurons. *Physiological Reviews*, **57**, 574–658.

Hayward, J. N., Murgas, K., Pavasuthipaisit, K., Perez-Lopez, F. R., & Sofroniew, M. V. (1977). Temporal patterns of vasopressin release following electrical stimulation of the amygdala and the neuroendocrine pathway in the monkey. *Neuroendocrinology*, **23**, 61–75.

Hendrickson, A. E., Wagoner, N., & Cowan, W. M. (1972). An autoradiographic and electron microscopic study of retino–hypothalamic connections. *Zeitschrift für Zellforschung und mikroskopische Anatomie*, **135**, 1–26.

Hild, W. (1951). Vergleichende Untersuchungen über Neurosekretion im Zwischenhirn von Amphibien und Reptilien. *Zeitschrift für Anatomie und Entwicklungsgeschichte*, **115**, 459–479.

Hosoya, Y., & Matsushita, M. (1979). Identification and distribution of the spinal and hypophyseal projection neurons in the paraventricular nucleus of the rat. A light and electron microscopic study with the horseradish peroxidase method. *Experimental Brain Research*, **35**, 315–331.

Jacobowitz, D. M., & Palkovits, M. (1974). Topographic atlas of catecholamine and acetylcholinesterase-containing neurons in the rat brain. I. Forebrain (Telencephalon, Diencephalon). *Journal of Comparative Neurology*, **157**, 13–28.

Jenkins, J. S., Mather, H. M., & Ang, V. (1980). Vasopressin in human cerebrospinal fluid. *Journal of Clinical Endocrinology and Metabolism*, **50**, 364–367.

Kendler, K. S., Weitzman, R. E., & Fisher, D. A. (1978). The effect of pain on plasma arginine vasopressin concentrations in man. *Clinical Endocrinology*, **8**, 89–94.

Kling, A. (1972). Effects of amygdalectomy on social-affective behavior in non-human primates. *In* B. E. Eleftheriou (Ed.), *The Neurobiology of the Amygdala*, pp. 511–536. New York: Plenum.

Kluever, H., & Bucy, P. (1939). Preliminary analysis of functions of the temporal lobe in monkeys. *Archives of Neurology and Psychiatry*, **42**, 979–1000.

Koizumi, K., & Yamashita, H. (1978). Influence of atrial stretch receptors on hypothalamic neurosecretory neurones. *Journal of Physiology*, **285**, 341–358.

König, J.F.R., & Klippel, R. A. (1963). *The rat brain. A stereotaxic atlas of the forebrain and lower parts of the brain stem*. Baltimore: Williams & Wilkins.

Kucera, P., & Favrod, P. (1979). Suprachiasmatic nucleus projection to mesencephalic central grey in the woodmouse (Apodemus sylvaticus L.). *Neuroscience*, **4**, 1705–1716.

Legait, H., & Legait, E. (1957). Les voies extra-hypophysaires des noyaux neurosécrétoires hypothalamiques chez les batraciens et les reptiles. *Acta Anatomica*, **30**, 429–443.

Legros, J.-J., & Gilot, P. (1979). Vasopressin and memory in the human. *In* A. M. Gotto, Jr., E. J. Peck, Jr., & A. E. Boyd, III (Eds.), *Brain Peptides: A New Endocrinology*, pp. 347–364. Amsterdam: Elsevier North Holland.

Ljungdahl, Å., Hökfelt, T., & Nilsson, G. (1978). Distribution of substance P-like immunoreactivity in the central nervous system of the rat. I. Cell bodies and nerve terminals. *Neuroscience*, **3**, 861–943.

Loewy, A. D., & McKellar, S. (1980). The neuroanatomical basis of central cardiovascular control. *Federation Proceedings*, **39**, 2495–2503.

Miczek, K. A., & Grossmann, S. P. (1972). Effects of septal lesions on inter- and intraspecific aggression in rats. *Journal of Comparative and Physiological Psychology*, **79**, 37–45.

Miley, W. M., & Baenninger, R. (1972). Inhibition and facilitation of interspecies aggression in septal lesioned rats. *Physiology and Behavior*, **9**, 379–384.

Möhring, J., Arbogast, R., Düsing, R., Glänzer, K., Kintz, J., Liard, J.-F., Maciel, J. A., Montani, J. P., & Schoun, J. (1980). Vasopressor role of vasopressin in hypertension. *In* W.

Wuttke, A. Weindl, K. Voigt, & R. R. Dries (Eds.), *Ferring Symposium on Brain and Pituitary Peptides*, pp. 157–167. Basel: Karger.

Moore, R. Y. (1973). Retinohypothalamic projection in mammals: A comparative study. *Brain Research, 49,* 403–409.

Moore, R. Y., & Eichler, V. B. (1972). Loss of a circadian adrenal corticosterone rhythm following suprachiasmatic lesions in the rat. *Brain Research, 42,* 201–206.

Moore, R. Y., & Lenn, N. J. (1972). A retinohypothalamic projection in the rat. *Journal of Comparative Neurology, 146,* 1–14.

Morris, R., Salt, T. E., Sofroniew, M. V., & Hill, R. G. (1980). Actions of microiontophoretically applied oxytocin, and immunohistochemical localization of oxytocin, vasopressin and neurophysin in the rat caudal medulla. *Neuroscience Letters, 18,* 163–168.

Nieuwenhuys, R., Voogd, J., & van Huijzen, C. (1979). *The Human Central Nervous System: A Synopsis and Atlas.* Berlin: Springer.

Nishino, H., Koizumi, K., & Brooks, C. McC. (1976). The role of suprachiasmatic nuclei of the hypothalamus in the production of circadian rhythm. *Brain Research, 112,* 45–59.

Nosaka, S., Yamamoto, T., & Yasunaga, K. (1979). Localization of vagal cardioinhibitory preganglionic neurons within rat brain stem. *Journal of Comparative Neurology, 186,* 79–92.

Oades, R. D. (1979). Search and attention: Interactions of the hippocampal–septal axis, adrenocortical and gonadal hormones. *Neuroscience and Biobehavioral Reviews, 3,* 31–48.

Ono, T., Nishino, H., Sasaka, K., Muramoto, K., Yano, I., & Simpson, A. (1978). Paraventricular nucleus connections to spinal cord and pituitary. *Neuroscience Letters, 10,* 141–146.

Palkovits, M., & Jacobowitz, D. M. (1974). Topographic atlas of catecholamine and acetylcholinesterase-containing neurons in the rat brain. II. Hindbrain (Mesencephalon, Rhombencephalon). *Journal of Comparative Neurology, 157,* 29–42.

Palkovits, M., Mezey, E., & Záborszky, L. (1979). Neuroanatomical evidences for direct neural connections between the brain stem baroreceptor centers and the forebrain areas involved in the neural regulation of blood pressure. *In* P. Meyer & H. Schmitt (Eds.), *Nervous System and Hypertension*, pp. 18–30. Paris: Wiley Flammarion.

Palkovits, M., & Záborszky, L. (1977). Neuroanatomy of central cardiovascular control. Nucleus tractus solitarii: Afferent and efferent neuronal connections in relation to the baroreceptor reflex arc. *In* W. De Jong, A. P. Provoost, & A. P. Shapiro (Eds.), *Hypertension and Brain Mechanisms: Progress in Brain Research,* Vol. 47, pp. 9–34. Amsterdam: Elsevier.

Palkovits, M., Záborszky, L., & Ambach, G. (1974). Accessory neurosecretory cell groups in the rat hypothalamus. *Acta Morphologica Academiae Scientiarum Hungaricae, 22,* 21–33.

Parry, H. B., & Livett, B. G. (1973). A new hypothalamic pathway to the median eminence containing neurophysin and its hypertrophy in sheep with natural scrapie. *Nature, 242,* 63–65.

Pavel, S. (1971). Evidence for the ependymal origin of arginine vasotocin in the bovine pineal gland. *Endocrinology, 89,* 613–614.

Peterson, G. M., Watkins, W. B., & Moore, R. Y. (1980). The suprachiasmatic hypothalamic nuclei of the rat. VI. Vasopressin neurons and circadian rhythmicity. *Behavioral and Neural Biology, 29,* 236–245.

Peterson, R. P. (1966). Magnocellular neurosecretory centers in the rat hypothalamus. *Journal of Comparative Neurology, 128,* 181–190.

Pfeifer, W. D., & Bookin, H. B. (1978). Vasopressin antagonizes retrograde amnesia in rats following electroconvulsive shock. *Pharmacology, Biochemistry and Behavior, 9,* 261–263.

Pickering, B. T. (1976). The molecules of neurosecretion: Their formation, transport, and release. *In* M. A. Corner & D. F. Swaab (Eds.), *Perspectives in Brain Research: Progress in Brain Research,* Vol. 45, pp. 161–179. Amsterdam: Elsevier.

Pittman, Q. J., Blume, H. W., & Renaud, L. P. (1978a). Electrophysiological indications that individual hypothalamic neurons innervate both median eminence and neurohypophysis. *Brain Research, 157,* 364–368.

Pittman, Q. J., Blume, H. W., & Renaud, L. P. (1978b). Afferent and efferent connections of putative peptidergic neurons of the paraventricular nucleus (PVN). *Society for Neuroscience Abstracts*, **4**, 352.

Pittman, Q. J., Blume, H. W., & Renaud, L. P. (1981). Connections of the hypothalamic paraventricular nucleus with the neurohypophysis, median eminence, amygdala, lateral septum and midbrain periaqueductal gray: An electrophysiological study in the rat. *Brain Research*, **215**, 15–28.

Rapoport, S. I., Klee, W. A., Pettigrew, K. D., & Ohno, K. (1980). Entry of opioid peptides into the central nervous system. *Science*, **207**, 84–86.

Reaves, T. A., & Hayward, J. N. (1979). Immunocytochemical identification of the vasopressinergic and oxytocinergic neurons in the hypothalamus of the cat. *Cell and Tissue Research*, **196**, 117–122.

Reinharz, A. C., Czernichow, P., & Vallotton, M. B. (1974). Neurophysin-like protein in bovine pineal gland. *Journal of Endocrinology*, **62**, 35–44.

Renaud, L. P. (1976). Influence of amygdala stimulation on the activity of identified tuberoinfundibular neurones in the rat hypothalamus. *Journal of Physiology*, **260**, 237–252.

Rexed, B. (1952). The cytoarchitectonic organization of the spinal cord in the cat. *Journal of Comparative Neurology*, **96**, 415–496.

Robinson, A. G., & Zimmerman, E. A. (1973). Cerebrospinal fluid and ependymal neurophysin. *Journal of Clinical Investigation*, **52**, 1260–1267.

Roche, K. E., & Leshner, A. I. (1979). ACTH and vasopressin treatments immediately after a defeat increase future submissiveness in male mice. *Science*, **204**, 1343–1344.

Rosenbloom, A. A., & Fisher, D. A. (1975). Arginine vasotocin in the rabbit subcommissural organ. *Endocrinology*, **96**, 1038–1039.

Rusak, B., & Zucker, I. (1979). Neural regulation of circadian rhythms. *Physiological Reviews*, **59**, 449–526.

Sakai, K., Touret, D., Salvert, D., Leger, L., & Jouvet, M. (1977). Afferent projections to the cat locus coeruleus as visualized by the horseradish peroxidase technique. *Brain Research*, **119**, 21–41.

Saper, C. B., Loewy, A. D., Swanson, L. W., & Cowan, W. M. (1976). Direct hypothalamo–autonomic connections. *Brain Research*, **117**, 305–312.

Sawyer, W. H. (1977). Evolution of active neurohypophysial principles among the vertebrates. *American Zoologist*, **17**, 727–737.

Scharrer, E. (1951). Neurosecretion. A relationship between the paraphysis and the paraventricular nucleus in the garter snake (Thamnophis sp.). *Biological Bulletin*, **1**, 106–113.

Schober, F. (1978). Darstellung der neurosekretorischen hypothalamo-rhombenzephalen Verbindung bei der Ratte durch retrograden axonalen Transport von Meerrettich-Peroxidase. *Acta Biologica et Medica Germanica*, **37**, 165–167.

Sherlock, D. A., Field, P. M., & Raisman, G. (1975). Retrograde transport of horseradish peroxidase in the magnocellular neurosecretory system of the rat. *Brain Research*, **88**, 403–414.

Silverman, A. J., & Zimmerman, E. A. (1975). Ultrastructural immunocytochemical localization of neurophysin and vasopressin in the median eminence and posterior pituitary of the guinea pig. *Cell and Tissue Research*, **159**, 291–301.

Sofroniew, M. V. (1979). Immunoreactive β-Endorphin and ACTH in the same neurons of the hypothalamic arcuate nucleus in the rat. *American Journal of Anatomy*, **154**, 283–289.

Sofroniew, M. V. (1980). Projections from vasopressin, oxytocin, and neurophysin neurons to neural targets in the rat and human. *Journal of Histochemistry and Cytochemistry*, **28**, 475–478.

Sofroniew, M. V., & Glasmann, W. (1981). Golgi-like immunoperoxidase staining of hypothalamic magnocellular neurons that contain vasopressin, oxytocin or neurophysin in the rat. *Neuroscience*, **6**, 619–643.

Sofroniew, M. V., Madler, M., Müller, O. A., & Scriba, P. C. (1978). A method for the consistent production of high quality antisera to small peptide hormones. *Fresenius Zeitschrift für Analytische Chemie*, **290**, 163.

Sofroniew, M. V., & Schrell, U. (1980). Hypothalamic neurons projecting to the rat caudal medulla oblongata, examined by immunoperoxidase staining of retrogradely transported horseradish peroxidase. *Neuroscience Letters*, **19**, 257–263.

Sofroniew, M. V., & Schrell, U. (1981). Evidence for a direct projection from oxytocin and vasopressin neurons in the hypothalamic paraventricular nucleus to the medulla oblongata: Immunohistochemical visualization of both the horseradish peroxidase transported and the peptide produced by the same neurons. *Neuroscience Letters*, **22**, 211–217.

Sofroniew, M. V., Schrell, U., Glasmann, W., Weindl, A., & Wetzstein, R. (1980). Hypothalamic accessory magnocellular vasopressin, oxytocin and neurophysin neurons projecting to the neurohypophysis in the rat. *Society for Neuroscience Abstracts*, **6**, 456.

Sofroniew, M. V., & Weindl, A. (1978a). Projections from the parvocellular vasopressin- and neurophysin-containing neurons of the suprachiasmatic nucleus. *American Journal of Anatomy*, **153**, 391–429.

Sofroniew, M. V., & Weindl, A. (1978b). Extrahypothalamic neurophysin-containing perikarya, fiber pathways and fiber clusters in the rat brain. *Endocrinology*, **102**, 334–337.

Sofroniew, M. V., & Weindl, A. (1980). Identification of parvocellular vasopressin and neurophysin neurons in the suprachiasmatic nucleus of a variety of mammals including primates. *Journal of Comparative Neurology*, **193**, 659–675.

Sofroniew, M. V., Weindl, A., Schinko, I., & Wetzstein, R. (1979a). The distribution of vasopressin-, oxytocin-, and neurophysin-producing neurons in the guinea pig brain. I. The classical hypothalamo–neurohypophyseal system. *Cell and Tissue Research*, **196**, 367–384.

Sofroniew, M. V., Weindl, A., Schrell, U., & Wetzstein, R. (1981). Immunohistochemistry of vasopressin, oxytocin and neurophysin in the hypothalamus and extrahypothalamic regions of the human and primate brain. *Acta Histochemica*, (Supplementum XXIV), 79–95.

Sofroniew, M. V., Weindl, A., & Wetzstein, R. (1977). Immunoperoxidase staining of vasopressin in the rat median eminence following adrenalectomy and steroid substitution. *Acta Endocrinologica*, (Supplementum 212), 72. (Abstract)

Sokol, H. W., Zimmerman, E. A., Sawyer, W. H., & Robinson, A. G. (1976). The hypothalamic–neurohypophysial system of the rat: Localization and quantitation of neurophysin by light microscopic immunocytochemistry in normal rats and in Brattleboro rats deficient in vasopressin and a neurophysin. *Endocrinology*, **98**, 1176–1188.

Steinbusch, H.W.M. (1981). Distribution of serotonin-immunoreactivity in the central nervous system of the rat-cell bodies and terminals. *Neuroscience*, **6**, 557–618.

Stephan, F. K., & Zucker, I. (1972). Circadian rhythms in drinking behavior and locomotor activity of rats are eliminated by hypothalamic lesions. *Proceedings of the National Academy of Sciences, U.S.A.*, **69**, 1583–1586.

Stephan, F. K., & Zucker, I. (1974). Endocrine and neural mediation of the effects of constant light on water intake of rats. *Neuroendocrinology*, **14**, 44–60.

Sterba, G. (1974). Ascending neurosecretory pathways of the peptidergic type. *In* F. Knowles & L. Vollrath (Eds.), *Neurosecretion—the Final Neuroendocrine Pathway*, pp. 38–47. Berlin: Springer.

Sternberger, L. A. (1974). *Immunocytochemistry*. Englewood Cliffs, N.J.: Prentice-Hall.

Stillman, M. A., Recht, L. D., Rosario, S. L., Seif, S. M., Robinson, A. G., & Zimmerman, E. A. (1977). The effects of adrenalectomy and glucocorticoid replacement on vasopressin and vasopressin–neurophysin in the zona externa of the median eminence of the rat. *Endocrinology*, **101**, 42–49.

Swaab, D. F., Nijveldt, F., & Pool, C. W. (1975). Distribution of oxytocin and vasopressin in the rat supraoptic and paraventricular nucleus. *Journal of Endocrinology*, **67**, 461–462.

Swanson, L. W. (1977). Immunohistochemical evidence for a neurophysin-containing autonomic pathway arising in the paraventricular nucleus of the hypothalamus. *Brain Research*, **128**, 346–353.

Swanson, L. W., & Kuypers, H.G.J.M. (1980). The paraventricular nucleus of the hypothalamus: Cytoarchitectonic subdivisions and organization of projections to the pituitary, dorsal vagal complex, and spinal cord as demonstrated by retrograde fluorescence double-labeling methods. *Journal of Comparative Neurology*, **194**, 555–570.

Swanson, L. W., & McKellar, S. (1979). The distribution of oxytocin- and neurophysin-stained fibers in the spinal cord of the rat and monkey. *Journal of Comparative Neurology*, **188**, 87–106.

Tanaka, M., de Kloet, E. R., de Wied, D., & Versteeg, D.H.G. (1977). Arginine⁸-vasopressin affects catecholamine metabolism in specific brain nuclei. *Life Sciences*, **20**, 1799–1808.

Thompson, R. (1978). Localization of a "passive avoidance memory system" in the white rat. *Physiological Psychology*, **6**, 263–274.

Tindal, J. S., Knaggs, G. S., & Turvey, A. (1967). The afferent path of the milk-ejection reflex in the brain of the guinea pig. *Journal of Endocrinology*, **38**, 337–349.

Tindal, J. S., Knaggs, G. S., & Turvey, A. (1968). Preferential release of oxytocin from the neurohypophysis after electrical stimulation of the afferent path of the milk-ejection reflex in the brain of the guinea pig. *Journal of Endocrinology*, **40**, 205–214.

Van Bogaert, A., Wellens, D., Van Bogaert, P. P., Martin, J. J., & De Wilde, A. (1976). Characteristics of hypotension elicited by electrical stimulation of the lateral hypothalamus in anesthetized dogs. *Archives internationales de Physiologie et de Biochimie*, **84**, 35–43.

van den Pol, A. N., & Powley, T. (1979). A fine-grained anatomical analysis of the role of the rat suprachiasmatic nucleus in circadian rhythms of feeding and drinking. *Brain Research*, **160**, 307–326.

Vandesande, F., De Mey, J., & Dierickx, K. (1974). Identification of neurophysin producing cells. I. The origin of the neurophysin-like substance-containing nerve fibres of the external region of the median eminence of the rat. *Cell and Tissue Research*, **151**, 187–200.

Vandesande, F., & Dierickx, K. (1975). Identification of the vasopressin producing and of the oxytocin producing neurons in the hypothalamic magnocellular neurosecretory system of the rat. *Cell and Tissue Research*, **164**, 153–162.

Vandesande, F., & Dierickx, K. (1979). The activated hypothalamic magnocellular neurosecretory system and the one neuron—one neurohypophysial hormone concept. *Cell and Tissue Research*, **200**, 29–33.

Vandesande, F., Dierickx, K., & De Mey, J. (1975a). Identification of the vasopressin–neurophysin II and the oxytocin–neurophysin I producing neurons in the bovine hypothalamus. *Cell and Tissue Research*, **156**, 189–200.

Vandesande, F., Dierickx, K., & De Mey, J. (1975b). Identification of the vasopressin–neurophysin producing neurons of the rat suprachiasmatic nuclei. *Cell and Tissue Research*, **156**, 377–380.

Vandesande, F., Dierickx, K., & De Mey, J. (1977). The origin of the vasopressinergic and oxytocinergic fibers of the external region of the median eminence of the rat hypophysis. *Cell and Tissue Research*, **180**, 443–452.

van Dyke, H. B., Chow, B. F., Greep, R. O., & Rothen, A. (1942). The isolation of a protein from the pars neuralis of the ox pituitary with constant oxytocic, pressor and diuresis-inhibiting activities. *Journal of Pharmacology*, **74**, 190–209.

van Leeuwen, F. W., & Swaab, D. F. (1977). Specific immunoelectronmicroscopic localization of vasopressin and oxytocin in the neurohypophysis of the rat. *Cell and Tissue Research*, **177**, 493–501.

van Leeuwen, F. W., Swaab, D. F., & de Raay, C. (1978). Immunoelectronmicroscopic localization of vasopressin in the rat suprachiasmatic nucleus. *Cell and Tissue Research*, **193**, 1–10.

van Wimersma Greidanus, Tj. B., Croiset, G., Bakker, E., & Bouman, H. (1979). Amygdaloid lesions block the effect of neuropeptides (vasopressin, ACTH₄₋₁₀) on avoidance behavior. *Physiology and Behavior*, **22**, 291–295.

van Wimersma Greidanus, Tj. B., & Versteeg, D.H.G. (1980). Neurohypophysial hormones; their role in endocrine function and behavioral homeostasis. *In* Ch.B. Nemeroff & A. J. Dunn (Eds.), *Behavioral Neurobiology*. Jamaica, N.Y.: Spectrum.

Versteeg, D.H.G., de Kloet, E. R., van Wimersma Greidanus, Tj. B., & de Wied, D. (1979). Vasopressin modulates the activity of catecholamine-containing neurons in specific brain regions. *Neuroscience Letters*, **11**, 69–73.

Watson, S. J., Richard, C. W., & Barchas, J. D. (1978). Adrenocorticotropin in rat brain: Immunocytochemical localization in cells and axons. *Science*, **200**, 1180–1182.

Weindl, A., & Sofroniew, M. V. (1976). Demonstration of extrahypothalamic peptide secreting neurons. A morphologic contribution to the investigation of psychotropic effects of neurohormones. *Pharmakopsychiatrie*, **9**, 226–234.

Weindl, A., & Sofroniew, M. V. (1978). Neurohormones and circumventricular organs. *In* D. E. Scott, G. P. Kozlowski, & A. Weindl (Eds.), *Brain–Endocrine Interaction III: Neural Hormones and Reproduction*, pp. 117–137. Basel: Karger.

Yamashita, H., & Koizumi, K. (1979). Influence of carotid and aortic baroreceptors on neurosecretory neurons in supraoptic nuclei. *Brain Research*, **170**, 259–277.

Yates, F. E., & Maran, J. W. (1974). Stimulation and inhibition of adrenocorticotropin release. *In* R. O. Greep, & E. B. Astwood (Eds.), *Handbook of Physiology, Section 7: Endocrinology IV*, pp. 367–404. Washington, D.C.: American Physiological Society.

Zemlan, F. P., Kow, L.-M., Morrell, J. I., & Pfaff, D. W. (1979). Descending tracts of the lateral columns of the rat spinal cord: A study using the horseradish peroxidase and silver impregnation techniques. *Journal of Anatomy*, **128**, 489–512.

Zimmerman, E. A. (1976). Localization of hypothalamic hormones by immunocytochemical techniques. *In* L. Martini, & W. F. Ganong (Eds.), *Frontiers in Neuroendocrinology* (Vol. 4), pp. 25–62. New York: Raven.

Zimmerman, E. A., & Antunes, J. L. (1976). Organization of the hypothalamic–pituitary system: Current concepts from immunohistochemical studies. *Journal of Histochemistry and Cytochemistry*, **24**, 807–815.

Zimmerman, E. A., Carmel, P. W., Husain, M. K., Ferin, M., Tannenbaum, M., Frantz, A. G., & Robinson, A. G. (1973). Vasopressin and neurophysin: High concentrations in monkey hypophyseal portal blood. *Science*, **182**, 925–927.

CHAPTER 17

Avoidance Conditioning and Endocrine Function in Brattleboro Rats

William H. Bailey and Jay M. Weiss

The Brattleboro strain of rat, genetically defective so that it produces little or no vasopressin, was used to examine the importance of endogenous vasopressin in modulating avoidance performance. These rats exhibit hypothalamic diabetes insipidus: Rats homozygous for this trait (DI) have a total deficiency of vasopressin, whereas heterozygous rats (HE) demonstrate only a partial vasopressin deficiency.

In this chapter, we review the physiological and hormonal characteristics of Brattleboro rats that are relevant to the interpretation of their performance on avoidance tasks. Besides the genetically determined deficiency of vasopressin in DI rats, other endocrine abnormalities are known to exist, including: (1) reduced pituitary–adrenal response to mild stressors; (2) increased release of oxytocin into the circulation; (3) physiological and structural changes in kidney that develop as a consequence of chronic polyuria; and (4) a deficiency of growth hormone. These abnormalities are reversible with vasopressin replacement therapy except for the deficiency of growth hormone, which therefore may also be genetic in origin.

On avoidance tasks, DI rats are reported to perform somewhat more poorly than HE rats. Homozygous Brattleboro rats acquire an active shuttlebox avoidance slightly more slowly than do HE rats, a finding confirmed in our experiments. We also found, as have others, that in passive

371

avoidance DI rats exhibit shorter passive avoidance latencies (poorer performance) than do HE rats. We did not find, however, that DI rats were totally unable to demonstrate passive avoidance 24 hr or more after the training trial, as reported by other investigators (Bohus, van Wimersma Greidanus, & de Wied, 1975; de Wied, Bohus & van Wimersma Greidanus, 1975). Differences in activity level are not the cause of this difference in passive avoidance, as HE and DI rats exhibited similar activity and behavior in open-field tests. Reduced activity and heightened fearfulness of Brattleboro rats (both DI and HE) observed during open-field tests does, however, provide an explanation for the longer latencies (superior performance) displayed by these animals in comparison to normal Long-Evans (LE) rats on passive avoidance tests. The poor passive avoidance of DI rats relative to HE rats in these studies is probably related to their deficiency of vasopressin, but other factors secondary to a deficiency of vasopressin, such as a reduced pituitary–adrenal response to stress, an increased release of oxytocin, and altered electrolyte balance, cannot be discounted. In a related study with Brattleboro rats, we demonstrated that the behavioral effect of adrenocorticotropin (ACTH) on passive avoidance cannot be attributed to the release of endogenous vasopressin, as the $ACTH_{4-10}$ analog did significantly affect the passive avoidance of DI rats that are entirely devoid of vasopressin. The differences between Brattleboro and normal LE rats may involve genetic discontinuities unrelated to those affecting vasopressin; it is also possible that such behavioral characteristics of the Brattleboro strain reflect nutritional inadequacies, particularly during early life, that are associated with deficiencies of vasopressin.

I. INTRODUCTION

The pioneering work of Selye (1950) on the role of the pituitary–adrenal system as part of an organism's response to stress has stimulated psychologists and neuroendocrinologists to investigate whether pituitary and adrenal hormones affect the behavior as well as the physiology of animals in stressful situations. Research on this problem has focused on the relationship of various pituitary–adrenal hormones to the performance of rats on a variety of aversively motivated conditioning tasks. Of all the hormones associated with the pituitary and the adrenal glands, particular attention has been given to adrenocorticosteroids and adrenomedullary catecholamines (not to be reviewed in this chapter) and to the pituitary peptides, adrenocorticotropic hormone (ACTH), melanocyte-stimulating hormone (MSH), and vasopressin.

Studies of these pituitary peptides have been quite consistent in showing that they improve, or at least support, behavior in aversively motivated tasks. For example, studies by de Wied and his colleagues have shown that the removal of the pituitary or the adenohypophysis alone severely impairs the acquisition of active avoidance responses, and that this behavioral impairment can be prevented by the administration of ACTH and MSH as well as a number of analogs of the ACTH molecule (de Wied, 1969; de Wied, Witter, & Lande, 1970; Greven & de Wied, 1973). Similarly, the rapid extinction of active and passive avoidance characteristic of hypophysectomized rats as well as the expected performance decline of normal rats during extinction can be prevented by the administration of these same peptides (Dempsey, Kastin, & Schally,

1972; de Wied, 1966; de Wied & Bohus, 1966; Guth, Seward, & Levine, 1971).

The foregoing data suggest that endogenous hormones of the anterior pituitary influence the learning and performance of aversively motivated behaviors. However, these studies by no means exclude a role for posterior pituitary hormones in aversively motivated behavior. Neurohypophysectomy, which does not remove the anterior pituitary but does remove the intermediate and posterior lobes of the pituitary, fails to affect acquisition of shuttlebox avoidance but does produce rapid extinction (de Wied, 1965). This effect could be due to a deficiency in the intermediate lobe peptide, MSH, or to a reduction in the posterior lobe hormones, oxytocin and vasopressin.

Without denying a possible role for MSH or oxytocin, the importance of vasopressin to normal extinction processes has been demonstrated by the finding that intraventricular injections of specific antibodies to vasopressin immediately after the training trial dramatically reduce performance on subsequent passive avoidance retention tests (van Wimersma Greidanus & de Wied, 1976). It is to the behavioral effects of the hormone, vasopressin, that the present chapter is directed.

Recent studies of rats of the Brattleboro strain have been carried out in the hope of shedding light on the role of vasopressin in avoidance responding. These rats, derived from Long-Evans (LE) stock, inherit hypothalamic diabetes insipidus; rats homozygous for this strain (DI) have a total deficiency of vasopressin, whereas heterozygous rats (HE) demonstrate only a partial vasopressin deficiency (Valtin, Sawyer, & Sokol, 1965). In this chapter, we will consider some of the behavioral characteristics of Brattleboro rats. The chapter begins with a review of the physiological and endocrine characteristics of Brattleboro rats to provide a proper background for the interpretation of behavioral performance of Brattleboro rats in avoidance tasks. It is suggested that all differences between DI and HE rats, or differences between DI and LE rats cannot be categorically attributed to the inability of DI rats to synthesize vasopressin. In the discussion of almost all behavioral studies of DI rats, it is implicitly assumed that any peculiarities in their performance relative to HE or normal rats reflects the absence of vasopressin at specific receptors, most likely in the central nervous system (CNS). Under this assumption, differences between DI rats and controls are attributable to what we define as a primary consequence of vasopressin deficiency. We point out in this review, however, that chronic vasopressin deficiency in the DI rat inevitably leads to alterations in electrolyte balance, which in turn affects physiological systems throughout the body, and the alterations in these physiological systems may account for some of the behavioral characteristics of DI rats. Such indirect physiological changes as loss of body potassium, increased oxytocin and mineralocorticoid secretion, decreased sensitivity of the adrenal cortex to ACTH, etc., we therefore define as secondary consequences of vasopressin deficiency. Indeed, it may well be that the far-reaching secondary consequences of the vasopressin deficiency of DI rats will prove to be of greater importance to the understanding of their avoidance capabilities than the primary deficiency of vasopressin itself. We shall also present behavioral experiments that evaluate the significance of the vasopressin deficiency of Brattleboro rats to performance on standard avoidance tasks.

II. PHYSIOLOGICAL AND HORMONAL CHARACTERISTICS OF BRATTLEBORO RATS RELEVANT TO THE INTERPRETATION OF AVOIDANCE EXPERIMENTS

The Brattleboro strain of rat is of LE ancestry and is studied by physiologists and endocrinologists because of the heritability, in this strain, of hypothalamic diabetes insipidus. The defect appears to be an autosomal recessive trait occurring with homozygosity of recessive genes at either locus of a single pair of loci, and may be associated in some pedigrees with albinism and the frequent occurrence of runts, stillbirths, and newborn mortality (Saul, Garrity, Benirschke, & Valtin, 1968). Assays of the hypothalamus, extrahypothalamic brain structures, and pituitary in DI rats homozygous for this trait indicate that these structures do not contain biologically active or immunoreactive vasopressin (Dogterom, Snijdewint, & Buijs, 1978; Miller & Moses, 1971; Valtin et al., 1965); in addition, their urine output is 16 times that of normal rats (Swaab, Boer, & Nolten, 1973). Heterozygous rats have levels of vasopressin in the hypothalamus and pituitary that are two-thirds to one-seventh of those found in normal rats, indicating a partial impairment in the synthesis (Valtin et al., 1965) and release of vasopressin (Miller & Moses, 1971; Moses & Miller, 1970). Despite the partial vasopressin defect in HE rats, their urine volume is only one-half to two times greater than normal, and serum osmolarities are not different from normal (Moses & Miller, 1970; Swaab et al., 1973).

It has been reported that replacement therapy with vasopressin is not quickly effective in restoring normal function in vasopressin-deficient rats, an observation that appears to compromise the view that a lack of vasopressin is the outstanding characteristic of DI rats. Harrington and Valtin (1965) reported that 28 days of vasopressin treatment was required before DI rats produced urine as concentrated as normal rats. The most plausible explanation for the refractoriness of DI rats to vasopressin therapy is that their kidneys have been damaged by a chronic deficiency of potassium, which occurs as a result of chronic polyuria. Möhring, Dauda, Haack, Homsy, Kohrs, and Möhring (1972) have shown that DI rats have below normal levels of potassium in serum (hypokalemia) and this produces kaliopenic nephropathy that increases in severity with age. Altered kidney function is also reflected by the high blood levels of renin and angiotensin II found in DI rats (Hoffman, Ganten, Schelling, Phillips, Schmid, & Ganten, 1978; Möhring, Kohrs, Möhring, Petri, Homsy, & Haack, 1978). This may in part account for the hypertension reported in DI rats (Hall, Ayachi, & Hall, 1973).

Although, in these investigations, no explanation was offered for the potassium deficiency of DI rats, it is almost certainly caused by the loss of salts associated with the large volumes of urine excreted and, perhaps, by changes in the extrarenal distribution of sodium, potassium, and water associated with the relative dominance of adrenal mineralocorticoids in the absence of vasopressin (Friedman, Scherrer, Nikashima, & Friedman, 1958). In rats with diabetes insipidus, hypokalemia, excessive renal loss of potassium (Möhring et al., 1972), slight elevations of sodium in plasma (Valtin & Schroeder, 1964), and diminished extracellular fluid volume (Harrington & Valtin, 1968) suggest aldosteronism or the relative dominance of similar mineralocorticoids (Tepperman, 1968). Nevertheless, in DI rats, the level of aldosterone in plasma is reduced (Möhring et al., 1978) and the levels of deoxycorticosterone and 18-hydroxydeoxycorticosterone in the adrenal are greatly diminished (Vinson, Goddard, & Whitehouse, 1977).

Although DI rats are not known to have any genetic hypothalamic–pituitary deficiency other than that affecting vasopressin, some studies suggest that a deficient release of, or response to, adenohypophyseal hormones may also be present in these animals. Most important to studies of avoidance behavior is the observation that Brattleboro rats as well as neurohypophysectomized rats show a diminished pituitary–adrenal response to weak, but not to strong stressors. For instance, DI rats show normal increases in plasma corticosterone after histamine, hypoxia, or large doses of ACTH, but show a diminished response to noise or to brief ether exposures (Yates, Russell, Dallman, Hedge, McCann, & Dhariwal, 1971). Arimura, Saito, Bowers, and Schally (1967) found no difference between HE and DI rats in their corticosterone response to ether, histamine, vasopressin, or acetylcholine, but in response to an intraperitoneal (i.p.) injection of saline, DI rats did not secrete as much corticosterone as did normals. McCann, Antunes-Rodrigues, Naller, and Valtin (1966) found that restraint and etherization for blood sampling elicited subnormal corticosterone responses from DI rats 1 min after the initiation of stress. Similar results were obtained on later testing even after dehydration was reduced with several weeks of pitressin tannate treatment (McCann et al., 1966).

Whereas these ·observations indicate a reduced adenohypophyseal responsiveness of Brattleboro rats, it appears likely that the change is directly related to a deficiency of vasopressin. Recent *in vitro* experiments have shown that the vasopressin deficiency of DI rats produces a decreased release of corticotropin releasing factor (CRF) and a reduced responsiveness to CRF. Both of these effects are reversible by the addition of small amounts of vasopressin (Buckingham & Leach, 1979a,b; Krieger & Liotta, 1977). Another factor contributing to the diminished corticosterone response of DI rats to some stressors is the reduced sensitivity of their adrenals to small quantities of ACTH. This, too, might be a primary consequence of a deficiency of vasopressin, but the finding that chronic administration of vasopressin, beginning at 4 days of age to both normal and DI rats, leads to similar corticosterone responses as adults (Wiley, Pearlmutter, & Miller, 1974) does not rule out the reduced adrenal responsiveness as a secondary consequence of vasopressin deficiency. However, it should be noted that, if both normal and DI rats are given chronic vasopressin treatment, they show a smaller corticosterone response to stress than do untreated rats, possibly because of habituation of the pituitary–adrenal axis to repeated stimulation from the injections. This would explain why McCann et al. (1966) failed to observe a reversal of the low steroid response of DI rats when they administered vasopressin, as they did not use a vasopressin-treated control group for comparison.

The extreme unlikelihood that the origin of the reduced pituitary–adrenal responsiveness of DI rats is attributable to a genetic defect in the pituitary–adrenal system itself rather than to primary or secondary effects of the vasopressin deficiency of DI rats is further underscored by the finding that deficiencies of vasopressin produced by neurohypophysectomy are also known to lead to a subnormal corticosteroid response to weak stresses (Arimura, Yamaguchi, Yoshimura, Imazeki, & Itoh, 1965; de Wied, 1968; Miller, Yueh-Chien, Wiley, & Hewitt, 1974).

In addition to the lack of vasopressin, DI rats show pituitary levels of oxytocin that are about one-third that of normal animals. However, this decrease is brought about by the continual release of oxytocin from the pituitary in response to the mild but chronic state of dehydration found in these animals, and it is reversed by vasopressin therapy (Valtin et al., 1965). Elevated levels of oxytocin are also found in the plasma and several brain regions of DI rats as compared with HE rats (Dogterom, van Wimersma Greidanus, & Swaab, 1977; Dogterom et al., 1978). This acceleration of oxytocin synthesis and release, apparently a secondary effect of vasopressin deficiency of DI rats, may prove to be relevant to the avoidance performance of these rats. Recent studies have shown that intraventricular or peripheral administration of oxytocin produces effects on extinction of passive avoidance and pole jumping that are opposite to those obtained with the administration of vasopressin (Bohus, Kovács, & de Wied, 1978a; Bohus, Urban, van Wimersma Greidanus, & de Wied, 1978b). In addition, intraventricular injections of antibodies to vasopressin or oxytocin have opposite effects on the extinction of passive avoidance (de Wied & Bohus, 1979).

A further pituitary deficiency of the DI rat that may be of significance to avoidance studies is that of growth hormone. DI rats are reported to be smaller than similar aged normal or HE rats (Valtin et al., 1965). Chronic vasopressin therapy to DI rats for 2 months after weaning increased body weight relative to peanut oil-injected DI rats but did not bring their weight up to normal nor did it increase tail length. Growth hormone treatment, however, did normalize both body weight and tail length after 5 months, without affecting water intake (Sokol, 1973). As thyroid function is reported to be normal in DI rats (Galton, Valtin, & Johnson, 1966), a genetic abnormality of growth hormone secretion may be responsible for the smaller size of DI rats. Another possibility is that pre- and postnatal malnutrition associated with an unusual drain of the mother's metabolic resources and body salts by DI rats and, to a lesser extent, by the HE littermates could produce irreversible stunting of growth and lowered growth hormone levels (Nitzan & Wilber, 1974). Regardless of the cause of the stunted growth of DI rats, there is, as yet, no evidence that the defect is the result of the deficit in vasopressin, although such evidence may in time be found.

To summarize this overview of the endocrine physiology of the Brattleboro strain, the evidence to date indicates that the principal endocrine abnormality found in DI rats is a deficit in vasopressin. This causes DI rats to show secondary increases and decreases in the activity of other hormonal systems, a fact not often appreciated by those who claim that the behavioral characteristics of these rats are solely attributable to the deficiency of vasopressin per se. A reduced pituitary–adrenal response to mild stress seems to be characteristic of DI rats, and this may be attributed to the reduced responsiveness of hypothalamic, pituitary, and adrenal tissues in the absence of vasopressin. DI rats are hypo-kalemic; this and other alterations in electrolyte distribution are in part related to polyuria and to the relative dominance of mineralocorticoid hormones and, in the absence of vasopressin, affect the kidney and other tissues. Also, the synthesis and release of oxytocin is increased in DI rats. All of the foregoing changes can be normalized to varying extents by vasopressin administration. Finally, DI rats show stunted growth, which is likely to be caused by a deficiency of growth hormone. This, however, may be due to a genetic factor. HE rats often have been employed as normal controls for DI rats in physiological studies, even though they are known to have a slight defect in the synthesis and release of vasopressin. The use of the HE rat as a control in behavioral studies is perhaps even more important as it is not known what genetic factors other than those responsible for a vasopressin deficiency may be peculiar to this strain. Thus, until further data are available on the significance of a slight disturbance in vasopressin synthesis capability or other characteristics of the HE rat, it is desirable to use both HE and normal rats as controls whenever possible.

III. BEHAVIORAL STUDIES OF BRATTLEBORO RATS

A. Active Avoidance

Early studies suggested that manipulation of pituitary hormones affected the acquisition and extinction of shuttlebox avoidance in rats. The work of de Wied and others pointed to an involvement of anterior pituitary hormones in the acquisition of shuttlebox avoidance based upon the slow acquisition of this conditioned avoidance response by adenohypophysectomized and hypophysectomized rats, and the normal acquisition performance by neurohypophysectomized rats deficient in vasopressin and oxytocin (see Introduction). Although these latter neurohypophyseal hormones seemed not to be necessary for the acquisition of shuttlebox avoidance, vasopressin appeared to contribute to the maintenance of avoidance responding, as evidenced by the observation that neurohypophysectomized rats showed rapid extinction of the shuttle response, which is reversed by injections of pitressin tannate and lysine vasopressin (de Wied, 1965).

Considering the aforementioned findings, one might expect that DI Brattleboro rats with a total deficiency of vasopressin would, like neurohypophysectomized rats, show normal acquisition of shuttlebox avoidance, but accelerated extinction. However, DI rats are reported to learn this task more slowly than either HE or normal Wistar rats (Bohus et al., 1975). Similar findings have been presented in two other studies (Celestian, Carey, & Miller, 1975; Miller, Barranda, Dean, & Brush, 1976). In our own studies two of eight DI rats were able to reach a criterion of 80% avoidance on 3 consecutive days during 14 days of training (10 trials/day), whereas most HE rats (five of seven) reached this criterion. The DI rats did, however, make the same number of avoidance responses (Fig. 1), consecutive responses, and intertrial responses during the first 100 trials (before any animal reached criterion) as did HE rats. The reason that more DI rats did not reach the 80% avoidance criterion was because their performance deteriorated, did not continue to improve, or became quite variable after 10 or 11 days. A decline in the performance of DI rats toward the end of training was also observed by Celestian et al. (1975) and Miller et al. (1976). The drop in performance may indicate that the less robust and weaker DI rats (muscular weakness is a symptom of potassium deficiency and adrenocortical insufficiency) were not able to withstand the exertion and stress associated with daily active avoidance training over long periods of time.

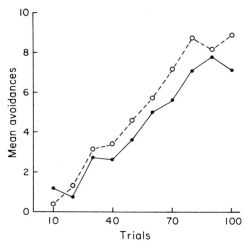

Fig. 1. Acquisition of conditioned avoidance by Brattleboro rats that have a total (DI, indicated by ●——●) or partial (HE, depicted by o---o) hereditary hypothalamic deficiency of vasopressin. For DI, $n = 8$; for HE, $n = 7$. The conditioned stimulus (CS) was a 1000.0 Hz tone of 80.0 dB intensity presented for 5 sec prior to a 0.6 mA shock. Ten trials were given each day.

Other data on the performance of DI rats during extinction prove to be contradictory. Bohus et al. (1975) claimed that avoidance responding by DI rats declined significantly during extinction of shuttlebox avoidance, a finding compatible with the results of studies with neurohypophysectomized rats (de Wied, 1965). In contrast, Celestian et al. (1975) reported that DI rats made significantly more avoidance responses during extinction than did either HE or normal LE rats. Miller et al. (1976) observed that the lower rate of avoidance by DI rats in comparison with HE or LE rats during extinction was confounded by differences in performance between DI, HE, and LE rats at the end of training. When extinction performance was adjusted for avoidance proficiency at the end of training by covariance analysis, no significant differences between DI, HE, or LE rats were found. Such considerations may explain why, in the report by Bohus et al. (1975), the differences between DI and HE rats during extinction of shuttlebox avoidance were large as compared with the differences between these rats observed during extinction of pole-jump avoidance. In the former task, there were marked differences between the acquisition performance of DI and HE rats, but in the latter task only minimal differences in acquisition performance were noted.

Therefore, to summarize the available evidence, it appears that the acquisition of active avoidance by DI rats is somewhat impaired relative to HE rats. However, because in large part this impairment appears most pronounced after 10 or 11 days of training, it is likely that fatigue and weakness resulting from hypokalemia and mild adrenocortical insufficiency contribute to this effect. The data are less consistent with regard to the performance of DI rats during extinction of active avoidance, but for the most part indicate that DI rats do not extinguish significantly faster than do HE rats.

B. Passive Avoidance

Effects of pituitary hormones are found in passive avoidance as well as in active avoidance tasks. Exogenously administered $ACTH_{4-10}$ and vasopressin prolong extinction (i.e., improve performance) of passive avoidance responding (Bohus, Gispen, & de Wied, 1973; Thompson & de Wied, 1973). Conversely, the hypophysectomized rat extinguishes faster than does the normal rat in passive avoidance (Weiss, McEwen, Silver, & Kalkut, 1969, 1970). As DI rats are similar to a greater or lesser extent to hypophysectomized rats in that they have reductions of vasopressin, ACTH, or corticosteroids, and probably growth hormone, it might be expected that DI rats would show rapid extinction of passive avoidance.

It has indeed been found that DI rats show poorer retention of passive avoidance than do normal rats (Bohus et al., 1975; de Wied et al., 1975). However, the magnitude of the reported effects has been surprising. These studies reported that DI rats did not exhibit any passive avoidance whatsoever 24 hr or more after training. This absolute avoidance deficit of DI rats was observed despite the use of footshocks of high intensity or long duration, whereas HE rats demonstrated a marked passive avoidance response for as long as 5 days after the training trial. These data were interpreted to indicate a complete impairment of memory on the part of vasopressin deficient DI rats (de Wied et al., 1975).

Using a slightly different testing procedure, we have not found such severe passive avoidance deficits in DI rats of either sex (Bailey & Weiss, 1979). In our experiment, 7-month-old DI and HE rats were tested for passive avoidance in the step-through apparatus described by Weiss et al. (1970) and shown in Fig. 2. On Day 1, the rat was administered a 2.0 sec, 1.0 mA ac scrambled grid shock after entering the large compartment. Avoidance of the large compartment was measured by the time required for the rat to reenter the compartment on single tests of passive avoidance given on Days 1, 2, 3, and 4. On Day 3 and Day 4 of testing, the sawdust under the grid floor was removed, the clear Plexiglas walls were covered with white paper, and the grid floor was covered with white paper over cardboard, a procedure used to increase the sensitivity of the passive avoidance test to differences in hormonal status by reducing fear stimuli (Weiss et al., 1969).

Under these testing conditions, postshock passive avoidance latencies signif-icantly greater than those observed on the training trial were observed in DI

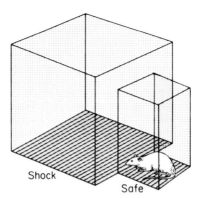

Fig. 2.   Step-through apparatus used in passive avoidance tests. The rat is placed into the safe compartment on the training trial. After entering the shock compartment, the rat is given a single footshock and removed 15 sec later. On subsequent tests, the latency to reenter the shock compartment is taken as the measure of retention of the passive avoidance response. (From Weiss et al., 1970.)

and HE groups of each sex on all 4 days of testing (Table 1). The data on the first 2 days of testing show that DI rats, although totally deficient in vasopressin, demonstrate passive avoidance behavior. The passive avoidance of DI rats is, nonetheless, overall significantly poorer than that demonstrated by HE rats.

The short passive avoidance latencies of DI rats cannot be explained by the hypothesis that their smaller size relative to HE rats is associated with a reduced sensitivity to shock, as lighter rats have been found to be more sensitive to footshock than are heavier ones (Gibbs, Sechzer, Smith, & Weiss, 1973). Nonetheless, to be sure that weight differences per se could not explain the differences in passive avoidance observed here, we tested two groups of normal LE female rats, one similar in weight to DI females, the other similar in weight to HE females. No significant differences, however, were observed in the passive avoidance performance of LE females of these different weight ranges on any day of testing.

Although the passive avoidance of DI rats was poorer than that of HE rats in this experiment, as was also reported by de Wied et al. (1975) and Bohus et al. (1975), the fact that the DI rats did exhibit retention of passive avoidance over long time intervals does not support the hypothesis of these authors that vasopressin plays a necessary role in memory processes. Furthermore, the variety and severity of the secondary consequences of diabetes insipidus not directly attributable to a primary deficiency of vasopressin at a CNS site, which were reviewed in Part II, urge caution in attributing all behavioral differences between DI and HE rats to a deficiency of vasopressin itself.

Several other consequences of diabetes insipidus may provide an explanation for the relatively poor passive behavior of DI rats. For example, a reduced

TABLE 1

Passive Avoidance of Brattleboro and Long-Evans Rats

Sex	Strain	n	Preshock latency (sec)	Median avoidance latency (sec) Postshock			
				1	2	3[a]	4[a]
Male	DI[d]	8	6.0	360.0[c]	196.0[b]	48.0[c]	143.0[b]
	HE[e]	7	7.0	360.0[b]	360.0[b]	360.0[b]	360.0[b]
Female	DI[f]	11	3.0	90.0[c]	83.0[c]	21.0[c]	15.0[b]
	HE[g]	11	4.0	360.0[c]	298.0[c]	360.0[c]	277.0[c]
	LE[h]	24	4.0	136.2	76.4	63.2	97.6

(Male 3[a] and 4[a] comparisons: .02[i]; Female 1 comparison: .02[i]; Female 2 comparison: .02[i]; Female 3[a] and 4[a] comparisons: .02[i])

[a] Tests on which "fear" stimuli were reduced.
[b] p < .01 pre- versus postshock comparisons made with Wilcoxon tests (one-tailed).
[c] p < .005 pre- versus postshock comparisons made with Wilcoxon tests (one-tailed).
[d] 270 ± 11 body weight in grams.
[e] 363 ± 23 body weight in grams.
[f] 219 ± 5 body weight in grams.
[g] 160 ± 4 body weight in grams.
[h] 186 ± 4 (n = 12); 230 ± 5 (n = 12) body weight in grams.
[i] All statistical comparisons were made with Mann-Whitney tests (two-tailed). (Data from Bailey & Weiss, 1979.)

pituitary–adrenal response to footshock on the training trial might be associated with poor passive avoidance responding (Weiss et al., 1969, 1970), and DI rats, like rats depleted of ACTH in the brain after neonatal injections of monosodium glutamate or adenohypophysectomized rats, are reported to show a subnormal analgesic response to cold water stress (Bodnar, Zimmerman, Nilaver, Mansour, Thomas, Kelly, & Glusman, 1980; Glusman, Bodnar, Kelly, Sirio, Stern, & Zimmerman, 1979). Another possibility is that the elevated levels of oxytocin in plasma and brain of DI rats (Dogterom et al., 1977, 1978; Valtin et al., 1965) associated with chronic dehydration might adversely affect passive avoidance conditioning, because the administration of oxytocin to normal rats has effects opposite to those of vasopressin, including the attenuation of passive avoidance behavior (Bohus, Urban, van Wimersma Greidanus, & de Wied, 1978c; Kovács, Vécsei, & Telegdy, 1976). The attractiveness of this possibility, however, is considerably diminished by a report that the intraventricular injection of antibodies to oxytocin does not improve passive avoidance behavior of DI rats, as was observed with normal rats (Bohus et al., 1978b). These data suggest that high levels of oxytocin in the cerebrospinal fluid of DI rats are not directly responsible for their poor passive avoidance performance. Nevertheless, these data are consistent with the hypothesis that oxytocin might affect passive avoidance in normal animals by competing for vasopressin sites and thereby reducing the facilitative effects of the latter peptide.

C. Activity in the Open Field

As the foregoing section makes clear, all studies, including our own, demonstrate that DI rats return to a compartment in which they received footshock sooner than do HE rats. Although differences in body weight do not seem to be related to this difference in performance, it is still possible that differences in activity could account for the findings. If DI rats show stronger activity tendencies than HE rats, then they are more likely to enter the shock compartment on extinction trials. In this regard, Endröczi (1972) reported that activity levels in a maze-type open field were inversely related to the rate of learning of a passive avoidance response. To evaluate the possibility that differences in passive avoidance performance of DI and HE rats were related to differences in their characteristic levels of activity, we observed their behavior during a single open-field test. The behavior of the rat in response to a brief exposure to an open area, particularly ambulation, may be interpreted as indicating either activity, emotionality (Whimby & Denenberg, 1967), or fearfulness (Gray, 1971).

In this experiment, 3-month-old HE and DI rats were compared with each other and with age-matched normal LE rats. Each rat was placed in the center of a 120 cm² box placed at one end of the animal colony room and the number of 24 cm squares crossed and rearings on hind legs were recorded during 3 min of observation. The results are summarized in Fig. 3.

Comparisons of the effects of genotype and sex on crossings and rearings were made by analysis of variance. Genotype but not sex differences were significant [F (2,39) = 15.42, p < .01] with respect to crossings. Newman-Keuls comparisons indicated that DI and HE rats did not differ between themselves, but each made significantly fewer crossings than did LE rats. No significant genotypic or sex differences in measures of rearing were observed. An analysis of the crossings by male rats of each group during each of the 3 min of the test (only 3-min totals were taken on female rats) indicated that overall differences were related to a decline in the activity of Brattleboro groups in the second and third min, whereas a high constant level of activity was exhibited by LE rats during the entire test.

The similar activity level displayed by DI and HE rats during this and other open-field tests (Bohus et al., 1975) suggests that the marked difference in their passive avoidance behavior that we observed in our previous experiment cannot be explained by differences between DI and HE rats in their activity levels. However, both DI and HE rats showed lower activity than did normal LE rats, so that activity differences may indeed be reflected in passive avoidance differences between the Brattleboro rats in general and normal LE rats. It is interesting to note that the Brattleboro rats in the open field froze or remained quiet when approached at the end of the testing period, in contrast to LE rats that ran away from the experimenter's hand. Gray and others have argued that fear inhibits exploration in tests such as those used here (Gray, 1971; Montgomery, 1955; Russell, 1973). Thus, Brattleboro rats may be more fearful, and/

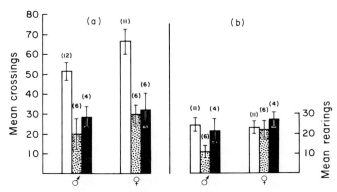

Fig. 3. Open-field activity of Brattleboro and Long-Evans rats. The bars represent (a) mean crossings and (b) rearings (± SEM) observed in these groups during a 3-min test. Long-Evans rats (white bar), rats heterozygous for hereditary diabetes insipidus (shaded bar), and rats homozygous for hereditary diabetes insipidus (black bar) are shown. Numbers in parentheses indicate the number of animals included in each mean.

or show diminished active avoidance behavior, than do normal LE rats. Whatever the nature of the difference between the response of LE and Brattleboro
rats to the open field, it is important to remember that several differences
unrelated to those responsible for the defect in vasopressin metabolism may be
involved, including other genetic differences, differences in early environment,
life history (Denenberg, 1967), and nutritional status.

D. Do Behaviorally Active Peptides Act Through the Release of Endogenous Vasopressin?

As has been pointed out previously, deficits in avoidance performed such
as the short passive avoidance latencies of DI rats described in Section III-B
may be related either to a deficiency of vasopressin or to changes in the activity
of other hormonal systems in the absence of vasopressin. For instance, if the
release of ACTH is impaired in DI rats, short passive avoidance latencies might
also be expected. Weiss et al. (1969, 1970) suggested that the relatively short
passive avoidance latencies in hypophysectomized rats may be related to a
deficiency of ACTH. A straightforward method of determining the hormonal
basis for the poor passive avoidance behavior of the DI rat would be to contrast
the effects of ACTH and vasopressin treatment. Whichever hormone increases
the latencies of DI rats to the level of HE rats would presumably indicate the
relevant hormone deficiency. However, this experiment has in a sense already
been done in normal rats. It has been shown that administration of $ACTH_{4-10}$,
MSH, and vasopressin all improve passive avoidance responding in normal rats
(Ader & de Wied, 1972; Bohus, Ader, & de Wied, 1972; Dempsey et al., 1972;
Guth et al., 1971), so it is conceivable that all of these hormones could raise
the passive avoidance latencies of DI rats to the level of HE rats. One possible
explanation for the commonality of effect of these peptides is that they facilitate
passive avoidance under certain conditions via increased arousal. Such a mechanism would not be peptide-specific, as other compounds that act peripherally
to increase arousal also improve passive avoidance (e.g., 4-hydroxy-ampheta-
mine, Martinez, Vasquez, Rigter, Messing, Jensen, Liang, & McGaugh, 1980;
epinephrine, and norepinephrine, Gold & van Buskirk, 1976).

Although the data discussed hardly provide a definitive answer to the proposed
problem, they do bring us to one of the focal points of this chapter. Another
explanation of how several pituitary hormones, including ACTH and vasopressin, have a similar effect on avoidance responding is that these hormones
influence behavior through a common mechanism. Specifically, it is possible
that only one hormone or peptide sequence actually affects avoidance behavior,
whereas the other hormones affect behavior by stimulating the release of endogenous stores of the active hormone or by potentiating its effects. A good

candidate for the primary role is vasopressin. In this case, the shorter duration of action of exogenously administered ACTH noted in some avoidance studies could be accounted for by the limited quantity of vasopressin available for endogenous release by ACTH. In addition, reports that (1) $ACTH_{4-10}$ and desglycinamide lysine vasopressin facilitate the release of antidiuretic material, thought to be vasopressin, into the blood of rats during passive avoidance tests (Thompson & de Wied, 1973), and (2) the behaviorally active oxytocin fragment cyclo(Leu-Gly) increases plasma levels of arginine vasopressin in the rat (de Wied, Bohus, van Ree, Urban, & van Wimersma Greidanus, 1977) are consistent with this thesis.

To test for interactions between exogenously administered peptides and the release of endogenous vasopressin, we examined the effect of $ACTH_{4-10}$ on the passive avoidance of DI rats with the idea that the passive avoidance behavior of DI rats should not be affected by $ACTH_{4-10}$ if the release of endogenous vasopressin is absolutely necessary for ACTH to affect behavior (Bailey & Weiss, 1978). The passive avoidance test was carried out as previously described except that HE, DI, and normal LE rats were injected 1 hr before testing with $ACTH_{4-10}$ (30.0 µg/100.0 g) or acidified saline vehicle, and the shock level was lowered to 0.5 mA.

As shown in Fig. 4, group differences and injection effects were most evident on Day 3 and Day 4 of testing when the stimulus conditions of the shock chamber were changed ("fear reduced" condition). $ACTH_{4-10}$ significantly increased the passive avoidance latencies of LE rats ($p < .02$), as was expected.

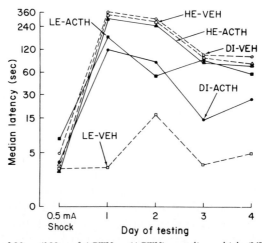

Fig. 4. Effect of 30 µg/100 g of $ACTH_{4-10}$ (ACTH), or saline vehicle (VEH) on the passive avoidance latencies of normal Long-Evans rats (LE) and rats with total (DI) or partial (HE) hereditary deficiency of vasopressin. Note that the testing apparatus was modified on Day 3 and Day 4 to reduce fear cues. (From Bailey & Weiss, 1978.)

However, the administration of $ACTH_{4-10}$ significantly *decreased* the latencies of DI rats [Day 3 ($p < .05$)] while not affecting the latencies of HE rats. The hypothesis that $ACTH_{4-10}$ requires the release of endogenous vasopressin for its behavioral potency was clearly contradicted by the significant action of $ACTH_{4-10}$ on the performance of DI rats. This finding, therefore, is more in keeping with new data demonstrating that significant amounts of arginine vasopressin or MSH above control levels are not released into the bloodstream by ordinary avoidance conditioning or by any but the most severe footshock or other stressors (Husain, Manger, Rock, Weiss, & Frantz, 1979; van Wimersma Greidanus, 1979). However, when interpreting these data, it should be kept in mind that the quantity of neuropeptide released within the brain, which was not measured in these experiments, may well show a correlation with avoidance performance.

Perhaps of most interest in this experiment was the finding that the passive avoidance of both DI and HE rats that were injected with vehicle were significantly greater than those of vehicle-injected normal LE rats on every trial ($p < .02$). On the basis of previous studies that have used exogenously administered vasopressin, it might be expected that Brattleboro rats would show poorer rather than better passive avoidance than would normal animals, because they have total or partial impairments of vasopressin metabolism. Our unexpected data therefore require explanation.

One category under which explanations would fall might be called the "sensitivity to footshock" hypothesis. For example, it could be argued that the superior passive avoidance of Brattleboro rats merely reflects the greater sensitivity to footshock of smaller rats (Gibbs et al., 1973), and the Brattleboro rats are generally smaller than LE rats of the same age. However, the data from our study, as well as those from other experiments, do not support this interpretation (Bohus et al., 1975; Celestian et al., 1975; de Wied et al., 1975). An examination of the individual passive avoidance latencies of HE, DI, and LE rats in the present experiment indicated that performance was not a simple function of body weight. Also, we have shown that the passive avoidance latencies of HE rats are longer than those of LE rats even when the groups are matched for body weight (cf. Section III-B-1, Table 1). Thus, differences in body size per se are not sufficient to account for the differences in the passive avoidance performance of Brattleboro and LE rats.

It is possible, however, that Brattleboro rats are more sensitive to footshock than are LE rats for other reasons. The total or partial vasopressin deficiency of these animals may affect descending oxytocin- or vasopressin-containing fibers that synapse in the spinal cord and could, in this way, produce increased sensitivity to nociceptive stimuli (Nilaver, Zimmerman, Wilkins, Michaels, Hoffman, & Silverman, 1980). Bondar et al. (1980) found that DI rats have lower thresholds for shock-elicited flinch–jump responses than do LE rats of the same size and that daily injections of arginine vasopressin or desamino-D-arginine vasopressin reverse this effect. However, other studies report that DI and HE rats do not differ from one another (Bohus et al., 1975; Celestian et al., 1975) or from LE rats (Celestian et al., 1975) in their immediate behavioral response to footshock. Whether differences in testing procedures are responsible for these discrepancies is not clear, but the observations by de Wied and Gispen (1976) that DI rats have lower thresholds for painful thermal stimulation than do HE rats suggest that this question deserves further investigation.

A second type of explanation can be called the arousal hypothesis. This explanation for the superior passive avoidance of Brattleboro rats is suggested by comparisons of Brattleboro and LE rats on open-field tests. In comparison with LE rats, both HE and DI Brattleboro rats were considerably less active in an open field (cf. Section III-C), indicating that Brattleboro rats (HE and DI) are more generally aroused than LE rats (Gray, 1971), and therefore become more fearful in a shock situation than do LE rats. Such a difference would explain why the Brattleboro rats displayed superior passive avoidance behavior. Because other investigators have not compared the passive avoidance performance of Brattleboro and normal LE rats, the generality of this finding cannot be judged at this time. However, within the Brattleboro strain, poorer performance on the part of DI rats relative to HE rats has been observed in three studies (Bailey & Weiss, 1978; Bohus et al., 1975; de Wied et al., 1975).

This possibility that Brattleboro rats were more aroused or fearful than LE rats not only explains their longer passive avoidance latencies relative to LE rats, but also explains why $ACTH_{4-10}$ had opposite effects on the passive avoidance responding of Brattleboro and LE rats. The relationship between arousal–excitability and performance on a variety of tasks, including passive avoidance (Gold & van Buskirk, 1976), has been described by an inverted U-shaped function (Duffy, 1962). In other words, performance of passive avoidance will improve (latencies become longer) as arousal increases from a very low level, but then performance will decline when arousal becomes very high. As ACTH is thought to increase arousal or excitability (Korányi, Endröczi, Lissák, & Szepes, 1967; Weiss et al., 1969), the effect of ACTH on passive avoidance conditioning will depend upon the arousal or excitability of the animal as influenced by experimental conditions. If these conditions establish a low arousal baseline, ACTH will increase arousal from this low level and improve avoidance performance (increase latencies). But if such conditions have already produced high arousal, the addition of ACTH may not affect or even interfere with performance, producing no change or shorter latencies, respectively. There is, in fact, considerable evidence to support this conclusion. Most passive avoidance studies do indeed report that ACTH or MSH produce longer passive avoidance latencies (see Dempsey et al., 1972; Guth et al., 1971; Thompson & de Wied, 1973). However, it has been noted that these peptides produce shorter latencies when (1) high shock levels are used in the passive avoidance test (Gold & van Buskirk, 1976), or (2) the experiment is conducted with hypoactive animals (Korányi et al., 1967). Similar interactions of ACTH and MSH with high shock levels or hypoactive animals have been reported for other avoidance tasks (Ley & Corson, 1971a,b; Stratton & Kastin, 1974). Thus, when animals are themselves highly aroused or are placed in conditions that are highly arousing, such as a high shock condition, they respond to $ACTH_{4-10}$ with decreased rather than with increased latencies. Consistent with this formulation, as well as with the view that ACTH increases arousal or excitability, are findings that high doses of ACTH will decrease rather than increase passive avoidance latencies (Gold & van Buskirk, 1976).

With regard to the finding that $ACTH_{4-10}$ was more effective in reducing the passive avoidance of DI than of HE rats, it is consistent with the above formulation that DI rats are basically more aroused than HE rats, and for this reason show more clearly the overarousing effects of $ACTH_{4-10}$. Supporting this aspect of the arousal hypothesis are findings that peripheral sympathetic activity, one measure of arousal, is considerably greater in DI than in HE rats (Wooten, Hansen, & Lamprecht, 1975). We also found that the passive avoidance latencies of DI rats are significantly shorter than those of HE rats when a footshock of 1.0 mA is used (Section III-B) but not when a shock of 0.5 mA is administered (this section), a finding consistent with the idea that an increase in arousal produced by more intense shock leads to fearfulness greater than that which is optimal for passive avoidance.

The above considerations indicate that the effect of $ACTH_{4-10}$ on the passive avoidance of vasopressin deficient DI rats is not in principle different from the effect of $ACTH_{4-10}$ on normal LE rats. Indeed, we speculate that if the performance of HE, DI, and LE rats had been adjusted so that the passive avoidance latencies of vehicle-treated rats were equal, then $ACTH_{4-10}$ would have increased or decreased their latencies depending upon the level of performance of vehicle-treated animals. Thus, the results of this study, together with the observation that vasopressin improves the avoidance of hypophysectomized rats that are totally deficient in ACTH (Bohus et al., 1973; de Wied & Bohus, 1966), strongly suggests that the effects of ACTH and vasopressin on avoidance need not depend upon, or even require, the presence of the other hormone. These results are thus in agreement with those of de Wied and others who have suggested that vasopressin and ACTH do not affect avoidance in the same way (Ader & de Wied, 1972; Lissák & Bohus, 1972) or have a common site of action (van Wimersma Greidanus, Bohus, & de Wied, 1974). In normal rats, however, the release of endogenous vasopressin within the brain may contribute to the improved passive avoidance produced by $ACTH_{4-10}$.

Although it is reasonable to consider that differences between LE and Brattleboro rats are related to hereditary deficiencies of vasopressin in Brattleboro rats, the marked differences in the performance of HE and LE rats in this experiment indicate that other differences not involving vasopressin may be involved. One possibility, for instance, is that the behavior of Brattleboro rats is affected by undernutrition during the suckling period following birth. Partial or total diabetes insipidus of Brattleboro pups may make it difficult, if not impossible, for a HE mother (DI females are not used for breeding purposes) to provide sufficient quantities of milk to either DI or HE pups in her litter; moreover, the quality of milk from Brattleboro mothers might be expected to be affected by the potassium deficiency known to occur in these rats (Möhring et al., 1974). Supporting the contention that the behavior of Brattleboro rats is altered by undernutrition in early life are studies on rats with pre- or postnatal undernutrition showing that these rats, like Brattleboro rats, are smaller (Lynch, 1976; Smart, Dobbing, Adlard, Lynch, & Sands, 1973), less active in the open field (Levitsky & Barnes, 1970), display greater passive avoidance (Levitsky & Barnes, 1970; Lissák & Bohus, 1972; Smart et al., 1973; Smart & Dobbing, 1972) and appear to be more frightened than controls in stressful situations (Levitsky & Barnes, 1970). Undernourished and DI rats are also similar in that their adrenals have been found to release less corticosterone when stimulated by ACTH (Shoemaker, 1973; Wiley et al., 1974). Because of these similarities between Brattleboro rats and rats that are undernourished in early life, it would be interesting to determine by cross-fostering studies whether passive avoidance or other behaviors of Brattleboro rats could be altered by providing sufficient quantities of milk from normal rats.

IV. GENERAL CONCLUSION

Rats of the Brattleboro strain exhibit total or partial deficiencies of vasopressin. The performance of these rats on avoidance tasks is of interest because of the possible influence of vasopressin on behavioral and physiological reactions to stressful situations in normal rats. In addition, animal studies and observations with humans have suggested that vasopressin may favorably influence memory processes (cf. relevant chapters in this volume). As is evident from this review, however, it is difficult to determine the behavioral and physiological significance of vasopressin in normal animals by studying animals with a total genetic lesion of the vasopressin–hypothalamic–neurohypophyseal system. The problems inherent in this approach are obvious when one considers the numerous and widespread consequences of chronic vasopressin deficiency in rats with diabetes insipidus. For example, we do not know whether the reduced passive avoidance performance of rats with diabetes insipidus is attributable to the absence of vasopressin at a specific receptor in the brain or merely an indirect consequence of a potassium deficiency that develops with the voiding of copious volumes of urine. Future studies of the Brattleboro rat may yet yield new and exciting information about the contribution of vasopressin to learning, memory, and behavior. The key to such developments requires, however, a greater understanding of the secondary consequences of vasopressin deficiency.

REFERENCES

Ader, R., & de Wied, D. (1972). Effects of lysine vasopressin on passive avoidance learning. *Psychonomic Science*, **29**, 46–48.

Arimura, A., Saito, T., Bowers, C. Y., & Schally, A. V. (1967). Pituitary–adrenal activation in rats with hereditary hypothalamic diabetes insipidus. *Acta Endocrinologica*, **54**, 155–165.

Arimura, A., Yamaguchi, T., Yoshimura, K., Imazeki, T., & Itoh, S. (1965). Role of the neurohypophysis in the release of adrenocorticotrophic hormone in the rat. *Japanese Journal of Physiology*, **15**, 278–295.

Bailey, W. H., & Weiss, J. M. (1978). Effect of $ACTH_{4-10}$ on passive avoidance of rats lacking vasopressin (Brattleboro strain). *Hormones and Behavior*, **10**, 22–29.

Bailey, W. H., & Weiss, J. M. (1979). Evaluation of a 'memory deficit' in vasopressin-deficient rats. *Brain Research*, **162**, 174–178.

Bodnar, R. J., Zimmerman, E. A., Nilaver, G., Mansour, A., Thomas, L. W., Kelly, D. D., & Glusman, M. (1980). Dissociation of cold-water swim and morphine analgesia in Brattleboro rats with diabetes insipidus. *Life Sciences* **26**, 1581–1590.

Bohus, B., Ader, R., & de Wied, D. (1972). Effects of vasopressin on active and passive avoidance behavior. *Hormones and Behavior*, **3**, 191–197.

Bohus, B., Gispen, W. H., & de Wied, D. (1973). Effect of lysine vasopressin and $ACTH_{4-10}$ on conditioned avoidance behavior of hypophysectomized rats. *Neuroendocrinology*, **11**, 137–143.

Bohus, B., Kovács, G. L., & de Wied, D. (1978a). Oxytocin, vasopressin and memory: Opposite effects on consolidation and retrieval processes. *Brain Research*, **157**, 414–417.

Bohus, B., Kovács, G. L., & de Wied, D. (1978b). Attenuation of memory processes by oxytocin. Paper presented at the 9th Congress of the International Society of Neuroendocrinology, August 22–24, Dublin, Ireland.

Bohus, B., Urban, I., van Wimersma Greidanus, Tj. B., & de Wied, D. (1978c). Opposite effects of oxytocin and vasopressin on avoidance behavior and hippocampal theta rhythm in the rat. *Neuropharmacology*, **17**, 239–247.

Bohus, B., van Wimersma Greidanus, Tj. B., & de Wied, D. (1975). Behavioral and endocrine responses of rats with hereditary hypothalamic diabetes insipidus (Brattleboro strain). *Physiology and Behavior*, **14**, 609–615.

Buckingham, J. C., & Leach, J. H. (1979a). Corticotropin secretion in the Brattleboro rat. *Journal of Endocrinology*, **81**, 126P.

Buckingham, J. C., & Leach, J. H. (1979b). Vasopressin and hypothalamo-pituitary-adrenocorticotrophic activity. *Journal of Physiology*, **296**, 87P.

Celestian, J. F., Carey, R. J., & Miller, M. (1975). Unimpaired maintenance of a conditioned avoidance response in the rat with diabetes insipidus. *Physiology and Behavior*, **15**, 707–711.

Dempsey, G. L., Kastin, A. J., & Schally, A. V. (1972). The effects of MSH on a restricted passive avoidance response. *Hormones and Behavior*, **3**, 333–337.

Denenberg, V. H. (1967). Stimulation in infancy, emotional reactivity, and exploratory behavior. *In* D. H. Glass (Ed.), *Biology and Behavior: Neurophysiology and Emotion*, pp. 161–189. New York: Rockefeller University Press.

de Wied, D. (1964). Influence of anterior pituitary on avoidance learning and escape behavior. *American Journal of Physiology*, **207**, 255–259.

de Wied, D. (1965). The influence of the posterior and intermediate lobe of the pituitary and pituitary peptides on the maintenance of a conditioned avoidance response in rats. *International Journal of Neuropharmacology*, **4**, 157–167.

de Wied, D. (1966). Inhibitory effect of ACTH and related peptides on extinction of conditioned avoidance behavior in rats. *Proceedings of the Society for Experimental Biology and Medicine*, **122**, 28–32.

de Wied, D. (1968). Influence of vasopressin and of a crude CRF preparation on pituitary ACTH-release in posterior-lobectomized rats. *Neuroendocrinology*, **3**, 129–135.

de Wied, D. (1969). Effects of peptide hormones on behavior. *In* W. F. Ganong & L. Martini (Eds.), *Frontiers in Neuroendocrinology*, pp. 91–140. London and New York: Oxford University Press.

de Wied, D., & Bohus, B. (1966). Long term and short term effects on retention of a conditioned avoidance response in rats by treatment with long acting pitressin and α-MSH. *Nature (London)*, **212**, 1484–1486.

de Wied, D., & Bohus, B. (1979). Modulation of memory processes by neuropeptides of hypothalamic-neurohypophyseal origin. *In* M.A.B. Brazier (Ed.), *Brain Mechanisms in Memory and Learning: From the Single Neuron to Man*, pp. 139–149. New York: Raven.

de Wied, D., Bohus, B., van Ree, J. M., Urban, I., & van Wimersma Greidanus, Tj. B. (1977). Neurohypophyseal hormones and behavior. *In* A. M. Moses & L. Share (Eds.), *Proceedings of the International Conference on the Neurohypophysis*, pp. 201–210. Basel: Karger.

de Wied, D., Bohus, B., & van Wimersma Greidanus, Tj. B. (1975). Memory deficit in rats with diabetes insipidus. *Brain Research*, **85**, 152–156.

de Wied, D., & Gispen, W. H. (1976). Impaired development of tolerance to morphine analgesia in rats with hereditary diabetes insipidus. *Psychopharmacologia*, **46**, 27–29.

de Wied, D., Witter, A., & Lande, S. (1970). Anterior pituitary peptides and avoidance acquisition of hypophysectomized rats. *In* D. de Wied & J.A.W.M. Weijnen (Eds.), *Pituitary, Adrenal and Brain: Progress in Brain Research*, Vol. 32, pp. 213–220. Amsterdam: Elsevier.

Dogterom, J., Snijdewint, F.G.M., & Buijs, R. M. (1978). The distribution of vasopressin and oxytocin in the rat brain. *Neuroscience Letters*, **9**, 341–346.

Dogterom, J., van Wimersma Greidanus, Tj. B., & Swaab, D. F. (1977). Evidence for the release of vasopressin and oxytocin into cerebrospinal fluid: Measurements in plasma and CSF of intact and hypophysectomized rats. *Neuroendocrinology*, **24**, 108–118.

Duffy, E. (1962). *Activation and Behavior*. New York: Wiley.

Endröczi, E. (1972). Limbic system, learning and pituitary-adrenal function. Budapest: Akadémiai Kiadó.

Friedman, S. M., Scherrer, H. F., Nikashima, M., & Friedman, C. L. (1958). Extrarenal factors in diabetes insipidus in the rat. *American Journal of Physiology*, **192**, 401–404.

Galton, V. A., Valtin, H., & Johnson, D. G. (1966). Thyroid function in the absence of vasopressin. *Endocrinology*, **78**, 1224–1229.

Gibbs, J., Sechzer, J. A., Smith, G. P., Conners, R., & Weiss, J. M. (1973). Behavioral responsiveness of adrenalectomized, hypophysectomized and intact rats to electric shock. *Journal of Comparative and Physiological Psychology*, **82**, 165–169.

Glusman, M., Bodnar, R. J., Kelly, D. D., Sirio, C., Stern, J., & Zimmerman, E. Z. (1979). Attenuation of stress-induced analgesia by anterior hypophysectomy in the rat. *Society for Neuroscience Abstracts*, **5**, 609.

Gold, P. E., & van Buskirk, R. (1976). Effects of posttrial hormone injections on memory processes. *Hormones and Behavior*, **7**, 509–517.

Gray, J. (1971). *The Psychology of Fear and Stress*. New York: McGraw-Hill.

Greven, H. M., & de Wied, D. (1973). The influence of peptides derived from corticotropin (ACTH) on performance: Structure activity studies. *In* E. Zimmerman, W. H. Gispen, B. H.

Marks, & D. de Wied (Eds.), *Drug Effects on Neuroendocrine Regulation: Progress in Brain Research*, Vol. 39, pp. 429–442. Amsterdam: Elsevier.

Guth, S., Seward, J. P., & Levine, S. (1971). Differential manipulation of passive avoidance by exogenous ACTH. *Hormones and Behavior*, **2**, 127–138.

Hall, C. E., Ayachi, S., & Hall, O. (1973). Spontaneous hypertension in rats with hereditary hypothalamic diabetes insipidus (Brattleboro strain). *Texas Reports on Biology and Medicine*, **31**, 471–487.

Harrington, A. R., & Valtin, H. (1965). Vasopressin effect on urinary concentration in rats with hereditary hypothalamic diabetes insipidus. *Proceedings of the Society for Experimental Biology and Medicine*, **118**, 448–450.

Harrington, A. R., & Valtin, H. (1968). Impaired urinary concentration after vasopressin and its gradual correction in hypothalamic diabetes insipidus. *Journal of Clinical Investigation*, **47**, 502–510.

Hoffman, W. E., Ganten, U., Schelling, P., Phillips, M. I., Schmid, P. G., & Ganten, D. (1978). The renin and isorenin-angiotensin system in rats with heridatary hypothalamic diabetes insipidus. *Neuropharmacology*, **17**, 919–923.

Husain, M. K., Manger, W. M., Rock, T. W., Weiss, R. J., & Frantz, A. G. (1979). Vasopressin release due to manual restraint in the rat: Role of body compression and comparison with other stressful stimuli. *Endocrinology*, **10**, 641–644.

Korányi, L., Endröczi, E., Lissák, K., & Szepes, E. (1967). The effect of ACTH on behavioral processes motivated by fear in mice. *Physiology and Behavior*, **2**, 439–445.

Kovács, G. L., Vécsei, L., & Telegdy, G. (1978). Opposite action of oxytocin to vasopressin in passive avoidance in rats. *Physiology and Behavior*, **20**, 801–802.

Krieger, D. T., & Liotta, A. (1977). Pituitary ACTH responsiveness in the vasopressin deficient rat. *Life Sciences*, **20**, 327–336.

Levitsky, D. A., & Barnes, R. H. (1970). Effects of early malnutrition on the reaction of adult rats to aversive stimuli. *Nature (London)*, **225**, 468–469.

Ley, K. F., & Corson, J. A. (1971a). ACTH: Differential effects on avoidance and discrimination. *Experientia*, **27**, 958–959.

Ley, K. F., & Corson, J. A. (1971b). Effects of ACTH, adrenalectomy and time of day on emotional activity of the rat. *Behavioral Biology*, **9**, 111–115.

Lissák, K., & Bohus, B. (1972). Pituitary hormones and avoidance behavior of the rat. *International Journal of Psychobiology*, **2**, 103–115.

Lynch, A. (1976). Passive avoidance behavior and response thresholds in adult male rats after early postnatal undernutrition. *Physiology and Behavior*, **16**, 27–32.

Martinez, J. L., Jr., Vasquez, B. J., Rigter, H., Messing, R. B., Jensen, R. A., Liang, K. C., & McGaugh, J. L. (1980). Attenuation of amphetamine-induced enhancement of learning by adrenal demedullation. *Brain Research*, **195**, 433–443.

McCann, S. M., Antunes-Rodrigues, J., Nallar, R., & Valtin, H. (1966). Pituitary adrenal function in the absence of vasopressin. *Endocrinology*, **79**, 1058–1064.

Miller, M., Barranda, E. G., Dean, M. C., & Brush, F. R. (1976). Does the rat with hereditary hypothalamic diabetes insipidus have impaired avoidance learning and/or performance? *Pharmacology, Biochemistry and Behavior, Supplement 1*, **5**, 35–40.

Miller, M., & Moses, A. M. (1971). Radioimmunoassay of urinary antidiuretic hormone with application to study of the Brattleboro rat. *Endocrinology*, **88**, 1389–1396.

Miller, R. E., Yueh-Chien, H., Wiley, M. K., & Hewitt, R. (1974). Anterior hypophysial function in the posterior-hypophysectomized rat: Normal regulation of the adrenal system. *Neuroendocrinology*, **14**, 233–250.

Möhring, J., Dauda, G., Haack, D., Homsy, E., Kohrs, G., & Möhring, B. (1972). Increased potassium intake and kaliopenic nephropathy in rats with genetic diabetes insipidus. *Life Sciences (Part I)*, **11**, 679–683.

Möhring, J., Kohrs, G., Möhring, B., Petri, M., Homsy, E., & Haack, D. (1978). Effects of prolonged vasopressin treatment in Brattleboro rats with diabetes insipidus. *American Journal of Physiology*, **234**, F106–F111.

Montgomery, K. C. (1955). The relation between fear induced by novel stimulation and exploratory behavior. *Journal of Comparative and Physiological Psychology*, **48**, 254–260.

Moses, A. M., & Miller, M. (1970). Accumulation and release of pituitary vasopressin in rats heterozygous for hypothalamic diabetes insipidus. *Endocrinology*, **86**, 34–41.

Nilaver, G., Zimmerman, E. A., Wilkins, J., Michaels, J., Hoffman, D., & Silverman, A.-J. (1980). Magnocellular hypothalamic projections to the lower brain stem and spinal cord of the rat. *Neuroendocrinology*, **30**, 150–158.

Nitzan, M., & Wilber, J. F. (1974). Effect of postnatal malnutrition on plasma proteins and growth hormone in the rat. *Hormone Research*, **5**, 167–172.

Russell, P. A. (1973). Relationships between exploring behavior and fear. *British Journal of Psychology*, **64**, 417–433.

Saul, G. B., Garrity, E. B., Benirschke, K., & Valtin, H. (1968). Inherited hypothalamic diabetes insipidus in the Brattleboro strain of rats. *Journal of Heredity*, **59**, 113–117.

Schultz, H., Kovács, G. L., & Telegdy, G. (1976). The effect of oxytocin and vasopressin on avoidance in rats. In E. Endröczi (Ed.), *Cellular and Molecular Bases of Neuroendocrine Processes*, pp. 555–564. Budapest: Akadémiai Kiadó.

Selye, H. (1950). *Stress*. Montreal: Acta Medical Publishers.

Shoemaker, W. J. (1973). Early undernutrition and brain neurotransmitters. In R. O. Scow (Ed.), *Endocrinology (Proceedings of the 4th International Congress of Endocrinology)*, pp. 1327–1338. Amsterdam: Excerpta Medica.

Smart, J. L., & Dobbing, J. (1972). Vulnerability of developing brain: IV. Passive avoidance behavior in young rats following maternal undernutrition. *Developmental Psychobiology*, **5**, 129–136.

Smart, J. L., Dobbing, J., Adlard, B.P.F., Lynch, A., & Sands, J. (1973). Vulnerability of developing brain: Relative effects of growth restriction during fetal and suckling periods on behavior in young rats. *Journal of Nutrition*, **103**, 1327–1338.

Sokol, H. W. (1973). Skeletal elongation stimulated by exogenous growth hormone, but not vasopressin, in growth-retarded rats having diabetes insipidus. Program of the 55th meeting of the Endocrine Society (Abstract).

Stratton, L. O., & Kastin, A. J. (1974). Avoidance learning at two levels of shock in rats receiving MSH. *Hormones and Behavior*, **5**, 149–155.

Swaab, D. F., Boer, G. J., & Nolton, J.W.L. (1973). The hypothalamic-neurohypophyseal system (HNS) of the Brattleboro rat. *Acta Endocrinologia*, **80**, 80.

Takahashi, H., Daughday, W. H., & Kipins, D. M. (1971). Regulation of immunoreactive growth hormone secretion in male rats. *Endocrinology*, **88**, 909–917.

Tepperman, J. (1968). *Metabolic and Endocrine Physiology* (Second Edition), Chicago, Illinois: Year Book Medical Publishers.

Thompson, E. A., & de Wied, D. (1973). The relationship between the antidiuretic activity of rat eye plexus blood and passive avoidance behavior. *Physiology and Behavior*, **11**, 377–380.

Valtin, H., Sawyer, W. H., & Sokol, H. W. (1965). Neurohypophyseal principles in rats homozygous and heterozygous for hypothalamic diabetes insipidus (Brattleboro strain). *Endocrinology*, **77**, 701–706.

Valtin, H., & Schroeder, H. A. (1964). Familial hypothalamic diabetes insipidus in rats (Brattleboro strain). *American Journal of Physiology*, **206**, 425–430.

van Wimersma Greidanus, Tj. B. (1979). Neuropeptides and avoidance behavior; with special reference to the effects of vasopressin, ACTH, and MSH on memory processes. In R. Collu, A. Barbeau, J. Rochefort, & J. R. Ducharme (Eds.), *Central Nervous System Effects of Hypothalamic Hormones and Other Peptides*, pp. 177–187. New York: Raven.

van Wimersma Greidanus, Tj. B., Bohus, B., & de Wied, D. (1974). Differential localization of lysine vasopressin and $ACTH_{4-10}$ on avoidance behavior: A study in rats bearing lesions in the parafascicular nuclei. *Neuroendocrinology*, **14**, 280–288.

van Wimersma Greidanus, Tj. B., & de Wied, D. (1976). Modulation of passive-avoidance behavior of rats by intracerebroventricular administration of antivasopressin serum. *Behavioral Biology*, **18**, 325–333.

Vinson, G. P., Goddard, C., & Whitehouse, B. J. (1977). Steroid profiles formed by adrenocortical tissue from rats with hereditary diabetes insipidus (Brattleboro strain) and from normal female Wistar rats under different conditions of stimulation. *Journal of Endocrinology*, **75**, 31P–32P.

Weiss, J. M., McEwen, B. S., Silva, M.T.A., & Kalkut, M. F. (1969). Pituitary-adrenal influences on fear responding. *Science*, **163**, 197–199.

Weiss, J. M., McEwen, B. S., Silva, M.T.A., & Kalkut, M. F. (1970). Pituitary-adrenal alterations and fear responding. *American Journal of Physiology*, **218**, 864–868.

Whimby, A. E., & Denenberg, V. H. (1967). Two independent behavioral dimensions in open field performance. *Journal of Comparative and Physiological Psychology*, **63**, 500–504.

Wiley, M. K., Pearlmutter, A. F., & Miller, R. E. (1974). Decreased adrenal sensitivity to ACTH in the vasopressin-deficient (Brattleboro rat). *Neuroendocrinology*, **14**, 280–288.

Wooten, G., Hanson, T., & Lamprecht, F. (1975). Elevated serum dopamine-beta-hydroxylase activity in rats with inherited diabetes insipidus. *Journal of Neural Transmission*, **36**, 107–112.

Yates, F. E., Russell, S. M., Dallman, M. F., Hedge, G. A., McCann, S. M., & Dhariwal, A.P.S. (1971). Potentiation by vasopressin of corticotropin release induced by corticotropin-releasing factor. *Endocrinology*, **88**, 3–15.

CHAPTER *18*

Vasopressin, Oxytocin, and Dependence on Opiates

Jan M. van Ree and David de Wied

Learning and memory processes have been implicated in development of tolerance to and physical dependence on opiates. Accordingly, both these opiate-induced changes and learning and memory processes are modulated by neurohypophyseal hormones and their fragments. These neuropeptides also affect heroin self-administration behavior. Desglycinamide–arginine–vasopressin decreased heroin intake of rats during acquisition of this behavior and of heroin addicts during methadone detoxification. It is suggested that neuropeptides related to neurohypophyseal hormones may influence opiate-induced adaptive changes via alterations in brain catecholamine systems.

I. INTRODUCTION

Learning and memory processes are essential elements in an animal's ability to adapt to environmental changes and invading noxious stimuli. Repeated

397

ENDOGENOUS PEPTIDES
AND LEARNING AND MEMORY PROCESSES

exposure to opiates is accompanied by alterations in a variety of homeostatic mechanisms in the organism resulting in tolerance and physical dependence and eventually in a state of behavioral dependence (abuse). Accordingly, learning and memory processes have been implicated in these opiate-induced adaptive changes. Evidence has been presented that learning and memory processes in both experimental animals and men are affected by treatment with the neurohypophyseal hormones, vasopressin and oxytocin, and fragments of these peptides (for references see Chapter 19 by van Wimersma Greidanus, Bohus, & de Wied). Thus, these peptides could also be involved in the development of opiate-induced adaptive changes. The present chapter summarizes the data obtained so far concerning the interaction of neurohypophyseal hormones and their fragments on the development of opiate tolerance, physical dependence, and opiate self-administration behavior.

II. OPIATE ACTION AND LEARNING–MEMORY PROCESSES

A. Tolerance and Physical Dependence

Tolerance is said to be demonstrated when the effect of a fixed dose of a drug is shown to be less than that of the first dose administered, and when this reduced effect can be overcome by increasing the dose. Thus, tolerance is due to adaptive changes in the organism that are reflected in altered dose response curves (Kalant, 1978). The development of tolerance is certainly not based on a single process. Many changes induced by drug administration contribute to some degree to the amount of tolerance. Various hypotheses have been offered to explain the mechanism underlying the development of tolerance (see Chapter 23 by Kesner & Baker). One of these hypotheses considers the development of tolerance to be a process that is analogous to learning and memory. Two different elaborations of this concept have been proposed and both are supported by experimental evidence. First, the display of tolerance seems to be at least partly dependent upon environmental stimuli associated with drug administration. The role of such environmental signals in tolerance development has been explained by reference to Pavlovian conditioning principles (Siegel, 1976, 1978). Second, the process of tolerance development can be regarded as consisting of adaptations that may be similar to those involved in learning and memory processes. The cellular functions adapt, as it were, to a new environment, including presence of the drug. This adaptation compensates for drug-induced changes. Thus the cells "remember" their experience with the drug, because their response to the drug is changed by repeated exposure. Evidence

supporting this learning concept of tolerance development is provided by experiments showing that a variety of protein synthesis inhibitors attenuate both
the consolidation of conditioned behavior and the development of tolerance
to morphine (Cohen, Keats, Krivoy, & Ungar, 1965; Cox, Ginsburg, & Osman,
1968; Cox & Osman, 1970; Feinberg & Cochin, 1977; Geller, Robustelli,
Barondes, Cohen, & Jarvik, 1969; Jarvik, 1972; Smith, Karmin, & Gavitt,
1966). Moreover, learning as well as development of tolerance can be retarded
by cerebral insult such as electroconvulsive shock, electrical stimulation of the
frontal cortex, or stimulation of the periaqueductal grey, when these events are
administered shortly after either the learning trial or morphine (Kesner &
Priano, 1977; Kesner, Priano, & Dewitt, 1976; Stolerman, Bunker, Johnson,
Jarvik, Krivoy, & Zimmermann, 1976).

Physical dependence can be defined as an altered state induced by repeated
administration of a drug and recognized by a specific pattern of disturbances
following withdrawal of the drug. Most of the hypotheses formulated to explain
tolerance suggest that the development of physical dependence covaries with
the development of tolerance, although there are exceptions (Kalant, 1978).
This similarity concerns the development but not the expression of physical
dependence and tolerance. These expressions (e.g., withdrawal symptoms and
decreased drug efficacy) may be modified in ways quite different from the
development of tolerance and physical dependence (Way, 1978).

B. Dependence (Abuse)

Self-administration of certain psychoactive drugs may lead to a state of drug
dependence characterized by the drug user's behavior leading specifically to
further administration of the drug, even at the expense of other behaviors
(Kalant, Engel, Goldberg, Griffiths, Jaffe, Krasnegor, Mello, Mendelson,
Thompson, & van Ree, 1978a). Severe degrees of dependence are commonly
labeled as addiction, particularly in clinical practice. The fact that addictive
behavior is continued implies that the drug is a reinforcer. Obviously, this
reinforcing action is the common denominator for the occurrence of abuse
with various drugs, and can be analyzed in self-administration experiments in
laboratory animals as well as in humans. The self-administration procedure has
been shown to be useful for establishing the reinforcing efficacy of drugs and
also the variables that interfere with drug taking behavior (Kalant et al., 1978a;
van Ree, 1979).

A variety of factors can be implicated in the mechanisms determining, in
a particular individual, whether or not a drug injection gains and maintains
control over behavior to such an extent that a state of dependence is achieved.
Among these factors are learning and memory processes. It is well known that

a drug user has to learn how to obtain and to handle the drug, and to remember the pleasant and unpleasant effects of the drug during the first and subsequent experiences with it. In fact, an addict learns and remembers how to select the reinforcing effects from all other influences of the drug. These reinforcing effects are suggested to be the prominent incentive of drug seeking behavior. Experimentally, the implication of learning and memory processes in the reinforcing efficacy of drugs can reliably be studied by analyzing acquisition and maintenance of drug self-administration in laboratory animals.

III. NEUROHYPOPHYSEAL HORMONES AND OPIATE TOLERANCE–PHYSICAL DEPENDENCE

A. Development of Tolerance

The involvement of neurohypophyseal hormones in learning and memory processes and the learning hypothesis of tolerance development, led to studies on the influence of these hormones and related peptides on morphine tolerance and physical dependence. Krivoy et al. (1974) reported that the fragment of vasopressin (desglycinamide[9]) lysine[8]-vasopressin (DG-LVP) that has only little pressor and antidiuretic activity facilitated development of resistance to the antinociceptive action of morphine. Other studies showed that DG-LVP injected directly into the nucleus linearis intermedius raphé accelerated the development of tolerance to morphine-induced behavioral changes in freely moving cats (Cools, Broekkamp, Gieles, Megens, & Mortiaux, 1977). A possible physiological role for vasopressin in tolerance development was suggested by two series of experiments. In the first, an impaired development of tolerance to morphine analgesia was found in rats of the Brattleboro strain with hereditary diabetes insipidus. These animals lack the ability to synthesize vasopressin. Treatment of Brattleboro rats with either arginine[8]-vasopressin or DG-LVP immediately after the daily morphine injection improved the impaired development of tolerance of these rats toward that of normal control rats (de Wied & Gispen, 1976). In the second series of experiments, vasopressin antiserum was injected intracerebroventricularly in normal rats, which were given repeated morphine injections. Treatment with this antiserum transiently inactivates vasopressin that is present in the brain. It appeared that development of tolerance to the antinociceptive action of morphine in normal rats was inhibited by treatment with vasopressin antiserum, whether it was given after the first morphine injection or prior to the second injection. The development of tolerance was inferred from the response of the animals to the second injection of morphine (van Wimersma Greidanus, Tjon Kan Fat-Bronstein, & van Ree, 1978; van Wimersma Greidanus, van Ree, & Versteeg, 1980). Oxytocin antiserum was not effective in this respect (Tjon Kan Fat-Bronstein, unpublished observations). Thus, in terms of memory processes, vasopressin antiserum may affect tolerance development at the level of consolidation as well as retrieval of information. Interestingly, treatment with vasopressin antiserum prior to the first injection did not influence subsequent tolerance development, suggesting that a possible acute interaction of vasopressin with morphine action did not contribute to the effect on tolerance development (Tjon Kon Fat-Bronstein, unpublished observations). The effects of vasopressin antiserum and the effect of DG-LVP, a peptide with reduced peripheral activity, indicate that brain vasopressin may be a modulator for the process of development of tolerance to morphine. Other peptides related to neurohypophyseal hormones may also be important in this respect. In fact, the C-terminal tripeptide of oxytocin, prolyl-leucyl-glycinamide (PLG) has been shown to facilitate development of tolerance to the antinociceptive action of intracerebroventricularly injected β-endorphin (van Ree, de Wied, Bradbury, Hulme, Smyth, & Snell, 1976). Moreover, this neuropeptide accelerated development of morphine tolerance in both rats and mice (Contreras & Takemori, 1980; Székely, Miglécz, Dunai-Kovács, Tarnawa, Ronai, Graf, & Bajusz, 1979).

B. Development of Physical Dependence

The action of neurohypophyseal hormones and their fragments on the development of physical dependence to morphine was also studied. Apparently, development of physical dependence as assessed with different tests was facilitated in rats treated with either (desglycinamide^9) arginine^8-vasopressin (DG-AVP) or oxytocin (van Ree & de Wied, 1976; van Ree & de Wied, 1977a). The degree of physical dependence in rats repeatedly treated with morphine was measured by the body weight loss and temperature decrease following naloxone injection. Whether or not peptide treatment was given on the test day did not substantially influence the results, suggesting that it was the process of development, rather than the expression of physical dependence that was affected by these neuropeptides. Oxytocin was much more potent than DG-AVP in facilitating physical dependence on morphine. Structure–activity relationship studies revealed that the covalent ring structures of vasopressin and oxytocin were inactive in this respect, whereas the C-terminal tripeptides of the hormones, PAG (prolyl-argyl-glycinamide) and PLG respectively, were as active as the parent hormones. The observation that PLG was effective in facilitating development of physical dependence on morphine has led to a number of studies (see Table 1). The data so far are somewhat conflicting, although the different test procedures used may account for some of the observed discrepancies. In contrast to development of tolerance to and physical dependence on morphine, the ultimate degree of these phenomena was not affected

TABLE 1

Neurohypophyseal Hormones and Development of Morphine Tolerance and Physical Dependence

Peptide	Test	Animals	Tolerance & physical dependence		Reference
			Development	Ultimate degree	
DG-LVP	Antinociception	Mice	+		Krivoy et al., 1974
DG-LVP	Antinociception	DI-rats	+		de Wied & Gispen, 1976
DG-AVP/OXY/PLG	Withdrawal signs[a]	Rats	+	0	van Ree & de Wied, 1976, 1977a
DG-AVP/OXY/PLG	Antinociception	Rats	+		van Ree & de Wied, 1976, 1977a
PLG	Antinociception	Rats	+		van Ree et al., 1976
DG-LVP	Behavioral changes	Cats	+		Cools et al., 1977
VP/OXY	Antinociception	Rats		0	Schmidt et al., 1978
Z-Pro-D-Leu	Withdrawal signs[a]	Mice	−		Walter et al., 1978
PLG	Withdrawal signs[a]	Mice	−		Walter et al., 1979
OXY	Withdrawal signs[a]	Mice	+		Walter et al., 1979
PLG	Antinociception	Rats	0		Mucha & Kalant, 1979
PLG	Antinociception	Mice	−		Bhargava et al., 1980
PLG	Antinociception	Rats	+		Székely et al., 1979
PLG	Withdrawal signs[a]	Mice	+		Székely et al., 1979
PLG	Antinociception	Mice	+		Contreras & Takemori, 1980
PLG	Withdrawal signs[a]	Mice	+		Contreras & Takemori, 1980

[a] Naloxone-precipitated withdrawal signs: body weight loss, hypothermia; not affected: jumping response.
+: Facilitation, −: attenuation, 0: no change.

by vasopressin or oxytocin (Schmidt, Holaday, Loh, & Way, 1978; van Ree & de Wied, 1976; van Ree & de Wied, 1977a). Although the effectiveness of PLG in rats could not be replicated in one study (Mucha & Kalant, 1979), others confirmed the finding that PLG facilitates development of physical dependence (Contreras & Takemori, 1980; Székely et al., 1979).

Replacement of an L-amino acid residue in a neuropeptide by its D-enantiomer may yield a peptide with a different activity or even effects opposite to those of the original entity (Bohus & de Wied, 1966; de Wied, Witter, & Greven, 1975; Ramachandran, 1973). For this reason, the dipeptide Z-Pro-D-Leu was prepared and tested for effects on morphine tolerance and physical dependence. Indeed, it was observed that this peptide prevented the development of tolerance to and physical dependence on morphine when implanted as a pellet in mice (Walter, Ritzmann, Bhargava, Rainbow, Flexner, & Krivoy, 1978). Subsequently, the same procedure was used to show that PLG, cyclo-(leucylglycine), and other structurally related peptides also prevent the development of physical dependence on morphine, whereas oxytocin accelerates this development (Bhargava, Walter, & Ritzmann, 1980; Walter, Ritzmann, Bhargava, & Flexner, 1979). This finding is in contradiction to those reported by others with respect to the action of PLG (Contreras & Takemori, 1980; Székely et al., 1979; van Ree & de Wied, 1976; van Ree & de Wied, 1977a). This discrepancy may be related to differences in experimental conditions, but more likely it is due to the dose of PLG used. In fact, the attenuating effect of PLG on development of tolerance and physical dependence in mice was observed with a dose of this neuropeptide that was approximately 1500 times higher than the dose that was effective in accelerating development of tolerance and physical dependence in the same species (Bhargava et al., 1980; Contreras & Takemori, 1980). The action of high doses of PLG may be explained by the antagonistic effect of these doses of PLG on acute morphine action (e.g., antinociception and catalepsy; Chiu & Mishira, 1979; Kastin, Olson, Ehrensing, Berzas, Schally, & Coy, 1979).

C. Possible Mode of Action

These findings suggest that neurohypophyseal hormones and their fragments are involved in the adaptive brain processes that are activated by morphine injection and are also involved in the development of tolerance and physical dependence. This influence is not limited to morphine, because vasopressin and oxytocin affect development of tolerance to and physical dependence on ethanol as well (Hoffman, Ritzmann, & Tabakoff, 1979; Hoffman, Ritzmann, Walter, & Tabakoff, 1978; Kalant, Mucha, & Niesink, 1978; Rigter & Crabbe, 1980). Although both vasopressin and PLG are effective in facilitating morphine tolerance and physical dependence, it is not clear whether these neuropeptides interact with the same or with different substrates. Moreover, the mode of action of these peptides in accelerating tolerance development might be quite different from their involvement in physical dependence. Tolerance development was affected when PLG was administered 1 hr before the morphine injection, but not when given immediately prior to morphine injection (Székely et al., 1979). In contrast, vasopressin injection was effective even when given after the morphine injection (de Wied & Gispen, 1976; Krivoy et al., 1974). Because α-MSH induces an effect opposite to that of PLG, it has been suggested that the action of PLG is mediated by its suppressive effect on α-MSH release (Székely et al., 1979); however, such an action is very unlikely for vasopressin. Considering that the development of tolerance and physical dependence is the result of several processes, it is conceivable that one of these processes is affected by vasopressin, whereas other processes are preferentially affected by PLG. In addition, it must be kept in mind that many environmental cues may function as conditioned stimuli in the process of tolerance development (Siegel, 1976, 1978). These environmental stimuli may either attenuate or augment the effectiveness of the peptides, and the impact of these cues may, in turn, be affected by the neurohypophyseal hormones. These factors may explain some of the variability of the data collected to date.

Evidence has been presented for a specific role of dopamine in morphine tolerance (Clouet & Iwatsubo, 1975; Lal, 1973; Takemori, 1975). Interestingly, the most prominent effect of PLG on catecholamine metabolism is an increase in dopaminergic activity especially in the nigrostriatal system as assessed by the method of α-methyl-p-tyrosine-induced decrease in dopamine content (Versteeg, Tanaka, de Kloet, van Ree, & de Wied, 1978). Also, other data point to an interaction of PLG with dopaminergic transmission in the brain such as an increase in dopamine synthesis or potentiation of apomorphine action (Friedman, Friedman, & Gershon, 1973; Kostrzewa, Kastin, & Sobrian, 1978; Plotnikoff & Kastin, 1976; Schulz, Kovács, & Telegdy, 1979). Thus, the influence of PLG on morphine tolerance may be mediated at least partially by its effect on brain dopamine. Whether vasopressin and related peptides also act via dopamine in this respect is as yet unknown. Vasopressin affects striatal dopamine, but, in addition, has marked facilitatory effects on noradrenaline transmission in limbic structures (Tanaka, de Kloet, de Wied, & Versteeg, 1977). Both of these effects of vasopressin may be of physiological significance, because after intracerebroventricular administration of vasopressin antiserum in normal rats and in rats with hereditary diabetes insipidus, opposite effects were noted (Versteeg, de Kloet, van Wimersma Greidanus, & de Wied, 1979; Versteeg, Tanaka, & de Kloet, 1978).

From studies of the structural requirements of neurohypophyseal hormones for facilitating or impairing memory processes, it was concluded that the ring structure of the hormone is important for modulating consolidation processes, whereas the C-terminal part seems to be more involved with retrieval processes. The amnesic effect of oxytocin is mainly a feature of the whole molecule (Bohus, 1980; de Wied & Bohus, 1979; van Ree, Bohus, Versteeg, & de Wied, 1978). Thus, one might speculate that PLG predominantly affects retrieval processes in development of tolerance and physical dependence. These effects may be mediated by dopamine transmission in the brain. Vasopressin may have a similar mode of action but, in addition, may affect consolidation processes concerned with the development of tolerance. These may possibly be mediated by noradrenergic activity in the brain (van Ree et al., 1978a).

IV. NEUROHYPOPHYSEAL HORMONES
AND OPIATE DEPENDENCE (ABUSE)

A. Heroin Self-Administration

As outlined earlier, drug self-administration behavior is particularly useful for establishing the reinforcing efficacy of drugs and the variables that interfere with drug taking behavior. To investigate the action of neurohypophyseal principles on this behavior in rats, intravenous heroin self-administration was selected, because, with this drug, self-injection behavior develops relatively quickly and is rather reproducible under standard conditions (van Ree & de Wied, 1980a; van Ree, Slangen, & de Wied, 1978). The experiments focused on the acquisition of this behavior, because neuropeptides related to neurohypophyseal hormones have been implicated in learning and memory processes. In a 5-day, 6 hr/day, test procedure, subcutaneous treatment with DG-AVP reduced heroin self-administration in opiate-naive rats. The effect of the peptide was clearly present after some days of testing (van Ree & de Wied, 1977b). The effect of DG-AVP was long-lasting, because the behavior was also attenuated in the second phase, whereas DG-AVP was given only in the first phase of testing. Further experiments suggested that this action of DG-AVP is a central effect, as a much lower dose of this neuropeptide was required when it was administered intracerebroventricularly (van Ree & de Wied, 1977c). Application of vasopressin antiserum via this route markedly stimulated heroin self-administration when compared to the behavior of rats treated with control serum. Thus, vasopressin may be physiologically involved in this behavior. Antiserum-containing antibodies against oxytocin or human growth hormone did not influence the behavior (van Ree & de Wied, 1977c).

Structure–activity relationship studies showed that the ring structure of vasopressin is important for attenuating heroin self-administration behavior. Oxytocin tended to facilitate the behavior. A more pronounced facilitating effect was observed with PLG, whereas the covalent ring structure of oxytocin appeared to be inactive (van Ree & de Wied, 1977b). Both (desglycinamide)-vasotocin containing the tail and (desglycinamide)-oxypressin containing the ring structure of vasopressin, attenuated heroin self-administration to a similar degree as that observed after treatment with either the tail or the ring structure of vasopressin (van Ree & de Wied, 1980a). Because DG-AVP markedly attenuated heroin self-administration, the influence of DG-AVP was studied in human addicts in an outpatient clinic, during the initial phase of methadone detoxification therapy (van Beek-Verbeek, Fraenkel, Geerlings, van Ree, & de Wied, 1979; van Ree, 1980). The outcome of this pilot experiment showed that sublingual application of DG-AVP facilitated methadone detoxification as could be inferred from the longer time course of attending the clinic, and from the lower percentage of urine samples with detectable morphine in patients treated with DG-AVP as compared to those receiving placebo. Moreover, the medical attendant rated the methadone detoxification of DG-AVP treated patients as more successful than that of the patients who received placebo. The beneficial effect of DG-AVP was present not only during treatment with this neuropeptide, but also in the period following cessation of DG-AVP administration. These preliminary findings in humans need to be extended, but they suggest that the human responds to DG-AVP in a way similar to experimental animals.

B. Possible Mode of Action

Because the amount of drug taken can serve as an index of the reinforcing efficacy of the reinforcer i.e., drug injection (van Ree et al., 1978b), it was postulated that DG-AVP attenuates and PLG enhances the reinforcing efficacy of heroin. This hypothesis is supported by data on intracranial electrical self-stimulating behavior. This measure is widely used to investigate the significance of certain brain structures with respect to reward (Wauquier & Rolls, 1976). Administration of DG-AVP decreased, while administration of PLG enhanced self-stimulation behavior elicited from electrodes implanted in the ventral tegmental medial substantia nigra area. This brain area contains the cell bodies of the mesolimbic and mesocortical dopaminergic pathways (Dorsa & van Ree, 1979). Subsequently, it was observed that the development of self-administration of fentanyl directly into the ventral tegmental area was attenuated and facilitated, respectively, by subcutaneous treatment with DG-AVP and PLG (van Ree & de Wied, 1980b). Thus, these peptides may influence heroin self-administration by modifying transmission in mesolimbic dopaminergic systems. This idea is consistent with the assumption that dopaminergic systems in the brain are critically involved in reward and in the reinforcing effects of opiates (Wise, 1978). The interference of PLG with dopaminergic activity in the brain may be related to the effects of the neuropeptide on heroin self-administration and electrical self-stimulation behavior. With respect to DG-AVP, the picture is more complicated. Because vasopressin increases noradrenergic transmission, and noradrenergic activity inhibits dopaminergic action, especially in self-stimulation (Koob, Balcom, & Meyerhoff, 1976; Phillips & Nikaido, 1975), DG-AVP may decrease the reward value of stimuli via its effect on brain noradrenaline.

It is rather difficult to reconcile the influence of vasopressin on heroin self-administration with that on memory consolidation. Nevertheless, the characteristics of the effectiveness of vasopressin are similar in both test procedures (de Wied, 1976; van Ree et al., 1978a). It may be the case that DG-AVP is only effective during acquisition of the behavior or when the behavior is changed in response to variation in the reinforcement or environmental cues. This hypothesis is supported by findings showing that DG-AVP does not reduce morphine self-administration in monkeys physically dependent on morphine and having a long history of self-administration (Mello & Mendelson, 1979). In addition, DG-AVP interacts with electrical self-stimulation at low but not at high current intensities (Dorsa & van Ree, 1979). Thus, the effect of DG-AVP on reward mechanisms depends on the degree of reinforcement control over behavior. This fits well with the assumption that neuropeptides exert their effects on behavioral adaptation by modulation of ongoing activity in the brain (Barchas, Akil, Elliott, Holman, & Watson, 1978).

V. CONCLUDING REMARKS

The data reviewed here indicate that neuropeptides related to neurohypophyseal hormones affect opiate-induced changes in homeostatic mechanisms and responses in the central nervous system. Learning and memory mechanisms have been implicated in adaptive processes elicited by repeated administration of opiates such as development of tolerance and physical dependence. Interestingly, tolerance, physical dependence development, and learning and memory processes are all modulated by neurohypophyseal peptides. However, the data, so far, are too limited to draw definite conclusions concerning the similarities and dissimilarities with respect to the mode of action of these peptides in both phenomena.

Reward mechanisms triggered by heroin during acquisition of drug seeking behavior may also be modulated by neurohypophyseal hormones and their fragments. Most probably, this influence of the peptides is dissociated from their involvement in development of tolerance and physical dependence. Neither tolerance nor physical dependence is an essential condition for drug self-administration (Kalant et al., 1978a; van Ree et al., 1978b). Thus, disturbances in brain and/or neurohypophyseal hormone-containing systems, or in the generation of neuropeptides from these hormones, may lead to a state in which drug seeking behavior may be elicited. Treatment of these disturbances may eventually decrease the probability that addictive behavior develops and may attenuate the reinforcing efficacy of heroin in the addict resulting in decreased drug seeking behavior. The beneficial effect of DG-AVP on methadone detoxification of some heroin addicts justifies detailed analysis of the activity of these systems during drug induced changes in experimental animals, as well as in human addicts.

REFERENCES

Barchas, J. D., Akil, H., Elliott, G. R., Holman, R. B., & Watson, S. J. (1978). Behavioral neurochemistry: Neuroregulators and behavioral states. *Science, 200,* 964–973.

Bhargava, H. N., Walter, R., & Ritzmann, R. F. (1980). Development of narcotic tolerance and physical dependence: Effects of Pro-Leu-Gly-NH$_2$ and cyclo (Leu-Gly). *Pharmacology, Biochemistry and Behavior, 12,* 73–77.

Bohus, B. (1980). Vasopressin, oxytocin, and memory: Effects on consolidation and retrieval processes. *Acta Psychologica Belgica, 80.*

Bohus, B., & de Wied, D. (1966). Inhibitory and facilitatory effect of two related peptides on extinction of avoidance behavior. *Science, 153,* 318–320.

Chiu, S., & Mishira, R. K. (1979). Antagonism of morphine induced catalepsy by L-Prolyl-L-Leucyl-Glycinamide. *European Journal of Pharmacology, 53,* 119–125.

Clouet, D. H., & Iwatsubo, K. (1975). Mechanism of tolerance to and dependence on narcotic analgesic drugs. *Annual Review of Pharmacology, 15,* 49–71.

Cohen, M., Keats, A. S., Krivoy, W. A., & Ungar, G. (1965). Effect of actinomycin D on morphine tolerance. *Proceedings of the Society for Experimental Biology and Medicine, 119,* 381–384.

Contreras, P. C., & Takemori, A. E. (1980). The effects of prolyl-leucyl-glycinamide on morphine tolerance and dependence. *Federation Proceedings, 39,* 845.

Cools, A. R., Broekkamp, C.L.E., Gieles, L.C.M., Megens, A., & Mortiaux H.J.G.M. (1977). Site of action of development of partial tolerance to morphine in cats. *Psychoneuroendocrinology, 2,* 17–33.

Cox, B. M., & Osman, O. H., (1970). Inhibition of the development of tolerance to morphine in rats by drugs which inhibit ribonucleic acid or protein synthesis. *British Journal of Pharmacology, 38,* 157–170.

Cox, B. M., Ginsburg, M., & Osman, O. H. (1968). Acute tolerance of narcotic analgesic drugs in rats. *British Journal of Pharmacology, 33,* 245–256.

de Wied, D. (1976). Behavioral effects of intraventricularly administered vasopressin and vasopressin fragments. *Life Sciences, 19,* 685–690.

de Wied, D., & Bohus, B. (1979). Modulation of memory processes by neuropeptides of hypothalamic–neurohypophyseal origin. In M.A.B. Brazier (Ed.), *Brain Mechanisms in Memory and Learning: From the Single Neuron to Man,* pp. 139–145. New York: Raven.

de Wied, D., & Gispen, W. H. (1976). Impaired development of tolerance to morphine analgesia in rats with hereditary diabetes insipidus. *Psychopharmacologia, 46,* 27–29.

de Wied, D., Witter, A., & Greven, H. M. (1975). Behaviorally active ACTH analogs. *Biochemical Pharmacology, 24,* 1463–1468.

Dorsa, D. M., & van Ree, J. M. (1979). Modulation of substantia nigra self-stimulation by neuropeptides related to neurohypophyseal hormones. *Brain Research*, **172**, 367–371.

Feinberg, M. P., & Cochin, J. (1977). Studies on tolerance II. The effect of timing on inhibition of tolerance to morphine by cycloheximide. *Journal of Pharmacology and Experimental Therapeutics*, **203**, 332–339.

Friedman, E., Friedman, J., & Gershon, S. (1973). Dopamine synthesis: Stimulation by a hypothalamic factor. *Science*, **182**, 831–832.

Geller, A., Robustelli, E., Barondes, S. H., Cohen, H. D., & Jarvik, M. E. (1969). Impaired performance by posttrial injections of cycloheximide in a passive avoidance task. *Psychopharmacologia*, **14**, 371–376.

Hoffman, P. L., Ritzmann, R. F., Walter, R., & Tabakoff, B. (1978). Arginine vasopressin maintains ethanol tolerance. *Nature*, **276**, 614–616.

Hoffman, P. L., Ritzmann, R. F., & Tabakoff, B. (1979). The influence of arginine vasopressin and oxytocin on ethanol dependence and tolerance. *In* M. Galanter (Ed.), *Currents in Alcoholism* V, pp. 5–16. New York: Grune & Stratton.

Jarvik, M. E. (1972). Effects of chemical and physical treatments on learning and memory. *Annual Review of Psychology*, **23**, 457–486.

Kalant, H. (1978). Behavioral criteria for tolerance and physical dependence. *In* J. Fishman (Ed.), *The Bases of Addiction*, pp. 199–220. Berlin: Dahlem Konferenzen.

Kalant, H., Engle, J. A., Goldberg, L., Giffiths, R. R., Jaffe, J. H., Krasnegor, N. A., Mello, N. K., Mendelson, J. H., Thompson, T., & van Ree, J. M. (1978a). Behavioral aspects of addiction. *In* J. Fishman (Ed.), *The Bases of Addiction*, pp. 463–496. Berlin: Dahlem Konferenzen.

Kalant, H., Mucha, R. F., & Niesink, R. (1978b). Effects of vasopressin and oxytocin fragments on ethanol tolerance. *Proceedings Canadian Biology Society*, **21**, 71.

Kastin, A. J., Olson, R. D., Ehrensing, R. H., Berzas, M. C., Schally, A. V., & Coy, D. H. (1979). MIF-I's differential actions as an opiate antagonist. *Pharmacology, Biochemistry and Behavior*, **11**, 721–723.

Kesner, R. P., & Priano, D. J. (1977). Time-dependent disruptive effects of periaqueductal gray stimulation on development of morphine tolerance. *Behavioral Biology*, **21**, 462–469.

Kesner, R. P., Priano, D. J., & Dewitt, J. R. (1976). Time-dependent disruption of morphine tolerance by electroconvulsive shock and frontal cortical stimulation. *Science*, **194**, 1079–1081.

Koob, G. F., Balcom, G. J., & Meyerhoff, J. L. (1976). Increases in intracranial self-stimulation in the posterior hypothalamus following unilateral lesions in the locus coeruleus. *Brain Research*, **101**, 554–560.

Kostrzewa, R. M., Kastin, A. J., & Sobrian, S. K. (1978). Potentiation of apomorphine action in rats by L-prolyl-L-leucyl-glycine amide. *Pharmacology, Biochemistry and Behavior*, **9**, 375–378.

Krivoy, W. A., Zimmermann, E., & Lande, S. (1974). Facilitation of development of resistance to morphine analgesia by desglycinamide^9-lysine vasopressin. *Proceedings of the National Academy of Sciences, U.S.A.*, **71**, 1852–1856.

Lal, H. (1973). Narcotic dependence, narcotic action, and dopamine receptors. *Life Sciences*, **17**, 483–496.

Mello, N. K. & Mendelson, J. H. (1979). Effects of the neuropeptide DG-AVP on morphine and food self-administration by dependent rhesus monkey. *Pharmacology, Biochemistry and Behavior*, **10**, 415–419.

Mucha, R. F., & Kalant, H. (1979). Failure of prolyl-leucyl-glycinamide to alter analgesia measured by the Takemori test in morphine-pretreated rats. *Journal of Pharmacy and Pharmacology*, **31**, 572–573.

Phillips, A. G., & Nikaido, R. S. (1975). Disruption of brain stimulation induced feeding by dopamine receptor blockade. *Nature (London)*, **258**, 750–751.

Plotnikoff, N. P., & Kastin, A. J. (1976). Neuropharmacology of hypothalamic releasing factors. *Biochemical Pharmacology*, 25, 363–365.

Ramachandran, J. (1973). The structure and function of adrenocorticotropin. In C. H. Li (Ed.), *Hormonal Proteins and Peptides*, pp. 1–28. New York: Academic Press.

Rigter, H., & Crabbe, J. C. (1980). Neurohypophyseal peptides and ethanol. In D. de Wied and P. van Keep (Eds.), *Hormones and the Brain*, pp. 263–275. Lancaster: MTP Press.

Schmidt, W. K., Holaday, J. W., Loh, H. H., & Way, E. L. (1978). Failure of vasopressin and oxytocin to antagonize acute morphine antinociception or facilitate narcotic tolerance development. *Life Sciences*, 23, 151–158.

Schulz, H., Kovacs, G. L., & Telegdy, G. (1979). Action of posterior pituitary neuropeptides on the nigrostriatal dopaminergic system. *European Journal of Pharmacology*, 57, 185–190.

Siegel, S. (1976). Morphine analgesic tolerance: Its situation specificity supports a Pavlovian conditioning model. *Science*, 193, 323–325.

Siegel, S. (1978). A Pavlovian conditioning analysis of morphine tolerance. In N. A. Krasnegor (Ed.), *Behavioral Tolerance: Research and Treatment Implications*, pp. 27–53. Rockville, Md.: NIDA Research Monograph 18, U.S. Government Printing Office.

Smith, A. A., Karmin, M., & Gavitt, J. (1966). Blocking effect of puromycin, ethanol, and chloroform on the development of tolerance to an opiate. *Biochemical Pharmacology*, 15, 1877–1879.

Stolerman, I. P., Bunker, P., Johnson, C. A., Jarvik, M. E., Krivoy, W. A. & Zimmermann, E. (1976). Attenuation of morphine tolerance development by electroconvulsive shock in mice. *Neuropharmacology*, 15, 309–313.

Székely, J. I., Miglécz, E., Dunai-Kovacs, Z., Tarnawa, I., Rónai, A. Z., Gráf, L., & Bajusz, S. (1979). Attenuation of morphine tolerance and dependence by α-melanocyte-stimulating hormone (α-MSH). *Life Sciences*, 24, 1931–1938.

Takemori, A. E. (1975). Neurochemical bases for narcotic tolerance and dependence. *Biochemical Pharmacology*, 24, 2121–2126.

Tanaka, M., de Kloet, E. R., de Wied, D., & Versteeg, D.H.G. (1977). Arginine-8-vasopressin affects catecholamine metabolism in specific brain nuclei. *Life Sciences*, 20, 1799–1808.

van Beek-Verbeek, G., Fraenkel, P. J., Geerlings, J. M., van Ree, J. M., & de Wied, D. (1979). Desglycinamide-arginine-vasopressin in methadone detoxification of heroin addicts. *Lancet*, 2, 738–739.

van Ree, J. M. (1979). Reinforcing stimulus properties of drugs. *Neuropharmacology*, 18, 963–969.

van Ree, J. M. (1980). Neurohypophyseal hormones and addiction. In D. de Wied & P. van Keep (Eds.), *Hormones and the Brain*, pp. 167–173. Lancaster: MTP Press.

van Ree, J. M., & de Wied, D. (1976). Prolyl-leucyl-glycinamide (PLG) facilitates morphine dependence. *Life Sciences*, 19, 1331–1340.

van Ree, J. M., & de Wied, D. (1977a). Effect of neurohypophyseal hormones on morphine dependence. *Psychoneuroendocrinology*, 2, 35–41.

van Ree, J. M., & de Wied, D. (1977b). Modulation of heroin self-administration by neurohypophyseal principles. *European Journal of Pharmacology*, 43, 199–202.

van Ree, J. M., & de Wied, D. (1977c). Heroin self-administration is under control of vasopressin. *Life Sciences*, 2, 315–320.

van Ree, J. M., & de Wied, D. (1980a). Brain peptides and psychoactive drug effects. In Israel, Claser, Kalant, Popham, Schmidt, & Smart (Eds.), *Research Advances in Alcohol and Drug Problems VI*. New York: Plenum.

van Ree, J. M., & de Wied, D. (1980b). Involvement of neurohypophyseal peptides in drug-mediated adaptive responses. *Pharmacology, Biochemistry and Behavior*, 13, (Supplement 3), 257–263.

van Ree, J. M., de Wied, D., Bradbury, A. F., Hulme, E. C., Smyth, D. G., & Snell, C. R. (1976). Induction of tolerance to the analgesic action of lipotropin C-fragment. *Nature*, 264, 792–794.

van Ree, J. M., Bohus, B., Versteeg, D.H.G., & de Wied, D. (1978a). Neurohypophyseal principles and memory processes. *Biochemical Pharmacology, 27,* 1793–1800.

van Ree, J. M., Slangen, J. L., & de Wied, D. (1978b). Intravenous self-administration of drugs in rats. *Journal of Pharmacology and Experimental Therapeutics, 204,* 547–557.

van Wimersma Greidanus, Tj. B., Tjon Kon Fat-Bronstein, H., & van Ree, J. M. (1978). Antisera to pituitary hormones modulate development of tolerance to morphine. *In* J. M. van Ree, & L. Terenius (Eds.), *Characteristics and Function of Opioids,* pp. 73–74. Amsterdam: Elsevier North Holland.

van Wimersma Greidanus, Tj. B., van Ree, J. M., & Versteeg, D.H.G. (1980). Neurohypophyseal peptides and avoidance behavior: The involvement of vasopressin and oxytocin in memory processes. *In* C. Ajmone Marsan, & W. Z. Traczyk (Eds.), *Neuropeptides and Neural Transmission,* pp. 293–300. New York: Raven.

Versteeg, D.H.G., Tanaka, M., & de Kloet, E. R. (1978). Catecholamine concentrations and turnover in discrete regions of the brain of the homozygous Brattleboro rat deficient in vasopressin. *Endocrinology, 103,* 1654–1661.

Versteeg, D.H.G., Tanaka, M., de Kloet, E. R., van Ree, J. M., & de Wied, D. (1978). Prolyl-leucyl-glycinamide (PLG): Regional effects on α-MPT-induced catecholamine disappearance in rat brain. *Brain Research, 143,* 561–566.

Versteeg, D.H.G., de Kloet, E. R., van Wimersma Greidanus, Tj. B., & de Wied, D. (1979). Vasopressin modulates the activity of catecholamine containing neurons in specific brain regions. *Neuroscience Letters, 11,* 69–73.

Walter, R., Ritzmann, R. F., Bhargava, H. N., Rainbow, Th. C., Flexner, L. B., & Krivoy, W. A. (1978). Inhibition by Z-Pro-D-Leu of development of tolerance to and physical dependence on morphine in mice. *Proceedings of the National Academy of Sciences, U.S.A., 75,* 4573–4576.

Walter, R., Ritzmann, R. F., Bhargava, H. N., & Flexner, L. B. (1979). Prolyl-leucylglycinamide, cyclo (leucylglycine), and derivatives block development of physical dependence on morphine in mice. *Proceedings of the National Academy of Sciences, U.S.A., 76,* 518–520.

Wauquier, A., & Rolls, E. T. (1976). *Brain-stimulation Reward.* Amsterdam: North Holland.

Way, E. L. (1978). Common and selective mechanisms in drug dependence. *In* J. Fishman (Ed.), *The Bases of Addiction,* pp. 333–352. Berlin: Dahlem Konferenzen.

Wise, R. A. (1978). Catecholamine theories of reward: A critical review. *Brain Research, 152,* 215–247.

CHAPTER *19*

Vasopressin and Oxytocin in Learning and Memory

Tjeerd B. van Wimersma Greidanus, Béla Bohus, and David de Wied

The neurohypophyseal peptide vasopressin facilitates memory in numerous aversive and appetitive tasks, and reduces or reverses the effects of various amnesic agents. In general, the neurohypophyseal peptide oxytocin affects behavior in an opposite direction to vasopressin. Structure–activity studies of the effects of these peptides have shown that their behavioral and endocrine effects may be dissociated. Furthermore, it may be that vasopressin and oxytocin are precursor molecules for highly active neuropeptides that differentially affect consolidation and retrieval of memory. This interpretation is also supported by data showing that: (1) extremely low intracerebroventricular doses of vasopressin (25 pg), which approach physiological levels, are behaviorally active; (2) removal of the posterior pituitary impairs avoidance behavior in a way that is reversible by vasopressin administration; (3) *diabetes insipidus* rats, naturally deficient in vasopressin, have impaired memory function that is corrected with vasopressin; (4) intracerebroventricular injection of vasopressin antiserum impairs passive avoidance retention, whereas intracerebroventricular injection of oxytocin antiserum improves retention. These results also suggest that the site of action of these behaviorally active neurohypophyseal peptides is in the central nervous system, and that they may act via a direct hypothalamic–limbic neurosecretory pathway.

413

ENDOGENOUS PEPTIDES
AND LEARNING AND MEMORY PROCESSES

I. INTRODUCTION

Vasopressin and oxytocin, two peptide hormones of hypothalamic origin that are secreted into the bloodstream by the posterior lobe (*pars nervosa*) of the pituitary, exert pronounced effects on behavior by direct action on the brain. In addition to these hypothalamic–neurohypophyseal pathways, an extrahypothalamic peptidergic neuronal system also exists. The fibers of this system terminate in various limbic and midbrain structures (Buijs, 1978; Buijs, Swaab, Dogterom, & van Leeuwen, 1978; Sofroniew & Weindl, 1978a). These neurohypophyseal nonapeptides consist of a ring structure of six amino acids and a C-terminal tail portion of three amino acids. The ring structure is closed by a disulfide bridge between two cysteine residues in Positions 1 and 6.

Behavioral research on these peptides began when it was observed that pitressin, a crude extract of posterior pituitary tissue containing considerable amounts of vasopressin, induced rats to maintain a high response level during extinction of conditioned avoidance behavior (de Wied & Bohus, 1966). This inhibitory action on extinction of a conditioned avoidance response (CAR) was observed in various active avoidance situations and could also be established by a single injection of synthetic vasopressin. In fact, a single subcutaneous (s.c.) injection of vasopressin resulted in a long-term, dose-dependent inhibition of extinction of the CAR, which lasted long after the disappearance of the injected peptide (de Wied, 1971; de Wied, Bohus, & van Wimersma Greidanus, 1974; van Wimersma Greidanus, Bohus, & de Wied, 1973). This suggested that vasopressin triggers a long-term effect on the maintenance of a learned response, probably by facilitation of memory processes.

In addition to active avoidance behavior, passive (inhibitory) avoidance retention is affected by vasopressin (Bohus, Ader, & de Wied, 1972; Kovács, Vécsei, & Telegdy, 1978; Krejĉi, Kupková, Metys, Barth, & Jost, 1979). Vasopressin facilitates the retention of a sexually motivated T-maze choice behavior in the male rat (Bohus, 1977), delays extinction of an appetitive discrimination task (Hostetter, Jubb, & Kozlowski, 1977), affects approach behavior to an imprinting stimulus (Martin & van Wimersma Greidanus, 1979) and delays the postcastration decline in copulatory behavior of male rats (Bohus, 1979a). Accordingly, the behavioral effects are not restricted to aversively motivated responses. The majority of these effects may be explained by influences on memory processes. The findings that vasopressin is able to prevent or reverse amnesia induced by electroconvulsive shock, CO_2-inhalation, pentylenetetrazol seizures, or the protein synthesis inhibitor puromycin (Bookin & Pfeifer, 1977; Lande, Flexner, & Flexner, 1972; Pfeifer & Bookin, 1978; Rigter, van Riezen, & de Wied, 1974) also support a memory hypothesis.

It is generally accepted that memory processes consist of a consolidation (storage) phase that follows acquisition, and a retrieval phase during which the stored information is recalled whenever a number of cues, present at acquisition, reoccur. It was proposed (McGaugh, 1966) that treatments given shortly after learning affect consolidation processes only, because specific influences on motivation, attention, perception, and motor activity can be excluded. In addition, postlearning treatments affect memory processes in a time-dependent manner: The shorter the interval between learning and treatment, the more profound the facilitation or attenuation of the later retention of the behavior (McGaugh, Zornetzer, Gold, & Landfield, 1972). Treatments given prior to the retention test, usually 24 hr or more after learning, are considered to influence retrieval of memory.

Our observations in a pole-jumping active avoidance situation suggested that subcutaneously administered vasopressin affected the maintenance of the response during extinction in a time-dependent manner: Lysine[8]-vasopressin (LVP) increased resistance to extinction when it was given immediately after the last acquisition session. The peptide was much less effective when the training–treatment period lasted longer than 3 hr (de Wied, 1971). These findings have recently been replicated in the rat in a passive avoidance paradigm using intracerebroventricular (i.c.v.) administration of arginine[8]-vasopressin (AVP). Administration of AVP in the amount of 1 ng immediately but not 6 hr after learning resulted in enhanced retention of passive avoidance behavior. Postponing the treatment for 3 hr after learning led to a slight facilitation of later retention. AVP given shortly before the retention test also facilitated passive avoidance retention (Bohus, Kovács, & de Wied, 1978a). These observations suggest that both consolidation and retrieval of memory are affected by vasopressin. The memory hypothesis can be generalized across aversive and appetitive conditions, as posttraining vasopressin administration facilitated retention of a T-maze choice behavior (Bohus, 1977) and postcastration copulatory behavior in male rats (Bohus, 1979a).

Our initial observations suggested that peripherally administered oxytocin mimicked the effect of vasopressin on avoidance extinction (de Wied & Gispen, 1977). However, Schulz, Kovács, & Telegdy (1974) and Kovács, Vécsei, and Telegdy (1978) found that oxytocin affected active and passive avoidance behavior in an opposite direction to vasopressin. Also, following i.c.v. administration of the peptides, we observed an effect of oxytocin that was opposite to that of vasopressin when doses between 0.1 and 1.0 ng were administered immediately after training in a passive avoidance task. These amounts of oxytocin attenuated the retention of the response. Following higher doses, this effect disappeared or an opposite effect was observed. In active avoidance behavior, i.c.v. injection of oxytocin (1.0 ng) only slightly affected acquisition and extinction (Bohus, Urban, van Wimersma Greidanus, & de Wied, 1978b). Oxytocin may be a naturally occurring amnesic neuropeptide. One ng of the peptide attenuated later retention of a passive avoidance response in a time-dependent manner if administered i.c.v. after learning. Oxytocin also impaired retention performance when administered prior to the retention test (Bohus et al., 1978b). These observations suggested that oxytocin affects both consolidation and retrieval of memory. It is not yet known whether this peptide is amnesic in appetitive situations.

II. STUDIES OF STRUCTURE–ACTIVITY RELATIONSHIPS

Shortly after the recognition that vasopressin was the active peptide in posterior pituitary extract (pitressin), which caused a long-term maintenance of an avoidance response (de Wied, 1971), it was found that the behavioral and endocrine effects of the peptides could be dissociated. Removal of the C-terminal glycinamide residue did not affect behavioral activity, but abolished the endocrine properties of LVP (de Wied, Greven, Lande, & Witter, 1972).

Detailed studies of the structural requirements for delaying extinction of an active avoidance response by single s.c. injections of vasopressin or its fragments have been performed by Walter, van Ree, and de Wied (1978). For these studies, a pole-jumping avoidance paradigm was used. The rats were trained to avoid an electric footshock administered through the grid floor of the conditioning box, by jumping onto a pole that was situated vertically in the center of the box. Onset of light above the pole for 5 sec prior to footshock was used as a conditioned stimulus (CS). As soon as the rat jumped on the pole, the CS or the CS/US (footshock) combination was terminated. The rats were trained for 3 days (10 trials on each day) to acquire the conditioned response. Rats that made seven or more avoidance responses during the third acquisition session were injected with the peptides immediately after training. Extinction sessions were run 24, 48, and 120 hr later on the same schedule as that used during acquisition, except that footshock was not supplied if the rat did not jump onto the pole within 5 sec. The CS was terminated after 5 sec.

It appeared that certain amino acid residues in the ring structure of arginine[8]-vasopressin (AVP) are more essential for the expression of an effect on avoidance extinction than in Positions 8 and 9 of the C-terminal linear position of the peptide molecule. Residues in Positions 2, 3, and 5 were critical. For example, an amino acid residue with an aromatic side chain in Position 3 was critical for high activity in this behavioral test. If a phenylalanine is replaced by less aromatic residues, such as thiamine or isoleucine, activity declines. Replacement of tyrosine[2] by alanine also dramatically reduced the activity. The presence of asparagine in Position 5 is essential for the behavioral activity. It was suggested that, similar to the endocrine effects, Position 3 may be critical for receptor binding, and the side chains of residues in Positions 2 and 5 are important for the activation of a receptor. The C-terminal dipeptide portion of AVP and oxytocin were almost inactive in this test situation.

Because the structure–activity studies in the pole-jumping situation failed to show any effects of oxytocin (which are opposite to the findings attained with vasopressin), we decided to investigate the structural requirements for the facilitatory and amnesic action of neurohypophyseal peptides in a memory test. The main conclusion of these studies is that vasopressin and oxytocin may serve as precursor molecules for highly active neuropeptides that may differentially affect consolidation and retrieval processes (Bohus, Kovács, Greven, & de Wied, 1978c; de Wied & Bohus, 1979).

In these experiments, fragments and analogs of AVP and oxytocin (OXT) were administered i.c.v. in a dose of 1 ng either immediately after training in a passive avoidance task (Ader, Weijnen, & Moleman, 1972) or before the first retention test. Briefly, this test uses the innate preference of rats for darkness over light. The animals were adapted to the dark chamber and, following four pretraining trials (approach dark from light), they received a single learning trial by applying an unavoidable electric footshock in the dark compartment. Retention of the passive avoidance response (avoidance of the dark) was tested at 24 and 48 hr after the learning trial.

AVP contains two active sites in the molecule that facilitate consolidation of memory. One active site is located in the covalent ring structure, pressinamide. The other site is present in the C-terminal linear portion of the molecule. Both Pro-Arg-Gly-NH$_2$ and Lys-Gly-NH$_2$ were active. Practically the same holds for the influence of AVP on retrieval of memory. Although pressinamide was not effective, elongation of the ring structure with proline (AVP$_{1-7}$) results in a highly active peptide. The second site in the C-terminus of the molecule was also active. The amnesic effect of the oxytocin molecule was absent when the peptide is shortened from the C-terminus. OXT$_{1-8}$ was still active, but OXT$_{1-7}$ and OXT$_{1-6}$ affected consolidation and retrieval of memory differently from OXT. One-tenth of one nanogram of these peptides were amnesic, but 1.0 ng induced a facilitation of memory. The C-terminal linear portion of OXT did not affect consolidation processes, but both Pro-Leu-Gly-NH$_2$ and Leu-Gly-NH$_2$ facilitated retrieval. It is worthwhile to mention that structural requirements for reversal of experimental amnesias by vasopressin and fragments (Flexner et al., 1977; Rigter & Crabbe, 1979; Walter, Hoffman, Flexner, & Flexner, 1975) are practically identical to those for "retrieval-type" effects.

Structure–activity studies suggest that fragments of the neurohypophyseal peptides are behaviorally active and that the information content of oxytocin fragments may differ from that of the parent molecule. The differences in the behavioral activity of OXT in the pole-jumping and passive avoidance experiments may be due to a different fate of the peptide following systemic or intraventricular administration. One cannot, however, exclude the possibility that the performance of the rats during pole-jumping extinction and passive avoidance retention does reflect different memory processes.

III. THE PHYSIOLOGICAL ROLE OF ENDOGENOUS NEUROHYPOPHYSEAL HORMONES IN MEMORY PROCESSES

The suggestion that vasopressin may be physiologically involved in memory processes was evidenced when picogram amounts were found to affect behavior after i.c.v. administration. In fact, as little as 25 pg of AVP injected into a lateral ventricle of a rat brain inhibited extinction of a CAR (de Wied, 1976). This amount approaches the physiological levels of vasopressin in cerebrospinal fluid (CSF).

A. Posterior Lobectomy

Experiments dating from 1964, in which the posterior lobe of the pituitary had been removed from rats, pointed to an essential role of this "gland" in avoidance behavior (de Wied, 1965). Posterior lobectomized rats were found to display a faster extinction of a CAR as compared to sham-operated animals. This derangement could be restored by treatment with pitressin. Also, the disturbed behavior of totally hypophysectomized rats could be restored by treatment with pitressin or with synthetic vasopressin preparations (Bohus, Gispen, & de Wied, 1973; de Wied, 1964). These data suggested an important role for vasopressin in brain processes involved in avoidance behavior. However, recent findings of vasopressin-containing neurons running directly toward limbic system structures (Buijs, 1978; Buijs et al., 1978), which are supposed to be the site(s) of action of vasopressin on active avoidance behavior (van Wimersma Greidanus et al., 1973; van Wimersma Greidanus, Bohus, & de Wied, 1974; van Wimersma Greidanus, Bohus, & de Wied, 1975a; van Wimersma Greidanus & de Wied, 1976a), make it necessary to reevaluate the findings with posterior lobectomized or totally hypophysectomized animals.

B. Diabetes Insipidus Rats

Convincing evidence for a physiological involvement of vasopressin in brain processes related to memory was obtained from experiments with Brattleboro rats. A homozygous variant of this strain has hereditary hypothalamic diabetes insipidus (HO-DI) because it lacks the ability to synthesize vasopressin, due to a mutation of a single pair of autosomal loci (Valtin, 1967; Valtin &

Schroeder, 1964). Because this topic is extensively discussed by Bailey and Weiss in Chapter 17, only the main facts obtained from our experiments with these animals will be mentioned here.

Memory function of HO-DI rats is impaired in one-trial passive avoidance when retention is tested 24 hr or more after training. Vasopressin (AVP or DG-LVP), administered immediately after the single acquisition trial, restores the disturbed passive avoidance behavior (de Wied, Bohus, & van Wimersma Greidanus, 1975). This favors the hypothesis that memory processes are disturbed in the absence of vasopressin. However, HO-DI rats are able to acquire fear-motivated responses in multiple-trial paradigms (shuttlebox or pole-jump) (Bohus, van Wimersma Greidanus, & de Wied, 1975; Celestian, Carey, & Miller, 1975). Furthermore, complete retention of passive avoidance behavior is obtained in HO-DI rats when retention is tested shortly after the learning trial. Accordingly, deficits in the maintenance of avoidance behaviors cannot be due to learning impairment, suggesting that memory is impaired in the absence of vasopressin. These data were partly confirmed by Bailey and Weiss (1979), who reported poorer passive avoidance behavior in HO-DI rats as compared with heterozygous diabetes insipidus (HE-DI) animals. However, under some of the conditions employed by these authors, no differences between HO-DI and HE-DI animals were found.

Whether or not the absence of vasopressin is the primary cause of the abnormalities of HO-DI rats must still be determined. Low concentrations of OXT in the pituitaries of HO-DI rats and high levels of this hormone in their plasma have been reported (Dogterom, van Wimersma Greidanus, & Swaab, 1977; van Wimersma Greidanus & de Wied, 1977). We have observed that oxytocin antiserum does not substantially influence the behavior of HO-DI rats (Bohus, 1979b). This indicates that oxytocin is most probably active only in the presence of vasopressin.

C. Intracerebroventricular Administration of Antisera to Neurohypophyseal Hormones

Another strategy for investigating the physiological significance of endogenous neurohypophyseal hormones in brain processes related to memory, is to effectively neutralize bioavailable peptides in the brain by i.c.v. administration of specific antisera. Administration of vasopressin antiserum into one lateral ventricle of the brain, immediately after training, induced a marked deficit in one-trial passive avoidance, when the rats were tested 6 hr or more later. No interference with the behavior was found when retention was tested less than 2 hr after administration of the vasopressin antiserum. Testing at 3 and 4 hr resulted in intermediate levels of retention (van Wimersma Greidanus & de

Wied, 1976b). Intravenous injection of a hundred times as much vasopressin antiserum, which effectively removed the peptide from the circulation, as indicated by the virtual absence of vasopressin in the urine and by a marked diuresis, did not affect passive avoidance behavior. These results indicate the importance of centrally available vasopressin in relation to avoidance behavior, and in particular to memory consolidation (van Wimersma Greidanus, Dogterom, & de Wied, 1975b).

If the injection of anti-vasopressin serum was postponed for up to 2 hr after training, the treatment still resulted in a marked disturbance of passive avoidance behavior, which is in agreement with the idea that processes of memory consolidation last for several hours. Moreover, administration of antivasopressin serum 1 hr prior to the retention test resulted in passive avoidance deficits. This supports the hypothesis that vasopressin is also involved in retrieval processes (van Wimersma Greidanus & de Wied, 1976b).

Interestingly, the effect of neutralization of centrally available vasopressin by i.c.v. injection of antisera can be overcome by systemic administration of vasopressin. Administration of 2 μl antivasopressin serum (i.c.v.) in a dilution of 1:10 results in a significant reduction of passive avoidance latencies, which can be normalized by s.c. administration of vasopressin in an amount of approximately 15 μg. This is true for treatment with antiserum and peptide after the learning trial, as well as prior to the retention test (van Wimersma Greidanus & van Egmond, unpublished data).

Further evidence for the role of central vasopressin in avoidance behavior was obtained from experiments using multiple-trial learning paradigms. Rats treated with vasopressin antiserum (i.c.v.), during acquisition of a CAR tended to have a slower rate of acquisition, as compared with control rats, but both groups reached the learning criterion. However, extinction of the avoidance response was faster in the rats injected with vasopressin antiserum, although the treatment was performed during acquisition and was discontinued during extinction (van Wimersma Greidanus, Bohus, & de Wied, 1975c). These results reinforce the notion that learning can take place in the absence of vasopressin, or at least when reduced amounts of this peptide are available in the brain but that, under these conditions, the maintenance of the behavior is disturbed.

Neutralization of bioavailable OXT in the brain by i.c.v. administration of a specific antiserum, immediately after the training trial (Bohus et al., 1978a), or 1 hr before the first extinction session, induced longer passive avoidance latencies than those found in controls. Rats that received antioxytocin serum before each acquisition session in the pole-jump avoidance situation, showed a weak but significant increase in resistance to extinction of the CAR. These results again suggest opposing effects of OXT and vasopressin on memory processes.

Neutralization of centrally available vasopressin by i.c.v. administration of antivasopressin serum also interferes with the development of tolerance to morphine (van Wimersma Greidanus, Tjon Kon Fat-Bronstein, & van Ree, 1978). Studies with rats in a footshock test paradigm revealed that administration of antivasopressin serum delays the development of tolerance to morphine. Control rats showed an increasing responsiveness to footshock during subsequent morphine injections, presumably reflecting the development of tolerance. Rats given both morphine and antivasopressin serum showed the same responsiveness to footshock in subsequent sessions. The development of tolerance to morphine was thus inhibited. Time gradient studies suggest that endogenous, centrally available vasopressin may be physiologically involved in the development of tolerance to morphine in a manner that is comparable to its role in memory processes (van Wimersma Greidanus et al., 1978).

IV. HOW DO VASOPRESSIN AND OXYTOCIN REACH THEIR SITE(S) OF ACTION?

Morphological and functional evidence has accumulated to suggest that hypothalamic hormones are released from neural tissue into the fluid of the cerebral ventricles. Recent studies indicate that neurosecretory axons run to the ependyma of the ventricles (Leonhardt, 1974; Rodriguez, 1970; Sterba, 1974a,b; Vigh-Teichmann & Vigh, 1974; Vorherr, Bradbury, Houghoughi, & Kleeman, 1968; Wittkowski, 1968). It has been suggested that neurohumors are released directly into CSF from these terminals (Heller & Zaidi, 1974; Sterba, 1974a).

Vasopressin and/or vasopressin and OXT have been shown to be present in the CSF of rats (Dogterom et al., 1977; Dogterom, van Wimersma Greidanus, & de Wied, 1978; van Wimersma Greidanus, Croiset, Goedemans, & Dogterom, 1979); rabbits (Heller, Hasan, & Saifi, 1968; Unger, Schwarzberg, & Schulz, 1974); dogs (Dogterom et al., 1977); and humans (Dogterom et al., 1978; Gupta, 1969; Pavel, 1970). In fact in rats, after hypophysectomy, vasopressin in the CSF is slightly higher than in the controls. This may suggest a compensatory release of this hormone into the ventricular system in association with decreased blood levels. However, to what extent the anesthesia, which was used in these experiments, affects vasopressin levels in CSF remains to be determined. How the hormones are channeled from the CSF to effector sites is not clear. Perhaps different pools of ependymal cells or some circumventricular organs act as mediators in this respect (Sterba, 1974b). However, it is doubtful whether vasopressin present in the CSF is of importance for the behavioral effect of the peptide. No correlation could be found between behavioral performance and vasopressin levels when measured by radioimmunoassay in peripheral blood, nor was there a significant difference between the levels of vasopressin in the CSF collected from the lateral ventricle (van Wimersma Greidanus et al., 1970) or from the cisterna magna (van Wimersma Greidanus, unpublished data) immediately after retention of a passive avoidance response. The absence of a direct correlation between plasma or CSF levels of vasopressin and behavioral performance suggests an independent release of neurohypophyseal principles into the blood, the CSF, and/or the brain. In fact, direct hypothalamic–limbic neurosecretory pathways exist by which neurohypophyseal peptides could affect the limbic system directly. Extrahypothalamic neurosecretory projections to the choroid plexus, circumventricular organs, and limbic system structures have been shown (Buijs, 1978, Buijs et al., 1978; Kozlowski, Brownfield, & Hostetter, 1978; Sterba, 1974a, 1978; Weindl & Sofroniew 1978a, 1979). Vasopressin- and oxytocin-containing pathways have been traced from the paraventricular nucleus toward the ventral hippocampus, the amygdaloid nuclei, the substantia nigra, the substantia grisea, and other extrahypothalamic structures. (Buijs, 1978, Weindl & Sofroniew, 1978b). Furthermore, vasopressin–neurophysin pathways between the suprachiasmatic nucleus and the dorsal thalamus, and between the lateral septum and the lateral habenulae have been demonstrated (Buijs, 1978; Sofroniew & Weindl, 1978b). Neurophysin–vasopressin-containing terminals have also been found in the medial and lateral septal nuclei, in the periventricular and dorsolateral areas of the thalamus, in parts of the amygdaloid complex, and in the habenular region by Kozlowski et al. (1978). Neurophysin-containing fibers have been observed to course within the stria terminalis (Kozlowski et al., 1978; Weindl & Sofroniew, 1978b, 1979). Recent observations with microinjection of AVP and OXT into limbic–midbrain areas showed that AVP facilitates memory consolidation when injected into the dentate gyri of the hippocampus, dorsal septum, and dorsal raphé nucleus. Oxytocin, on the other hand, caused an attenuation of memory consolidation when injected into the dentate gyri or dorsal raphé nucleus. Accordingly, effective sites of injection overlay those brain areas where vasopressinergic or oxytocinergic nerve fiber endings have been traced (Kovács, Bohus, & Versteeg, 1980). It may well be that the release of neurosecretory peptides from the peptidergic nerve endings is of physiological importance in modulating brain processes related to behavioral adaptation. The effectiveness of i.c.v. administered peptides or their antibodies may be due to passive or active penetration into brain sites that contain peptidergic terminals.

V. PERSPECTIVE

The finding that neurohypophyseal peptides influence brain processes related to memory may be regarded as a breakthrough in neuroendocrine research. The fact that vasopressin and oxytocin and/or their fragments exert opposite effects in this respect suggests a modulatory mechanism, and it is tempting to assume that disturbances in this mechanism could result in alterations in brain function and consequently in behavioral changes. In particular, some derangements in memory function may be caused by a dysfunction of the neu-rohypophyseal system. It is conceivable that some disorders of a psychopath-ological nature may be the result of disturbances in the equilibrium between bioavailable neurohypophyseal entities in the brain. This may be illustrated by the various clinical effects of vasopressin or vasopressin analogs and by the correlation between neurophysin levels and certain physiological performances (Legros, Gilot, Seron, Claessens, Adam, Moeglen, Audibert, & Berchier, 1978).

REFERENCES

Ader, R., Weijnen, J.A.W.M., & Moleman, P. (1972). Retention of a passive avoidance response as a function of the intensity and duration of electric shock. *Psychonomic Science*, **26**, 125–128.

Bailey, W. H., & Weiss, J. M. (1979). Evaluation of a 'memory deficit' in vasopressin-deficient rats. *Brain Research*, **162**, 174–178.

Bohus, B. (1977). Effect of desglycinamide-lysine vasopressin (DG-LVP) on sexually motivated T-maze behavior of the male rat. *Hormones and Behavior*, **8**, 52–61.

Bohus, B. (1979a). Neuropeptide influences on sexual and reproductive behavior. *In* L. Zichella, & P. Pancheri (Eds.), *Psychoneuroendocrinology in Reproduction: Developments in Endocrinology* (Vol. 5), pp. 111–120. Amsterdam: Elsevier, North Holland.

Bohus, B. (1979b). Inappropriate synthesis and release of vasopressin in rats: Behavioral conse-quences and effects of neuropeptides. *Neuroscience Letters*, **Supplement 3**, 329.

Bohus, B., Ader, R., & de Wied, D. (1972). Effects of vasopressin on active and passive avoidance behavior. *Hormones and Behavior*, **3**, 191–197.

Bohus, B., Gispen, W. H., & de Wied, D. (1973). Effect of lysine vasopressin and $ACTH_{4-10}$ on conditioned avoidance behavior of hypophysectomized rats. *Neuroendocrinology*, **11**, 137–143.

Bohus, B., Kovács, G. L., & de Wied, D. (1978a). Oxytocin, vasopressin and memory: Opposite effects on consolidation and retrieval processes. *Brain Research*, **157**, 414–417.

Bohus, B., Kovács, G. L., Greven, H., & de Wied, D. (1978c). Memory effects of arginine-vasopressin (AVP) and oxytocin (OXT): Structural requirements. *Neuroscience Letters*, **Supple-ment 1**, 571.

Bohus, B., Urban, I., van Wimersma Greidanus, Tj. B., & de Wied, D. (1978b). Opposite effects of oxytocin and vasopressin on avoidance behaviour and hippocampal theta rhythm in the rat. *Neuropharmacology*, **17**, 239–247.

Bohus, B., van Wimersma Greidanus, Tj. B., & de Wied, D. (1975). Behavioral and endocrine response of rats with hereditary hypothalamic diabetes insipidus (Brattleboro strain). *Physiology and Behavior*, **14**, 609–615.

Bookin, H. B., & Pfeifer, W. D. (1977). Effect of lysine vasopressin on pentylenetetrazol-induced retrograde amnesia in rats. *Pharmacology, Biochemistry and Behavior*, **7**, 51–54.

Buijs, R. M. (1978). Intra- and extrahypothalamic vasopressin and oxytocin pathways in the rat. Pathways to the limbic system, medulla oblongata and spinal cord. *Cell and Tissue Research*, **192**, 423–435.

Buijs, R. M., Swaab, D. F., Dogterom, J., & van Leeuwen, F. W. (1978). Intra- and extra-hypothalamic vasopressin and oxytocin pathways in the rat. *Cell and Tissue Research*, **186**, 423–433.

Celestian, J. F., Carey, R. J., & Miller, M. (1975). Unimpaired maintenance of a conditioned avoidance response in the rat with diabetes insipidus. *Physiology and Behavior*, **15**, 707–711.

de Wied, D. (1964). Influence of anterior pituitary on avoidance and escape behavior. *American Journal of Physiology*, **207**, 255–259.

de Wied, D. (1965). The influence of the posterior and intermediate lobe of the pituitary and pituitary peptides on the maintenance of a conditioned avoidance response in rats. *International Journal of Neuropharmacology*, **4**, 157–167.

de Wied, D. (1971). Long term effect of vasopressin on the maintenance of a conditioned avoidance response in rats. *Nature (London)*, **232**, 58–60.

de Wied, D. (1976). Behavioral effects of intraventricularly administered vasopressin and vasopressin fragments. *Life Sciences*, **19**, 685–690.

de Wied, D., & Bohus, B. (1966). Long term and short term effects on retention of a conditioned avoidance response in rats by treatment with long acting pitressin and α-MSH. *Nature (London)*, **212**, 1484–1486.

de Wied, D., & Bohus, B. (1979). Modulation of memory processes by neuropeptides of hypothalamic–neurohypophyseal origin. *In* M.A.B. Brazier (Ed.), *Brain Mechanisms in Memory and Learning: From the Single Neuron to Man*, pp. 139–149. New York: Raven.

de Wied, D., Bohus, B., & van Wimersma Greidanus, Tj. B. (1974). The hypothalamo-neurohypophyseal system and the preservation of conditioned avoidance behavior in rats. *In* D. F. Swaab & J. P. Schade (Eds.), *Integrative Hypothalamic Activity: Progress in Brain Research*, **41**, pp. 417–428. Amsterdam: Elsevier North Holland.

de Wied, D., & Bohus, B., & van Wimersma Greidanus, Tj. B. (1975). Memory deficit in rats with hereditary diabetes insipidus. *Brain Research*, **85**, 152–156.

de Wied, D., & Gispen, W. H. (1977). Behavioral effects of peptides. *In* H. Gainer (Ed.), *Peptides in Neurobiology*, pp. 397–448. New York: Plenum.

de Wied, D., Greven, H. M., Lande, S., & Witter, A. (1972). Dissociation of the behavioural and endocrine effects of lysine vasopressin by tryptic digestion. *British Journal of Pharmacology*, **45**, 118–122.

Dogterom, J., van Wimersma Greidanus, Tj. B., & de Wied, D. (1978). Vasopressin in cerebrospinal fluid and plasma of man, dog, and rat. *American Journal of Physiology*, **234**, E463–E467.

Dogterom, J., van Wimersma Greidanus, Tj. B., & Swaab, D. F. (1977). Evidence for the release of vasopressin and oxytocin into cerebrospinal fluid: Measurements in plasma and CSF of intact and hypophysectomized rats. *Neuroendocrinology*, **24**, 108–118.

Flexner, J. B., Flexner, L. B., Hoffman, P. L., & Walter, R. (1977). Dose–response relationships in attenuation of puromycin-induced amnesia by neurohypophyseal peptides. *Brain Research*, **134**, 139–144.

Gupta, K. K. (1969). Antidiuretic hormone in cerebrospinal fluid. *Lancet*, **1**, 581.

Heller, H., Hasan, S. H., & Saifi, A. Q. (1968). Antidiuretic activity in the cerebrospinal fluid. *Journal of Endocrinology*, **41**, 273–280.

Heller, H., & Zaidi, S.M.A. (1974). The problem of neurohypophyseal secretion into the cerebrospinal fluid: Antidiuretic activity in the liquor and choroid plexus. *In* A. Mitro (Ed.), *Ependyma and Neurohormonal Regulation*, pp. 229–250. Bratislava: Veda.

Hostetter, G., Jubb, S. L., & Kozlowski, G. P. (1977). Vasopressin affects the behavior of rats in a positively-rewarded discrimination task. *Life Sciences*, **21**, 1323–1328.

Kovács, G. L., Bohus, B., & Versteeg, D.H.G. (1980). The interaction of posterior pituitary neuropeptides with monoaminergic neurotransmission: Significance in learning and memory processes. P. S. McConnell, G. J. Boer, H. J. Romjin, N. E. van de Poll, & M. A. Corner. *Adaptive Capabilities of the Nervous System: Progress in Brain Research.* **53**, 123–140.

Kovács, G. L., Véscei, L., & Telegdy, G. (1978). Opposite action of oxytocin to vasopressin in passive avoidance behavior in rats. *Physiology and Behavior*, **20**, 801–802.

Koslowski, G. P., Brownfield, M. S., & Hostetter, G. (1978). Neurosecretory supply to extra-hypothalamic structures: Choroid plexus, circumventricular organs and limbic system. *In* W.

Bargmann, A. Oksche, A. Polenov, & B. Scharrer (Eds.), *Neurosecretion and Neuroendocrine Activity: Evolution Structure and Function*, pp. 217–227. Berlin: Springer-Verlag.

Krejĉi, I., Kupkova, B., Metys, J., Barth, T., & Jost, K. (1979). Vasopressin analogs: Sedative properties and passive avoidance behavior in rats. *European Journal of Pharmacology*, **56**, 347–353.

Lande, S., Flexner, J. B., & Flexner, L. B. (1972). Effect of corticotropin and desglycinamide[9]-lysine vasopressin on suppression of memory by puromycin. *Proceedings of National Academy of Sciences, U.S.A.*, **69**, 558–560.

Legros, J. J., Gilot, P., Seron, X., Claessens, J., Adam, A., Moeglen, J. M., Audibert, A., & Berchier, P. (1978). Influence of vasopressin on learning and memory. *Lancet*, **1**, 41–42.

Leonhardt, H. (1974). Ependymstrukturen im Dienst des Stofftransportes zwischen Ventrikelliquor und Hirnsubstanz. *In* A. Mitro (Ed.), *Ependyma and Neurohormonal Regulation*, pp. 29–75a. Bratislava: Veda.

Martin, J. T., & van Wimersma Greidanus, Tj. B. (1979). Imprinting behavior: Influence of vasopressin and ACTH analogues. *Psychoneuroendocrinology*, **3**, 261–269.

McGaugh, J. L. (1966). Time-dependent processes in memory storage. *Science*, **153**, 1351–1358.

McGaugh, J. L., Zornetzer, S. F., Gold, P. E., & Landfield, P. W. (1972). Modification of memory systems: Some neurobiological aspects. *Quarterly Reviews of Biophysics*, **5**, 163–186.

Pavel, S. (1970). Tentative identification of arginine vasotocin in human cerebrospinal fluid. *Journal of Clinical Endocrinology*, **31**, 369–371.

Pfeifer, W. D., & Bookin, H. B. (1978). Vasopressin antagonizes retrograde amnesia in rats following electroconvulsive shock. *Pharmacology, Biochemistry and Behavior*, **9**, 261–263.

Rigter, H., & Crabbe, J. C. (1979). Modulation of memory by pituitary hormones and related peptides. *Vitamins and Hormones*, **37**, 153–241.

Rigter, H., van Riezen, H., & de Wied, D. (1974). The effects of ACTH- and vasopressin-analogues on CO_2-induced retrograde amnesia in rats. *Physiology and Behavior*, **13**, 381–388.

Rodriguez, E. M. (1970). Morphological and functional relationship between the hypothalamo-neurohypophyseal system and cerebrospinal fluid. *In* W. Bargmann & B. Scharrer (Eds.), *Aspects of Neuroendocrinology*, pp. 352–365. Berlin: Springer-Verlag.

Schulz, H., Kovács, G. L., & Telegdy, G. (1974). Effect of physiological doses of vasopressin and oxytocin on avoidance and exploratory behaviour in rats. *Acta Physiologica Academiae Scientiarum Hungaricae*, **45**, 211–215.

Sofroniew, M. V., & Weindl, A. (1978a). Extrahypothalamic neurophysin-containing perikarya, fiber pathways and fiber clusters in the rat brain. *Endocrinology*, **102**, 334–337.

Sofroniew, M. V., & Weindl, A. (1978b). Projections from the parvocellular vasopressin- and neurophysin-containing neurons of the suprachiasmatic nucleus. *American Journal of Anatomy*, **153**, 391–430.

Sterba, G. (1974a). Ascending neurosecretory pathways of the peptidergic type. *In* F. Knowles & L. Vollrath (Eds.), *Neurosecretion: The Final Neuroendocrine Pathway*, pp. 38–47. Berlin: Springer-Verlag.

Sterba, G. (1974b). Cerebrospinal fluid and hormones. *In* A. Mitro (Eds.), *Ependyma and Neurohormonal Regulation*, pp. 143–179. Bratislava: Veda.

Sterba, G. (1978). Oxytocinergic extrahypothalamic neurosecretory system of the vertebrates and memory processes. *In* W. Bargmann, A. Oksche, A. Polenov, & B. Scharrer (Eds.), *Neurosecretion and Neuroendocrine Activity: Evolution, Structure and Function*, pp. 293–299. Berlin: Springer-Verlag.

Unger, H., Schwarzberg, H., & Schulz, H. (1974). The vasopressin and oxytocin content in the cerebrospinal fluid of rabbits under changed conditions. *In* A. Mitro (Ed.), *Ependyma and Neurohormonal Regulation*, pp. 251–259. Bratislava: Veda.

Valtin, H. (1967). Hereditary hypothalamic diabetes insipidus in rats (Brattleboro strain). A useful experimental model. *American Journal of Medicine*, **42**, 814–827.

Valtin, H., & Schroeder, H. A. (1964). Familial hypothalamic diabetes insipidus in rats (Brattleboro strain). *American Journal of Physiology*, **206**, 425–430.

van Wimersma Greidanus, Tj. B., Bohus, B., & de Wied, D. (1973). Effects of peptide hormones on behaviour. *International Congress Series*, **273**, 197–201. Amsterdam: Excerpta Medica.

van Wimersma Greidanus, Tj. B., Bohus, B., & de Wied, D. (1974). Differential localization of the influence of lysine vasopressin and of ACTH$_{4-10}$ on avoidance behavior: A study in rats bearing lesions in the parafascicular nuclei. *Neuroendocrinology*, **14**, 280–288.

van Wimersma Greidanus, Tj. B., Bohus, B., & de Wied, D. (1975a). CNS sites of action of ACTH, MSH and vasopressin in relation to avoidance behavior. *In* W. E. Stumpf & L. D. Grant (Eds.), *Anatomical Neuroendocrinology*, pp. 284–289. Basel: Karger.

van Wimersma Greidanus, Tj. B., Bohus, B., & de Wied, D. (1975c). The role of vasopressin in memory processes. *In* W. H. Gispen, Tj. B. van Wimersma Greidanus, B. Bohus, & D. de Wied (Eds.), *Hormones, Homeostasis and the Brain. Progress in Brain Research* (Vol. 42), pp. 135–141. Amsterdam: Elsevier North Holland.

van Wimersma Greidanus, Tj. B., Croiset, G., Goedemans, H., & Dogterom, J. (1979). Vasopressin levels in peripheral blood and cerebrospinal fluid during passive and active avoidance behavior in rats. *Hormones and Behavior*, **12**, 103–111.

van Wimersma Greidanus, Tj. B., & de Wied, D. (1976a). Dorsal hippocampus: A site of action of neuropeptides on avoidance behavior? *Pharmacology, Biochemistry and Behavior*, (Supplement 1), **5**, 29–33.

van Wimersma Greidanus, Tj. B., & de Wied, D. (1976b). Modulation of passive avoidance behavior of rats by intracerebroventricular administration of anti-vasopressin serum. *Behavioral Biology*, **18**, 325–333.

van Wimersma Greidanus, Tj. B., & de Wied, D. (1977). The physiology of the neurohypophyseal system and its relation to memory processes. *In* A. N. Davison (Ed.), *Biochemical Correlates of Brain Structure and Function*, pp. 215–248. London: Academic Press.

van Wimersma Greidanus, Tj. B., Dogterom, J., & de Wied, D. (1975b). Intraventricular administration of anti-vasopressin serum inhibits memory consolidation in rats. *Life Sciences*, **16**, 637–644.

van Wimersma Greidanus, Tj. B., Tjon Kon Fat-Bronstein, H., & van Ree, J. M. (1978). Antisera to pituitary hormones modulate development of tolerance to morphine. *In* J. M. van Ree & L. Terenius (Eds.), *Characteristics and Function of Opioids*, pp. 73–74. Amsterdam: Elsevier North Holland.

Vigh-Teichmann, I., & Vigh, B. (1974). Correlation of CSF contacting neuronal elements to neurosecretory and ependymosecretory structures. *In* A. Mitro (Ed.), *Ependyma and Neurohormonal Regulation*, pp. 281–295. Bratislava: Veda.

Vorherr, H., Bradbury, M.W.B., Houghoughi, M., & Kleeman, C. R. (1968). Antidiuretic hormone in cerebrospinal fluid during endogenous and exogenous changes in its blood level. *Endocrinology*, **83**, 246–250.

Walter, R., Hoffman, P. L., Flexner, J. B., & Flexner, L. B. (1975). Neurohypophyseal hormones, analogs and fragments: Their effect on puromycin-induced amnesia. *Proceedings of the National Academy of Sciences, U.S.A.*, **72**, 4180–4184.

Walter, R., van Ree, J. M., & de Wied, D. (1978). Modification of conditioned behavior of rats by neurohypophyseal hormones and analogues. *Proceedings of the National Academy of Sciences, U.S.A.*, **75**, 2493–2496.

Weindl, A., & Sofroniew, M. V. (1976). Demonstration of extrahypothalamic peptide secreting neurons. A morphologic contribution to the investigation of psychotropic effects of neurohormones. *Pharmakopsychiatrie*, **9**, 226–234.

Weindl, A., & Sofroniew, M. V. (1978a). The functional morphology of vascular and neuronal efferent connections of neuroendocrine systems. *Drug Research*, **28**, 1264–1268.

Weindl, A., & Sofroniew, M. V. (1978b). Neurohormones and circumventricular organs. *In* D. E. Scott, G. P. Kozlowski, & A. Weindl (Eds.), *Brain-Endocrine Interaction III. Neural Hormones and Reproduction*, pp. 117–137. Basel: Karger.

Wittkowski, W. (1968). Elektronenmikroskopische Studien zur intraventrikularen Neurosekretion in den Recessus infundibularis der Maus. *Zeitschrift für Zellforschung und Mikrokopische Anatomie*, **92**, 207–216.

PART IV

Opiates

CHAPTER 20

Opiate Modulation of Memory

Rita B. Messing, Robert A. Jensen, Beatriz J. Vasquez, Joe L. Martinez, Jr., Vina R. Spiehler, and James L. McGaugh

Retention performance of rats is changed following posttraining administration of morphine, naloxone, or naltrexone, suggesting that endogenous opioid systems are involved in memory storage processes. Furthermore, the memory modulatory effects of these drugs occur only within a narrow dose range, suggesting that these effects are mediated by μ-receptors. The qualitative effects of these drugs on memory are highly time-dependent. Morphine, naloxone, and naltrexone all improve retention performance when given in equally divided doses immediately and 30 min after training. However, a single administration of morphine immediately posttraining, or of naloxone 30 min after training, impairs retention performance. The similar effects of an opiate agonist and an opiate antagonist on memory may be due to their common action of displacing endogenous peptides from receptor sites. The time-dependency of these drug effects may be caused by drug interactions with endogenous neurohumors that change as a function of time after a stressful experience. Alternatively, it is proposed that a single neurochemical manipulation may result in different functional effects if it occurs at different times in the process of memory consolidation.

431

ENDOGENOUS PEPTIDES
AND LEARNING AND MEMORY PROCESSES

I. INTRODUCTION

A. Role of Neurohumors
in Memory Storage Processes

Substances that alter mood, attention, motivation, or arousal may also alter memory storage processes. These include monoamine neurotransmitters and peptide hormones and neuromodulators.

However, the strategy of elucidating the precise role of a neurohumor in memory storage by studying the effects on memory of drugs that alter it has resulted in a great deal of puzzling data. The problem is that similar drug-induced alterations in an endogenous neurohumoral system can cause opposite effects on retention performance. For example, posttraining administration of diethyldithiocarbamate, which blocks monoamine synthesis by inhibiting aromatic amino acid decarboxylase, can either facilitate or impair subsequent retention performance (see e.g., Haycock, van Buskirk, & McGaugh, 1976). Similarly, an increase in catecholaminergic neurotransmission caused by posttraining administration of amphetamine or epinephrine can either facilitate or impair subsequent retention performance (Gold & van Buskirk, 1978; Haycock, van Buskirk, & Gold, 1977). Opposite effects on retention performance caused by posttraining administration of the same drug also occur following peptide administration. Thus, both ACTH (Gold & van Buskirk, 1976) and Substance P (Huston & Stäubli, 1979) can either enhance or impair memory following posttraining administration.

These opposite effects on performance produced by the same drugs can be attributed to several variables, such as the strain or species of animal, the task, and the site of action. Thus, for instance, Substance P has different effects if it is injected into the lateral hypothalamus, or into the amygdala (Huston & Stäubli, 1979). It is therefore possible that some seemingly contradictory data on drug effects following systemic administration can be explained by multiple sites of drug action. However, this explanation is not sufficient to cover all cases, because stimulation of a single brain area can also enhance or impair memory (Destrade, Soumireu-Mourat, & Cardo, 1973; Gold, Hankins, Edwards, Chester, & McGaugh, 1975).

Interactions between drug administration and level of reinforcement (e.g., footshock intensity) are also thought to be responsible for many puzzling data. This is because stressful experiences such as footshocks result in the release of varying amounts of several neurohumors in the brain and periphery, and pharmacological manipulations are superimposed upon these naturally occurring processes. Because there is probably an optimal balance of circulating and brain neurohumors for memory storage, it is not strange that biphasic dose–response curves, interactions between motivational level and drug dose, or species differences are quite common.

Nevertheless, these considerations clearly indicate that the elucidation of the mechanism of action of any neurohumoral substance in memory processes is a complicated problem.

B. Modulation of Memory Storage Processes by Opioid Systems

Endogenous opioid peptides are released when an animal is subjected to any of a wide variety of stressors (for a review see Bodnar, Kelly, Brutus, & Glusman, 1980), or to conditioned fear-inducing stimuli (Chance, White, Krynock, & Rosecrans, 1978; Fanselow & Bolles, 1979). These peptides are in turn thought to play roles in the control of physiological mechanisms that constitute the defensive response of an organism to stress (for a review see Amir, Brown, & Amit, 1980). Endogenous opioids are also thought to be important mediators of behavioral adaptations to stress and, as such, have been implicated in the acquisition of many learned behaviors: the conditioned emotional response (CER), preference for signaled versus unsignaled shock, conditioned taste aversion (CTA) and learned helplessness (for a review see Riley, Zellner, & Duncan, 1980).

Similarly, work from several laboratories has demonstrated that opiate agonists, antagonists, and endogenous opioid peptides can influence memory storage and retrieval for aversive events. We initially found (Messing, et al., 1978) that naloxone enhances and morphine impairs memory. The general conclusion that agonists impair and antagonists enhance memory is in agreement with the work of some (Castellano, 1975; Gallagher & Kapp, 1978; Izquierdo, 1979; Izquierdo, Paiva, & Elisabetsky, 1980; Martinez & Rigter, 1980) and in disagreement with the work of others (Belluzzi & Stein, 1977; Castellano, 1980; Mondadori & Waser, 1979, Stäubli & Huston, 1980). Furthermore, subsequent research in our own laboratory has convinced us that both opiate agonists and antagonists can enhance and impair memory consolidation processes, just as is the case with drugs that act on other neurochemical systems. This work is described in detail in the following sections.

II. EFFECTS OF OPIATE AGONISTS AND ANTAGONISTS ON RETENTION OF INHIBITORY AND ACTIVE AVOIDANCE LEARNING

A. Description of the Tasks

Almost all of our work with opiate drug effects on memory has been done using a one-trial, shock-motivated, inhibitory avoidance step-through task. The reasons for the use of this task are well-known (McGaugh & Herz, 1972); briefly stated, they are that the time of initial learning is short and well defined, and that drug injections can be given after the learning experience, when they cannot influence the animal's immediate perception of that experience.

In these experiments, male rats were placed in the lighted part of a trough-shaped alleyway, facing away from the door leading to a dark compartment. When the rat turned around, the door was opened and the rat was allowed to step through; the door was then closed and a mild footshock was delivered (500.0 or 750.0 μA; 0.5 sec). After receiving the footshock, rats were removed from the apparatus and given intraperitoneal (i.p.) drug injections. Each rat was again placed in the apparatus, and latency to step through into the dark compartment was measured 24 hr later. If a rat failed to cross within 600 sec for the 750.0 μA footshock, or 200 sec for the 500.0 μA footshock, the retention trial was terminated.

In the active avoidance experiment, rats received eight acquisition trials on Day 1, and eight retention trials 24 hr later on Day 2. Drug injections were given i.p. before and/or after the eight acquisition trials. At the start of a trial, the rat was placed in the larger, dark compartment of the same straight alley apparatus used in the inhibitory avoidance task, facing the door to the smaller, lighted compartment. The door was opened and the trial was terminated when the rat entered the smaller compartment. An avoidance response was recorded if the rat entered the smaller compartment within 10 sec, and if no avoidance response was made, a 640.0 μA footshock was administered. If the rat did not escape to the smaller compartment within 30 sec after the onset of shock (i.e., 40 sec from the beginning of the trial), the trial was terminated, and the rat was placed in the smaller compartment and retained there during the 30 sec intertrial interval. For each trial, latency to enter the smaller compartment was recorded.

B. Effects of Morphine, Naloxone, and Naltrexone

The effect of morphine, given immediately posttraining, in the inhibitory avoidance tasks is shown in Fig. 1 (Messing et al., 1978). Low to moderate doses of morphine (1.0 or 3.0 mg/kg) significantly impaired retention, as measured by the percentage of rats exhibiting perfect retention scores. Higher doses had no effect.

We next reasoned that if opioid systems are involved in memory storage, administration of an antagonist should also affect retention. We therefore administered naloxone to rats following inhibitory avoidance training. However, because the duration of action of naloxone is short (Tallarida, Harakal, Maslow, Geller, & Adler, 1978), we administered two posttraining doses of the drug, 30 min apart. In two experiments, naloxone did significantly enhance retention (Figs. 2a and 3a) (Messing, Jensen, Martinez, Spiehler, Vasquez, Soumireu-Mourat, Liang, & McGaugh, 1979). As with morphine, the effect was seen

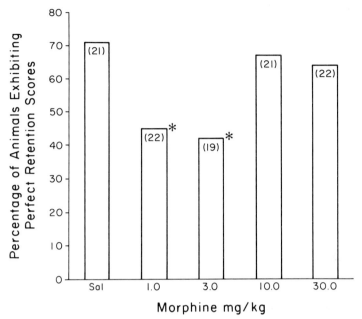

Fig. 1.   Percentage ceiling retention latencies of F344 rats after morphine administration in a one-trial inhibitory avoidance task using 750 μA/0.5-sec footshock. Rats were given posttraining injections (1.0 ml/kg, i.p.) and tested for retention 24 hr later. The number of animals in each group are shown in parentheses. * $p < .02$ compared using a χ^2-test ($df = 1$) with Yates' correction. (Data from Messing et al., 1978.)

with a moderate dose of naloxone (a total of 1.0 mg/kg), whereas a higher dose had no effect. Also, the effect of naloxone could be antagonized by concomitant administration of morphine (Fig. 2b).

One anomalous result did occur in these experiments: If naloxone was administered in a single dose, 30 min after training, an impairment in retention performance occurred (Fig. 3b).

In a second series of experiments, we confirmed the facilitatory effect of naloxone on retention, using an active avoidance task. Again, the improvement in performance occurred at less than the maximum dose investigated, and higher doses had no effect (Fig. 4a) (Messing et al., 1979). Furthermore, the effect could not be attributed to a naloxone-induced decrease in light aversion, because naloxone also facilitated retention performance when rats ran from a lighted to a dark compartment (Fig. 5b). However, the dose of naloxone needed to improve performance on the retention test was much higher than the effective dose for inhibitory avoidance. We have hypothesized that more naloxone is necessary to facilitate retention performance in the active avoidance task because the animal is administered a greater amount of shock, resulting in a larger amount of endorphin release.

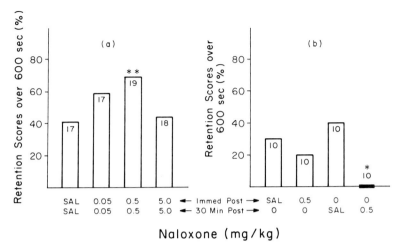

Fig. 2. Percentage ceiling retention latencies of rats after naloxone administration in a one-trial inhibitory avoidance task using 750 μA shock. Rats were given posttraining injections and tested for retention 24 hr later. (a) Rats were given either naloxone or 0.9% NaCl in equally divided doses immediately and 30 min after training. (b) Rats were given naloxone or 0.9% NaCl either immediately or 30 min after training. * $p < .05$, ** $p < .01$ compared to the appropriate saline-injected control group using a χ^2-test ($df = 1$) with Yates' correction. (Data from Messing et al., 1979.)

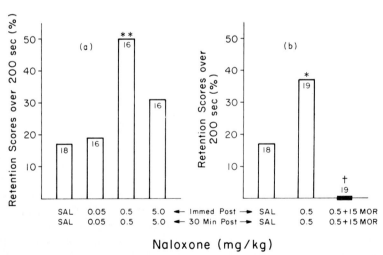

Fig. 3. Percentage ceiling retention latencies of rats after naloxone or combined naloxone and morphine administration in a one-trial inhibitory avoidance task using 500 μA shock. † $p < .1$, * $p < .05$, ** $p < .01$ compared to the saline-injected control group, using a χ^2-test ($df = 1$) with Yates' correction. (Data from Messing et al., 1979.)

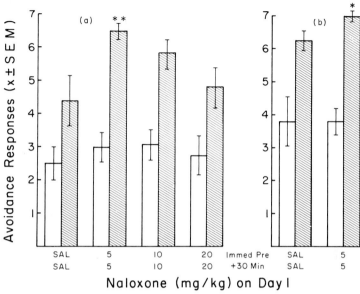

Fig. 4. Number of avoidances during training and retention as a function of naloxone administration. Rats were injected with 0.9% NaCl on Day 1 (training; white bar) and given eight active avoidance trials; after training they were injected again with 0.9% NaCl or naloxone. On Day 2 (retention; shaded bar) they were given an additional eight active avoidance trials. (a) Rats were subjected to footshock on the darker side of the apparatus, and ran to the lighted side. (b) The light was moved from the safe compartment of the compartment in which rats were administered shock, so that rats ran to the darker side to avoid shock. ** $p = .03$, * $p = .08$ compared to the saline-injected control group, using a Student's t-test (two-tailed). (Data from Messing et al., 1979.)

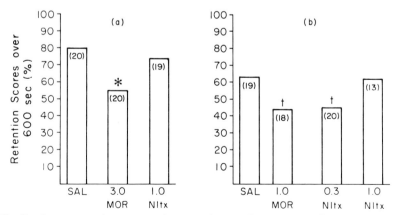

Fig. 5. Percentage ceiling retention latencies after a single posttraining administration of morphine (MOR) or naltrexone (Nltx) in a one-trial inhibitory avoidance task using 750 μA shock. Doses are in mg/kg. † $p < .1$, * $p < .01$, compared to the saline-injected control group, using a χ^2-test ($df = 1$) with Yates' correction.

As with inhibitory avoidance, the effects of naloxone on retention of an active avoidance task are highly time-dependent: Naloxone had an effect only when administered immediately before training and 30 min later. Other schedules of administration (such as two posttraining injections, a single pretraining injection, one injection 15 min prior to training combined with an injection immediately posttraining) had no effect. Unlike inhibitory avoidance, the facilitatory effect of naloxone was dependent upon pretraining drug administration. However, by itself, this had no effect on either learning or retention. We have hypothesized that pretraining naloxone administration is necessary because the training session takes about 10 min, during which time memory consolidation is also occurring.

Therefore, in the next series of studies we decided to investigate the time-dependency of opiate influences on memory.

The first thing we did was to evaluate the effect on retention of posttraining naltrexone administration in the inhibitory avoidance task. We had reasoned that two posttraining administrations of naloxone were necessary because of the short duration of action of this drug. Therefore, it was expected that naltrexone, a longer-acting antagonist, would facilitate performance after a single posttraining administration. In two experiments, naltrexone failed to have a significant effect on retention (Fig. 5). We did, however, replicate memory impairment with 3.0 mg/kg of morphine (the 1.0 mg/kg dose did not produce a statistically significant amnesic effect) (Fig. 5).

The next experiment was an investigation of the effects on memory of morphine and naltrexone given in a dose schedule similar to the one already used for naloxone. That is, morphine or naltrexone was administered to rats in equally divided doses, immediately and 30 min posttraining. The results of these experiments are shown in Fig. 6. Panel (a) shows that if morphine is administered according to this schedule, it enhances memory, just as had previously been found for naloxone. Moreover, it does so at a low dose, and not at a higher dose. Naltrexone also enhances memory when injected twice over a 30 min period. Moreover, it does so at a dose that is below the effective dose of naloxone. Thus, naltrexone, which has a lower rate of metabolism than does naloxone, is also more potent. However, enhancement by each of the three drugs, all of which are metabolized at different rates, depends upon the dual administration procedure.

III. INTERPRETATION OF THE RESULTS

These data raise three questions:

1. What is the reason for the U-shaped dose–response curves?
2. Why are the effects of these drugs on memory so highly time-dependent?

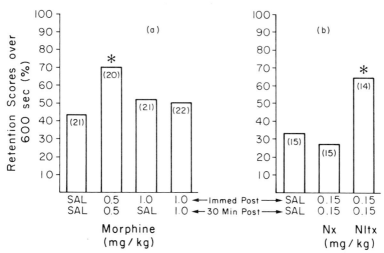

Fig. 6. Percentage ceiling retention latencies after two posttraining injections of (a) morphine (MOR) and (b) naloxone (Nx) or naltrexone (Nltx) in a one-trial inhibitory avoidance task using 750 μA shock. * $p = .03$ compared to the saline-injected control groups, using a χ^2-test ($df = 1$) with Yates' correction.

In particular, why can morphine and naloxone each produce memory enhancement and impairment in rats at similar doses and levels of footshock, when only the time of injection is varied?

3. Why are the effects of morphine, an agonist, similar to those of the opiate antagonists naloxone and naltrexone?

Some possible answers to these questions are proposed in the following.

A falling off in the response of an organism to a high dose of an agonist drug is most easily explained by receptor fatigue, or tachyphylaxis. However, such an explanation is inadequate in this case, because naloxone is a relatively "pure" antagonist, the doses of naloxone and morphine that have memory modulatory effects are relatively low, and the effective dose range is narrow. A more plausible explanation has to do with the concept of multiple types of opiate receptors (Kosterlitz, Lord, Paterson, & Waterfield, 1980). At low concentrations, morphine, naloxone, and naltrexone are presumed to interact primarily with the hypothesized μ-receptor, whereas, at higher concentrations, they also have effects on other receptor types. At high doses, when other types of drug–receptor interactions occur, the memory modulatory effects of these drugs may be masked. It is interesting, in this context, that others have also found that dose–response curves for naloxone are sometimes U-shaped (Jacob & Ramabadran, 1978; Pert & Walter, 1976).

The second question has to do with the time-dependence of these drug effects: Both μ-receptor agonists and antagonists modulate memory, but the direction of the effects is determined by the time after training and number of drug administrations. The usual explanations for the ability of a single drug to both impair and enhance memory will not suffice in these experiments. Thus, footshock appears to be irrelevant, because the same drug can enhance or impair when footshock is not varied. Similarly, dose is not the primary determinant of the opposite effects of these drugs, since the doses of morphine and naloxone used to attain memory enhancement or impairment are similar. This same argument makes it equally improbable that these effects depend on a type of anatomical specificity whereby one brain region mediates memory-enhancing effects of naloxone and memory-impairing effects of morphine (e.g., the amygdala; Gallagher & Kapp, 1978), whereas another region mediates memory-enhancing effects of morphine and memory-impairing effects of nal-oxone (e.g., periventricular areas)(Belluzzi & Stein, 1977). Because the doses used to achieve memory effects are similar whether enhancement or impairment of memory is observed, it is unlikely that the anatomical distribution of a given drug is markedly different when the drug enhances or impairs memory.

However, levels of released endogenous neurohumors, including opioids, change as the time after stress is varied. There is an accumulating body of evidence that endogenous opioid peptides are released after stressful experiences, including electroshock (Bodnar et al., 1980), and that these endogenous sub-stances bind to opiate receptors (Pert & Bowie, 1979). Thus, the interactions of exogenously administered opiate receptor ligands with receptors will differ as a function of time after a stressful experience. The other side of the coin, since endogenous and exogenous ligands are thought to compete for the same receptor sites, is that the degree to which exogenous drugs interfere with binding of endogenous ligands to receptors will also vary with the time of drug administration.

Finally, even if these considerations are ignored and it is assumed that a drug such as morphine exerts the same physiological effect no matter when after training the drug injections occur, the memory modulatory effect still may vary with time. Because memory consolidation occurs over a considerable time period (McGaugh & Herz, 1972), the same physiological or neurochemical manipulation may result in a different functional effect, depending upon the time in the memory consolidation process it occurs.

The third question is perhaps the most difficult one to speculate about: Why do morphine, naloxone, and naltrexone enhance retention performance under similar conditions?

One simple answer may be that neither naloxone nor naltrexone is a "pure" antagonist (Jacob & Michaud, 1976; Sawynok, 1979) and it is the "agonist" action of these drugs that is responsible for their memory-enhancing effect.

Another explanation, similar to the first, but more sophisticated, has to do with the "intrinsic agonist activity" of drugs (see Hollenberg, 1978). Certain models of ligand–receptor–effector interactions postulate that a receptor may interact with more than one effector (e.g. with a cyclase and an ionophore). Furthermore, it has been postulated that the allosteric regulation of receptor conformation by ligands is different for different ligands. Thus, ligands vary in their intrinsic agonist activities because each ligand induces a different change in receptor conformation, which in turn affects receptor affinities for effectors. However, it is possible that endogenous opioid peptides have different intrinsic agonist activities from alkaloid agonists and antagonists; that is, the neuro-chemical consequences of opioid peptide–receptor interactions may not be the same as neurochemical effects of alkaloid–receptor interactions. Therefore, under certain circumstances, morphine and antagonists may exert similar effects (or antagonists may appear to have "agonist" actions) because both morphine and antagonists displace endogenous ligands from receptors, and it is this action that is reponsible for the memory-enhancing effects of these drugs.

IV. CONCLUSIONS

Work from our own and other laboratories has established that opioid systems play a role in memory consolidation. However, just as with the memory mod-ulatory effects of other neurohumors, the effects of opiate drugs are difficult to schematize simply. In the case of opiates, as well as drugs that interact with other neurochemical systems, the following factors are important for under-standing drug effects on memory:

1. Drugs may interact with more than one receptor. This means that dose–response functions may be complex.

2. Exogenously administered drugs and endogenous ligands may differ in their "intrinsic agonist activities." This implies that it may not always be useful to classify drugs as agonists or antagonists.

3. The qualitative effect of a drug on memory varies with the time of drug administration. It may, therefore, be the case that the same biochemical events can either enhance or impair memory, depending upon when in the process of memory storage they occur.

ACKNOWLEDGMENTS

This research was supported by Research Grants MH 12526 (to JLMcG, JLMJr, RAJ, RBM, and BJV) and AG 00538 (to JLMcG, JLMJr, RAJ, and RBM) from the United States Public Health Service, and BNS 76- 17370 (to JLMcG) from the National Science Foundation.

REFERENCES

Amir, S., Brown, Z. W., & Amit, Z. (1980). The role of endorphins in stress: Evidence and speculations. *Neuroscience & Biobehavioral Reviews,* **4,** 77–86.

Belluzzi, J. D., & Stein, L. (1977). Enkephalin- and morphine-induced facilitation of long-term memory. *Society for Neuroscience Abstracts,* **3,** 230.

Bodnar, R. J., Kelly, D. D., Brutus, M., & Glusman, M. (1980). Stress-induced analgesia: Neural and hormonal determinants. *Neuroscience & Biobehavioral Reviews,* **4,** 87–100.

Castellano, C. (1975). Effects of morphine and heroin on discrimination learning and consolidation in mice. *Psychopharmacologia,* **42,** 235–242.

Castellano, C. (1980). Dose-dependent effects of heroin on memory in two inbred strains of mice. *Psychopharmacology,* **67,** 235–239.

Chance, W. T., White, A. C., Krynock, G. M., & Rosecrans, J. A. (1978). Conditioned fear-induced antinociception and decreased binding of ^{3}H-N-leu-enkephalin to rat brain. *Brain Research,* **141,** 371–374.

Destrade, C., Soumireu-Mourat, B., & Cardo, B. (1973). Effects of post-trial hippocampal stimulation on acquisition of operant behavior in the mouse. *Behavioral Biology,* **8,** 713–724.

Fanselow, M. S., & Bolles, R. C. (1979). Triggering of the endorphin analgesic reaction by a cue previously associated with shock: reversal by naloxone. *Bulletin of the Psychonomic Society,* **14,** 88–90.

Gallagher, M., & Kapp, B. S. (1978). Manipulation of opiate activity in the amygdala alters memory processes. *Life Sciences,* **23,** 1973–1978.

Gold, P. E., Hankins, L., Rose, M. E., Chester, J., & McGaugh, J. L. (1975). Memory interference and facilitation with post-trial amygdala stimulation: Effect on memory varies with footshock level. *Brain Research,* **86,** 509–513.

Gold, P. E., & van Buskirk, R. (1976). Enhancement and impairment of memory processes with post-trial injections of adrenocorticotrophic hormone. *Behavioral Biology,* **16,** 387–399.

Gold, P. E., & van Buskirk, R. (1978). Effects of α- and β-adrenergic receptor antagonists on post-trial epinephrine modulation of memory: Relationship to post-training brain norepinephrine concentrations. *Behavioral Biology,* **24,** 168–184.

Haycock, J. W., van Buskirk, R., & Gold, P. E. (1977). Effects on retention of post-training amphetamine injections in mice: Interaction with pre-training experience. *Psychopharmacology,* **54,** 21–24.

Haycock, J. W., van Buskirk, R., & McGaugh, J. L. (1976). Facilitation of retention performance in mice by post-training diethyldithiocarbamate. *Pharmacology, Biochemistry & Behavior,* **5,** 525–528.

Hollenberg, M. D. (1978). Receptor models and the action of neurotransmitters and hormones. *In* H. I. Yamamura, S. J. Enna, M. J. Kuhar (Eds.), *Neurotransmitter Receptor Binding,* pp. 13–39. New York: Raven.

Huston, S. P., & Stäubli, U. (1979). Post-trial injection of substance P into lateral hypothalamus and amygdala, respectively, facilitates and impairs learning. *Behavioral and Neural Biology,* **27,** 244–248.

Izquierdo, I. (1979). Effect of naloxone on various forms of memory in the rat: Possible role of endogenous opiate mechanisms in memory consolidation. *Psychopharmacology,* **66,** 199–203.

Izquierdo, I., Paiva, A.C.M., & Elisabetsky, E. (1980). Post-training intraperitoneal injections of β-endorphin cause retrograde amnesia for two different tasks in rats. *Behavioral and Neural Biology,* **28,** 246–251.

Jacob, J.J.C., & Ramabadran, K. (1978). Enhancement of a nociceptive reaction by opioid antagonists in mice. *British Journal of Pharmacology,* **64,** 91–98.

Jacob, J. J., & Michaud, G. M. (1976). Production par la naloxone d'effets inverses de ceux de la morphine chez le chein éveillé. *Archives Internationale de la Pharmacodynamique*, **222**, 332–340.

Kosterlitz, H. W., Lord, J.A.H., Paterson, S. J., & Waterfield, A. A. (1980). Effects of changes in the structure of enkephalins and of narcotic analgesic drugs on their interactions with μ- and δ-receptors. *British Journal of Pharmacology*, **68**, 333–342.

Martinez, J. L., Jr., & Rigter, J. R. (1980). Endorphins alter acquisition and consolidation of an inhibitory avoidance response. *Neuroscience Letters*, **19**, 197–201.

McGaugh, J. L., & Herz, M. J. (Eds.) (1972). *Memory Consolidation*. San Francisco, California: Albion Publishing.

Messing, R. B., Jensen, R. A., Martinez, J. L., Jr., Spiehler, V. R., Vasquez, B. J., Soumireu-Mourat, B., Liang, K. C., & McGaugh, J. L. (1979). Naloxone enhancement of memory. *Behavioral and Neural Biology*, **27**, 266–273.

Messing, R. B., Jensen, R. A., Martinez, J. L., Jr., Vasquez, B. J., Soumireu-Mourat, B., & McGaugh, J. L. (1978). *7th International Congress of Pharmacology Abstracts*, 560.

Mondadori, C., & Waser, P. G. (1979). Facilitation of memory processing by posttrial morphine: Possible involvement of reinforcement mechanisms? *Psychopharmacology*, **63**, 297–300.

Pert, A., & Walter, M. (1976). Comparison between naloxone reversal of morphine and electrical stimulation induced analgesia in the rat. *Life Sciences*, **19**, 1023–1032.

Riley, A. W., Zellner, D. A., & Duncan, H. J. (1980). The role of endorphins in animal learning and behavior. *Neuroscience & Biobehavioral Reviews*, **4**, 69–76.

Sawynok, J., Pinsky, C., & La Bella, F. S. (1979). Specificity of naloxone as an opiate antagonist. *Life Sciences*, **25**, 1621–1632.

Stäubli, U., & Huston, J. P. (1980). Avoidance learning enhanced by posttrial morphine injection. *Behavioral and Neural Biology*, **28**, 487–490.

Tallarida, R. J., Harakal, C., Maslow, J., Geller, E. B., & Adler, M. W. (1978). The relationship between pharmacokinetics and pharmacodynamic action as applied to *in vivo* pA$_2$: Application to the analgesic effect of morphine. *Journal of Pharmacology and Experimental Therapeutics*, **206**, 38–45.

CHAPTER 21

Influence of Amygdala Opiate-Sensitive Mechanisms, Fear Motivated Responses, and Memory Processes for Aversive Experiences

Michela Gallagher and Bruce S. Kapp

Intracranial injections of opiate agents into the amygdala complex of rats altered retention of passive avoidance conditioning. Administration of the opiate agonist, levorphanol, immediately following conditioning, impaired retention measured 24 hr later. This effect was observed to be dose-dependent, stereospecific, and blocked by concurrent administration of the opiate antagonist naloxone. In addition, administration of naloxone only immediately following conditioning produced a dose-dependent facilitation of retention. The effects of both levorphanol and naloxone administration were time-dependent because administration of either agent 6 hr following conditioning did not significantly alter retention of passive avoidance conditioning. These results support the hypothesis of a role for amygdala opioid peptides in modulating memory processes for an aversive experience.

The results of other research are presented demonstrating that opiate injections into the central nucleus of the amygdala complex alter the acquisition of a fear motivated classically conditioned heart-rate response in rabbits. The conditioning procedure consisted of 15 presentations of the conditioned stimulus (CS) alone followed by 20 paired presentations of the conditioned and unconditioned stimuli (US). The CS was a 5 sec tone; the US was a 200 msec eyelid shock. Administration of levorphanol (5.0 nmol) into the central nucleus attenuated the magnitude of the conditioned decelerative heart-rate response. The effect of opiate agonist administration was

445

ENDOGENOUS PEPTIDES
AND LEARNING AND MEMORY PROCESSES

observed to be stereospecific and blocked by concurrent opiate antagonist administration. Conversely, naloxone (2.5 nmol) administration into the central nucleus increased the magnitude of the conditioned heart-rate response.

The effects of opiate manipulation within the central nucleus of the amygdala on aversively conditioned heart-rate responding may be due to alterations in the arousal of fear. This interpretation is consistent with other research demonstrating that manipulations of the central nucleus or of opiate-sensitive mechanisms within the central nucleus alter fear-like behaviors. The possibility that opiate-sensitive mechanisms within the central nucleus provide a common substrate for the effects of opiates on memory processes and the regulation of emotional states is discussed.

I. INTRODUCTION

Specific regions within the limbic system are among those brain areas that possess high densities of opiate receptors and high concentrations of endorphins. To investigate the possible contribution of endorphins within the limbic system to memory processes, we focused our research on the amygdala complex, a brain region that possesses large quantities of opiate receptors and the pentapeptide enkephalins (Atweh & Kuhar, 1977; Elde, Hökfelt, Johansson, & Terenius, 1976; Gros, Pradelles, Humbert, Dray, Le Gal La Salle, & Ben-Ari, 1978; Hiller, Pearson, & Simon, 1973; Kuhar, Pert, & Snyder, 1973; Simantov, Kuhar, Uhl, & Snyder, 1977). Our choice of the amygdala was guided by the results of a number of investigations using a variety of aversive conditioning procedures that have demonstrated that postconditioning electrical stimulation or chemical manipulation within this group of nuclei affects subsequent retention (Gallagher, Kapp, Musty, & Driscoll, 1977; Gold, Hankins, & Rose, 1977; Gold, Rose, Hankins, & Spanis, 1976; Handwerker, Gold, & McGaugh, 1974; Kesner, Berman, Burton, & Hankins, 1975). Based on this evidence, we have investigated the effects of manipulating opiate activity within the amygdala of rats on retention of passive avoidance conditioning (Gallagher & Kapp, 1978).

II. ALTERATION OF MEMORY PROCESSES BY AMYGDALA–OPIATE MANIPULATIONS

Rats were surgically prepared with bilateral cannulae positioned at the dorsal surface of the amygdala. Following surgery, 1 wk later animals received passive avoidance conditioning. Each rat was placed into a lighted compartment of a two-compartment apparatus. As the animal stepped into an adjacent dark compartment, a door connecting the two compartments was closed, the animal received a footshock (1.0 mA/2 sec) and was immediately removed from the apparatus. All intracerebral (i.c.) injections were administered bilaterally in a

0.5 μl volume either immediately or 6 hr following this conditioning trial. Retention of passive avoidance conditioning was measured 24 hr after the conditioning trial by replacing the animal in the lighted compartment and recording the latency for entry into the dark compartment. Increased latency in entering the dark compartment during the retention test compared to the conditioning trial provided a measure of retention of the conditioning experience. The opiate agents used in this study were the agonist levorphanol, its inactive enantiomer dextrorphan and the opiate antagonist naloxone. The vehicle was a Kreb's Ringer phosphate solution.

Retention test latency data are presented in Fig. 1. Statistical analysis revealed that animals injected with levorphanol into the amygdala immediately after the conditioning trial demonstrated significant retention deficits when compared to the vehicle injected and unoperated control animals (Fig. 1a). Conversely, postconditioning administration of naloxone significantly increased retention latencies (Fig. 1b). The effects of these opiate agents on retention appear to be mediated by opiate receptors. As can be observed in Fig. 1c, the effect obtained with levorphanol was stereospecific, as the inactive enantiomer of levorphanol, dextrorphan, did not significantly alter retention. In addition, the severe retention deficit produced by levorphanol (5.0 nmol) was blocked by a relatively low dose of naloxone (1.25 nmol). These results are in agreement with the pharmacological characterization of opiate receptors in brain tissue. Opiate binding sites in brain are stereospecific and exhibit high affinity binding for both agonist and antagonist opiate compounds (Pert & Snyder, 1973; Simon, Hiller, & Edelman, 1973).

The possibility that endogenous activation of opiate receptors within the amygdala is normally involved in memory processes for an aversive experience is further suggested by the effects of naloxone. The increased retention produced by naloxone administration suggests that opioid peptide systems within this region are active for a period of time following conditioning. According to this interpretation, the increased retention produced by opiate antagonist administration may reflect the effects of blocking endogenously mediated activity at opiate receptors.

Because these opiate agents were administered after the conditioning trial, their effects on retention cannot be attributed to drug induced interference with the processing of information during the conditioning trial. In addition, due to the short-acting effects of the opiate agents used, it is unlikely that opiates administered immediately after training alter retention by affecting performance during retention testing 24 hr later. Indeed, this interpretation is further supported by evidence demonstrating that the effects of postconditioning opiate administration are time-dependent. Groups of animals that received injections of either levorphanol or naloxone into the amygdala 6 hr after conditioning exhibited retention latencies that did not differ significantly from those of the

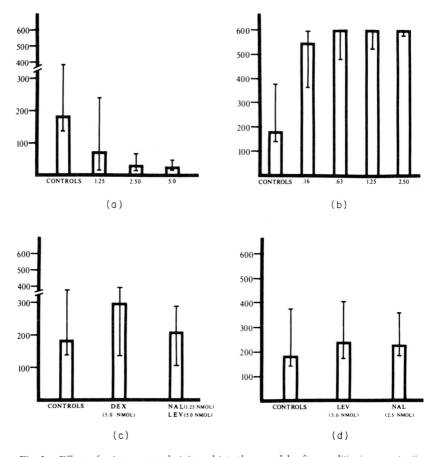

Fig. 1. Effects of opiate agents administered into the amygdala after conditioning on retention of passive avoidance. Bars represent median retention test latencies for groups of animals. The interquartile range for each group is indicated. Each group was composed of 7–10 animals. Control groups consisted of an unoperated group and a vehicle injected group that did not differ significantly from one another (Mann-Whitney U test, two-tailed). (a) Injections of levorphanol (nmol). A Kruskal-Wallis one-way analysis of variance performed on these groups revealed a significant difference ($p < .001$). A Mann-Whitney U test (two-tailed) performed between the levorphanol 5.0 nmol group and the pooled control data (CONTROLS) revealed a significant difference ($p < .002$). (b) Injections of naloxone (nmol). A Kruskal-Wallis one-way analysis of variance performed on these groups revealed a significant difference ($p < .005$). A Mann-Whitney U test (two-tailed) performed between the naloxone 2.5 nmol group and the control group revealed a significant difference ($p < .02$). (c) Stereospecificity and reversibility of levorphanol. Results of independent Mann-Whitney U tests revealed that neither the group receiving dextrorphan (DEX, 5.0 nmol) nor the group receiving combined injections of levorphanol and naloxone (LEV, 5.0 nmol + NAL, 1.25 nmol) differed significantly from the controls. (d) Drug administration 6 hr after conditioning. Independent Mann-Whitney U tests revealed that neither the levorphanol group (LEV, 5.0 nmol) nor the naloxone group (NAL, 2.5 nmol) that were injected 6 hr after conditioning differed significantly from controls. (From Gallagher & Kapp, 1978.)

448

control groups (Fig. 1d). Therefore, the effects of manipulating opiate activity within the amygdala after conditioning may best be attributed to drug induced alterations in neural processes that occur for a period of time after an aversive experience and that enable recently acquired information to be utilized at a later time.

Finally, we have provided some evidence that the effects of amygdala injection of opiate agonists and antagonists on memory processes are not due to spread of the injection to adjacent structures. Groups of animals that received bilateral injections of either levorphanol (5.0 nmol) or naloxone (2.5 nmol) into the basal ganglia approximately 1.0 mm dorsal to the amygdala complex exhibited normal retention of passive avoidance conditioning (Gallagher & Kapp, 1978). Because high concentrations of endorphins and opiate receptors have been observed in the basal ganglia (Atweh & Kuhar, 1977; Elde et al., 1976; Simantov et al., 1977), these results also indicate that not all opiate-sensitive systems in the brain are involved in memory processes for passive avoidance conditioning. In further support of this interpretation, Kesner (personal communication 1978) has observed that opiate administration into another opiate-sensitive area, the periaqueductal grey region of the midbrain, does not alter memory processes for aversive conditioning in rats.

That opiate-sensitive systems are involved in memory processes is supported by additional research demonstrating that postconditioning systemic opiate administration alters retention of avoidance conditioning (Castellano, 1975; Messing, Jensen, Martinez, Spiehler, Vasquez, Soumireu-Mourat, Liang, & McGaugh, 1979). In agreement with our findings, these studies have reported that opiate agonist administration immediately following conditioning decreases retention for both active and passive avoidance conditioning. In these studies, this effect was blocked by concurrent administration of the antagonist naloxone. Furthermore, Messing et al. (1979) reported that systemic administration of naloxone alone facilitates retention of both passive and active avoidance conditioning. Similar to our results with opiate administration into the amygdala, the effects of opiate agents on memory processes in these studies have been uniformly observed to be time-dependent.

Although our results as well as those of Castellano (1975) and Messing et al. (1979) demonstrate that opiate agonist administration decreases retention of avoidance conditioning, other investigators have reported increased retention of avoidance conditioning following opiate agonist administration (Belluzzi & Stein, 1977; Jensen, Martinez, Messing, Spiehler, Vasquez, Soumireu-Mourat, Liang, & McGaugh, 1978; Mondadori & Waser, 1979). Mondadori and Waser observed increased retention of avoidance conditioning following postconditioning systemic administration of morphine in mice at doses of 40.0 and 100.0 mg/kg. Similar results in rats have been reported by Belluzzi and Stein (1977) and by Jensen et al. (1978) following intraventricular (i.v.) opiate agonist administration. Belluzzi and Stein observed increased retention of avoidance conditioning when morphine (20.0 μg) or enkephalin (200.0 μg) were administered i.v. after passive avoidance conditioning. Jensen et al. (1978) observed that postconditioning i.v. administration of a low dose of morphine (3.0 μg) decreased retention of passive avoidance conditioning, whereas administration of a high dose of morphine (40.0 μg) produced increased retention. However, the available evidence makes it difficult to assess whether opiate agonist administration in some instances facilitates and in others impairs retention of avoidance conditioning through a common mechanism of action. Indeed, studies (Belluzzi & Stein, 1977; Jensen et al., 1978; Mondadori & Waser, 1979) that have reported facilitation of retention following administration of relatively high doses of morphine have not assessed whether this effect is altered by opiate antagonist administration. In addition, no evidence has been provided that opiate agonist facilitation of retention exhibits stereospecificity. In light of other evidence that morphine produces pharmacological effects following intracerebral administration that are not stereospecific or blocked by naloxone (Firemark & Weitzman, 1979; Jaquet, Klee, Rice, Iijima, & Minamikawa, 1977), such tests would be useful in determining whether opiate agonists produce facilitation and impairment of retention through mechanisms exhibiting similar pharmacological properties. Therefore, although systemic or i.v. injections of opiates have been observed, in some instances, to facilitate retention of avoidance conditioning, evidence that this effect is mediated by an opiate-sensitive system that exhibits the pharmacological properties of opiate receptors has not been provided. The results of studies in which the effects of opiate administration on memory processes for aversive conditioning are characteristic of mediation by opiate receptor mechanisms consistently indicate that increased activity at opiate receptors following avoidance conditioning impairs subsequent retention (Castellano, 1975; Gallagher & Kapp, 1978; Messing et al., 1979). These results implicate a role for opiate-sensitive receptor mechanisms in the establishment of enduring memories for aversive experiences. Furthermore, the results of our research indicate that an opiate-sensitive system within the amygdala complex provides an important site for the effects of opiate receptor agonists and antagonists on memory processes. Recent neuroanatomical descriptions of endorphin distribution within the amygdala have guided our more recent investigations of this system.

III. AMYGDALA ENKEPHALINS: A FOCUS FOR RESEARCH ON THE CENTRAL NUCLEUS

The amygdala complex is a heterogeneous structure composed of a number of different nuclei, each possessing unique neuroanatomical projections to and from other brain systems. Furthermore, these nuclei have been observed to possess different relative concentrations of a wide variety of neurochemicals. The enkephalins, in particular, exhibit a relatively localized distribution within the amygdala complex. Studies using immunohistochemical and radioimmunoassay techniques have reported particularly high densities of enkephalin immunoreactively within the central nucleus of the amygdala complex (Gros et al., 1978; Sar, Stumpf, Miller, Chang, & Cuatracasas, 1978; Simantov et al., 1977). Because the cannula placements in our previously described research were positioned at the dorsal surface of the amygdala complex, with the majority of placements near the central nucleus (see Fig. 2), the possibility exists that the observed effects were due to manipulation of the enkephalin system within the central nucleus. Several lines of evidence suggest that the opiate-sensitive system within the central nucleus may provide an important substrate for the effects of intra-amygdala opiate injections on retention of avoidance conditioning.

First, several studies have reported that damage to the central nucleus impairs the acquisition of both passive and active avoidance conditioned responding in rats (Grossman, Grossman, & Walsh, 1975; McIntyre & Stein, 1973; Werka, Skär, & Ursin, 1978). In interpreting these effects, it has been proposed that central nucleus lesions may interfere with brain mechanisms that are normally involved in the arousal of fear (Werka et al., 1978). Indeed, a recent investigation by Werka et al. demonstrated that lesions largely confined to the central nucleus reduce fear-like behaviors in rats. In their study, separate groups of animals received bilateral lesions of either the central or basolateral nuclei or of the insular cortex lateral to the amygdala complex. Using a variety of open-field measures generally held to be indices of fear or emotionality, Werka et al. observed that animals with lesions of the central nucleus, when compared to other lesion or surgical control groups, exhibited highly significant increases in (1) open-field activity; (2) open-field rearings; (3) time spent in the center of the open field; and (4) total number of entries into the open field from the home cage. The basolateral lesion, cortical lesion, and control groups did not differ from one another on any of these measures. These effects of central nucleus lesions led to the conclusion that the central nucleus plays an important role in the expression of fear, and is consistent with earlier studies demonstrating that large amygdala lesions decrease emotional reactivity, particularly fear-like behaviors in response to threatening stimuli (Blanchard & Blanchard, 1972; Goddard, 1964).

A second line of evidence implicating the amygdala complex in emotional responding has demonstrated that electrical stimulation of sites within the central nucleus elicits somatic and autonomic responses indicative of fear in a number of species (Fernandez de Molina & Hunsperger, 1962; Hilton & Zbrozyna, 1963; Ursin & Kaada, 1960; Wood, Schoteluis, Frost, & Baldwin, 1958; Zbrozyna, 1972). For example, as described by Ursin and Kaada (1960), stimulation of a restricted zone within the amygdala that includes the central nucleus and anterior portion of the lateral nucleus elicits a fear response in cats. This

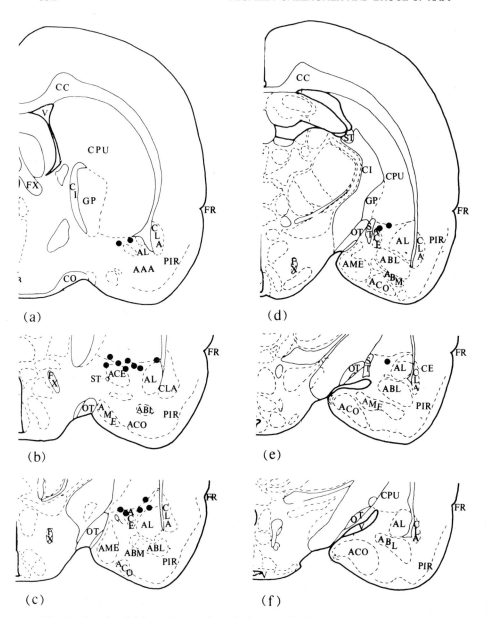

Fig. 2. Rostrocaudal brain sections through the amygdala from 1.2 mm anterior to bregma to 1.4 mm posterior to bregma according to the Pellegrino and Cushman (1967) atlas. Cannula tip placements (●) are schematically represented for the group that received levorphanol injections (5.0 nmol) immediately after conditioning. The distribution of cannula placements for this group was representative of other groups included in the experiment. Abbreviations used: AAA, anterior amygdaloid area; ABL, basolateral amygdaloid nucleus; ACE, central amygdaloid nucleus; ACO,

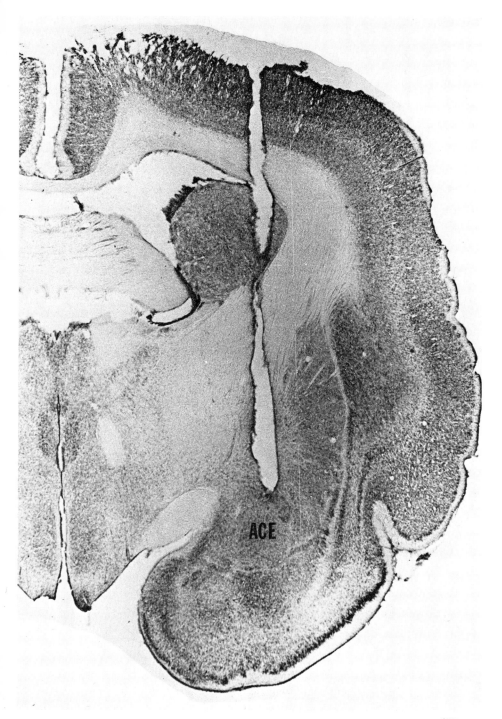

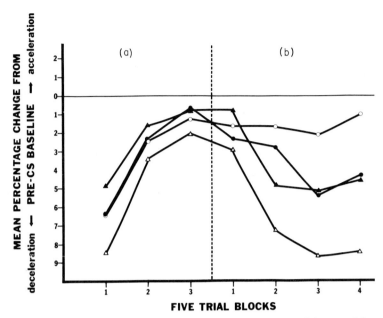

Fig. 4. Effects of opiate manipulations within the central nucleus of the amygdala on the acquisition of a classically conditioned heart-rate response in rabbits. (a) CS alone; (b) CS-US presentations. Each group was composed of eight animals. Unoperated, ●——●; vehicle, ▲——▲; levorphanol (5.0 nmol), ○——○; naloxone (2.5 nmol), △——△.

group, a group injected with vehicle, a group injected with 5.0 nmol of levorphanol, and a group injected with 2.5 nmol of naloxone. All injections were administered bilaterally in a 1.0 μl volume of vehicle approximately 5 min prior to the onset of the conditioning session.

The conditioning session consisted of 15 presentations of the conditioned stimulus (CS) followed by 20 paired conditioning trials in which the onset of the unconditioned stimulus (US) was coincident with the offset of the CS. The CS was a 5 sec tone (1000 Hz, 92 dB). The US consisted of a 500 msec, 2.0 mA eyelid shock. The heart-rate response to each CS presentation was computed as a percentage change in heart rate to the 5 sec tone compared to the 5 sec baseline period immediately preceding the tone onset.

All groups exhibited a bradycardia orienting response to the initial presentations of the CS alone, which habituated over the course of the 15 CS alone presentations (Fig. 4). Analysis of variance comparing the heart-rate responses to the CS during the 15 CS only trials revealed no significant differences among the groups, indicating that administration of opiates into the central nucleus does not alter this component of the orienting response in rabbits nor the course of its habituation. However, a two-way analysis of variance conducted on the

20 paired conditioning trials revealed a significant Groups effect [F (3,28) = 8.23, $p < .001$], and a significant Groups x Trials interaction [F (57,514) = 1.62, $p < .005$]. Whereas subsequent statistical analysis revealed that the unoperated and vehicle groups did not differ significantly from one another, comparisons performed between the drug injected groups and vehicle group revealed that levorphanol significantly decreased the magnitude of the conditioned response [F (1,14) = 4.70, $p < .05$], whereas naloxone significantly increased the magnitude of the conditioned decelerative heart-rate response [F (1,14) = 7.62, $p < .02$]. Two additional groups, one injected with the inactive enantiomer of levorphanol, dextrorphan (5.0 nmol) and the other receiving a combined injection of levorphanol (5.0 nmol) and naloxone (2.5 nmol) did not significantly differ from the vehicle group. Therefore, these results are consistent with the interpretation that administration of opiate agents into the central nucleus of the rabbit amygdala alters the acquisition of aversively conditioned heart-rate responding by altering activity at opiate receptors. Preliminary results from research now in progress indicates that comparable effects are not observed when opiates are injected into other regions of the amygdala.

Several observations render it unlikely that the effects of opiate administration into the central nucleus of the amygdala on heart-rate conditioning are due to gross alterations in sensory or motor functions. First, drug-injected groups in our experiment exhibited normal bradycardia orienting responses to the CS alone presentations indicating that the animals were responsive to the tone that was subsequently used during conditioning. Second, Rodgers (1978) reported that opiate administration into the central nucleus in rats does not alter pain sensitivity. Finally, opiate administration into the amygdala does not alter motor functions in rats (Pert, DeWald, Liao, & Sivit, 1979). However, in light of the evidence already presented supporting a role for the central nucleus and for opioid peptides in this region, in the regulation of responses elicited by fear-provoking stimuli, these results are consistent with the notion that opiate manipulation within the central nucleus in rabbits affects the acquisition of conditioned heart-rate responses by altering the arousal of fear.

V. THE INFLUENCE OF AMYGDALA–OPIATE MECHANISMS ON FEAR-MOTIVATED RESPONSES AND MEMORY: A COMMON NEURAL SUBSTRATE IN THE CENTRAL NUCLEUS?

In returning to the evidence described earlier in this chapter demonstrating that postconditioning opiate manipulation within the amygdala complex alters memory processes for an aversive experience, we raised the possibility that the effects were due to alterations in central nucleus function. This proposal is supported by (1) the location of injection sites in our investigation (Fig. 2); (2) the presence of high concentrations of enkephalins within the central nucleus (Gros et al., 1978; Sar et al., 1978; Simantov et al., 1977); and (3) the results of other investigations demonstrating that postconditioning electrical stimulation at electrode sites in the region of the central nucleus alters retention of avoidance conditioning in rats (Gold et al., 1976, 1977). Furthermore, we have reviewed evidence indicating that the central nucleus of the amygdala and the opiate receptor population therein may be part of a system that is involved in the regulation of emotional reactivity; specifically, the arousal of fear-like responses. Indeed, the observations that (1) electrical stimulation within this region induces fear-like behaviors (Fernandez de Molina & Hunsperger, 1962; Hilton & Zbrozyna, 1963; Wood et al., 1958; Zbrozyna, 1972); (2) lesions of the central nucleus reduce fear-like behaviors (Werka et al., 1978) and produce deficits in aversively conditioned responding (Grossman et al., 1975; Kapp et al., 1979; McIntyre & Stein, 1973); and (3) opiate administration within the region of the central nucleus reduces fear-like behaviors (File & Rogers, 1979) and alters aversively conditioned heart-rate responding are consistent with this interpretation. In summary, the evidence suggests that the central nucleus, as well as the opioid peptide system within this region, contributes to both memory processes and the regulation of emotional states.

A basis for understanding the possible relationship between these functions for an opiate-sensitive system in the amygdala may be provided within a theoretical framework proposed by Kety (1970, 1976) and Gold and McGaugh (1975). For example, Kety (1976) has proposed that physiological processes sensitive to the motivational or reinforcing properties of an experience may contribute to memory processes. Specifically, he states that physiological systems associated with the arousal of emotional states may serve an adaptive role in regulating the establishment of an enduring memory. Such mechanisms would insure that information accompanied by the arousal of significant positive or negative affective responses would be preserved, while information accompanied by an insignificant arousal of such responses would be less likely to be retained. Within this theoretical framework, the effects of posttraining opiate manipulations within the amygdala, and particularly the central nucleus, on memory may be functionally related to changes in the emotional response elicited by the aversive experience. Such alterations, according to Kety's (1976) proposal, could influence the establishment of an enduring memory and would be reflected in altered retention. Although this interpretation is admittedly speculative, it is consistent with the literature reviewed in this chapter. Only additional investigations into the neural substrates of emotion and learning will determine the validity of this theoretical position and whether an amygdala–opioid system functions in memory processes by regulating affective states elicited by aversive experiences.

ACKNOWLEDGMENTS

This research was supported by NIH grant R03 01577-01, NIMH grant R01 32292-01 and a Research Scientist Development Award K02 MH00118 (NIMH) to B.S.K.

REFERENCES

Atweh, S., & Kuhar, M. J. (1977). Autoradiographic localization of opiate receptors in rat brain. III. The telencephalon. *Brain Research*, **134**, 393–405.

Belluzzi, J. D., & Stein, L. (1977). Enkephalin- and morphine-induced facilitation of long-term memory. *Society for Neuroscience Abstracts*, **3**, 230.

Blanchard, D. C., & Blanchard, R. J. (1972). Innate and conditioned reactions to threat in rats with amygdala lesions. *Journal of Comparative and Physiological Psychology*, **81**, 281–290.

Castellano, C. (1975). Effects of morphine and heroin on discrimination learning and consolidation in mice. *Psychopharmacologia*, **42**, 235–242.

Elde, H. T., Johansson, O., & Terenius, L. (1976). Immunohistological studies using antibodies to leucine-enkephalin: Initial observations on the nervous system of the rat. *Neuroscience*, **1**, 349–351.

Evans, M. H. (1976). Stimulation of rabbit hypothalamus: caudal projections to respiratory and cardiovascular centers. *Journal of Physiology*, **260**, 205–222.

Evans, M. H. (1978). Potentiation of a cardioinhibitory reflex by hypothalamic stimulation in the rabbit. *Brain Research*, **154**, 331–343.

Fernandez de Molina, A., & Hunsperger, R. W. (1962). Organization of the subcortical system governing defence and flight reactions in the cat. *Journal of Physiology*, **160**, 200–213.

File, S. E., & Hyde, J.R.G. (1978). Can social interactions be used to measure anxiety? *British Journal of Pharmacology*, **62**, 19–25.

File, S. E., & Hyde, J.R.G. (1979). A test of anxiety that distinguishes between the actions of benzodiazepines and those of other minor tranquilizers and stimulants. *Pharmacology, Biochemistry and Behavior*, **11**, 65–69.

File, S. E., & Rodgers, R. J. (1979). Partial anxiolytic action of morphine sulphate following microinjection into the central nucleus of the amygdala in rats. *Pharmacology, Biochemistry and Behavior*, **11**, 313–318.

Firemark, H. M., & Weitzman, R. E. (1979). Effects of β-endorphin, morphine and naloxone on arginine vasopressin secretion and the electroencephalogram. *Neuroscience*, **4**, 1895–1902.

Gallagher, M., & Kapp, B. S. (1978). Manipulation of opiate activity in the amygdala alters memory processes. *Life Sciences*, **23**, 1973–1978.

Gallagher, M., Kapp, B. S., Musty, R. E., & Driscoll, P. A. (1977). Memory formation: Evidence for a specific neurochemical system in the amygdala. *Science*, **198**, 423–425.

Goddard, G. V. (1964). Amygdaloid stimulation and learning in the rat. *Journal of Comparative and Physiological Psychology*, **58**, 23–30.

Gold, P. E., Hankins, L. L., & Rose, R. P. (1977). Time-dependent post-trial changes in localization of amnestic electrical-stimulation sites within amygdala in rats. *Behavioral Biology*, **20**, 32–40.

Gold, P. E., & McGaugh, J. L. (1975). A single-trace, two-process view of memory storage processes. In D. Deutsch and J. A. Deutsch (Eds.), *Short-Term Memory*, pp. 356–378. New York: Academic Press.

Gold, P. E., Rose, R. P., Hankins, L. L., & Spanis, C. (1976). Impaired retention of visual discriminated escape training produced by subseizure amygdala stimulation. *Brain Research*, **118**, 73–85.

Gros, C., Pradelles, P., Humbert, J., Dray, F., Le Gal La Salle, G., & Ben-Ari, Y. (1978). Regional distribution of met-enkephalin within the amygdaloid complex and bed nucleus of the stria terminalis. *Neuroscience Letters*, **10**, 193–196.

Grossman, S. P., Grossman, L., & Walsh, L. (1975). Functional organization of the rat amygdala with respect to avoidance behavior. *Journal of Comparative and Psychological Psychology*, **88**, 829–850.

Handwerker, M. J., Gold, P. E., & McGaugh, J. L. (1974). Impairment of active avoidance learning with posttraining amygdala stimulation. *Brain Research*, **75**, 324–327.

Hiller, J. M., Pearson, J., & Simon, E. J. (1973). Distribution of stereospecific binding of the potent narcotic analgesic etorphine in the human brain: Predominance in the limbic system. *Research Communications in Chemical Pathology and Pharmacology*, **6**, 1052–1062.

Hilton, S. M., & Zbrozyna, A. W. (1963). Amygdaloid region for defence reactions and its efferent pathway to the brain stem. *Journal of Physiology*, **165**, 160–173.

Jacquet, Y. F., Klee, W. A., Rice, K. C., Iijima, I., & Minamikawa, J. (1977). Stereospecific and nonstereospecific effects of $(+) - $ and $(-) - $ morphine: Evidence for a new class of receptors? *Science*, **198**, 842–845.

Jensen, R. A., Martinez, J. L., Jr., Messing, R. B., Spiehler, V., Vasquez, B. J., Soumireu-Mourat, B., Liang, K. C., and McGaugh, J. L. (1978). Morphine and naloxone alter memory in the rat. *Society for Neuroscience Abstracts*, **4**, 260.

Kapp, B. S., Frysinger, R. C., Gallagher, M., & Haselton, J. R. (1980). Amygdala central nucleus lesions: Effect on heart rate conditioning in the rabbit. *Physiology and Behavior*, **23**, 1109–1117.

Kesner, R. P., Berman, R. F., Burton, B., & Hankins, W. G. (1975). Effects of electrical stimulation of amygdala upon neophobia and taste aversion. *Behavioral Biology*, **13**, 349–358.

Kety, S. (1970). The biogenic amines in the central nervous system: Their possible roles in arousal, emotion, and learning. *In* F. O. Schmitt (Ed.), *The Neurosciences: Second Study Program*, pp. 324–336. New York: Rockefeller University Press.

Kety, S. S. (1976). Biological concomitants of affective states and their possible role in memory processes. *In* M. R. Rosenzweig and E. L. Bennett (Eds.), *Neural Mechanisms of Learning and Memory*, pp. 321–326. Cambridge, Massachusetts: MIT Press.

Krettek, J. E., & Price, J. L. (1978). A description of the amygdaloid complex in the rat and cat with observations on intra-amygdaloid axonal connections. *Journal of Comparative Neurology*, **178**, 255–280.

Kuhar, M. J., Pert, C. B., & Snyder, S. H. (1973). Regional distribution of opiate receptor binding in monkey and human brain. *Nature (London)*, **245**, 447–450.

McIntyre, M., & Stein, D. G. (1973). Differential effects of one vs. two stage amygdaloid lesions on activity, exploratory and avoidance behavior in the albino rat. *Behavioral Biology*, **9**, 451–465.

Messing, R. B., Jensen, R. A., Martinez, J. L., Jr., Spiehler, V. R., Vasquez, B. J., Soumireu-Mourat, B., Liang, K. C., & McGaugh, J. L. (1979). Naloxone enhancement of memory. *Behavioral and Neural Biology*, **27**, 266–275.

Mondadori, C., & Waser, P. G. (1979). Facilitation os memory processing by post-trial morphine: Possible involvement of reinforcement mechanisms? *Psychopharmacology*, **63**, 297–300.

Pellegrino, L. J., & Cushman, A. J. (1967). *A Stereotaxic Atlas of the Rat Brain*. New York: Appleton.

Pert, A., De Wald, L. A., Liao, H., & Sivit, C. (1979). Effects of opiates and opioid peptides on motor behaviors: Sites and mechanisms of action. *In* E. Usdin, W. E. Bunney, & N. S. Kline (Eds.), *Endorphins in Mental Health Research*, pp. 45–61. London and New York: Oxford University Press.

Pert, C. B., & Snyder, S. H. (1973). Opiate receptor: Demonstration in nervous tissue. *Science*, **179**, 1011–1014.

Rodgers, R. J. (1978). Influence of intra-amygdaloid opiate injections on shock thresholds, tail flick latencies and open field behavior in rats. *Brain Research*, **153**, 211–216.

Rodgers, R. J., & Deacon, R.M.J. (1979). Effect of naloxone on the behavior of rats exposed to a novel environment. *Psychopharmacology*, **65**, 103–105.

Sar, M., Stumpf, W. E., Miller, R. J., Chang, K. J., & Cuatrecasas, P. (1978). Immunohistochemical localization of enkephalin in rat brain and spinal cord. *Journal of Comparative Neurology*, **182**, 17–38.

Simantov, R., Kuhar, M. J., Uhl, G. R., & Snyder, S. H. (1977). Opioid peptide enkephalin: Immunohistochemical mapping in rat central nervous system. *Proceedings of the National Academy of Sciences U.S.A.*, **74**, 2167–2171.

Simon, E. J., Hiller, J. M., & Edelman, I. (1973). Stereospecific binding of the potent narcotic analgesic [³H] etorphine to rat-brain homogenate. *Proceedings of the National Academy of Sciences U.S.A.*, **70**, 1947–1949.

Urban, I., & Richard, P. (1972). *A Stereotaxic Atlas of the New Zealand Rabbit's Brain*. Springfield, Massachusetts: Thomas.

Ursin, H., & Kaada, B. R. (1960). Functional localization within the amygdaloid complex in the cat. *Electroencephalography and Clinical Neurophysiology*, **12**, 1–20.

von Frisch, O. (1966). Herzfrequenzänderung bei Drückreaktion junger Nestflücher. *Zeitschrift für Tierpsychologie*, **23**, 497–500.

Werka, T., Skär, J., & Ursin, H. (1978). Exploration and avoidance in rats with lesions in amygdala and piriform cortex. *Journal of Comparative and Physiological Psychology*, **92**, 672–681.

Wood, C. D., Schottelius, B., Frost, L. L., and Baldwin, M. (1958). Localization within the amygdaloid complex of anaesthetized animals. *Neurology*, **8**, 477–480.

Zbrozyna, A. W. (1972). The organization of the defence reaction elicited from the amygdala and its connections. *In* E. B. Eleftheriou (Ed.), *The Neurobiology of the Amygdala*, pp. 597–606. New York: Plenum.

CHAPTER 22

Changes in Brain Peptide Systems and Altered Learning and Memory Processes in Aged Animals

Robert A. Jensen, Rita B. Messing, Joe L. Martinez, Jr.,
Beatriz J. Vasquez, Vina R. Spiehler,
and James L. McGaugh

The nature of the deficit in cognitive processes in aged individuals is not well understood. However, it appears that aged humans and infrahumans show similar neuroanatomical, neurochemical, and behavioral pathologies and that the senescent rodent may be a good model for the study of the neurobiological basis of impaired memory in aged individuals. One established finding is that aged animals show poor retention of an inhibitory (passive) avoidance response. Although initial learning appears to be relatively undisturbed in aged animals, deficits in retention performance occur after a much shorter training–testing interval in aged animals compared to their younger controls.

There is substantial research evidence that changes in cholinergic systems in the central nervous system (CNS) may underlie some deficits in memory. However, changes in hormonal systems may also be important. Treatment with vasopressin appears to attenuate some aspects of age-related memory deficits, but administration of adrenocorticotropic hormone (ACTH) fragments appears to be relatively ineffective, although these peptides may help to improve mood and attention.

Alteration in opioid systems may contribute to changes in learning and memory processes in aged animals. The effect of administration of the opiate receptor antagonist naloxone may be

463

ENDOGENOUS PEPTIDES
AND LEARNING AND MEMORY PROCESSES

different in young and aged rats. In addition, there are regionally specific differences in opiate receptor concentrations and apparent binding affinities in young and old rats. When considered as a whole, these findings suggest that the study of alterations in peptide systems in aged animals may be a promising avenue of research for understanding the disorders of memory seen in the aged.

I. INTRODUCTION

It is well documented that elderly people frequently complain about impaired cognitive abilities, especially impaired memory. Although some of these complaints may be the result of an expectation that a poor memory comes with age, data from laboratory studies support the idea that aging affects many aspects of mental functioning (for reviews see Botwinick, 1978; Botwinick & Storandt, 1974; Craik, 1977; Thatcher, 1976). Usually, the cognitive decline related to normal aging does not significantly impair an aged individual's day to day existence. However, in approximately 10% of the elderly, and in about 30% of those over 80, there is a severe and progressive decay of memory function and cognitive processes in general. A substantial body of research evidence has been gathered concerning the etiology of the memory deficits seen in the aged, but, as yet, no clear consensus about the nature of these memory problems has emerged (see Arenberg & Robertson-Tchabo, 1977; Craik, 1977; Fozard, 1980).

In our laboratory, we have adopted a research strategy that emphasizes the use of pharmacological tools to study animal models of behavioral senescence (see Jensen, Messing, Martinez, Vasquez, & McGaugh, 1980b; Jensen, Messing, Spiehler, Martinez, Vasquez, & McGaugh, 1980c). Young and aged laboratory rats are trained in a variety of learning and memory tasks and the effects of drug treatments on acquisition and retention are assessed. Because some of the drug treatments have quite specific actions, differences in drug effects between young and aged animals may provide information about the nature of those changes in the brain that lead to memory deficits with age.

The value of the rodent model for studies of memory and aging can be seen in the finding that aged rats, 2 to 3 years old, exhibit many of the same CNS pathologies as do aged humans. In addition, most of these changes are not general degeneration but appear to be quite specific and are confined to discrete brain regions and cell types (see Brizzee, Ordy, Knox, & Jirge, 1980). For example, in the aged rat and aged human, changes such as astroglial hypertrophy (Landfield, Rose, Sandles, Wohlstadter, & Lynch, 1977), degenerative alterations in the membranes of pyramidal cells and Purkinje cells, and a loss of dendritic spines from cortical pyramidal cells (Brizzee, Ordy, Knox, & Jirge, 1980; Feldman, 1976) are seen. Also, other changes such as the accumulation of lipofuscin pigment (Brizzee, Ordy, & Kaack, 1974), the development of

neurofibrillary tangles (Knox, 1979; Wisniewski & Terry, 1973), and the appearance of senile plaques (Terry & Wisniewski, 1974) have been documented
in aged laboratory animals and in humans.

II. AMINE NEUROTRANSMITTERS AND MEMORY ALTERATIONS IN AGED RATS

There are substantial age-related changes in neurotransmitter metabolism in
aged humans and other animals that may underlie some of the memory deficits
associated with aging. As with neuroanatomical alterations, most neurochemical
alterations that occur with aging are quite specific, and there are complex
regionally specific patterns of change that cannot be accounted for by a simple
decline in cerebral metabolic capacity. McGeer and McGeer (1976) studied
changes with age in the activities of the enzymes tyrosine hydroxylase, DOPA
decarboxylase, glutamic decarboxylase, choline acetyltransferase, and acetylcholinesterase in 56 brain areas of human subjects aged 5–50 years. They
reported decreases in enzyme activities in some brain regions, but not in others.
Furthermore, those regions in which activities were reduced were different for
the various enzymes. Regionally specific age-related changes were also found
for the enzymes monoamine oxidase and catechol-O-methyl-transferase (see
Domino, Dren, & Giardina, 1978).

The importance of changes in catecholaminergic systems is pointed out by
differences in amphetamine effects on memory in aged and young animals.
Doty and Doty (1966) found that amphetamine administration enhanced retention performance of aged rats when given as long as 1 to 4 hr after training.
However, in younger rats, amphetamine was effective in facilitating memory
consolidation only when given shortly after training, and was ineffective when
given 1 to 4 hr after training.

There are data suggesting that alterations in the cholinergic system may also
be important in mediating some of the changes in cognitive processes seen with
advanced age. The decrease in choline acetyltransferase activity reported by
McGeer and McGeer (1976) has been replicated by other researchers (Perry,
Perry, Gibson, Blessed, & Tomlinson, 1977) and very large decreases in the
activity of this enzyme are seen in the brains of elderly patients with Alzheimer
dementia, a condition in which there is a profound and progressive deterioration
of memory function (Perry, Tomlinson, Blessed, Bergmann, Gibson, & Perry,
1978; White, Goodhart, Kent, Hiley, Carrasco, Williams, & Bowen, 1977).
This decrease in enzyme activity is reported to be positively correlated with the
degree of cognitive impairment seen in these patients (Perry et al., 1978).

Bartus and his associates (for review see Bartus, 1980) have extensively studied
the role that changes in central cholinergic systems may play in mediating age-

related changes in learning and memory. They found that young monkeys, injected with the cholinergic receptor blocker, scopolamine, show a retention deficit that is very similar to that seen in aged monkeys. That is, progressively greater impairments were seen when longer training–retention test intervals were used (Bartus & Johnson, 1976). This dose-related deficit was attenuated by administration of the anticholinesterase, physostigmine. In addition, physostigmine administered to aged monkeys produced improvements in the performance of memory tasks that further suggest an involvement of cholinergic systems (Bartus, 1979).

These findings, along with others (see Bartus, 1980), suggested that it might be possible to decrease the rate at which memory impairments occur with age by increasing the amount of choline available in the diets of mature but not senescent animals. Bartus, Dean, Goas, and Lippa (1980) reported that 13-month-old mice raised on a choline deficient diet for 4.5 months showed impaired retention of an inhibitory avoidance response compared to mice given a choline-rich diet. The retention performance of the 13-month-old mice maintained on a choline-deficient diet was similar to that seen in 23-month-old mice, whereas the 13-month-old mice that were fed the choline-enriched diet performed comparably to 6-month-old mice. Although other explanations can be offered for these findings that are not related to changes in central cholinergic function, taken together, they do provide evidence supporting the idea that changes in cholinergic systems may be related to some aspects of age-related behavioral impairment.

III. CHANGES IN LEARNING AND MEMORY IN AGED RATS

Early in the research program of this laboratory, Gold and McGaugh (1975) reported that 24-month-old F344 rats show more rapid forgetting of an inhibitory avoidance response than do 2-month-old animals. When trained with a very low-level footshock (0.2 mA, 0.4 sec), the aged rats showed good retention 2 hr after training, but 6 hr after training, significant forgetting was already apparent. This basic finding has been confirmed with F344 rats by researchers in other laboratories (Bartus, 1980; Brizzee & Ordy, 1979; Ordy, Brizzee, Kaack, & Hansche, 1978), and with random-bred Sprague-Dawley rats (McNamara, Benignus, Benignus, & Miller, 1977), Wistar rats (Rigter, Martinez, & Crabbe, 1980), and C57B1/6j mice (Bartus et al., 1980).

We have extended this finding in a program of research aimed at characterizing the differences in retention between young and senescent rats so that a behavioral data base could be established for further pharmacological research. In the first phase of this study we found that both 3-month-old and 15-month-old F344 rats showed good retention of the inhibitory avoidance response when tested 1 day after training. However, when other groups of animals were tested 7 days after training, the 15-month-old rats showed a significant retention deficit (Jensen, Martinez, McGaugh, Messing, & Vasquez, 1980a; Vasquez, Martinez, Jensen, Rigter, Messing, & McGaugh, in press). Whether or not the accelerated forgetting seen in the older rats is the result of differences in the strength of initial learning or to some other factor is not presently clear (Rigter et al., 1980). However, Messing, Rigter, and Nickolson (submitted) obtained data indicating that initial learning in both young and aged rats is comparably resistant to disruption by amnesic treatments. This finding provides further evidence that acquisition performance is not substantially impaired in aged animals and again suggests that the deficit in these animals is related to an impairment of retention mechanisms. In addition, the 15-month-old rats that were used in this study cannot be considered aged, so the impaired retention performance of these animals indicates that the more rapid forgetting observed with the inhibitory avoidance task may appear at a younger age than some other behavioral measures and it may prove to be a useful indicator of impending deterioration in other areas of performance.

In contrast to the impaired performance seen in the older rats in the inhibitory avoidance task, in an active avoidance task, the 15-month-old rats showed acquisition and retention performance at 1, 3, or 7 days after training that was superior to that seen in the young animals. In a further study, young (3-month-old) and aged (24-month-old) male F344 rats were trained in a swim escape task. Young rats learned this task more efficiently than did the aged rats. Thus, the nature of the impairment in learning and memory in aged rats is task specific and is not simply a general decrement in behavioral performance.

IV. PEPTIDE MODULATION OF IMPAIRED MEMORY

There is a growing awareness that peptide systems, particularly the peptides of the pituitary, are important in memory processes (some representative reviews in this volume are: Bohus & de Wied; Gold & Delanoy; Koob & Bloom; Riccio & Concannon; Witter, Gispen, & de Wied). For example, the role of vasopressin in the modulation of memory is well established (see Bailey & Weiss and van Wimersma Greidanus, Bohus, & de Wied, this volume). A decrease in the amount of vasopressin produced by hypophysectomy or removal of the posterior portion of the pituitary produces behavioral impairments that can be attenuated by vasopressin administration (Bohus, Ader, & de Wied, 1972). Additionally, Brattleboro rats, homozygous for hereditary diabetes insipidus, lack the capacity to synthesize vasopressin, and these animals show an impaired performance in some learning and memory tasks (see Chapter 17 by Bailey & Weiss; see also de Wied, Bohus, Urban, & van Wimersma Greidanus, 1975). Both Brattleboro rats and aged rats show a substantial deficit in retention performance of an inhibitory avoidance task. Because there appear to be changes with age in the capacity of aged animals to synthesize and release vasopressin (Turkington & Everitt, 1976), the impaired memory performance of aged rats may be related to alterations in vasopressin systems.

To study this possibility, Cooper, McNamara, Thompson, and Marsh (1980) trained aged and young rats in an inhibitory avoidance task with low-level footshock. After a 1-wk training–retention–testing interval, some animals received a subcutaneous (s.c.) injection of lysine vasopressin 1 hr before the retention test. Other rats received a control injection. Vasopressin treatment significantly attenuated the retention deficit seen in the untreated aged animals. These data indicate that the aged animal may have impaired retrieval mechanisms and demonstrate that this deficit can be ameliorated by administration of vasopressin. Similarly, Cooper et al. (1980) showed that the recovery from a conditioned flavor aversion is more rapid in aged rats, and that this recovery can be significantly prolonged by treatment with lysine vasopressin. Although both young and aged rats showed vasopressin-induced improvement in retention of the flavor aversion, it appears that the aged rats benefitted more from the vasopressin treatment than did the younger animals. This finding again supports the hypothesis that decrements in retention in aged animals may result from an impairment in vasopressin-related mechanisms.

There are several studies of vasopressin effects in human subjects. In children suffering from diabetes insipidus, treatment with an analog of arginine vasopressin resulted in an "improved psychological status" as measured by a questionnaire given to the parents (Waggoner, Slonim, & Armstrong, 1978). In a very provocative study, Legros, Gilot, Seron, Claessena, Adam, Moeglen, Audibert, and Berchier (1978) gave 16 international units (IU) of lysine vasopressin as a nasal spray to elderly patients. They reported that the vasopressin produced a statistically significant improvement in performance of tests of attention, concentration, and memory. Although it is not clear whether these elderly patients were cognitively impaired, these results suggest that vasopressin may be potentially useful.

The anterior pituitary hormone ACTH also modulates learning and retrieval processes (see Bohus & de Wied, Chapter 3; Gold & Delanoy, Chapter 4; Martin, Chapter 5; Riccio & Concannon, Chapter 6; Soumireu-Mourat, Micheau, & Franc, Chapter 7). Several studies have been conducted to investigate the effects of administration of ACTH or one of its analogs in aged human subjects. Dornbush and Volavka (1976) compared several doses (15.0–60.0 mg) of $ACTH_{4-10}$ with control injections in normal elderly subjects in a cross-over design. They found a dose-related improvement in performance of a continuous reaction time test, but no improvement in measures of short-term memory or perceptual motor performance. Ferris, Sathananthan, Gershon, Clark, and Moshinsky (1976) gave either 15.0 or 30.0 mg of $ACTH_{4-10}$ or placebo to cognitively impaired elderly subjects in a cross-over design on different days. A battery of cognitive tests was administered before and after injection on each day. Following administration of 30.0 mg of $ACTH_{4-10}$, they observed a slowing of reaction time and also a slight improvement in the recall of visual memory as measured the next day. No other significant effects occurred.

In elderly subjects with some complaints of declining memory capacity, Will, Abuzzahab, and Zimmerman (1978) reported no consistent changes following administration of 15.0 mg $ACTH_{4-10}$. A similar lack of effect was also reported by Branconnier, Cole, and Gardos (1979). Thus, $ACTH_{4-10}$ does not seem to have the capacity to attenuate the memory impairment seen in aged subjects, although it may have some beneficial effects on mood and attentional processes.

One disadvantage of $ACTH_{4-10}$ treatment is that it must be administered by injection and its time of action in the body is quite short. The $ACTH_{4-9}$ analog ORG 2766, can be administered orally and it is more resistant to breakdown in the body. In research with laboratory animals, this peptide (H-Met[O]-Glu-His-Phe-D-Lys-Phe-OH) has been reported to be much more potent than $ACTH_{4-10}$ in delaying the extinction of pole-jump avoidance behavior (de Wied, 1974). For a review of the behavioral effects of this peptide, see Gaillard, Chapter 9, this volume. The practical advantage of oral administration of this peptide, and the possibility of giving it over extended time periods, makes it potentially very attractive for use with cognitively impaired patients. Unfortunately, the clinical effectiveness of this $ACTH_{4-9}$ analog in aged subjects has yet to be demonstrated and, if anything, the reported effects on memory-related processes seem to be somewhat weaker than those of $ACTH_{4-10}$ (Ferris, Reisberg, & Gershon, 1980). In one study, Branconnier et al. (1979) reported that this $ACTH_{4-9}$ analog was administered for 7 days to moderately impaired senile dementia patients, but no drug-related improvement was observed in a wide variety of measures. In a very comprehensive study, Ferris et al. (1980) gave several doses of the peptide to senile dementia patients in a double-blind multiple crossover design. Again no consistent effects were seen on measures of cognition. There were, however, some positive effects on the mood of the patients.

Thus, ACTH-related peptides appear to be less promising than vasopressin as therapeutic tools in the treatment of disorders of memory and cognitive function in the aged. However, the ACTH-related peptides may have substantial utility in treating some problems of attention and mood in some individuals.

V. OPIATE MODULATION OF MEMORY

Parallel with our research on memory in aged animals, our laboratory has conducted a program of research investigating the role that opioid systems may play in the modulation of memory storage processes. Initially, we found that two injections of naloxone given to young F344 rats immediately after training and 30 min later, enhanced retention performance of an inhibitory avoidance task (Jensen, Martinez, Messing, Spiehler, Vasquez, Soumireu-Mourat, Liang, & McGaugh, 1978; Messing, Jensen, Martinez, Spiehler, Vasquez, Soumireu-Mourat, Liang, & McGaugh, 1979). We concluded that this naloxone-induced facilitation is probably mediated by opiate receptor systems because naloxone's facilitatory effect is antagonized by treatment with morphine (Messing et al., 1979). For a complete discussion of naloxone and morphine effects on memory processes, see Chapter 20 by Messing, Jensen, Vasquez, Martinez, Spiehler, and McGaugh.

Because naloxone facilitated retention performance in young rats, we were interested in studying the effects of naloxone administration in aged rats to determine whether naloxone is capable of attenuating the memory deficit seen in these animals (Vasquez, Jensen, Messing, Martinez, Rigter & McGaugh, 1979). Three different tasks were used with 4- to 6-month-old and 24- to 26-month-old male F344 rats. They were inhibitory avoidance, active avoidance, and swim escape. In the inhibitory avoidance task, two footshock levels were used. These findings are presented in terms of percentage of change from the respective saline control baselines so that comparisons between the young and aged animals can more easily be made. No significant effects of naloxone were apparent in the young rats. Although trends can be seen, the lack of significant effects is probably due to the small size of the experimental groups used in this study. However, with the low-level footshock, 0.3 mg/kg naloxone significantly impaired the retention performance of the old rats when compared both to aged saline controls and to young rats given the same dose of naloxone. At a higher-level footshock, 0.3 mg/kg of naloxone tended to improve retention performance of both young and aged animals when compared to saline controls. This effect was significant only for the old rats in the second experiment. Thus, it appears that aged rats are more sensitive to the memory modulatory effects of naloxone than are young rats. Furthermore, in contrast to young rats, the direction of the drug effect on memory (i.e., enhancement or impairment) is dependent on the intensity of the footshock.

To determine whether these effects are general or confined only to the inhibitory avoidance task, young (5-month) and aged (26-month) rats were trained in the active avoidance task with a 640.0 µA footshock. Eight training trials were given on the first day and eight retention test trials were given 24 hr later. Naloxone was given immediately before training and again 30 min

TABLE 1

Effects of Naloxone and Morphine on Inhibitory Avoidance Retention in Young and Aged F344 Rats (Percent Change from Baseline (Saline) Retention Latencies)[a]

Drug and dose (mg/kg)	Young (4–6 mo)	Aged (24–26 mo)
Experiment 1: 500.0 μA, 0.5 sec footshock		
Saline	— ± 7.6 (11)	— ± 30.3 (9)
Naloxone 0.3	24.0 ± 26.7 (10)	−80.5 ± 14.4 (12)[b]
1.0	−1.1 ± 21.6 (10)	−39.7 ± 27.7 (8)
3.0	11.9 ± 29.8 (8)	−33.9 ± 27.7 (9)
Experiment 2: 500.0 μA, 0.5 sec footshock		
Saline	— ± 37.6 (7)	— ± 78.9 (8)
Naloxone 0.23	183.0 ± 107.0 (7)	−5.5 ± 79.5 (8)
1.0	700.0 ± 400.0 (7)	3.4 ± 91.1 (8)
Experiment 3: 750.0 μA, 0.5 sec footshock		
Saline	— ± 27.0 (7)	— ± 54.2 (7)
Naloxone 0.3	121.0 ± 66.8 (7)	144.5 ± 43.0 (7)[c]
1.0	40.9 ± 64.8 (7)	−31.6 ± 48.6 (6)
Experiment 4: 750.0 μA, 1.0 sec footshock		
Saline	— ± 19.6 (9)	— ± 92.2 (6)
Naloxone 0.3	8.8 ± 19.6 (9)	155.5 ± 96.9 (9)
Morphine 1.0	−23.8 ± 18.3 (9)	3.3 ± 91.6 (6)

[a] Naloxone was administered in equally divided doses immediately after training and 30 min later. Saline was administered immediately after training and 30 min later. Morphine was injected immediately after training followed by saline 30 min later. Retention was assessed 24 hr after training. The numbers in parentheses indicate the number of animals in each group.

[b] $p = .027$ compared to saline; $p = .002$ compared to young.

[c] $p = .58$ compared to saline.

later. The data shown in Table 2 show the effects of naloxone on acquisition, as no significant effects were observed on retention. Again, naloxone impaired the performance of old rats while having virtually no effect on the acquisition of the task in young rats. A dose of 1.5 mg/kg significantly impaired acquisition of the aged animals compared to their saline controls. However, a significant differential effect of the drug on young and aged animals was observed only at the 5.0 mg/kg dose.

In the swim escape task, a similar pattern of results was seen (Table 3). In two experiments, old rats given 1.0 mg/kg naloxone showed significantly longer latencies to reach the pole and escape from the water than did their saline controls, indicating that there was impaired retention performance in these rats. No significant effects of naloxone were observed in the younger rats. Morphine (3.0 mg/kg) also impaired retention performance of old, but not young, rats in this task. Similar effects of morphine and naloxone on memory have been

TABLE 2

Effect of Naloxone on Active Avoidance Acquisition Latencies in Young and Aged F344 Rats (Percent Change from Baseline (Saline) Acquisition Latencies)[a]

Drug and dose (mg/kg)	Young (5 mo)	Aged (26 mo)
Saline	— ± 10.3 (13)	— ± 12.8 (11)
Naloxone 1.5	11.3 ± 21.0 (12)	53.0 ± 22.0 (12)[b]
5.0	−23.4 ± 6.5 (13)	41.0 ± 19.0 (10)[c]
15.0	8.9 ± 11.6 (12)	27.7 ± 17.3 (10)

[a] Naloxone was administered immediately before training in the active avoidance task. The footshock was 640.0 μA and was given until the animal made an escape response. The interval between the beginning of the trial and the administration of footshock was 10 sec. The numbers in parentheses indicate the numbers of animals in each group.

[b] $p = .05$ compared to saline.

[c] $p < .01$ compared to young.

observed before in our laboratory, and this phenomenon is discussed in Chapter 20 by Messing et al.

Thus it appears that under some conditions, the response of aged rats to naloxone administration differs from that of young animals in memory tasks. Naloxone tends to impair learning and retention performance in aged rats, whereas it generally improves the performance of young rats (Messing et al., 1979). These findings also emphasize that the direction of the memory-modulating effects of naloxone are dependent on a number of different factors such as age, footshock level, and injection schedule (see Chapter 20 by Messing et al.).

VI. CHANGES WITH AGE IN OPIATE RECEPTOR SYSTEMS

We have conducted a series of studies in an effort to determine whether the differences in response to naloxone administration between young and aged animals might be related in some way to alterations in opiate receptor concentrations or binding affinities. We have completed measurements of the binding of tritiated dihydromorphine in a number of brain areas of young and aged male and female F344 rats.

In the female rats, we found that there are decreases in receptor densities in the thalamus and midbrain of the 26-month-old animals compared to the younger 4- to 5-month-old rats. Evidence for two binding sites was seen in the anterior cortex of the young rats, but the aged rats showed only a single binding site (Messing, Vasquez, Spiehler, Martinez, Jensen, Rigter, & McGaugh,

TABLE 3

Effects of Naloxone and Morphine on the Retention of a Swim-Escape Response in Young and Aged F344 Rats (Percent Change from Baseline (Saline) Retention Latencies)[a]

Drug and dose (mg/kg)	Young (4–6 mo)	Aged (24–26 mo)
Experiment 1		
Saline	— ± 9.5 (13)	—± 25.3 (10)
Naloxone 1.0	−21.4 ± 18.9 (11)	47.7 ± 31.3 (10)[b]
3.0	17.9 ± 17.5 (11)	130.9 ± 78.4 (10)
10.0	10.2 ± 24.8 (11)	73.0 ± 51.4 (9)
Experiment 2		
Saline	— ± 19.1 (10)	— ± 62.2 (3)
Naloxone 1.0	35.2 ± 22.9 (10)	211.0 ± 51.0 (6)[c,d]
3.0	7.1 ± 26.0 (10)	105.2 ± 48.5 (7)[c]

[a] Each rat was given two training trials 30 sec apart. Naloxone was administered in equally divided doses immediately after the second training trial and 30 min later. Saline was administered immediately after training and 30 min later. Morphine was injected immediately after training followed by saline 30 min later. Retention was assessed 24 hr after training. The numbers in parentheses indicate the number of animals in each group.
[b] $p < .01$ compared to young.
[c] $p < .05$ compared to saline.
[d] $p < .05$ compared to young.
[e] $p < .10$ compared to young.

1980). In male rats, substantial differences were seen in the pattern of change with age. The aged male rats had significantly lower receptor densities than their younger controls in the frontal poles, anterior cortex, and striatum. Interestingly, in the frontal poles, the decrease in receptor concentration was paralleled by a significant increase in the apparent affinity of the receptors for dihydromorphine (Messing, Vasquez, Samaniego, Jensen, Martinez, & McGaugh, 1981). These studies indicate that processes mediating aging in the brain may be different for males and females.

These findings led to a series of studies investigating functional changes in brain opioid systems in intact young and aged F344 rats (Jensen et al., 1980c). A subanalgesic dose of radioactively labeled etorphine was administered to young (5-month) and aged (26-month) male rats 8 min before flinch-jump testing (20 min before sacrifice). Specific binding was determined by dividing the amount of radioactivity (dpm/mg) in each brain region by the dpm/mg in the cerebellum, a brain region that contains no opiate receptors. We found significantly more specific labeled etorphine binding in the young animals in frontal poles, striatum, hypothalamus, hippocampus, thalamus, midbrain, and pons–medulla. Correlations were calculated between jump thresholds and labeled etorphine binding in different brain regions of the young and aged rats.

Overall, the data from young rats showed negative correlations between these measures, indicating the possibility of a direct relationship between the release of endogenous opioids (resulting in decreased etorphine binding) and changes in jump thresholds. Positive correlations were seen in the data from aged rats. Thus, it may be the case that stress-induced release of endogenous opioids is impaired in these animals or that opiate receptor ligand binding is qualitatively different in aged rats. Between young and aged rats, these correlations were found to be significantly different in the frontal poles and anterior cortex. These are two of the three brain regions in which we previously observed a decrease in opiate receptor concentrations using *in vitro* techniques. Thus, changes in some behaviors may be due to altered responsiveness of opioid receptor systems with age, or possibly to a decreased capacity of stress to activate opioid systems in the aged animal.

VII. CONCLUSIONS

Several conclusions can be drawn from these studies. First, animal models appear to be very useful for studies of the neurobiology of aging. Both human and other animals show similar deficits in memory processes for transient experiences or one-time-only events. The finding that aged animals are impaired in retention of an inhibitory avoidance response is well established by research in several laboratories. Cholinergic, catecholaminergic, vasopressin, and opioid systems all appear to be involved in alterations of memory processes in aged animals. Furthermore, pharmacological manipulations of these systems produce differential effects in young and aged animals. Thus, it is reasonable to hypothesize that alterations in brain systems subserved by these neurohumors may underlie some memory deficits in aged organisms. The precise nature of these biochemical alterations and their roles in memory processes is, at present, unclear. However, gaining an understanding of them will be a major task for memory research in the immediate future.

ACKNOWLEDGMENTS

This research was supported by USPHS Grants AG 00538 (J.L.McG., R.A.J., J.L.M., Jr., R.B.M.), MH 12526 (J.L.McG., R.A.J., J.L.M., Jr., R.B.M., B.J.V.), and NSF Grant BNS 76-17370 (J.L.McG.).

REFERENCES

Arenberg, D., & Robertson-Tchabo, E. A. (1977). Learning and aging, In J. E. Birren & W. K. Schaie (Eds.), *Handbook of the Psychology of Aging*, pp. 421–449. Princeton, New Jersey: Van Nostrand-Reinhold.

Bartus, R. T. (1979). Physostigmine and recent memory: Effects in young and aged non-human primates. *Science, 206,* 1087–1089.

Bartus, R. T. (1980). Cholinergic drug effects on memory and cognition in animals. *In* L. Poon (Ed.), *Aging in the 1980s. Psychological Issues,* pp. 163–180. Washington, D.C.: American Psychological Association.

Bartus, R. T., Dean, R. L., Goas, J. A., & Lippa, A. S. (1980). Age-related changes in passive avoidance retention and modulation with chronic dietary choline. *Science, 209,* 301–303.

Bartus, R. T., & Johnson, H. R. (1976). Short-term memory in the rhesus monkey. Disruption from the anticholinergic scopolamine. *Pharmacology, Biochemistry, & Behavior, 5,* 31–39.

Bohus, B., Ader, R., & de Wied, D. (1972). Effects of vasopressin on active and passive avoidance behavior. *Hormones & Behavior, 3,* 191–197.

Botwinick, J. (1978). *Aging and Behavior* (2nd ed.). New York: Springer.

Botwinick, J., & Storandt, M. (1974). *Memory-Related Functions and Age.* Springfield, Illinois: Thomas.

Branconnier, R. J., Cole, J. O., & Gardos, G. (1979). $ACTH_{4-10}$ in the amelioration of neuro-psychological symptomatology associated with senile organic brain syndrome. *Psychopharmacology, 61,* 161–165.

Brizzee, K. R., & Ordy, J. M. (1979). Age pigments, cell loss and hippocampal function. *Mechanisms of Ageing and Development, 9,* 143–162.

Brizzee, K. R., Ordy, J. M., & Kaack, B. (1974). Early appearance and regional differences in intraneuronal and extraneuronal lipofuscin accumulated with age in the brain of a nonhuman primate. *Journal of Gerontology, 29,* 366–381.

Brizzee, K. R., Ordy, J. M., Knox, C., & Jirge, S. K. (1980). Morphology and aging in the brain. *In* G. J. Maletta & F. J. Pirozzolo (Eds.), *The Aging Nervous System,* pp. 10–39. New York: Praeger.

Cooper, R. L., McNamara, M. C., Thompson, W. G., & Marsh, G. R. (1980). Vasopressin modulation of learning and memory in the rat. *In* L. Poon (Ed.), *Aging in the 1980s, Psychological Issues,* pp. 201–211. Washington, D.C.: American Psychological Association.

Craik, F.I.M. (1977). Age differences in human memory. *In* J. E. Birren & W. K. Schaie (Eds.), *Handbook of the Psychology of Aging,* pp. 384–420. Princeton, New Jersey: Van Nostrand-Reinhold.

de Wied, D. (1974). Pituitary-adrenal system hormones and behavior. *In* F. Schmitt & F. G. Worden (Eds.), *The Neurosciences. Third Study Program,* pp. 653–666. Cambridge, Massachusetts: MIT Press.

de Wied, D., Bohus, B., Urban, I., & van Wimersma Greidanus, Tj. B. (1975). Memory deficit in rats with diabetes insipidus. *Brain Research, 85,* 152–156.

Domino, D. F., Dren, A. T., & Giardina, W. J. (1978). Biochemical and neurotransmitter changes in the aging brain. *In* M. A. Lipton, A. DiMascio, & K. F. Killam (Eds.), *Psychopharmacology: A Generation of Progress.* New York: Raven.

Dornbush, R. L., & Volavka, J. (1976). $ACTH_{4-10}$: A study of toxicological and behavioral effects in an aging sample. *Neuropsychobiology, 2,* 350–360.

Doty, B. A., & Doty, L. A. (1966). Facilitative effects of amphetamine on avoidance conditioning in relation to age and problem difficulty. *Psychopharmacologia, 9,* 234–241.

Feldman, M. L. (1976). Aging changes in the morphology of cortical dendrite. *In* R. D. Terry & S. Gershon (Eds.), *The Neurobiology of Aging,* pp. 211–227. New York: Raven.

Ferris, S. H., Reisberg, B., & Gershon, S. (1980). Neuropeptide modulation of cognition and memory in humans. *In* L. Poon (Ed.), *Aging in the 1980s. Psychological Issues,* pp. 212–220. Washington, D.C.: American Psychological Association.

Ferris, S. H., Sathananthan, G., Gershon, S., Clark, C., & Moshinsky, J. (1976). Cognitive effects of $ACTH_{4-10}$ in the elderly. *Pharmacology, Biochemistry and Behavior, (Supplement 1), 5,* 73–78.

Fozard, J. L. (1980). The time for remembering. In L. Poon (Ed.), Aging in the 1980s. Psychological Issues, pp. 272–287. Washington, D.C.: American Psychological Association.

Gold, P. E., & McGaugh, J. L. (1975). Changes in learning and memory during aging. In J. M. Ordy & K. R. Brizzee (Eds.), Neurobiology of Aging. New York: Plenum.

Jensen, R. A., Martinez, Jr., J. L., McGaugh, J. L., Messing, R. B., & Vasquez, B. J. (1980a). The psychobiology of aging. In G. J. Maletta & F. J. Pirozzolo (Eds.), The Aging Nervous System, pp. 110–125. New York: Praeger.

Jensen, R. A., Martinez, Jr., J. L., Messing, R. B., Spiehler, V. R., Vasquez, B. J., Soumireu-Mourat, B., Liang, K. C., & McGaugh, J. L. (1978). Morphine and naloxone alter memory in the rat. Neuroscience Abstracts, 4, 260.

Jensen, R. A., Messing, R. B., Martinez, Jr., J. L., Vasquez, B. J., & McGaugh, J. L. (1980b). Opiate modulation of learning and memory in the rat. In L. Poon (Ed.), Aging in the 1980s. Psychological Issues, pp. 191–200. Washington, D.C.: American Psychological Association.

Jensen, R. A., Messign, R. B., Spiehler, V. R., Martinez, Jr., J. L., Vasquez, B. J., & McGaugh, J. L. (1980c). Memory, opiate receptors, and Aging, Peptides, (Supplement 1), 1, 197–201.

Knox, C. (1979). Morphology and permeability of the aging cerebral cortex in normotensive and hypertensive strains of rats. Anatomical Record, 193, 590–591.

Landfield, P. W., Rose, G., Sandles, L., Wohlstadter, T. C., & Lynch, G. (1977). Patterns of astroglial hypertrophy and neuronal degeneration in the hippocampus of aged, memory-deficient rats. Journal of Gerontology, 32, 3–12.

Legros, J. J., Gilot, P., Seron, X., Claessena, J., Adam, A., Moeglen, J. M., Audibert, A., & Berchier, P. (1978). Influence of vasopressin on learning and memory. Lancet, 1, 41–42.

McGeer, E., & McGeer, P. L. (1976). Neurotransmitter metabolism in the aging brain. In R. D. Terry & S. Gershon (Eds.), Neurobiology of Aging, pp. 389–403. New York: Raven.

McNamara, M. C., Benignus, G., Benignus, V. A., & Miller, A. T. (1977). Active and passive avoidance learning in rats as a function of age. Experimental Aging Research, 27, 266–275.

Messing, R. B., Jensen, R. A., Martinez, Jr., J. L., Spiehler, V. R., Vasquez, B. J., Soumireu-Mourat, B., Liang, K. C., & McGaugh, J. L. (1979). Naloxone enhancement of memory. Behavioral and Neural Biology, 27, 266–275.

Messing, R. B., Rigter, H., & Nickolson, V. (submitted). Memory consolidation in senescence: Effects of CO_2, amphetamine and morphine.

Messing, R. B., Vasquez, B. J., Samaniego, B., Jensen, R. A., Martinez, Jr., J. L., & McGaugh, J. L. (1981). Alterations in dihydromorphine binding in cerebral hemispheres of aged male rats. Journal of Neurochemistry, 36, 784–787.

Messing, R. B., Vasquez, B. J., Spiehler, V. R., Martinez, Jr., J. L., Jensen, R. A., Rigter, H., & McGaugh, J. L. (1980). ^{3}H-dihydromorphine binding in brain regions of young and aged rats. Life Sciences, 26, 921–927.

Ordy, J. M., Brizzee, K. R., Kaack, B., & Hansche, J. (1978). Age differences in short-term memory and cell loss in the cortex of the rat. Gerontology, 24, 276–285.

Perry, E. K., Perry, R. H., Gibson, P., Blessed, G., & Tomlinson, B. E. (1977). A cholinergic connection between normal aging and senile dementia in the human hippocampus. Neuroscience Letters, 6, 85–89.

Perry, E. K., Tomlinson, B. E., Blessed, G., Bergmann, K., Gibson, P. H., & Perry, R. H. (1978). Correlation of cholinergic abnormalities with senile plaques and mental test scores in senile dementia. British Medical Journal, 2, 1457–1459.

Rigter, H., Martinez, Jr., J. L., & Crabbe, Jr., J. C. (1980). Forgetting and other behavioral manifestations of aging. In D. Stein (Ed.), The Psychobiology of Aging: Problems and Perspectives, pp. 161–175. Amsterdam: Elsevier, North-Holland Publ.

Terry, R. D., & Wisniewski, M. (1974). Some structural and chemical aspects of the aging nervous system. Scandinavian Journal of Clinical Laboratory Investigation, 34, 13–15.

Thatcher, R. W. (1976). Electrophysiological correlates of animal and human memory. *In* R. D. Terry & S. Gershon (Eds.), *Neurobiology of Aging*, pp. 43–102. New York: Raven.

Turkington, M. R., & Everitt, A. V. (1976). The neurohypophysis and aging, with special reference to the antidiuretic hormone. *In* A. V. Everitt & J. A. Burgess (Eds.), *Hypothalamus, Pituitary, and Aging.* Springfield, Illinois: Thomas.

Vasquez, B. J., Jensen, R. A., Messing, R. B., Martinez, Jr., J. L., Rigter, H., & McGaugh, J. L. (1979). Naloxone impairs memory in aged rats. *The Pharmacologist*, **21**, 269.

Vasquez, B. J., Martinez, Jr., J. L., Jensen, R. A., Messing, R. B., Rigter, H., & McGaugh, J. L. Learning and memory in young and aged Fischer 344 rats. *Neurobiology of Aging*, in press.

Waggoner, R. W., Slonim, A. E., & Armstrong, S. H. (1978). Improved psychological status of children under DDAVP therapy for central diabetes insipidus. *American Journal of Psychiatry*, **135**, 361–362.

White, P., Goodhardt, M. J., Kent, J. P., Hiley, C. R., Carrasco, L. H., Williams, I. E., & Bowen, D. M. (1977). Neocortical cholinergic neurons in elderly people. *Lancet*, **1**, 668–671.

Will, J. C., Abuzzahab, F. S., & Zimmerman, R. L. (1978). The effects of $ACTH_{4-10}$ versus placebo in the memory of symptomatic geriatric volunteers. *Psychopharmacology*, **14**, 25–27.

Wisniewski, H. M., & Terry, R. D. (1973). Morphology of the aging brain, human and animal. *Progress in Brain Research*, **40**, 167–186.

CHAPTER *23*

A Two-Process Model of Opiate Tolerance

Raymond P. Kesner and Timothy B. Baker

This chapter evaluates and integrates recent findings concerning the development of opiate tolerance with respect to major behavioral and physiological tolerance models.

The behavioral phenomena of morphine tolerance can be explained by a two-process model, in which the importance of learning is emphasized. One process involves Pavlovian conditioning, with morphine serving as an unconditioned stimulus. Tolerance is viewed as a conditioned response that is discriminated on environmental cues accompanying drug administration. Such tolerance requires the regular pairing of drug with salient stimuli; it develops rapidly and has excellent long-term persistence. We propose that the development of Pavlovian tolerance is mediated by a negative endogenous opiate feedback circuit that reduces levels of endogenous opiate when organisms are exposed to stimuli that have previously been paired with the exogenous opiate.

The second hypothesized behavioral process of opiate tolerance involves drug-habituation, with morphine functioning as a conditioned stimulus. Drug-habituation tolerance is a function of iterative drug exposure in the absence of contiguous, salient environmental cues. Acquisition of this tolerance is less rapid than that of Pavlovian tolerance, develops with massed rather than with spaced drug administrations, and decays spontaneously after a decrease in opiate levels. Drug-habituation tolerance may conform to one of the traditional physiological models of opiate tolerance (e.g., derepression of one or more neurotransmitters). In conclusion, opiate tolerance appears to reflect at least two semiautonomous processes. Both processes are consistent with traditional models of behavior (Pavlovian learning and habituation), and each is mediated by different physiological substrates. Together, both processes can account for most of the behavioral phenomenon associated with the development of opiate tolerance.

479

ENDOGENOUS PEPTIDES
AND LEARNING AND MEMORY PROCESSES

I. INTRODUCTION

Recent research advances have stimulated considerable interest in behavioral and physiological models of morphine tolerance. Substantial amounts of experimental data have accumulated, but many findings are inconsistent or anomalous. One purpose of this chapter is to evaluate and to integrate recent literature with respect to major behavioral and physiological morphine tolerance models. A second purpose is to compare behavioral and physiological theories in terms of their postulates and in terms of their concordance with extant data. This comparison seems timely as results from a cross-fertilization of behavioral and physiological theories and techniques are beginning to appear. Research at this interface has produced evidence of congruity between behavioral tolerance models and the variation of important physiological indices (e.g., endogenous opiate levels). To accomplish our purposes, we will first define tolerance terms, then evaluate the relevance of behavioral or learning mechanisms to morphine tolerance development. In particular, we will concentrate on tolerance development to the analgesic or antinociceptive effects of morphine. This will be followed by a presentation of major behavioral models of tolerance, followed by a proposal of a new two-process model of physiological tolerance that is designed to account for recent behavioral observations.

II. TOLERANCE DEFINITIONS

The term *drug tolerance* generally refers to the phenomenon in which a drug dose exerts a decreasing effect with repeated administration, or a drug effect diminished through prior exposure can be reinstated with an increased dose (e.g., LeBlanc, Poulos, & Cappell, 1978). However, it has been observed that a more appropriate definition is that tolerance indicates a shift in the dose–response curve to the right (Kalant, 1978). In this chapter, *tolerance* per se is used in a way that is consistent with these definitions and the term carries little excess meaning (e.g., it does not specify a particular tolerance mechanism).

There is substantial consensus over general definitions of tolerance. However, confusion reigns over discussions concerning specific tolerance mechanisms or substrates, and procedurally derived tolerance labels. Evidence of such confusion is illustrated by Jones's (1978) compilation of over 20 different labels to refer to types of tolerance or tolerance phenomena including, behavioral, psychological, learned, behaviorally augmented, biochemical, metabolic, cellular, tissue, organic, dispositional, and pharmacological. Much of the dissension surrounding tolerance labels is due to the confusion of procedural and mechanistic definitions (Ferraro, 1978). Thus, we shall explicitly define the tolerance terms used in this chapter.

Most tolerance taxonomies include a label for types of tolerance resulting from physiological or organic changes that lead either to reduced levels of drug at target organs or receptor sites, or to insensitivity of receptor sites. Such tolerance has often been labeled *dispositional* or *physiological* tolerance and it may be due to altered metabolism, absorption, excretion, or distribution (Dews, 1978; Kalant, LeBlanc, & Gibbins, 1971). It has been suggested that dispositional or physiological tolerance should accrue because of drug exposure per se, and should be relatively unaffected by environmental context (LeBlanc et al., 1978). Another type of tolerance has been termed *behavioral* tolerance. This term has been used to refer to tolerance that is related to environmental context or learning contingencies. Because all of these terms refer to a *mechanism* or substrate of tolerance acquisition rather than to observed tolerance phenomena, we will refer to *dispositional, physiological,* or *behavioral (learning) mechanisms* of tolerance acquisition.

A number of authors have proposed that all forms of tolerance may share a common physiological substrate, reducing the need to distinguish among them (Jones, 1978). In addition, some authors have argued that, because any instance of tolerance is probably subserved by a variety of processes, it may be misleading to attribute instances of tolerance to particular mechanisms (Carder, 1978). However, Dews (1978) noted the advantages of attending to possible mechanisms of action in studying tolerance; "Such a course is surely better heuristically than defining behavioral tolerance as tolerance to the behavioral effects of a drug, which merely specifies the obvious dependent variable and provokes no further thought or experiment [1978, p. 21]." Furthermore, theories of drug tolerance that do not acknowledge the importance of environmental context, or learning (e.g., Collier, 1965) can neither explain nor predict results such as those reported by Siegel (1977).

The task of isolating and identifying processes contributing to tolerance development and maintenance is really one of experimental design and technique. That fact that tolerance appears to be a homogeneous phenomenon when studied with a particular experimental design (e.g., Jones, 1978) may reflect more on the design employed than on the phenomenon. Designs are available that do reflect the contributions of different mechanisms to tolerance development (Lê, Poulos, & Cappell, 1979; Siegel, 1977, 1978a; Siegel, Hinson, & Krank, 1978). Moreover, there is little reason to avoid behavioral or learning labels in discussions of tolerance mechanisms simply because all tolerance phenomena, at some level, have a physiological substrate. Such a reductionistic argument ignores the fact that all forms of learning undoubtedly possess a physiological substrate, which does not attenuate their predictive validity or theoretical utility.

In addition to terminology referring to mechanisms of tolerance development we will also operationally define terms to denote particular tolerance phenomena. Observed tolerance that results from a behavioral manipulation or a drug–context contingency will be referred to as *behaviorally augmented* (e.g., LeBlanc et al., 1978). Behaviorally-augmented tolerance will be contrasted with tolerance that appears to result from drug exposure per se (where there is no apparent behavioral contingency or manipulation), and it implies no particular mechanistic substrate. Finally, when we refer to a method of tolerance assessment, as opposed to putative mechanisms, we will use the term *behaviorally assessed* tolerance, or refer to the specific physiological index used to measure drug effects (e.g., blood ethanol checks).

III. MORPHINE TOLERANCE: BEHAVIORAL MODELS

There is ample evidence that environmental context, learning contingencies, and mnemonic interventions affect drug tolerance. Mitchell and his associates showed that maximal levels of morphine tolerance are obtained only when tolerance induction and testing are conducted in the same environment (Adams, Yeh, Woods, & Mitchell, 1969; Ferguson, Adams, & Mitchell, 1969; Kayan, Woods, & Mitchell, 1969). Others (Ferraro, Grilly, & Grisham, 1974; Manning, 1976; Schuster, Dockens, & Woods, 1966) have found that animals acquire levels of drug tolerance that reflect the extent to which previous drug exposure reduced appetitive, or increased aversive, consequences. Kesner, Priano, & DeWitt (1976) and Kesner & Priano (1977b) found that electroconvulsive shock, and electrical stimulation of frontal cortex or periaqueductal gray appeared to disrupt a morphine tolerance response just as it disrupts retention of other learned responses. In general, treatments that affect learning and memory processes exert similar effects on drug tolerance. LeBlanc, Matsunaga, and Kalant (1976) demonstrated that both frontal cortical ablation and cycloheximide treatments interfere with the development or expression of behaviorally assessed alcohol tolerance. Furthermore, serotonin depletion produced by *p*-chlorophenylalanine not only retards habituation to a variety of classes of environmental stimuli, but also attenuates alcohol and morphine tolerance development (Frankel, Khanna, LeBlanc, & Kalant, 1975; Way, Loh, & Shen, 1968).

What remains to be answered is which theoretical paradigm best accounts for environmental effects? A particular example of behaviorally augmented tolerance, the effect of context on tolerance to morphine antinociception, will be analyzed in an attempt to identify the tolerance theory that is most consistent with experimental observations.

A. The Pavlovian Model

Siegel is the principal proponent of the Pavlovian model and has explicated the model in an impressive series of papers (Siegel, 1975, 1977, 1978a, 1978c;

Siegel, et al., 1978). Theoretically, drugs can serve as unconditioned stimuli (USs) that produce direct drug effects, i.e., unconditioned responses (URs). Through an associative process, stimuli repeatedly paired with drug administration eventually elicit a response that is antagonistic, or counterdirectional to unconditioned drug effects. Thus, contextual cues that routinely accompany morphine exposure can be considered to be CSs that elicit drug compensatory CRs. This model explained earlier findings concerning the impact of environmental contingencies on morphine tolerance. Mitchell and his colleagues, for instance, obtained considerable evidence that maximal levels of analgesic morphine tolerance are obtained only when tolerance induction and testing are conducted in the same environment (Adams et al., 1969; Ferguson et al., 1969; Kayan et al., 1969). We will refer to this type of behaviorally augmented tolerance that arises from contextual associations as the *context effect*. Adams et al., (1969) for example, found that only animals receiving exposure to the tolerance test apparatus (a hot plate) under morphine evidenced significant levels of behaviorally assessed tolerance. Furthermore, this phenomenon was not simply due to test apparatus novelty because animals receiving morphine and test apparatus exposure at different times did not acquire tolerance. The Adams et al. (1969) data also argue against the notion that experience with the nociceptive stimulus or practice of the tolerance test response are necessary for the occurrence of the context effect. Temporally contiguous exposure to morphine and the test apparatus during tolerance development produced behaviorally augmented tolerance even when the hot plate was at ambient temperature during training (this precluded test response practice). These results suggest that exposure to the testing environment while under drug was the crucial factor in the development of the context effect. Adams et al. (1969) could not explain this "drug–test interaction" (actually a drug–test–environment interaction), but they speculated that it might be due to the concomitant occurrence of stress and morphine intoxication (cf. Bardo & Hughes, 1979; Carder, 1978).

Siegel and his co-workers extended these findings and proposed the Pavlovian paradigm as a model for the behavioral augmentation of tolerance to morphine. Siegel (1977) has shown that:

1. Behaviorally assessed tolerance to the analgesic effect of morphine is extinguished if animals are exposed to the tolerance test environment without the drug.

2. Preconditioning exposure to the injection regimen or the test environment retards tolerance acquisition, an effect analogous to latent inhibition.

3. Partial reinforcement results in attenuated tolerance.

These findings were obtained with either paw-pressure or hot plate measures of morphine analgesia. In a compelling demonstration that predrug signals or context are critical in eliciting behaviorally augmented tolerance, Siegel et al.

(1978) found that rats with consistent predrug signals developed greater morphine tolerance than did controls, even though predrug signals were not associated with the tolerance test apparatus or procedures. This finding reduces
the likelihood that context effects are due to a special relationship between test
environment exposure alone and drug effects. Rather, it suggests that behaviorally augmented tolerance to morphine may be elicited by a variety of
stimuli—the primary consideration being their contingent relationship with the
drug.

Not only is a Pavlovian model of tolerance acquisition consistent with Siegel's
research, but it is consistent with a considerable body of research demonstrating
the conditionability of a variety of drug effects. For instance, the ability of CSs
to elicit components of the abstinence syndrome (e.g., wet dog shakes), well
after physical dependence has waned, was shown by Wikler (1965), Wikler and
Pescor (1967), and Goldberg and Schuster (1970). On the other hand, Roffman,
Reddy, and Lal (1973) have shown that presentation of a bell–environment
complex CS, which was previously paired with morphine, attenuated morphine
withdrawal hypothermia for as long as 72 hr postwithdrawal. There is also
considerable evidence that, at least within particular training paradigms, tolerance conforms to operant conditioning principles (e.g., Ferraro, 1978; Schuster et al., 1966). In general, such research shows that the rate of acquisition
of behaviorally assessed tolerance depends, at least in part, on environmental–behavioral requirements impinging on the intoxicated organism. Tolerance
to drug effects is most likely if its occurrence allows an organism to reduce the
probability of aversive consequences, or enhance the likelihood of appetitive
consequences (cf. Corfield-Sumner & Stolerman, 1978, for a critical evaluation). A final source of information that is congruent, if not supportive, of an
associative explanation of tolerance development, derives from the previously
cited research showing that amnesic treatments interfere with tolerance just as
they interfere with other sorts of learning and memory tasks (Frankel et al.,
1975; Kesner et al., 1976; Kesner & Priano, 1977b; LeBlanc et al., 1976; Loh,
Shen, & Way, 1969; Way et al., 1968).

Some evidence, however, appears to be inconsistent with a Pavlovian model
of behaviorally augmented tolerance, and has served as a basis for the development of alternative tolerance models. First we will review specific critiques
of the Pavlovian model, and then discuss some of the major alternatives.

1. Dosage

The vast majority of data attesting to the explanatory prowess of the Pavlovian
tolerance model arise from studies employing low morphine doses, pyretic or
nociceptive tolerance measures, and the tolerance test apparatus as the CS.

Such procedural constraints are problematic because some tolerance studies using different parameters have produced evidence inconsistent with a Pavlovian model.

Most of the evidence pointing to the importance of location of morphine experience per se as a major contributor to behaviorally assessed tolerance derives from studies in which a 5.0 mg/kg morphine dose was employed (Adams et al., 1969; Ferguson et al., 1969; Kayan et al., 1969; Siegel, 1975, 1977, 1978a,b,c). Not only does the nearly exclusive use of a 5.0 mg/kg dose necessarily constrain the external validity of these experiments, but discrepant results have been produced by departures from this dose. For instance, Kayan et al. (1969) speculate that Cochin and Kornetsky's (1964) failure to find an effect due to location of morphine exposure was due to the high dose (20.0 mg/kg) employed in that study (Cochin & Kornetsky, 1964; Experiments 4, 5). Sherman (1979) found that an extinction manipulation failed to produce any diminution of tolerance when a 10.0 mg/kg dose was used. In addition, Sklar and Amit (1978) reported no difference in mortality incidence to a high dose of morphine as a function of environmental context. In this study, two groups of rats received increasing doses of morphine (15.0–105.0 mg/kg morphine) paired with a distinctive environment for 7 days. Then, one group had the distinctive environment paired with Ringer's solution for 7 days (extinction), while the other group remained in their home cages. On the test day, all rats were given 100.0 mg/kg morphine in the distinctive environment. Results showed no differences between animals that received extinction of the drug–environment pairing and those that did not. It is conceivable that high morphine doses might produce increased tolerance that is not attributable to environmental contingencies and that might overshadow any behaviorally augmented tolerance.

Siegel, Hinson, and Krank (1979) suggested that the reults of Sklar and Amit (1978) might have occurred because tolerance develops to certain lethal effects of morphine (those other than respiratory depression). In their own test of the effects of location on response to a high dose of morphine, Siegel et al. (1979) demonstrated that rats that had predrug signals extinguished (predrug signals were paired with saline) prior to receiving a lethal dose of morphine (300.0 mg/kg) died more quickly than rats for which predrug signals reliably predicted morphine. These results suggest that environmental context affects tolerance development at high, as well as at low, morphine doses. However, these results are merely suggestive, because the authors found a context effect for mortality latency, but not for incidence. Second, mortality latency was determined by failure to find a palpable heartbeat. Manual detection of a rat's heartbeat can be quite difficult, especially when it is weak and intermittent. The authors did not use an EKG measure for confirmation of death nor did they report estimates of interobserver reliability.

In an effort to obtain evidence of a context effect at a high morphine dose, Tiffany and Baker (1980) conducted two experiments using 10.0 mg/kg and 20.0 mg/kg morphine doses. In this research, a computerized flinch–jump measure was used to assess morphine-induced analgesia (Evans, 1961; Smith, Bowman, & Katz, 1978). This apparatus produces a voltage output that is a monotonic function of the force a rat exerts against a shock grid floor in response to shock (0.25–0.7 mA). In the first study, rats (N = 72) were assigned to one of six groups. During tolerance development, one group was given morphine sulfate (MS) (20.0 mg/kg) paired with the test (flinch–jump) environment for 5 days (F–J–MS), another group received morphine in their home cages (HC-MS), and a control group received NaCl (1.0 ml/kg) in both environments (SC, saline controls). In addition to these three groups, two groups were given flinch–jump tolerance pretests before tolerance development; one of these groups received morphine in the flinch–jump and the other in the home cage (PT, F–J-MS and PT, HC-MS groups, respectively). Pretesting was intended to disrupt tolerance acquisition because of latent inhibition. A final group had no pretesting, but did have morphine paired with both a highly distinctive plastic box environment and the flinch–jump apparatus during tolerance development (PB-MS).

Tolerance tests were begun on the day following tolerance development. Rats were given a 20.0 mg/kg morphine dose and their flinch–jump responses to shocks were assessed. All morphine experienced animals jumped significantly more than did SC rats, indicating a tolerance effect not due to environmental context (see Fig. 1). In addition, F–J-MS and HC-MS animals did not differ. Thus, in contrast to low-dosage studies (Siegel, 1977; Tiffany & Baker, 1980), simple exposure to the testing apparatus may not be sufficient to produce a context effect when using a higher dose. A context effect was obtained, however, because PB-MS rats jumped significantly more than did HC-MS animals on both test days. During both tolerance development and tolerance test sessions, these rats were given morphine in the plastic box environment and were later put in the flinch–jump apparatus. This treatment may have produced a more salient and effective CS, since the plastic box environment contained highly distinctive visual, auditory, and olfactory cues (Wagner, 1978).

Pretesting exposure to the flinch–jump apparatus produced no significant disruption of tolerance relative to nonpretested F–J-MS animals. This finding could also be explained by the lack of salience of the flinch–jump environment.

In a second high-dose study one group of rats had 20.0 mg/kg morphine paired with the plastic box environment during the 5 days of tolerance development (PB-MS). A second group received saline injections in the plastic box environment and morphine in the home cage (HC-MS). The last group received saline in both environments (SC). After the tolerance development phase, all rats received two tolerance tests in which they received morphine (10.0 mg/kg)

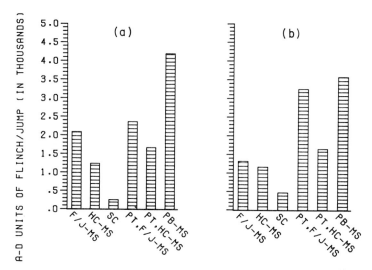

Fig. 1.   Mean flinch/jump amplitude in response to electric shock for (a) Test Day 1 and (b) Test Day 2 in A-D units. A-D units represent digitized units of force exerted against the shock grid floor in response to shock. Group designations reflect the following tolerance development treatment: F/J-MS, morphine in the flinch/jump test environment, and NaCl in the home cage; HC-MS, morphine in the home cage and NaCl in the flinch/jump environment; SC, NaCl in both the home cage and flinch/jump environments; and PB-MS, morphine in a complex environment consisting of a distinctive plastic box and the flinch/jump apparatus and NaCl in the home cage. The PT, F/J-MS and PT, HC-MS groups were treated the same as the F/J-MS and HC-MS groups, respectively, except that the PT groups received shock flinch/jump pretesting prior to tolerance development. All rats received morphine (20.0 mg/kg) on test days. (Reprinted with permission of American Psychological Association.)

in the plastic box environment before being tested in the flinch–jump apparatus. Results of the two tests again showed a significant reduction in morphine-induced analgesia that could be attributed to drug exposure per se: Both groups receiving morphine during tolerance development (PB-MS and HC-MS) jumped more than did saline controls. However, on the second test day, PB-MS rats did jump significantly more than HC-MS rats.

Taken together, these two experiments suggest that context effects can be obtained at a moderately high morphine dose. However, the magnitude of context effects at a high dose may be small relative to the magnitude of tolerance obtained that is unaffected by behavioral or contextual manipulations (e.g., tolerance attributable to dispositional mechanisms). Later, we shall present additional evidence showing that components of behaviorally assessed tolerance are differentially affected by learning or conditioning manipulations. In addition, these two studies suggest why other researchers using high doses may not have obtained context effects. Because the relative magnitude of context effects is small at high doses, some procedures that produce context effects at low doses may be ineffective at high doses. However, extremely advantageous conditioning procedures (e.g., a highly salient CS) can produce learning effects even at high doses.

Whereas context effects certainly appear to occur at more than one dose level, dose is only one procedural constraint that typifies Pavlovian research with morphine. Further research is required to determine the extent to which behavioral augmentation conforms to a Pavlovian paradigm when using other drugs, different dependent measures, and different behavioral manipulations. In a test of the generality of context effects, Lê et al., (1979) reported evidence of a context effect using a pyretic measure of ethanol tolerance.

2. Directionality of the CR

The problem of CR directionality in drug tolerance research arises in two forms. The first problem is that the Pavlovian tolerance paradigm requires that CRs be counterdirectional to drug URs. But, in most conditioning research, the CR is quite similar to the UR, at least in terms of general topography. Although it is unknown why drugs can produce drug antagonistic CRs, evidence of such drug-compensatory responding is abundant (cf. Hinson & Siegel, 1980; Siegel, 1978a). One hypothesis with respect to the directionality of CRs concerns the origin of physiological changes associated with the US. In most Pavlovian research, an external US impinges on the organism and the organism reacts to exteroceptive information concerning that stimulus. In drug conditioning, the drug US exerts agonistic effects by directly altering an organism's physiology; for instance, through receptor occupancy. After a drug exerts its agonist effects, an organism may react defensively in a compensatory fashion to mitigate drug-induced internal changes (Wikler, 1973). These defensive or preparatory reactions may be most analogous to US in the traditional Pavlovian paradigm: This would explain why cue-elicited CRs are isodirectional to these defensive reactions rather than to initial agonistic drug effects. Unlike the opponent-process theory of motivation (Solomon, 1977, 1980), this approach assumes that compensatory responses are more associable with drug-paired cues than are initial drug effects. It assumes that compensatory CRs are most likely to be acquired when USs directly alter internal processes. This tenet is, of course, consistent with the frequent observation of drug-compensatory responses to drug-paired cues. One problem with this theory is that the magnitude of drug-compensatory responses appears to be positively related to the number of drug exposures (e.g., Siegel, 1978a), an outcome inconsistent with the performance of URs.

Directionality of the CR to drug-paired cues poses a second problem for a Pavlovian paradigm: There are numerous examples of CRs that are isodirectional with direct drug effects. Siegel's Pavlovian conditioning explanation of morphine tolerance development is based upon the notion that drug-compensatory responses are elictied by cues previously paired with the drug. Evidence of compensatory CRs is fairly abundant (e.g., Hinson & Siegel, 1980; Lê et al., 1979; Siegel, 1975), yet, there is also copious evidence of isodirectional CRs. Many studies finding isodirectional CRs have measured pyretic effects of morphine. In fact, there are numerous conflicting reports of rats' conditioned or unconditioned pyretic responses to morphine: Some studies report CRs of hypothermia, hyperthermia, or a biphasic response (Cox, Ary, Chesarek, & Lomax, 1976; Eikelboom & Stewart, 1979; Gunne, 1960; Lal, Miksic, & Smith, 1976; Martin, Pryzbylik, & Spector, 1977; Miksic, Smith, Numan & Lal, 1975; Rosenfeld & Burks, 1977; Sherman, 1979; Siegel, 1978c; Stewart & Eikelboom, 1979; Thornhill, Hirst, & Gowdey, 1978). Such inconsistency is apparently due to the fact that pyretic response to morphine or to morphine-related cues varies as a function of dosage (Lal, Miksic, & Drawbaugh, 1978; Wilker, 1973), degree of restraint or activity (Martin et al., 1977), route of morphine administration (Lal et al., 1978), and duration of the pyretic assessment interval (Siegel, 1979). Moreover, Siegel (1979) has observed that studies showing CRs isodirectional to drug effects do not generally report evidence that tolerance was acquired. Because the association of drug-compensatory responses with tolerance is crucial to a Pavlovian model, it seems important to discover whether procedures that result in isodirectional CRs (e.g., Miksic et al., 1975) also produce tolerance. Although animals generally received extensive morphine exposure in these pyretic response studies (Lal et al., 1976; Miksic et al., 1975), certain procedural aspects might explain the absence of an acquired compensatory response. For instance, in the Miksic et al. (1975) study, rats were exposed to a 7-day baseline procedure that might have latently inhibited tolerance development.

3. Preconditioning US Exposure Effects

If an organism is administered a US prior to a CS–US pairing, learning will be blocked or attenuated. This is a relatively robust phenomenon and has been found in both taste mediated learning and traditional Pavlovian paradigms (Baker, 1976; Cannon, Berman, Baker, & Atkinson, 1975). The US preexposure effect is of interest as it involves a period of noncontingent drug administration followed by assessments of US potency. The similarity of this procedure to tolerance paradigms is obvious.

LeBlanc et al. (1978) report findings suggesting that the US preexposure effect in taste aversion learning may be due to a Kamin (1968) blocking effect. That is, during US preconditioning exposure, rats may associate the drug US with the injection ritual that regularly precedes drug administration. Therefore, taste aversions do not develop because drug effects are not "available" for association with a taste; drug effects are already paired with injection cues. In support of this notion, LeBlanc et al., (1978) reported that habituation to the injection ritual (latent inhibition) prior to drug preexposue reduced the disruption of taste aversion learning.

LeBlanc et al. (1978) noted the similarity of their findings to those supporting a Pavlovian model of behaviorally augmented tolerance (Siegel, 1978a). Using a hot plate assessment of morphine tolerance, Siegel (1977) showed that rats that have been habituated to the injection ritual acquire morphine tolerance more rapidly than do nonhabituated animals. He justifiably explains these results by suggesting that morphine tolerance is relatively more associable with tolerance test cues in injection habituated rats than in nonhabituated animals. If CS–US noncorrelation due to CS preexposure attenuates behaviorally assessed tolerance development, it seems reasonable that CS–US noncorrelation due to US preexposure should also block tolerance acquisition. Demonstration of a US preexposure effect, of course, would constitute evidence against any wholly nonassociative tolerance explanation as such an effect would occur in opposition or in addition to tolerance mechanisms that are activated by drug exposure per se (perhaps dispositional or physiological tolerance mechanisms).

The small amount of information that exists on the effects of morphine exposure prior to its explicit pairing with an environmental stimulus is either equivocal or somewhat contradictory to a Pavlovian model. For instance, Ferguson et al. (1969) administered morphine to one group of rats for 3 days without any concomitant tolerance testing (the putative CS). This group was then given 2 days of tolerance tests. A second group of animals received the same treatment except that they received saline in lieu of morphine over the first 3 days of the experiment. Ferguson et al. (1969) found no significant difference between the morphine pretreated rats and the group that received only morphine–test pairings. Siegel et al. (1978) also found that animals receiving morphine preexposure did not exhibit attenuated tolerance acquisition relative to controls. They offer reasons why a US preexposure effect was not obtained in their research, yet most of their reasons apply equally well to taste-mediated learning studies that produce potent US preexposure effects. Perhaps the US preexposure effect cannot be obtained in behaviorally assessed tolerance research because nonassociational substrates of tolerance (e.g., physiological tolerance) overwhelm the effects of drug–stimulus noncorrelation.

4. Criteria for Learned Tolerance

Corfield-Sumner and Stolerman (1978) acknowledge that procedural and contextual cues do affect tolerance, but argue that such effects do not necessarily indicate a learning mechanism. They note that the effects of contextual or procedural manipulations are small relative to maximum tolerance potential, and observe that the CSs in Siegel's research are "merely the cues present in the room to which the animals were moved when drugged" [p. 401]. These authors suggest that evidence should be obtained showing that tolerance is discriminated on particular *discrete* environmental events. They also note that it is possible that environmental manipulations do not affect a learned tolerance sui generis, but merely influence the rate of development of a general, un-differentiated tolerance (cf. LeBlanc et al., 1978). Based upon such concerns, Corfield-Sumner and Stolerman (1978) suggested that two major criteria be satisfied before accepting "learning" as an established route or contributor to tolerance:

1. identification of the specific learned responses purported to result in the appearance of tolerance; and
2. exclusion of factors such as enhanced metabolic rate that might also account for tolerance. Moreover, they observe that it would be difficult to unequivocally separate physiological and learning mechanisms of tolerance development.

Corfield-Sumner and Stolerman (1978) are justified in being interestd in the extent to which discrete, environmental events can serve as CSs for tolerance elicitation. However, Siegel has demonstrated a latent inhibition effect through preexposure to an injection ritual (Siegel, 1977) and this seems to be a fairly discrete cue that is not embedded in the tolerance test environment. Whereas it does seem important to discover whether tolerance can be elicited by a single, discrete drug-paired stimulus to examine the specificity of the associational process, a lack of such data does not lessen the present congruence between the Pavlovian paradigm and considerable basic research data (Hinson & Siegel, 1980; Lê et al., 1979; Siegel, 1975, 1977, 1978a,b,c; Siegel et al., 1978; Tiffany & Baker, 1980). Moreover, it now appears as though CSs in traditional Pav-lovian paradigms do not function as isolated, discrete stimuli, but are embedded in a contextual surround. It is this compound that elicits the full CR (Estes, 1979; Wagner, 1976, 1978). In addition, it seems unreasonable to assume that learning mechanisms must be dissociated from physiological changes involved in the tolerance test performance. As already noted, learning processes must necessarily be subserved by physiological changes.

B. Alternatives to the Pavlovian Model

We will now examine two theories that have been advanced as alternatives to a Pavlovian model; the novelty model and the stress-habituation model. These theories overlap considerably and some authors treat novelty and stress as equivalent concepts or factors. In this chapter, we shall evaluate a very narrow conceptualization of the novelty hypothesis and consider general stress theories under the stress-habituation heading.

1. The Novelty Hypothesis

Bardo and Hughes (1979), noting that nonstressful but distracting stimuli produce analgesia (Gardener & Lickleder, 1959), suggested that the context effect really represents habituation to novel aspects of the tolerance test apparatus or procedure. The authors posit that the performance of animals given analgesia tolerance tests in a novel environment reflects both the effects of novelty (which produces analgesia) and any dispositional or physiological drug tolerance that may have developed (which reduces analgesia).

In their first experiment, Bardo and Hughes (1979) found that rats given pretest morphine exposure in a hot plate environment showed more tolerance than rats given pretest morphine in a different environment. This, of course, replicates the context effect observed in many previous studies (Adams et al., 1969; Fergunsun et al., 1969; Kayan et al., 1969; Siegel, 1977). However, Bardo and Hughes also examined the paw-lick latencies of rats given saline during both the preexposure and tolerance tests. They found that among these animals, rats with pretest hot plate exposure had shorter paw-lick latencies than rats exposed to the hot plate for the first time. In a second experiment, Bardo and Hughes (1979) again found that animals given pretest pairings of saline in the tolerance test environment showed shorter paw-lick latencies than did rats given saline in a different environment. This, again, suggests a phenomenon topographically similar to drug tolerance (i.e., reduced paw-lick latencies). (It is important to note here that these effects are not due to increased practice of the test response as the hot plate was nonfunctional during pretests.) In addition, in their second experiment, Bardo and Hughes found that naloxone was ineffective in reducing the increment in analgesia produced by pretest hot plate exposure. However, naloxone was effective in reducing morphine induced analgesia. In summary, these results suggest that preexposure to a tolerance test environment results in a subsequent increased sensitivity to nociceptive stimulation in that environment and this effect is independent of drug history. Furthermore, Bardo and Hughes speculate that this sensitivity is not mediated by an endogenous opiate system, as it is not antagonized by naloxone.

Bardo and Hughes's data are not unique in their support of the hyperalgesic effect of environmental habituation. Kayan et al. (1969) found that hot plate testing experience resulted in decreased reaction times among animals with no morphine experience. In addition, Sherman (1979, Experiment 4) has produced data consistent with a novelty hypothesis. He found that rats with any pretest environment exposure showed shorter paw-lick latencies during tolerance tests—regardless of whether or not morphine was paired with such exposure.

The preceding findings suggest that, in particular circumstances, environmental exposure per se may sensitize an organism to nociceptive stimuli presented in that environment. This phenomenon does not, however, provide an adequate explanation of the robust context effects seen in morphine tolerance acquisition. First, a considerable body of research shows that environmental familiarity is an unreliable analgesic treatment. Cochin and Kornetsky (1964), for instance, found that rats with pretest hot plate practice (paired with distilled water) did not differ from rats naive to the hot plate environment when both groups were given hot plate tests under morphine (cf. Adams et al., 1969). Another major impediment to a novelty hypothesis arises from Siegel's research employing an extinction manipulation. Rats undergoing extinction trials have more test apparatus exposure than do rats with only morphine-paired test apparatus exposure, yet groups receiving the extinction manipulation evidence less morphine tolerance (i.e., they have slower reaction times) than nonextinguished rats, even though both groups have the same drug histories (e.g., Siegel, 1977, Experiment 2). In addition, in much context effect research (e.g., Siegel, 1978c), all animals receive equal exposure to test apparatus cues, but it is only those animals that have such exposure paired with morphine that display maximal levels of tolerance. Lastly, a novelty explanation cannot account for the elicitation of a drug antagonistic response by stimuli previously paired with drug (Lê et al., 1979; Siegel, 1975).

2. The Drug–Stress Habituation Model

This alternative to a Pavlovian model has been suggested by numerous authors but it remains poorly articulated and lacks a convincing modus operandi. Adams et al. (1969) were among the first to suggest a drug–stress interaction model of behavioral augmentation. However, neither these authors nor later writers formulated a model by which drug–stress pairings might produce an increment in behaviorally assessed drug tolerance.

In a recent formulation of the drug–stress hypothesis, Carder (1978) proposed that behaviorally augmented tolerance is produced by pairing drug with stress or arousal. Carder argues that results consistent with a Pavlovian model are equally compatible with a drug–stress theory. According to this theory, animals with an environment–drug (CS–US) pairing habituate to the drug–stress combination so that this effect gradually abates and an animal's behavior normalizes. This behavioral normalization is labeled *tolerance*.

To test his theory, Carder (1978) conducted a series of experiments in which tetrahydrocannabinol (THC) intoxication was paired with environmental cues or tasks dissimilar to the tolerance test. For instance, Carder exposed one group of rats to THC in the home cage, one group received THC in a novel experimental chamber, two additional groups received saline in one or the other of those two environments, and the final two groups received either THC or saline in an activated shock chamber. The tolerance test consisted of a swimming task under 5.0 mg/kg THC. No rats were exposed to the swimming task prior to tolerance testing. Carder discovered that rats exposed to the combinations of either THC in the novel environment or THC in the shock environment, had the fastest escape scores on the swimming task. Rats that were shocked without drug also showed faster escape times but not to the extent of animals with the drug–stress combinations. Thus, Carder found that drug exposure contiguous with a stressor or arousal agent produces greater behaviorally assessed tolerance than drug exposure, even if the stressor is dissimilar to the tolerance test.

Whereas the drug–stress habituation model may be germane to some tolerance phenomena, it does not provide an adequate explanation for the context effect. First, as previously noted, attributing tolerance to a habituation of stress or adaptation process lacks epistemological value since such a process, at least as currently presented, is imprecise. More importantly, there is substantial evidence antagonistic to that hypothesis.

Carder's (1978) research did not demonstrate that a drug–stress interaction resulted in complete display of behaviorally assessed tolerance on the swimming task as no rats received THC training doses in the swimming task environment. It is theoretically possible that such animals would display more tolerance than rats receiving THC paired with shock or the novel environment (although in this particular experiment a restriction in range renders this outcome unlikely). There is reason to anticipate such an outcome. Six years prior to Carder's (1978) report, Gebhart, Sherman, and Mitchell (1972) assessed whether the context effect depended upon a consistent morphine–context contingency, or upon a contingency between morphine and any stressor or arousal agent. They administered morphine in conjunction with pretest exposure to either the hot plate used in tolerance testing or to a stressful auditory stimulus, restraint, or to a swimming task. The results of two separate experiments clearly showed that only rats with pretest experience with morphine in the hot plate environment displayed behavioral augmentation. It is unlikely that these results were due to the inefficacy of the stressors employed, as Gebhart et al. (1972) found gastric pathology (type unspecified) in animals exposed to the auditory and restraint treatments. Given these results, it is unclear why Carder found an attenuation of THC effects through stress exposure. One possibility is that there are differences in the effects of morphine and THC. Another possibility is that commonalities between the pretest and test procedures (e.g., the drug administration ritual, handling) in the Carder experiment may have produced behavioral augmentation in the swimming task (Siegel, 1977).

Another problem with the drug–stress habituation model is that, like the novelty hypothesis, it is unable to account for the elicitation of drug compensatory responses when saline is administered in the presence of cues formerly paired with drug (Lê et al., 1979; Siegel, 1975).

3. The Drug-Habituation Model

Even though the Pavlovian model can account for a large number of behaviorally assessed tolerance effects of morphine including the *context* effect as well as latent inhibition, extinction, and drug compensatory effects, certain phenomena cannot be exclusively accounted for by a Pavlovian model. For example,

1. High doses of morphine often mask the context-dependent contribution to the development of morphine tolerance (Sherman, 1979; Sklar & Amit, 1978).

2. Treatments aimed at diminishing the importance of the environmental context (e.g., latent inhibition or extinction) attenuate, but do not abolish, the development of morphine tolerance (Siegel, 1977).

3. Procedural variables such as US preexposure have no effect on morphine tolerance development (Siegel et al., 1978).

4. Acquired morphine tolerance is not always persistent (Huidobro, Huidobro-Toro, & Way, 1976).

5. Morphine tolerance can be acquired even in nondistinctive environments (Kesner, Priano, & DeWitt, 1976).

It has often been suggested that morphine tolerance effects that cannot be explained by the Pavlovian model or one of its alternatives (the novelty or drug–stress habituation models) must therefore be subserved by physiological tolerance mechanisms. We propose, however, that in the absence of a distinctive environmental context, a drug-habituation model is consistent with adaptive changes observed during the development of morphine tolerance. Furthermore, behaviorally assessed tolerance within the drug-habituation model might be subserved by alterations in specific neurotransmitter systems. This proposed drug-habituation model differs from the drug–stress habituation model in that stress or arousal are not critical components of the model and the exogenous opiate serves as a CS rather than as a US. The drug-habituation model is not offered as an alternative to the Pavlovian model. Rather, the drug-habituation model attempts to account for opiate tolerance under conditions of minimal

influence of the environment and thus can be thought of as complementary
to the Pavlovian model.

Habituation is defined as a process characterized by a response decrement
with repeated presentations of a specific stimulus. It has been suggested (LeBlanc
& Cappell, 1977) that development of tolerance constitutes an adaptive process
similar to habituation. To support this LeBlanc and Cappell noted that serotonin
depletion by *p*-chlorophenylalanine not only retards habituation to a variety
of classes of environmental stimuli, but also attenuates alcohol and morphine
tolerance development (Frankel et al., 1975; Way et al., 1968). However, the
serotonergic system is involved in many behaviors, making it difficult to specifically relate habituation and the development of morphine tolerance.
Whereas LeBlanc and Kalant offer other parallels between tolerance and habituation, a more fruitful approach may be to determine whether some of the
important parametric characteristics that define habituation to external stimuli
(Thompson & Spencer, 1966) also apply to the development of tolerance. In
this section, we will discuss whether or not tolerance conforms to predictions
derived from a drug-habituation model. Whenever possible we will contrast
these predictions with those derived from other tolerance models.

There are a few important points that need to be made before a comparison
of tolerance and habituation can be undertaken. First, because of the long
duration of morphine action, the time required for tolerance development is
greater than that required for habituation to exteroceptive stimuli (e.g., tones,
lights). Second, for drug-habituation to occur, it is important to minimize the
salience of nondrug environmental cues to lessen the association of morphine
and exteroceptive cues, so that morphine will not function as a US. Unfortunately, most of the tolerance literature has not dealt with these issues, making
it very difficult to compare studies and to evaluate the results. Furthermore,
no one has systematically manipulated the parameters of habituation and examined their effects on the development of analgesic tolerance to morphine.

A basic requirement of habituation is that a particular stimulus elicits a
decreased response following iterative stimulation. The decrease in analgesic
response to repeated administration of morphine has been labeled *drug tolerance*
rather than habituation.

One parameter of habituation is that the more rapid the frequency of stimulation, the greater the amount of habituation (i.c., massed trials should result
in greater habituation than spaced trials). Using morphine doses of 5.0–10.0
mg/kg and time intervals in the order of hours–weeks, some studies indicate
that analgesic tolerance development is not influenced by frequency of drug
administration. However, other studies have yielded evidence that the shorter
the time interval between morphine injections, the greater the development
of tolerance (Ferguson et al., 1969; Kayan et al., 1969, Kayan & Mitchell,

1972a,b; Mushlin, Grell, & Cochin, 1976). For instance, Goldstein and Sheehan (1969) reported greater tolerance with massed compared to spaced injections
of levorphanol using running fits as the behavioral index. Because in most of
the above mentioned studies, not much attention has been paid to the effects
of environmental context, there appears to be a need to reassess the effects of
frequency on development of morphine tolerance in a situation in which
environmental contextual influences are minimized.

We performed a series of studies designed to investigate this question. Male
rats were injected with either a low (5.0 mg/kg) or high (15.0 mg/kg) dose of
morphine sulfate or saline once every 12 or 48 hr. All animals received 10
consecutive injections. After each injection, each animal remained in its home
cage for 30 min, then the animal was taken to an adjacent room with the same
level of illumination, and given a shock threshold test to determine sensitivity
to pain.

Each animal was placed in a small box with a grid floor. After a 60 sec
adaptation period, an ascending series of footshocks (starting with 0.5 mA
intensity and increased by 0.5 mA steps) was delivered until jump and squeal
responses were observed in response to three consecutive footshocks, or until
10.0 mA intensity was reached. Shocks were delivered via a constant current
scrambler for 0.5 sec; pulse repetition rate was 60 Hz. The intershock interval
was approximately 6 sec, but shocks were administered only when the animal
was making contact with the grid floor with all four paws.

Behavioral responses to each shock were observed. The shock intensities
required to initiate flinch, vocalization or avoidance, and achieve the criterion
of three consecutive vocalization or avoidance responses were used as dependent
measures. A flinch was defined as a response to shock characterized by any
detectable movement of any part of the animal. A vocalization response was
characterized by detection of audible vocalizations. An avoidance response was
characterized by shock-elicited rapid movement of the whole animal followed
by rapid running.

The mean shock intensity required for first vocalization across repeated tests
is shown in Figs. 2a, b, and c (data are depicted as percentages of control
values). The shorter the delay between treatments, the greater the development
of tolerance for both the 5.0 and 15.0 mg/kg dose levels. The data are clearly
congruent with the frequency parameters associated with habituation. It is
important to note that the Pavlovian model predicts the opposite, namely, that
spaced compared to massed trials should result in greater development of tolerance. Of course, as was previously noted, frequency parameters for habituation
and Pavlovian conditioning with exteroceptive stimuli were derived from research utilizing much shorter interstimulus intervals. To test the Pavlovian
model, new groups of animals were injected 10 times with 5.0 mg/kg morphine

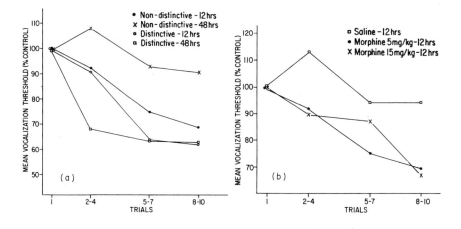

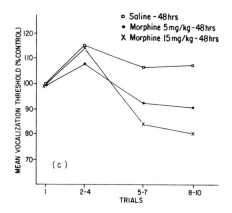

Fig. 2. (a) The effects of frequency of morphine (5.0 mg/kg) injections within distinctive and nondistinctive test environments upon mean shock intensity required for first vocalization. Data are depicted as a percentage of Test 1 value. Nondistinctive (12 hr) is indicated by •, nondistinctive (48 hr) by ×, distinctive (12 hr) by ○, distinctive (48 hr) by □.

The effects of 5.0 mg/kg (•) and 15.0 mg/kg (×) morphine once every (b) 12 hr and (c) 48 hr upon mean shock intensity required for first vocalization. Saline injections at every (b) 12 hr and (c) 48 hr are indicated by □. Data are depicted as a percentage of Test 1 value.

sulfate once every 12 or 48 hr and tested in a distinctive environment. For the distinctive environment groups, animals received their morphine injections in a novel, darkened room during tolerance development. They remained in the dark room for 30 min and were then tested for sensitivity to shock in a small box in the same dark room. The walls of the box were white with green fluorescent stripes. The animals were returned to their home cages in the animal colony 10 min after the shock test. The results are shown in Fig. 2a and clearly indicate that the longer the delay between treatments, the greater the development of tolerance. Thus, in a nondistinctive environment massed trials are more effective than spaced ones, supporting a habituation model; in a distinctive environment spaced trials are more effective than massed ones, supporting a Pavlovian model.

Another characteristic of habituation is that weak stimuli often produce the same or more rapid response diminution than do stronger stimuli. Considerable previous research shows that high morphine doses often produce as much, or more, tolerance than do low doses (Huidobro, Huidobro-Toro, & Way, 1976; Mucha, Kalant, & Linseman, 1979). However, in such research, initial differences in analgesic levels were not taken into account. This is of special importance when considering differential effects of high versus low dosage levels. Figures 2b and 2c depict the mean shock intensity required for first vocalization across repeated tests for 5.0 or 15.0 mg/kg morphine injected either every 12 or 48 hr. It can be seen that rats receiving 15.0 mg/kg show the same rate of tolerance development as do animals that received 5.0 mg/kg at both 12 or 48 hr injection frequencies. Thus, our failure to reject the null hypothesis with respect to the effect of dose appears to be consistent with a habituation model.

A third characteristic of habituation is that if the habituated stimulus is withheld, the response tends to recover over time (i.e., spontaneous recovery, lack of persistence, or lack of long-term retention). It has been shown that there is spontaneous recovery from analgesic tolerance to morphine after both single or multiple injections (Huidobro et al., 1976; Kesner, unpublished observations); and, with levorphanol-induced tolerance to running fits, there is spontaneous recovery over time (Goldstein & Sheehan, 1969). However, in Siegel's (1975, 1977) research, and that of Cochin and Kornetsky (1964) and Kornetsky and Bain (1968), no spontaneous recovery (loss of previous acquired tolerance) was observed. In fact, in the Cochin and Kornetsky (1964) research, analgesic tolerance was maintained for a period of 1 year.

To evaluate the apparent discrepancy between these studies, animals were given 5.0 mg/kg of morphine every 12 hr for 10 tests in either a nondistinctive or a distinctive environment. Two weeks after the last injection, animals were given a single 5.0 mg/kg morphine injection and again tested for analgesia in the same environment.

Greater tolerance development was seen for animals receiving morphine injections in the distinctive environment (Fig. 3). Moreover, these animals exhibited no spontaneous recovery of morphine effects after 2 weeks without morphine. However, there is spontaneous recovery or lack of persistence when animals are trained and tested in a nondistinctive environment.

The results shown in Fig. 3 suggest a resolution of the discrepancy concerning the persistence of morphine tolerance effects. Just as an exteroceptive stimulus may produce results consonant with habituation or Pavlovian paradigms depending upon the consistency of stimulus covariation (Wagner, 1976), morphine tolerance may resemble habituation or Pavlovian conditioning depending upon the availability of salient premorphine stimuli. Thus, when morphine

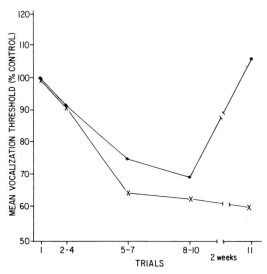

Fig. 3. The effects of a single morphine (5.0 mg/kg) injection 2 weeks following 10 morphine injections (5.0 mg/kg) once every 12 hr within distinctive (×) and nondistinctive (●) test environments upon mean shock intensity required for first vocalization. Data are depicted as a percentage of Test 1 value.

delivery is reliably signaled by a contextual cue, tolerance acquisition is rapid and little or no spontaneous recovery is evident. When morphine is delivered without reliable contingent cues, tolerance conforms to a drug-habituation model: There is a decrement in responsiveness and spontaneous recovery occurs. Response differences should be evident in the two different tolerance models, as in the Pavlovian model, morphine is serving as a US, whereas in the other model, morphine is merely an habituated stimulus, whose presentation has no signal function.

A fourth feature of habituation is that presentation of a previously unpaired stimulus results in recovery of the habituation response. In the case of a change in environmental context, this recovery has been labeled *dishabituation*. However, when an intense stimulus is presented, the recovery of the habituated response has been labeled *sensitization* and is independent of the habituation process (Groves & Thompson, 1970). Siegel (1975) has clearly demonstrated that a change in environmental context leads to attenuation of analgesic tolerance, suggesting recovery of the habituated response. It is of interest to note that in experiments carried out in the first author's laboratory, no change was observed in analgesic tolerance when rats made tolerant with 5.0 mg/kg of morphine in a nondistinctive environment were exposed to a distinctive environment.

There are a number of additional characteristics of the drug-habituation model that are shared with the Pavlovian model in that both make the same predictions. For example, a habituated or learned response to a given stimulus exhibits stimulus generalization of other stimuli (cross-tolerance). Animals exhibiting analgesic tolerance to morphine show cross-tolerance with β-endorphin, methionine-enkephalin, and electrical stimulation of the periaqueductal gray (Bläsig and Herz, 1976; Mayer and Hayes, 1975; Székely, Rónai, Dunai-Kovács, Miglécz, Bajusz, & Gráf, 1977).

As another example, if repeated series of habituation training trials and spontaneous recovery trials are given, habituation becomes successively more rapid. Kalant, LeBlanc, Gibbins, and Wilson (1978) report an accelerated development of tolerance to alcohol with repeated tolerance and spontaneous recovery sessions. Again, however, such an outcome is consistent with a Pavlovian model as reacquisition of a CR should be more rapid than the initial acquisition. Goldstein and Sheehan (1969) found no effect in development of tolerance to running fits with levorphanol using repeated tolerance and spontaneous recovery sessions. More data is needed to evaluate this parametric condition, especially with morphine.

In summary, in the absence of a distinct environmental context, the drug-habituation model of tolerance can account for the faster rate of tolerance development with massed compared to spaced morphine presentations, the lack of differential effect with different morphine dose levels, and the lack of persistence of morphine tolerance. Furthermore, the drug-habituation model can account for the presence of morphine tolerance (even though attenutated) when specific procedural manipulations block the environmental contribution to the development of tolerance. It appears that the Pavlovian model plus the drug-habituation model can account for most of the behavioral phenomena associated with opiate tolerance.

IV. BEHAVIORAL AND PHYSIOLOGICAL MODELS OF MORPHINE TOLERANCE: AN INTEGRATION

Considerable research evidence indicates that some tolerance phenomena are congruent with behavioral or learning models (e.g., Siegel, 1978), whereas others are not (e.g., the performance of HC-MS animals in Fig. 1; Cochin & Kornetsky, 1964; Sherman, 1979). Although learning theorists tend to emphasize behavioral augmentation phenomena (e.g., Siegel, 1976), researchers exploring physiological phenomena tend to stress dispositional and physiological tolerance mechanisms (e.g., Collier, 1965; Kosterlitz & Hughes, 1978). This tends to dichotomize the analysis of tolerance both at the behavioral and physiological levels. We propose a model of tolerance development that is an integration of behavioral and physiological data and theories.

Various physiological models have been advanced to account for drug tolerance and dependence: e.g., receptor occupancy (Axelrod, 1956; Seevers, 1958); receptor disuse (Collier, 1968; Sharpless & Jaffe, 1969); derepressant mechanisms (Shuster, 1971); immunological reaction (Cochin, 1972); new receptors (Collier, 1965); and an endogenous agonist model (Kosterlitz & Hughes, 1975). However, each model alone does not provide an adequate explanation for opiate tolerance effects, because many are faced with contrary evidence and none provides a role for contextual elicitation of a tolerance substrate. The new receptor model, for instance, is opposed by findings that opiate receptor sites do not increase in number following opiate tolerance induction (Klee & Streaty, 1974). The endogenous agonist model has difficulty accounting for increases in enkephalin levels during tolerance induction (Loh, Tseng, Holaday, & Wei, 1978).

In the proposed tolerance model we assume that exogenous opiates have a dual effect upon opiate receptors that mediate agonistic effects. It is assumed that derepression of one or more neurotransmitter systems (e.g., dopamine, serotonin, Substance P) or direct stimulatory effects on opiate-sensitive systems mediate processes associated with the drug-habituation model of tolerance development, while activation of a negative endogenous opiate feedback circuit mediates processes associated with the Pavlovian model of tolerance development.

We shall first elaborate on possible physiological mechanisms underlying the drug-habituation component of the two-process model of opiate tolerance followed by a more elaborate discussion of possible physiological mechanisms associated with the Pavlovian component of the two-process model.

It is assumed that this depression of neurotransmitter mechanism has properties that are consistent with the necessary short-term functional changes that presumably mediate the development of tolerance within nondistinctive environments according to the drug-habituation model. What are the models that describe the aforementioned physiological mechanism? There are at least three closely related models that operate sequentially to produce derepression of a specific neurotransmitter system. First is the *receptor occupancy model*, which hypothesizes that opiates remain associated with a specific receptor for very extended periods of time. Thus, each succeeding dose finds a smaller fraction of these receptors available for interaction, thereby producing tolerance (Axelrod, 1956; Seevers, 1958). Support for the prolonged binding of morphine comes from the observation that morphine can persist in detectable quantities for hours after a single injection (Hipps, Eveland, Meyer, Sherman, & Cicero, 1976; Misra, Mitchell, & Woods, 1971; Mullis, Perry, Finn, Stafford, & Sádee, 1979). Naloxone, a specific opiate antagonist injected during or after morphine injection disrupts the development of analgesic tolerance even when naloxone is administered 24 hr after an initial morphine injection (Kesner & Priano, 1977a; Martin, 1967; Mushlin & Cochin, 1976).

Second, there is the *receptor disuse model*, which hypothesizes that an opiate interrupts specific pathways by inhibiting release of a neurotransmitter. This results in a state of denervation with resultant supersensitivity of the postsynaptic effector and development of tolerance (Collier, 1968; Sharpless & Jaffe, 1969).

Third, there is the *derepressant model* proposed by Shuster (1971). He suggests that opiates inhibit an enzyme involved in the synthesis of a neurotransmitter. However, sustained repression of the enzyme eventually results in derepression of the enzyme, or it results in repression of an enzyme that destroys the neurotransmitter. The consequence is that neurotransmitter levels increase, counteracting the analgesic action of the opiate and thereby producing tolerance.

Goldstein and Goldstein (1961) have proposed a somewhat similar model involving enzyme induction. In both cases, whatever specific processes are involved, there is an increase in synthesis of the critical neurotransmitter that neutralizes the inhibitory effect of morphine, thereby initiating tolerance.

However, to identify a specific neurotransmitter as critical for development of analgesic tolerance, investigators have had to incorporate the receptor occupation and disuse models as a triggering system and the derepressant model as a subsequent homeostatic control mechanism. As an example of the possible interaction of a neurotransmitter system in tolerance development, we will illustrate the case for dopamine, even though other neurotransmitters like norepinephrine, acetylcholine, and serotonin might be similarly important (Way, 1972).

First, it is assumed that morphine induces analgesia by blockading postsynaptic receptors sensitive to dopamine. Support for this assumption comes from findings that:

1. Apomorphine, an agent that stimulates dopaminergic receptors (Anden, Rubenson, Fuxe, & Hökfelt, 1967) antagonized morphine-induced analgesia and other acute behavioral effects (Kuschinsky & Hornykiewiecz, 1974; Vanderwende & Spoerlein, 1973; Vedernikov, 1969).

2. Haloperidol, an agent that blocks dopaminergic receptors (Janssen, 1967), potentiates morphine-induced analgesia (Eidelberg & Erspamer, 1975).

3. Morphine produces an initial depletion of dopamine, especially in the striatum (Clouet, Jonsson, Ratner, Williams, & Gold, 1973; Clouet & Ratner, 1970; Gunne, Jonsson, & Fuxe, 1969).

Thus, the lower the level of dopamine at the putative dopamine receptor sites, the greater the analgesic effect of morphine.

It should be the case that analgesic tolerance is attenuated when agents that prevent the blockade of dopaminergic receptors are applied immediately after a morphine treatment. Kesner, Priano, and Gold (1977) found that apomorphine injected 5 min, but not 3 hr after an initial morphine injection disrupted the development of analgesic tolerance. Theoretically, the consequence of the

5 min apomorphine injection was the prevention of receptor blockade and, hence, prevention of denervation of disuse hypersensitivity. In the absence of such receptor hypersensitivity, there is no compensatory feedback to produce increased dopamine biosynthesis and release of dopamine (Hornykiewicz, 1972; Nybäck & Sedvall, 1968; Schwartz, Costentin, Martres, Protais, & Baudry, 1978). Increased dopamine synthesis should counteract the effects of subsequent injection of morphine resulting in less inhibition of dopaminergic postsynaptic receptors and thus less analgesia (i.e., tolerance to morphine induced analgesia). In support of this presumed process it has been shown that:

1. Acute morphine injections can lead to an initial depletion of dopamine in the striatum followed by an increased rate of dopamine biosynthesis in the same area (Clouet & Ratner, 1970; Clouet et al. 1973).

2. α-Methyl-p-tyrosine, a drug that blocks dopamine as well as norepinephrine synthesis (Levitt, Spector, Sjoerdsma, & Udenfriend, 1965; Spector, Sjoerdsma, & Udenfriend, 1965), injected simultaneously with morphine, blocks the development of morphine tolerance in mice (Marshall & Smith, 1973).

3. Chronic morphine injections produce an increase in the activity of tyrosine hydroxylase (Fukui, Shiomi, & Takagi, 1972).

4. Chronic treatment with morphine leads to a supersensitivity to dopamine agonists in that the behavioral and/or physiological effects of these agents are enhanced in morphine-treated rats (Iversen & Joyce, 1978; Puri & Lal, 1974; Smee & Overstreet, 1976).

If the dopamine hypothesis is correct, it should be the case that analgesic tolerance is attenuated when agents that prevent the enhanced synthesis of dopamine are applied after a morphine treatment. Kesner et al. (1977) found that α-methyl-p-tyrosine injected 5 min or 3 hr, but not 12 hr after an initial morphine injection disrupted the development of analgesic tolerance.

Thus, it appears that tolerance development depends on changes in dopamine biosynthesis and dopamine receptors. Similar data exist for the involvement of serotonin (Way, 1972), acetylcholine (Way & Bhargava, 1976), and the adenylate cyclase–cyclic 3′,5′adenosine monophosphate (cAMP) system (Collier & Francis, 1978).

In summary, it is hypothesized that exogenous opiates directly or indirectly stimulate the derepression of one or more neurotransmitter systems (e.g., dopamine, serotonin, acetylcholine, or norepinephrine) resulting in the development of morphine tolerance according to the laws of habituation that characterize the drug-habituation model.

In the presence of environmental cues or contexts that were previously paired with exogenous opiates, there is, in addition to the derepression of neurotransmitter systems, an activation of a negative endogenous opiate feedback circuit

(Kosterlitz & Hughes, 1975, 1978). This part of the model makes the following major assumptions: (1) changes in endogenous opiate activity can be elicited by environmental cues; (2) both exogenous opiates and tolerance test procedures produce an initial increase in endogenous opiate activity; (3) endogenous opiates potentiate exogenous opiate effects; and (4) naloxone produces incomplete antagonism of endogenous opiate effects.

We propose that endogenous opiate systems subserve Pavlovian opiate tolerance effects in the following way. When an animal is given exogenous opiate for the first time, the opiate elicits an increase in endogenous opiate release (Fu & Dewey, 1979; Weissman, Bergman, & Altstetter, 1978). This may be due to the effects of the stress involved in drug administration (Chesher & Chan, 1977; DeVries, Chance, Payne, & Rosecrans, 1979), or perhaps because endogenous opiate release is stimulated by the occupation of opiate receptor sites by exogenous opiates (Snyder, Pasternak, & Pert, 1975). In fact, if the tolerance test is itself stressful, it might augment the effects of morphine and enhance opiate effects. In any event, potent initial exogenous opiate effects are caused not only by exogenous opiates but also by increases in endogenous opiate release. There is considerable evidence that exogenous and endogenous opiates show cross-tolerance, produce similar behavioral effects, and may act synergistically (e.g., Bhargava, 1980; Browne & Segal, 1980; Morley, 1980; Urca, Frenk, Liebeskind, & Taylor, 1977; Vaught & Takemori, 1979a,b). If opiate doses are administered contingent with salient environmental cues, such cues are regularly paired with a state of excessive opiate availability (both endogenous and exogenous). Opiate-paired cues may eventually come to elicit a compensatory or preparatory response resulting in reductions in endogenous opiate release. Thus, we propose a cue-elicited negative endogenous opiate feedback system.

This model is consistent with observations that naloxone exerts relatively little pharmacological effect on drug-naive animals (e.g., Dykstra, 1980; El-Sobkey, Dostrovsky, & Wall, 1976; Goldstein, 1977). A number of explanations have been offered for this lack of effect including: (1) variability in endogenous opiate levels via diurnal rhythms (Davis, Buchsbaum, & Bunney, 1978; Frederickson, Burgis, & Edwards, 1977); (2) individual differences (Buchsbaum, Davis, & Bunney, 1977); (3) failure of naloxone to block those endogenous opiates or receptor sites critically involved in nociception (Lord, Waterfield, Hughes, & Kosterlitz, 1977; Kosterlitz & Hughes, 1978; Simantov & Snyder, 1976); or (4) maintenance of endogenous opiates on a standby basis and release or synthesis during particular events or in special contexts (Fanselow, 1979; Fanselow & Bolles, 1979a,b; Goldstein, 1977; Sherman & Liebeskind, 1980). Our model assumes that strong naloxone effects should not be observed in animals unless they have been subjected to a stressful procedure that results in the elicitation of endogenous opiates (Chesher & Chan, 1977).

The Pavlovian, or cue-elcitied, regulation of endogenous opiate changes is consistent with a variety of behaviorally augmented tolerance phenomena. The model predicts the prolonged maintenance of tolerance without further drug, given that the organism is not exposed to salient cues that had previously been paired with drug (Fig. 3) (Cochin & Kornetsky, 1964). Furthermore, it suggests that, in general, the rate of tolerance acquisition should be positively related to the salience or novelty of drug-paired cues (Fig. 1) (Siegel, 1977).

We have already presented evidence to support some of our assumptions: exogenous opiates act synergistically with, or potentiate, endogenous opiates, and exogenous opiate administration produces increased endogenous opiate activity. We will now address the remaining two assumptions: endogenous opiate activity may be cue-elicited and naloxone produces incomplete antagonism or blockade of exogenous opiate effects.

Data suggest that endogenous opiate activity is affected by contextual cues. For instance, stress-induced autoanalgesia may be caused by the release of endogenous opiates in an environment previously paired with stress. Numerous researchers have shown that stressors such as footshocks, centrifugal rotation, cold water swims, and even food deprivation produce analgesia (Akil, Madden, Patrick, & Barchas, 1976; Chance, Krynock, & Rosecrans, 1978; Chance, White, Krynock, & Rosecrans, 1978, 1979; Chesher & Chan, 1977; Hayes, Bennett, Newlon, & Mayer, 1978; Madden, Akil, Patrick, & Barchas, 1977). Moreover, autoanalgesia phenomena appear to be associated with alterations in opiate binding properties of brain tissue, suggesting an increase in endogenous opiate activity (Akil et al., 1976; Chance et al., 1978; Madden et al., 1977). Further evidence that endogenous opiate activity mediates autoanalgesia derives from studies in which stress-induced analgesia is blocked by naloxone (Akil et al., 1976; Bodnar, Kelly, Spiaggia, Ehrenberg, & Glusman, 1978; Chesher & Chan, 1977; Madden et al., 1977) and from studies demonstrating cross-tolerance between autoanalgesia and morphine (Chesher & Chan, 1977).

Some studies have failed to find evidence of naloxone blockade of autoanalgesia (Chance, 1980; Chance & Rosecrans, 1979b; Hayes et al., 1978), and some have found no evidence of cross-tolerance between morphine and stress-induced analgesia (Bodnar et al., 1978; Chance & Rosecrans, 1979a). Such discrepant findings may be due to the fact that antinociception is probably mediated, in part, by nonendorphinergic systems (Sherman & Liebeskind, 1980). Other explanations for these findings are:

1. Naloxone may have little effect on certain classes of endogenous opiates or opiate receptors involved in stress-induced analgesia (Chance & Rosecrans, 1979a,b; DeVries et al., 1979; Lord et al., 1977).

2. Context–stressor contingencies within particular experiments are not always conducive to recruitment of endogenous opiates (Chance, Krynock, & Rosecrans, 1980; Sherman & Liebeskind, 1980; Stewart & Eikelboom, 1979).

3. The pattern of stress administration may differentially affect affinity for those endogenous opiates blocked by naloxone (Chance et al., 1979b).

In any event, it is important to note that contextual cues can infleunce endogenous opiate activity—even if naloxone does not completely block endogenous opiate activity subserving autoanalgesia (Chance et al., 1978; DeVries et al., 1979; Fanselow, 1979; Fanselow & Bolles, 1979a,b).

Chance et al. (1978) provided evidence that is consistent with the notion that contextual cues may elicit alterations in endogenous opiate activity. These researchers exposed animals to 15 sec of shock per day for eight days. Shock administration procedures were invariant to produce a high context–shock correlation. Rats were given a tail-flick analgesia test after the eighth shock session, and then decapitated for determination of brain homogenate binding of radiolabeled enkephalin. Control animals were similarly treated except that the shock grid was not activated during their placement on it. Results showed that experimental animals had significantly longer tail-flick latencies than did controls, and experimental brain homogenates showed reduced binding of exogenous labeled enkephalin—presumably because receptors sites were occupied by endogenous opiates. Furthermore, the authors obtained a negative correlation between tail-flick latencies and binding levels of exogenous enkephalins in control animals. The authors interpret their results as showing that:

1. Stress produces a prolonged analgesia.
2. Stress-induced analgesia is due to greater receptor occupation by endogenous opiates.
3. Even among nonstressed rats, increased pain sensitivity is associated with reduced receptor occupation by endogenous opiates.

That the increased receptor binding by shocked animals was due to a context–shock association, rather than shock per se, is suggested by results of Chance et al. (1978). This research showed that autoanalgesia was discriminated on the particular cues present in the shock environment and was not produced by shock alone. Finally, if animals were exposed to a neutral context, stress-induced analgesia was attenuated.

Additional evidence of the associability of contextual cues with endogenous opiate activity can be found in research by Fanselow and Bolles (1979b), who compared the performance of rats that had received tone–shock pairings with rats that had received backward shock–tone pairings. All animals were tested for amount of freezing to a shock preceded by a tone. Results showed that tone–shock training was associated with decreased freezing to shock. However, if rats given tone–shock training were administered 4.0 mg/kg naloxone prior to the tone–shock test, they actually engaged in more freezing than did rats with shock–tone training. Thus, rats that had forward tone–shock pairings, but received no naloxone, froze less to the shock than animals that had received forward tone–shock pairings with naloxone, or animals that received backward shock–tone pairings with or without naloxone. The authors interpreted their results as showing that forward tone–shock training resulted in tone-elicited endogenous opiate release that rendered the shock less painful. This analgesic response was eliminated, however, if rats given tone–shock training were administered naloxone before the tone–shock test, presumably, naloxone blocked the effects of tone-elicited release of endorphins.

The results of this research are suggestive, but not conclusive. An alternative to the explanation offered by Fanselow and Bolles (1979b) is that reductions in freezing to shock occurred only if the freezing test was conducted in an identical manner to the training trials. Only tone–shock trained rats not receiving naloxone before the freezing test showed low levels of freezing to the shock. These were also the only rats to receive the same treatment during training and test trials. Rats trained with shock–tone pairings experienced a CS–US reversal during the test, whereas other rats were drugged (for the first time) before the freezing test. It is conceivable that such changes might have resulted in dishabituation.

Fanselow (1979) has reported additional evidence suggesting a link between endogenous opiate activity and a discrete environmental stimulus. Fanselow notes that it has been hypothesized that a signal preceding an aversive event allows an organism to marshal a preparatory response to that aversive event. This may be responsible for rats' preference for signaled, as opposed to unsignaled, shocks (Miller, Marlin, & Berk, 1977). Fanselow (1979) suggests that the basis of this preparatory response may be the activation of an endogenous opioid system. To test the notion that endogenous opiate release is involved in the preference for signaled shock, Fanselow administered naloxone to rats undergoing both signaled and unsignaled shock training procedures.

When rats were given naloxone prior to shock administration, they no longer showed a preference for signaled shock in a shuttlebox preference test. Only rats given preconditioning NaCl injections (in lieu of naloxone) exhibited preference for signaled shock. Apparently, naloxone administration prior to conditioning blocked the conditioned release of endogenous opiate and prevented the diminution of shock aversiveness. This, of course, eliminated the benefit of, and preference for a reliable shock signal. An alternative explanation is that naloxone increased shock aversiveness (cf. Fanselow & Bolles, 1979; Frederickson et al., 1977), resulting in the difference between the naloxone and saline groups. However, there is little evidence to support the notion that increased US aversiveness decreases preference for signaled shock.

At this point, we have reviewed evidence suggesting a link between endogneous opiate release and pain sensitivity, and have presented evidence that endogenous opiate activity may be elicited or affected by environmental stimuli.

Not only do stressors such as shock produce increases in endogenous opiate binding (as inferred from decreases in binding of exogenous ligand), but it is likely that shock-paired environmental cues also affect endogenous opiate release (Fanselow, 1979). The associability of contextual cues with endogenous opiate activity suggests that endogenous opiate release or activity might also be affected by contextual cues that share a contingent relationship with opiate administration or intoxication.

The final assumption of this model is that naloxone produces incomplete blockade of endogenous opiate effects. Our model assumes that contextual tolerance effects depend upon conditioned reductions in opiate release (it is, of course, possible that endogenous opiate activity could be reduced by an alternate mechanism such as increased peptidase activity; Morley, 1980). This does not, however, appear to explain the occurrence of compensatory responses in animals given vehicle in an environment previously paired with drug. If we consider response to nociceptive stimulation, animals showing compensatory responding are hyperalgesic relative to drug-naive animals (Siegel, 1978a). It has frequently been observed, however, that naloxone exerts modest effects on analgesia assessments (Akil & Watson, 1980). Whereas naloxone has been shown to increase nociceptive responsivity (cf. Jacob, Tremblay, & Colombel, 1974; Sawynok, Pinsky, & LaBella, 1979), such effects tend to be small and unreliable (Dykstra, 1980; Goldstein, Pryor, Otis, & Larson, 1976). If we assume that compensatory hyperalgesia is produced by reduced endogenous opiate release, and also that naloxone does not reliably produce hyperalgesia, we are faced with two alternatives. First, it may be that the modest effects produced by naloxone in drug-naive rats are comparable to the magnitude of cue-elicited compensatory responding seen in morphine treated animals (e.g., Siegel, 1975); Pavlovian tolerance effects account for only a portion of tolerance seen in low-dose studies (Tiffany & Baker, 1980). Another alternative is that naloxone provides incomplete blockade of endogenous opiate receptor sites. Indeed, there is mounting evidence to support this (Chance & Rosecrans, 1979b; Kosterlitz & Hughes, 1978; Lord et al., 1977; Rigter, Hannan, Messing, Martinez, Vasquez, Jensen, Veliquette, & Mcgaugh, 1980; Simantov & Snyder, 1976). Thus, compensatory responses may be mediated, at least in part, by reduced release of endogenous opiates that occupy receptor sites not blocked by naloxone.

In conclusion, behavioral phenomena of morphine tolerance can be explained by a two-process model, with each emphasizing the importance of specific learning processes. First, the drug-habituation model treats morphine as a conditioned stimulus. Drug-habituation tolerance is a function of iterative drug exposure in the absence of contiguous, salient environmental cues. Acquisition of this type of tolerance is more rapid with massed rather than spaced drug administrations, and decays spontaneously after a decrease in opiate levels.

Drug-habituation tolerance may conform to one of the traditional physiological models of opiate tolerance (e.g., derepression of one or more neurotransmitters).

Second, the Pavlovian model treats morphine as an US and views tolerance as a CR that is discriminated on environmental cues accompanying drug administration. This type of tolerance requires the regular pairing of drug with salient stimuli, it develops rapidly, and has excellent long-term persistence. We propose that the development of Pavlovian tolerance is mediated by a negative endogenous opiate feedback circuit that reduces levels of endogenous opiate when organisms are exposed to stimuli that had been previously paired with exogenous opiate.

Opiate tolerance appears to reflect at least two semiautonomous processes. Both processes are consistent with traditional models of behavior (Pavlovian learning and habituation) and each may be mediated by different physiological substrates. Together, the models can account for most of the behavioral phenomena associated with the development of opiate tolerance.

It is clear, however, that more research is required at the behavioral and physiological levels to determine the nature of the interaction between drug-habituation and Pavlovian tolerance acquisition processes, on the one hand, and interaction between the negative endogenous opiate feedback and derepression of neurotransmitter circuits, on the other hand. These interactions might be determined more easily if appropriate procedures are employed to separate habituation and classical conditioning processes using distinctive or nondistinctive environments, differing frequencies and intensities of opiate injections schedules, and short- and long-retention delays. Even with appropriate procedures, it is important to recognize that an observed tolerance phenomenon cannot be wholly ascribed to any single tolerance mechanism or process. Moreover, no adequate tolerance model can be constructed on the basis of data from one type of tolerance paradigm (e.g., analgesic tolerance). Any phenotypic example of analgesic tolerance is undoubtedly the product of diverse analgesic and drug processes (e.g., Lord et al., 1977; Rigter et al., 1980; Sherman & Liebeskind, 1980; Simantov & Snyder, 1976), and a similar complexity, no doubt, exists for other measures of tolerance.

REFERENCES

Adams, W. H., Yeh, S. Y., Woods, L. A., & Mitchell, C. L. (1969). Drug-test interaction as a factor in the development of tolerance to the analgesic effect of morphine. *Journal of Pharmacology and Experimental Therapeutics, 168,* 251–257.

Akil, H., Madden, J., Patrick, R. L., & Barchas, J. D. (1976). Stress-induced increase in endogenous opiate peptides: Concurrent analgesia and its partial reversal by naloxone. *In* H. W. Kosterlitz (Ed.), *Opiates and Endogenous Opioid Peptides.* Amsterdam: Elsevier.

Akil, H., & Watson, S. J. (1980). The role of endogenous opiates in pain control. *In* H. W. Kosterlitz & L. Y. Terenius (Eds.), *Pain and Society.* Weinheim: Verlag Chemie.

Andén, N. E., Rubenson, A., Fuxe, K., & Hokfelt, T. (1967). Evidence for dopamine receptor stimulation by apomorphine. *Journal of Pharmacy and Pharmacology*, 19, 627–629.

Axelrod, J. (1956). Possible mechanisms of tolerance to narcotic drugs. *Science*, 124, 263–264.

Baker, A. G. (1976). Learned irrelevance and learned helplessness: rats learn that stimuli, reinforcers, and responses are uncorrelated. *Journal of Experimental Psychology: Animal Behavior Process*, 2, 130–141.

Bardo, M. T., & Hughes, R. A. (1979). Exposure to a nonfunctional hot plate as a factor in the assessment of morphine-induced analgesia and analgesic tolerance in rats. *Pharmacology, Biochemistry and Behavior*, 10, 481–485.

Bhargava, H. N. (1980). Comparative effects of synthetic enkephalinamides and morphine on abstinence responses in morphine-dependent mice. *Pharmacology, Biochemistry & Behavior*, 12, 645–649.

Bläsig, J., and Herz, A. (1976). Tolerance and dependence induced by morphine-like pituitary peptides in rats. *Naunyn-Schmiederberg's Archives of Pharmacology*, 294, 297–300.

Bodnar, R. J., Kelly, D. D., Spiaggia, A., Ehrenberg, C., & Glusman, M. (1978). Dose-dependent reductions of analgesia induced by cold-water stress. *Pharmacology, Biochemistry and Behavior*, 8, 667–672.

Browne, R. G., & Segal, D. S. (1980). Behavioral activating effects of opiates and opioid peptides. *Biological Psychiatry*, 15, 77–86.

Buchsbaum, M. S., Davis, G. C., & Bunney, W. E. (1977). Naloxone alters pain perception and somatosensory evoked potentials in normal subjects. *Nature (London)*, 270, 620–622.

Cannon, D. S., Berman, R. F., Baker, T. B., & Atkinson, C. A. (1975). Effect of preconditioning unconditioned stimulus experience on learned taste aversions. *Journal of Experimental Psychology: Animal Behavior Processes*, 1, 270–284.

Carder, B. (1978). Environmental influences on morphine tolerance. In N. A. Krasnegor (Ed.), *Behavioral Tolerance: Research and Treatment Implications*, NIDA Research Monograph 18. Washington, D.C.: U.S. Government Printing Office.

Chance, W. T. (1980). Autoanalgesia: Opiate and non-opiate mechanisms. *Neuroscience and Biobehavioral Reviews*, 4, 55–67.

Chance, W. T., Krynock, G. M., & Rosecrans, J. A. (1978). Antinociception following lesion-induced hyperemotionality and conditioned fear. *Pain*, 4, 243–252.

Chance, W. T., Krynock, G. M., & Rosecrans, J. A. (1980). Investigation of pituitary influences on autoanalgesia. *Psychoneuroendocrinology*, 4, 199–205.

Chance, W. T., & Rosecrans, J. A. (1979a). Lack of cross-tolerance between morphine and autoanalgesia. *Pharmacology, Biochemistry and Behavior*, 11, 639–642.

Chance, W. T., & Rosecrans, J. A. (1979b). Lack of effect of naloxone on autoanalgesia. *Pharmacology, Biochemistry and Behavior*, 11, 643–646.

Chance, W. T., White, A. C., Krynock, G. M., & Rosecrans, J. A. (1978). Conditioned fear-induced antinociception and decreased binding of [^{3}H]N-Leu-enkephalin to rat brain. *Brain Research*, 141, 371–374.

Chance, W. T., White, A. C., Krynock, G. M., & Rosecrans, J. A. (1979). Autoanalgesia: Acquisition, blockade, and relationship to opiate binding. *European Journal of Pharmacology*, 58, 461–468.

Chesher, G. B., & Chan, B. (1977). Footshock induced analgesia in mice: Its reversal by naloxone and cross tolerance with morphine. *Life Sciences*, 21, 1569–1574.

Clouet, D. H., Jonsson, J. C., Ratner, M., Williams, N., & Gold, G. J. (1973). The effect of morphine on rat brain catecholamines: Turnover in vivo and uptake in isolated synaptosomes. In E. Usdin & S. H. Snyder (Eds.), *Frontiers in Catecholamine Research*, pp. 1039–1042. Oxford: Pergamon.

Clouet, D. H., & Ratner, M. (1970). Catecholamine biosynthesis in brains of rats treated with morphine. *Science*, 168, 854–856.

Cochin, J. (1972). Some aspects of tolerance to the narcotic analgesics. *In* J. M. Singh, L. H. Miller, & H. Lal (Eds.), *Drug Addiction*, Vol. 1, Experimental Pharmacology, pp. 365–375. Mt. Kisco, New York: Futura Publ.

Cochin, J., & Kornetsky, C. (1964). Development and loss of tolerance to morphine in the rat after single and multiple injections. *Journal of Pharmacology and Experimental Therapeutics*, **145**, 1–10.

Collier, H. O. J. (1965). A general theory of the genesis of drug dependence by induction of receptors. *Nature (London)*, **205**, 181–182.

Collier, H. O. J. (1968). Supersensitivity and dependence. *Nature (London)*, **220**, 228–231.

Collier, H. O. J., & Francis, D. L. (1978). A pharmacological analysis of opiate tolerance/dependence. *In* J. Fishman (Ed.), *The Bases of Addiction*, pp. 281–298. Berlin: Dahlem Konferenzen.

Corfield-Sumner, P. K., & Stolerman, I. P. (1978). Behavioral tolerance. *In* D. E. Blackman & D. J. Sanger (Eds.), *Contemporary Research in Behavioral Pharmacology*. New York: Plenum Press.

Cox, B. M., Ary, M., Chesarek, W., & Lomax, P. (1976). Morphine hyperthermia in the rat: An action on the central thermostats. *European Journal of Pharmacology*, **36**, 33–39.

Davis, G. C., Buchsbaum, M. S., & Bunney, W. E. (1978). Naloxone decreases diurnal variation in pain sensitivity and somatosensory evoked potentials. *Life Sciences*, **23**, 1449–1460.

DeVries, G. H., Chance, W. T., Payne, W. R., & Rosecrans, J. A. (1979). Effect of autoanalgesia on CNS enkephalin receptors. *Pharmacology, Biochemistry and Behavior*, **11**, 741–744.

Dews, P. B. (1978). Behavioral tolerance. *In* N. A. Krasnegor (Ed.), *Behavioral Tolerance: Research and Treatment Implications*, NIDA Research Monograph 18. Washington, D.C.: U.S. Government Printing Office.

Dykstra, L. A. (1980). Discrimination of electric shock: Effects of some opioid and nonopioid drugs. *Journal of Pharmacology and Experimental Therapeutics*, **213**, 234–240.

Eidelberg, E., & Erspamer, R. (1975). Dopaminergic mechanisms of opiate actions in brain. *Journal of Pharmacology and Experimental Therapeutics*, **192**, 50–57.

Eikelboom, R., & Stewart, J. (1979). Conditioned temperature effects using morphine as the unconditioned stimulus. *Psychopharmacology*, **61**, 31–38.

El-Sobkey, A., Dostrovsky, J. O., & Wall, P. D. (1976). Lack of effect of naloxone on pain perception in humans. *Nature (London)*, **263**, 783–784.

Estes, W. K. (1979). Cognitive processes in conditioning. *In* A. Dickinson & R. A. Boakes (Eds.), *Mechanisms of Learning and Motivation*. Hillsdale, New Jersey: Erlbaum.

Evans, W. O. (1961). A new technique for the investigation of some analgesic drugs on a reflexive behavior in the rat. *Psychopharmacologia*, **2**, 318–325.

Fanselow, M. S. (1979). Naloxone attenuates rats' preference for signaled shock. *Physiological Psychology*, **7**, 70–74.

Fanselow, M. S., & Bolles, R. C. (1979a). Naloxone and shock-elicited freezing in the rat. *Journal of Comparative and Physiological Psychology*, **93**, 736–744.

Fanselow, M. S., & Bolles, R. C. (1979b). Triggering of the endorphin analgesic reaction by a cue previously associated with shock: Reversal by naloxone. *Bulletin of the Psychonomic Society*, **14**, 88–90.

Ferguson, R. K., Adams, W. J., & Mitchell, C. L. (1969). Studies of tolerance development to morphine analgesia in rats tested on the hot plate. *European Journal of Pharmacology*, **8**, 83–92.

Ferraro, D. P. (1978). Behavioral tolerance to marihuana. *In* N. A. Krasnegor (Ed.), *Behavioral tolerance: Research and Treatment Implications*, NIDA Research Monograph 18. Washington, D.C.: U.S. Government Printing Office.

Ferraro, D. P., Grilly, D. M., & Grisham, M. G. (1974). Delta-9-tetrahydrocannabinol and delayed matching-to-sample in chimpanzees. *In* J. M. Singh & H. Lal (Eds.), *Drug Addiction*, Vol. 3. New York: Symposia Specialists.

Frankel, D., Khanna, J. M., LeBlanc, A. E., & Kalant, H. (1975). Effect of *p*-chlorophenylalanine on the acquisition of tolerance to ethanol and phenobarbital. *Psychopharmacology*, **44**, 247–252.

Frederickson, R. C. A., Burgis, B., & Edwards, J. D. (1977). Hyperalgesia induced by naloxone follows diurnal rhythm in responsivity to painful stimuli. *Science*, **198**, 756–758.

Fu, T. C., & Dewey, W. L. (1979). Morphine antinociception: Evidence for the release of endogenous substance(s). *Life Sciences*, **25**, 53–60.

Fukui, K., Shiomi, H., & Takagi, H. (1972). Effect of morphine on tyrosine hydroxylase activity in mouse brain. *European Journal of Pharmacology*, **19**, 123–125.

Gardner, W. J., & Licklider, J. C. R. (1959). Auditory analgesia in dental operations. *Journal of the American Dental Association*, **59**, 114–119.

Gebhart, G. F., Sherman, A. D., & Mitchell, C. L. (1972). The influence of stress on tolerance development to morphine in rats tested on the hot plate. *Archives Internationales de Pharmacodynamie et de Therapie*, **197**, 328–337.

Goldberg, S. R., & Schuster, C. R. (1970). Conditioned nalorphine-induced abstinence changes: Persistence in post-morphine dependent monkeys. *Journal of the Experimental Analysis of Behavior*, **14**, 33–46.

Goldstein, A., Pryor, G. T., Otis, L., & Larsen, F. (1976). On the role of endogenous opioid peptides: Failure of naloxone to influence shock escape threshold in the rat. *Life Sciences*, **18**, 599–604.

Goldstein, A. (1977). Future research on opioid peptides (endorphins): A preview. *In* K. Blum (Ed.), *Alcohol and Opiates: Neurochemical and Behavioral Mechanisms.* New York: Academic Press.

Goldstein, A., & Sheehan, P. (1969). Tolerance to opioid narcotics. I. Tolerance to the "running fit" caused by levorphanol in the mouse. *Journal of Pharmacology and Experimental Therapeutics*, **169**, 175–184.

Goldstein, D. B., & Goldstein, A. (1961). Possible role of enzyme inhibition and repression in drug tolerance and addiction. *Biochemistry and Pharmacology*, **8**, 48–49.

Groves, P. M., & Thompson, R. F. (1970). Habituation: A dual-process theory. *Psychological Review*, **77**, 419–450.

Gunne, L. M. (1960). The temperature response in rats during acute and chronic morphine administration: A study of morphine tolerance. *Archives Internationales de Pharmacodynamie et de Therapie*, **79**, 416–428.

Gunne, L. M., Jonsson, J., & Fuxe, K. (1969). Effects of morphine intoxication on brain catecholamine neurons. *European Journal of Pharmacology*, **5**, 338–342.

Hayes, R. L., Bennett, G. J., Newlon, P. G., & Mayer, D. J. (1978). Behavioral and physiological studies of non-narcotic analgesia in the rat elicited by certain environmental stimuli. *Brain Research*, **155**, 69–90.

Hinson, R. E., & Siegel, S. (1980). The contribution of Pavlovian conditioning to ethanol tolerance and dependence. *In* H. Rigter & J. C. Crabbe, Jr. (Eds.), *Alcohol Tolerance, Dependence, and Addiction*, pp. 181–187. Amsterdam: Elsevier/North-Holland.

Hipps, P. P., Eveland, M. R., Meyer, E. R., Sherman, W. R., & Cicero, T. J. (1976). Mass fragmentography of morphine: Relationship between brain levels and analgesic activity. *Journal of Pharmacology and Experimental Therapeutics*, **196**, 642–648.

Hornykiewicz, O. (1972). Dopamine and its physiological significance in brain function. *In* G. H. Bourne (Ed.), *Structure and Function of Nervous Tissue*, Vol. 4, pp. 367–415. New York: Academic Press.

Huidobro, F., Huidobro-Toro, J. P., & Way, E. L. (1976). Studies on tolerance development to single doses of morphine in mice. *Journal of Pharmacology and Experimental Therapeutics*, **198**, 318–329.

Iversen, S. D., & Joyce, E. M. (1978). Effects in the rat of chronic morphine treatment on the behavioral response to apomorphine. *British Journal of Pharmacology*, **62**, 390.

Jacob, J. J., Tremblay, E. C., & Colombel, M. C. (1974). Facilitations de reactions nociceptives par la naloxone chez la souris et chez le rat. *Psychopharmacologia*, **37**, 217–233.

Janssen, P. A. J. (1967). The pharmacology of haloperidol. *International Journal of Neuropsychiatry*, **3**, S-10–S-18.

Jones, R. T. (1978). Behavioral tolerance: Lessons learned from cannabis research. In N. A. Krasnegor (Ed.), *Behavioral Tolerance: Research and Treatment Implications*, NIDA Research Monograph 18. Washington, D.C.: U.S. Government Printing Office.

Kalant, H. (1978). Behavioral criteria for tolerance and physical dependence. In J. Fishman (Ed.), *The Bases of Addiction*. Berlin: Dahlem Konferenzen.

Kalant, H., LeBlanc, A. E., & Gibbins, R. J. (1971). Tolerance to, and dependence on, some non-opiate psychotropic drugs. *Pharmacological Review*, **23**, 135–191.

Kalant, H., LeBlanc, A. E., Gibbins, R. J., & Wilson, A. (1978). Accelerated development of tolerance during repeated cycles of ethanol exposure. *Psychopharmacology*, **60**, 59–65.

Kamin, L. J. (1968). Attention-like processes in classical conditioning. In M. R. Jones (Ed.), *Miami Symposium on the Prediction of Behavior: Aversive Stimulation*. Miami, Florida: University of Miami Press.

Kayan, S., & Mitchell, C. L. (1972a). The role of the dose-interval on the development of tolerance to morphine. *Archives International Pharmacodynamics*, **198**, 238–241.

Kayan, S., & Mitchell, C. L. (1972b). Studies on tolerance development to morphine: Effect of the dose-interval on the development of single dose tolerance. *Archives International Pharmacodynamics*, **199**, 407–414.

Kayan, S., Woods, L. A., & Mitchell, C. L. (1969). Experience as a factor in the development of tolerance to the analgesic effect of morphine. *European Journal of Pharmacology*, **6**, 333–339.

Kesner, R. P., & Priano, D. J. (1977a). Disruptive effects of naloxone on development of morphine tolerance. *Life Sciences*, **21**, 509–512.

Kesner, R. P., & Priano, D. J. (1977b). Time-dependent disruptive effects of periaqueductal gray stimulation on development of morphine tolerance. *Behavioral Biology*, **21**, 462–469.

Kesner, R. P., Priano, D. J., & DeWitt, J. R. (1976). Time-dependent disruption of morphine tolerance by electroconvulsive shock and frontal cortical stimulation. *Science*, **194**, 1079–1081.

Kesner, R. P., Priano, D. J., & Gold, T. (1977). Time-dependent disruptive effects of apomorphine and alpha-methyl-*p*-tyrosine on development of morphine tolerance. *Psychopharmacology*, **55**, 177–181.

Klee, W. A., & Streaty, R. A. (1974). Narcotic receptor sites in morphine-dependent rats. *Nature (London)*, **248**, 61–63.

Kornetsky, C., & Bain, G. (1968). Morphine: Single-dose tolerance. *Science*, **162**, 1011–1012.

Kosterlitz, H. W., & Hughes, J. (1975). Some thoughts on the significance of enkephalin, the endogenous ligand. *Life Sciences*, **17**, 91–96.

Kosterlitz, H. W., & Hughes, J. (1978). Endogenous opioid peptides. In J. Fishman (Ed.), *The Bases of Addiction*. Berlin: Dahlem Konferenzen.

Kuschinsky, K., & Hornykiewicz, O. (1974). Effects of morphine on striatal dopamine metabolism: Possible mechanism of its opposite effect on locomotor activity in rats and mice. *European Journal of Pharmacology*, **26**, 41–50.

Lal, H., Miksic, S., & Drawbaugh, R. (1978). Influence of environmental stimuli associated with narcotic administration on narcotic actions and dependence. In M. L. Adler, L. Manara, & R. Samanin (Eds.), *Factors Affecting the Action of Narcotics*. New York: Raven.

Lal, H., Miksic, S., & Smith, N. (1976). Naloxone antagonism of conditioned hyperthermia: An evidence for release of endogenous opiate. *Life Sciences*, **18**, 971–975.

Lê, A. D., Poulos, C. X., & Cappell, H. (1979). Conditioned tolerance to the hypothermic effect of ethyl alcohol. *Science*, **206**, 1109–1110.

LeBlanc, A. E., & Cappell, H. D. (1977). Tolerance as adaptation: Interactions with behavior and parallels to other adaptive processes. *In* K. Blum (Ed.), *Alcohol and Opiates: Neurochemical and Behavioral Mechanisms.* New York: Academic Press.

LeBlanc, A. E., Matsunaga, M., & Kalant, H. (1976). Effects of frontal polar cortical ablation and cycloheximide on ethanol tolerance in rats. *Pharmacology, Biochemistry and Behavior,* **4,** 175–179.

LeBlanc, A. E., Poulos, C. X., & Cappell, H. D. (1978). Tolerance as a behavioral phenomenon: Evidence from two experimental paradigms. *In* N. A. Krasnegor (Ed.), *Behavioral Tolerance: Research and Treatment Implications,* NIDA Research Monograph 18. Washington, D.C.: U.S. Government Printing Office.

Levitt, M., Spector, S., Sjoerdsma, S., & Udenfriend, S. (1965). Elucidation of the rate limiting step in norepinephrine biosynthesis in the perfused guinea pig heart. *Journal of Pharmacology and Experimental Therapeutics,* **148,** 1–8.

Loh, H. H., Shen, F. H., & Way, E. L. (1969). Inhibition of morphine tolerance and physical dependence development and brain serotonin synthesis by cycloheximide. *Biochemical Pharmacology,* **18,** 2711–2721.

Loh, H. H., Tseng, L. F., Holaday, J. W., & Wei, E. (1978). Endogenous peptides and opiate actions. *In* M. L. Adler, L. Manara, & R. Samanin (Eds.), *Factors Affecting the Action of Narcotics,* pp. 387–402. New York: Raven.

Lord, J. A. H., Waterfield, A. A., Hughes, J., & Kosterlitz, H. W. (1977). Endogenous opioid peptides: Multiple agonists and receptors. *Nature (London),* **267,** 495–499.

Madden, J., Akil, H., Patrick, R. L., & Barchas, J. D. (1977). Stress-induced parallel changes in central opioid levels and pain responsiveness in the rat. *Nature (London),* **265,** 358–360.

Manning, F. J. (1976). Chronic delta-9-tetrahydrocannabinol. Transient and lasting effects on avoidance behavior. *Pharmacology, Biochemistry and Behavior,* **4,** 17–21.

Marshall, I., & Smith, C. B. (1973). Development of tolerance in mice was blocked by simultaneous administration of alpha-methyl-tyrosine with morphine. *Pharmacologist,* **15,** 243.

Martin, G. E., Pryzbylik, A. T., & Spector, N. H. (1977). Restraint alters the effects of morphine and heroin on core temperature in the rat. *Pharmacology, Biochemistry and Behavior,* 7, 463–469.

Martin, W. R. (1967). Opioid antagonists. *Pharmacological Reviews,* **19,** 463–521.

Mayer, D. J., & Hayes, R. (1975). Stimulation-produced analgesia: Development of tolerance and cross-tolerance to morphine. *Science,* **188,** 941–943.

Miksic, S., Smith, N., Numan, R., & Lal, H. (1975). Acquisition and extinction of a conditioned hyperthermic response to a tone paired with morphine administration. *Neuropsychobiology,* **1,** 277–283.

Miller, R. R., Marlin, N. A., & Berk, A. M. (1977). Reliability and sources of control of preference for signaled shock. *Animal Learning and Behavior,* **5,** 303–308.

Misra, A. L., Mitchell, C. L., & Woods, L. A. (1971). Persistence of morphine in central nervous system of rats after a single injection and its bearing on tolerance. *Nature (London),* **232,** 48–50.

Morley, J. S. (1980). Structure-activity relationships of enkephalin-like peptides. *Annual Review of Pharmacology and Toxicology,* **20,** 81–110.

Mucha, R. F., Kalant, H., & Linseman, M. A. (1979). Quantitative relationships among measures of morphine tolerance and physical dependence in the rat. *Pharmacology, Biochemistry, and Behavior,* **10,** 397–405.

Mullis, K. B., Perry, D. C., Finn, A. M., Stafford, B., & Sádee, W. (1979). Morphine persistence in rat brain and serum after single doses. *Journal of Pharmacology and Experimental Therapeutics,* **208,** 228–231.

Mushlin, B. E., & Cochin, J. (1976). Tolerance to morphine in the rat: Its prevention by naloxone. *Life Sciences,* **18,** 797–802.

Mushlin, B. E., Grell, R., & Cochin, J. (1976). Studies on tolerance. I. The role of the interval between doses on the development of tolerance to morphine. *Journal of Pharmacology and Experimental Therapeutics,* **196,** 280–287.

Nybäck, H., & Sedvell, G. (1968). Effect of chloropromazine on accumulation and disappearance of catecholamines formed from tyrosine ^{14}C in brain. *Journal of Pharmacology and Experimental Therapeutics,* **162,** 294–301.

Puri, S. K., & Lal, H. (1974). Tolerance to the behavioural and neurochemical effects of haloperidol and morphine in rats chronically treated with morphine or haloperidol. *Naunyn Schmiedeberg's Archives of Pharmacology,* **282,** 155–170.

Rigter, H., Hannan, T. J., Messing, R. B., Martinez, Jr., J. L., Vasquez, B. J., Jensen, R. A., Veliquette, J., & McGaugh, J. L. (1980). Enkephalins interfere with acquisition of an active avoidance response. *Life Sciences,* **26,** 337–345.

Roffman, M., Reddy, C., & Lal, H. (1973). Control of morphine withdrawal hypothermia by conditioned stimuli. *Psychopharmacologia,* **29,** 197–201.

Rosenfeld, G. C., & Burks, T. F. (1977). Single-dose tolerance to morphine hypothermia in the rat: Differentiation of acute from long term tolerance. *Journal of Pharmacology and Experimental Therapeutics,* **180,** 136–143.

Sawynok, J., Pinsky, C., & LaBella, F. S. (1979). Minireview on the specificity of naloxone as an opiate antagonist. *Life Sciences,* **25,** 1621–1632.

Schuster, C. R., Dockens, W. S., & Woods, J. H. (1966). Behavioral variables affecting the development of amphetamine tolerance. *Journal of Pharmacology and Experimental Therapeutics,* **9,** 170–182.

Schwartz, J. C., Costentin, J., Martres, M. P., Protais, P., & Baudry, M. (1978). Modulation of receptor mechanisms in the CNS: Hyper- and hyposensitivity to catecholamines. *Neuropharmacology,* **17,** 665–685.

Seevers, M. H. (1958). Termination of drug action by tolerance development. *Federation Proceedings,* **17,** 1175–1181.

Sharpless, S., & Jaffe, J. (1969). Withdrawal phenomena as manifestations of disuse supersensitivity. *In* H. Steinberg (Ed.), *Scientific Basis of Drug Dependence,* pp. 67–76. New York: Grune and Stratton.

Sherman, J. E. (1979). The effects of conditioning and novelty on rats' analgesic and pyretic responses to morphine. *Learning and Motivation,* **10,** 383–418.

Sherman, J. E., & Liebeskind, J. C. (1980). An endorphinergic, centrifugal substrate of pain modulation: Recent findings, current concepts and complexities. *In* T. J. Bonica (Ed.), *Pain Proceedings of the Association for Research in Nervous and Mental Disease.* New York: Raven Press.

Shuster, L. (1971). Tolerance and physical dependence. *In* D. Clouet (Ed.), *Narcotic Drugs: Biochemical Pharmacology,* pp. 408–423. New York: Plenum.

Siegel, S. (1975). Evidence from rats that morphine tolerance is a learned response. *Journal of Comparative and Physiological Psychology,* **89,** 498–506.

Siegel, S. (1977). Morphine tolerance acquisition as an associative process. *Journal of Experimental Psychology: Animal Behavior Processes,* **3,** 1–13.

Siegel, S. (1978a). A Pavlovian conditioning analysis of morphine tolerance. *In* N. A. Krasnegor (Ed.), *Behavioral Tolerance: Research and Treatment Implications,* NIDA Research Monograph 18. Washington, D.C.: U.S. Government Printing Office.

Siegel, S. (1978b). Reply. *Science,* **200,** 344–345.

Siegel, S. (1978c). Tolerance to the hyperthermic effect of morphine in the rat is a learned response. *Journal of Comparative and Physiological Psychology,* **92,** 1137–1149.

Siegel, S. Critique of the Effects of Conditioning and Novelty on the Analgesic and Pyretic Responses to Morphine, by Jack E. Sherman. Unpublished manuscript, 1979.

Siegel, S., Hinson, R. E., & Krank, M. D. (1978). The role of predrug signals in morphine analgesic tolerance. *Journal of Experimental Psychology: Animal Behavior Processes,* **4,** 188–196.

Siegel, S., Hinson, R. E., & Krank, M. D. (1979). Modulation of tolerance to the lethal effects of morphine by extinction. *Behavioral and Neural Biology,* **25,** 257–262.

Simantov, R., & Snyder, S. H. (1976). Brain-pituitary opiate mechanisms: Pituitary opiate receptor binding, radioimmunoassays for methionine enkephalin and leucine enkephalin, and ³H-en-kephalin interactions with the opiate receptor. *In* H. W. Kosterlitz (Ed.), *Opiates and Endogenous Opioid Peptides,* pp. 41–48. Amsterdam: North Holland.

Sklar, L. S., & Amit, Z. (1978). Tolerance to high doses of morphine: Lack of evidence of learning. *Behavioral Biology,* **22,** 509–514.

Smee, M. L., & Overstreet, D. H. (1976). Alterations in effects of dopamine agonists and antagonists on general activity in rats following chronic morphine treatment. *Psychopharmacology,* **49,** 125–130.

Smith, R. F., Bowman, R. E., & Katz, J. (1978). Behavioral effects of exposure to halothane during early development in the rat: Sensitive period during pregnancy. *Anesthesiology,* **49,** 319–323.

Snyder, S. H., Pasternak, G. W., & Pert, C. B. (1975). Opiate receptor mechanisms. *In* L. L. Iversen, S. Iversen, & S. H. Snyder (Eds.), *Handbook of Psychopharmacology.* New York: Plenum.

Solomon, R. L. (1977). An opponent-process theory of acquired motivation: The affective dynamics of addiction. *In* J. D. Maser & M. E. P. Seligman (Eds.), *Psychopathology: Experimental Models.* San Francisco: Freeman.

Solomon, R. L. (1980). The opponent-process theory of acquired motivation: The cost of pleasure and the benefits of pain. *American Psychologist,* **35,** 691–712.

Spector, S., Sjoerdsma, F., & Udenfriend, S. (1965). Blockade of endogenous norepinephrine synthesis by alpha-methyl-tyrosine, an inhibitor of tyrosine hydroxylase. *Journal of Pharmacology and Experimental Therapeutics,* **147,** 86–102.

Stewart, J., & Eikelboom, R. (1979). Stress masks the hypothermic effect of naloxone in rats. *Life Sciences,* **25,** 1165–1172.

Székely, J. I., Rónai, A. Z., Dunai-Kovács, Z., Miglécz, E., Bajusz, S., & Gráf, L. (1977). Cross tolerance between morphine and β-endorphin in vivo. *Life Sciences,* **20,** 1259–1264.

Thompson, R. F., & Spencer, W. A. (1966). Habituation: A model phenomenon for the study of neuronal substrates of behavior. *Psychological Review,* **73,** 16–43.

Thornhill, J. A., Hirst, M., & Gowdey, C. W. (1978). Changes in the hyperthermic responses of rats to daily injections of morphine and the antagonism of the acute response by naloxone. *Canadian Journal of Physiology and Pharmacology,* **56,** 438–489.

Tiffany, S. T., & Baker, T. B. (1980). Morphine tolerance: Congruence with a Pavlovian paradigm. *Journal of Comparative and Physiological Psychology,* in press.

Urca, G., Frenk, H., Liebeskind, J. C., & Taylor, A. N. (1977). Morphine and enkephalin: Analgesic and epileptic properties. *Science,* **197,** 83–86.

Vanderwende, C., & Spoerlein, M. T. (1973). Role of dopaminergic receptors in morphine analgesia and tolerance. *Research Communications in Chemistry, Pathology, and Pharmacology,* **5,** 35–43.

Vaught, J. L., & Takemori, A. E. (1979a). Differential effects of leucine and methionine enkephalin on morphine-induced analgesia, acute tolerance and dependence. *Journal of Pharmacology and Experimental Therapeutics,* **208,** 86–90.

Vaught, J. L., & Takemori, A. E. (1979b). A further characterization of the differential effects of leucine enkephalin, methionine enkephalin and their analogs on morphine-induced analgesia. *Journal of Pharmacology and Experimental Therapeutics,* **211,** 280–283.

Vedernikov, Y. P. (1969). Interaction of amphetamine, apomorphine, and disulfiram with morphine and the role played by catecholamines in morphine analgesic action. *Archives of International Pharmacodynamics*, **182**, 59–64.

Wagner, A. R. (1976). Priming in STM: An information processing mechanism for self-generated or retrieval-generated depression in performance. *In* T. J. Tighe & R. N. Leaton (Eds.), *Habituation: Perspectives from Child Development, Animal Behavior and Neurophysiology*. Hillsdale, New Jersey: Erlbaum.

Wagner, A. R. (1978). Expectancies and the priming of STM. *In* S. H. Hulse, H. Fowler, & W. K. Honig (Eds.), *Cognitive Processes in Animal Behavior*. Hillsdale, New Jersey: Erlbaum.

Way, E. L. (1972). Brain neurohormones in morphine tolerance and dependence. *Pharmacology and the Future of Man*. (*Proceedings of the Fifth International Congress of Pharmacology, San Francisco*), Vol. 1, pp. 77–94.

Way, E. L., & Bhargava, H. M. (1976). Assessment of the role of acetylcholine in morphine analgesia, tolerance and physical dependence. *In* D. H. Ford & D. H. Clouet (Eds.), *Tissue Responses to Addictive Drugs*. New York: Spectrum Publications.

Way, E. L., Loh, H. H., & Shen, F. H. (1968). Morphine tolerance, physical dependence and the synthesis of brain 5-hydroxytryptamine. *Science*, **162**, 1290–1292.

Weissman, B. A., Bergman, F., & Altstetter, R. (1978). Characteristics and function of opioids. *International Narcotics Research Conference*, 73.

Wikler, A. (1965). Conditioning factors in opiate addiction and relapse. *In* D. M. Wilner & G. G. Kassenbaum (Eds.), *Narcotics*. New York: McGraw-Hill.

Wikler, A. (1973). Conditioning of successive adaptive responses to the initial effects of drugs. *Conditioned Reflex*, **8**, 193–210.

Wikler, A., & Pescor, F. T. (1967). Classical conditioning of a morphine-abstinence phenomenon, reinforcement of opioid drinking behavior and "relapse" in morphine-addicted rats. *Psychopharmacologia*, **10**, 255–284.

PART V

Other Neuropeptides

CHAPTER 24

Substance P and Its Effects on Learning and Memory

Joseph P. Huston and Ursula Stäubli

This chapter summarizes studies dealing with the effects of the undecapeptide Substance P on passive avoidance learning in rats. Substance P was injected posttrial into various brain sites with the following results. Amnesia was obtained with injection into the substantia nigra and amygdala. Learning was facilitated by injection into the medial nucleus of the septum and the lateral hypothalamus. Injection into the ventromedial nucleus of the hypothalamus had no influence on learning. Substance P mimics the effects of posttrial electrical stimulation of these brain sites on learning. It is hypothesized that Substance P enhances or hinders avoidance learning by virtue of posttrial reinforcing influences in the hypothalamus and the septum, and aversive or mixed effects in the amygdala and substantia nigra. Supporting evidence is provided by the demonstration that injection of Substance P into the lateral hypothalamus or medial septal nucleus positively reinforced T-maze learning, whereas injections into the amygdala or substantia nigra did not have reinforcing properties.

I. INTRODUCTION

Some of the properties of Substance P and its distribution in the brain (Huston & Stäubli, 1978; Huston & Stäubli, 1979; Stäubli & Huston, 1979a,

521

ENDOGENOUS PEPTIDES
AND LEARNING AND MEMORY PROCESSES

1980) will be reviewed, then a summary of our results of intracranial posttrial application of Substance P on learning will be presented.

Substance P was detected accidentally as an unidentified biologically active substance in extracts of equine intestine by von Euler & Gaddum in 1931. The analysis of this unknown material was performed with a "purified standard preparation simply referred to as 'P'," providing the name Substance P (SP) (von Euler, 1977). The isolation of SP from bovine hypothalamus (Chang & Leeman, 1970) and its identification as an undecapeptide (see Fig. 1) occurred only 40 years later (Chang, Leeman, & Niall, 1971). Its subsequent synthesis (Tregear, Niall, Potts, Leemann, & Chang, 1971) was a significant step forward in the research on SP, as the results of previous investigations were of limited value due to the lack of pure SP extracts.

The development of a specific radioimmunoassay for SP (Powell, Leeman, Tregear, Niall, & Potts, 1973), i.e., of SP-like immunoreactivity, prompted many studies dealing with the localization of SP in the central nervous system (CNS), although the distribution of SP in the nervous tissue had already been grossly determined by bioassay (Zetler, 1970). Radioimmunoassay can ideally provide a reliable measure of neural peptides in the low concentrations in which they normally occur. Although the widespread and uneven distribution of SP in the rat brain and spinal cord is now a well-established finding, a recent report has emphasized that SP antibodies cross-react with SP-fragments, and are relatively unspecific. Thus, there is some uncertainty about the exact distribution of SP (Ben-Ari, Pradelles, Gros, & Dray, 1979). The introduction of immunohistochemical procedures provided a technique for intraneuronal localization of SP-immunoreactive material in the CNS, making it possible to distinguish between SP cell bodies and terminals (Cuello & Kanazawa, 1978; Hökfelt, Kellerth, Nilsson, & Pernow, 1975b; Ljungdahl, Hökfelt, & Nilsson, 1978).

Substance P has been reported in most parts of the CNS. In the rat brain, SP is present in highest concentration in the substantia nigra (SN), medial amygdaloid nucleus, medial habenular nucleus, interpeduncular nucleus, and trigeminal nerve nuclei. It also is found in the caudate nucleus, globus pallidus,

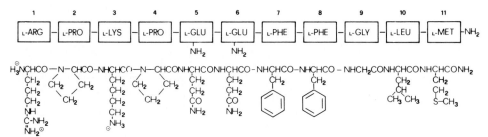

Fig. 1. Amino acid sequence and structure of Substance P.

nucleus accumbens, the septum, most hypothalamic nuclei, the raphé nuclei, the olfactory tubercle, the frontal cortex, and many other parts of the brain (Ben-Ari, Le Gal La Salle, & Kanazawa, 1977; Brownstein, Mroz, Kizer, Palkovits, & Leeman, 1976; Cuello & Kanazawa, 1978; Hökfelt, Kellerth, Nilsson, & Pernow, 1975a; Ljungdahl, Hökfelt, & Nilsson, 1978). This suggests that the neuronal distribution pattern of Substance P is as extensive as, for instance, that of the central catecholamine systems (Ljungdahl, Hökfelt, & Nilsson, 1978; Ljungdahl, Hökfelt, Nilsson, & Goldstein, 1978).

Based on lesion studies, a number of SP-containing projections have recently been proposed. The pathways that have so far been suggested for the rat CNS are presented in Fig. 2.

Results from studies examining the regional SP distribution in human brains (Cuello, Emson, Del Fiacco, Gale, Iversen, Jessel, Kanazawa, Paxinos, & Quik, 1978; Duffy & Powell, 1975; Zetler & Schlosser, 1955) indicate a similar pattern to that found in the rat, with the highest SP levels found in the SN, and moderate levels in the globus pallidus, caudate nucleus, and hypothalamus. A very small amount of SP is present in the cortex, and it is found in only a very low concentration or is absent in the cerebellum. Abnormally low levels of SP have been reported in the SN and globus pallidus in patients that died as a result of Huntington's chorea, probably as a consequence of degenerative changes in the basal ganglia that occur with this disease (Cuello et al., 1978; Gale, Bird, Spokes, Iversen, & Jessel, 1978; Kanazawa, Bird, Gale, Iversen, Jessel, Muramoto, Spokes, & Sutoo, 1979; Kanazawa, Bird, O'Connell, & Powell, 1977).

Neurochemical and electrophysiological studies suggest that SP serves as a neurotransmitter or as a neuromodulator. There is strong evidence for considering SP to be an excitatory agent in the brain as well as in the dorsal horn of the spinal cord (Otsuka & Konishi, 1977; Otsuka & Takahashi, 1977). Interactions of SP with other putative neurotransmitters are being actively investigated. For example, an interaction with catecholamines has been proposed, since SP-containing nerve terminals are present in moderate or high densities in various brain regions known to contain monoaminergic nerve terminals and cell bodies (Ljungdahl, Hökfelt, Nilsson, & Goldstein, 1978; Pickel, Tong, Reis, Leeman, & Miller, 1979). Injection of SP into various brain regions has been shown to influence such diverse neurotransmitters as dopamine (DA), noradrenaline (NE), acetylcholine (ACH), serotonin (5-HT), and γ-amino-butyric acid (GABA). For example, in the SN, exogenous SP leads to excitation of the nigroneostriatal DA neurons (Davies & Dray, 1976; Walker, Kemp, Yajima, Kitagawa, & Woodruff, 1976) and increases the spontaneous release of nigral 5-HT (Reubi, Emson, Jessel, & Iversen, 1978). In the locus coeruleus, SP was demonstrated to excite NE neurons (Guyenet & Aghajanian, 1977), whereas a decrease in hippocampal ACH turnover rate has been shown following

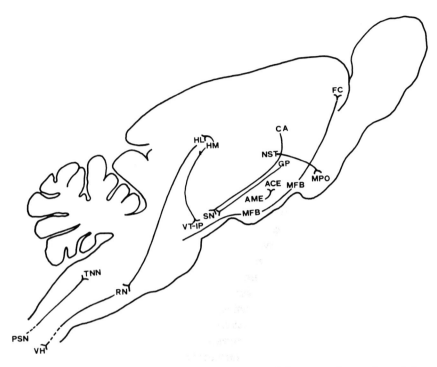

Fig. 2. Presumed Substance P pathways in the brain of the rat: caudate nucleus (CA) to substantia nigra (SN) (Brownstein, Mroz, Tappaz, & Leeman, 1977; Hong, Yang, Racagni, & Costa, 1977; Kanazawa, Emson, & Cuello, 1977; Mroz, Brownstein, & Leeman, 1977); substantia nigra to globus pallidus (GP) (Kanazawa, Bird, Gale, Iversen, Jessel, Muramoto, Spokes, & Sutoo, 1979); globus pallidus (GP) to substantia nigra (SN) (Jessel, Emson, Paxinos, & Cuello, 1978; Kanazawa, Emson, & Cuello, 1977; Kanazawa, Bird, Gale, Iversen, Jessel, Muramoto, Spokes, & Sutoo, 1979); medial nucleus of habenula (HM) to lateral nucleus of habenula (HL); medial nucleus of habenula to ventrotegmental area (VT) and interpeduncular nucleus (IP) (Cuello, Emson, Paxinos, & Jessel, 1978); lateral nucleus of habenula (HL) to dorsal raphé nucleus (RN) (Neckers, 1979); area posterior to interpeduncular nucleus to frontal cortex (FC) (via MFB) (Paxinos, Emson, & Cuello, 1978a); medial nucleus of amygdala (AME) to central nucleus of amygdala (ACE) (Emson, Jessel, Paxinos, & Cuello, 1978); interstitial nucleus of stria terminalis (NST) to medial preoptic area (MPO) (Paxinos, Emson, & Cuello, 1978b); raphé nuclei (RN) to ventral horn of spinal cord (VH) (Hökfelt, Johansson, Kellerth, Ljungdahl, Nilsson, Nygards, & Pernow, 1977); primary sensory neurons (PSN) to dorsal horn of spinal cord trigeminal nerve nuclei (TNN) (Hökfelt, Kellerth, Nilsson, & Pernow, 1975a; Takahashi & Otsuka, 1975).

intraseptal SP application (Malthe-Sorenssen, Cheney, & Costa, 1978; Wood, Cheney, & Costa, 1979). Interestingly, in the brain stem raphé nuclei, SP and 5-HT coexist in the same neurons (Chan-Palay, Jonsson, & Palay, 1978; Hökfelt, Ljungdahl, Steinbusch, Verhofstad, Nilsson, Brodin, Pernow, & Goldstein, 1978). Antagonism of SP by GABA was observed in the SN, where SP release was depressed by GABA and restored by the GABA antagonist picrotoxin (Jessel, 1978). It was therefore suggested that SP release in the substantia nigra is presynaptically controlled by GABA. In primary afferent nerve terminals in the trigeminal nerve nuclei, presynaptic enkephalin receptors located on SP terminals are possibly involved in modulating SP release (Jessel & Iversen, 1977). This association of SP terminals and enkephalin terminals is consistent with the concept, first proposed by Lembeck, that SP is a transmitter of primary afferent sensory neurons (Lembeck, 1953), and is thus, important in transmitting pain information (Henry, 1976). Although, at the level of the trigeminal brain-stem nuclei, considerable evidence points to a role of SP in nociception (e.g., Henry, 1980; Henry, Sessle, Lucier, & Hu, 1980); in "higher" brain structures, the relationship between SP and pain is tenuous. Some studies have reported analgesic effects in rats and mice following intraperitoneal and intraventricular SP administration (Malick & Goldstein, 1978; Stewart, Getto, Neldner, Reeve, Krivoy, & Zimmermann, 1976), whereas other studies fail to find an influence of SP on pain perception (Goldstein & Malick, 1979; Hayes & Tyers, 1979). In fact, SP has been proposed to function as a physiological tranquilizer in forebrain structures (Starr, James, & Gaytten, 1978).

Only a few studies have investigated the role of SP in behavior. In pigeons, SP given intracerebrally induced drinking (Evered, Fitzsimons, & DeCaro, 1977); whereas in rats it inhibited drinking behavior (DeCaro, Massi, & Micossi, 1978). Other studies concerned with the effects of exogenous SP upon motor behavior in rats have shown a dose-dependent behavioral activation (Rondeau, Jolicoeur, Belanger, & Barbeau, 1978), such as increased locomotion following injection of SP into SP-rich areas, including the ventral tegmental area and the substantia nigra (Diamond, Comaty, Sudakoff, Havdala, Walter, & Borison, 1979; Kelley & Iversen, 1978; Stinus, Kelley, & Iversen, 1978). Opposite effects, such as decreased locomotor activity, were reported in mice after in-traventricular and intraperitoneal injections of SP (Katz, 1979; Starr et al., 1978). Several studies have reported a unilateral activation of the nigroneostriatal dopaminergic system induced by SP injection into the SN as manifested by contraversive rotatory behavior (Diamond et al., 1979; James & Starr, 1977, 1979; Olpe & Koella, 1977). However, injection of SP into the SN, outside of the reticular portion, induced ipsiversive turning, suggesting an inhibitory action of SP on dopaminergic cell activity (James & Starr, 1979).

Two studies have shown an effect of SP on operant behavior. First, SP given intraventricularly in rabbits caused a decrease in bar pressing for sweetened water (Graeff & Arisawa, 1978). Second, self-stimulation rate in rats, via electrodes in the lateral hypothalamus (LH), was depressed following infusion of SP into the same site (Goldstein & Malick, 1977), suggesting a possible involvement of SP in reward processes.

In the only study other than ours dealing with the effects of SP on learning, Hecht, Oehme, Poppei, and Hecht (1979) found that intraperitoneal injection of SP (250 and 500 μg/kg) in rats disrupted learning to turn off an aversive stimulus that was conditioned to an acoustic CS. Unfortunately, because the injections were not given using the posttrial procedure, the results cannot be interpreted in terms of an influence on learning and memory, but could reflect nonspecific performance decrements.

II. POSTTRIAL INFLUENCES ON LEARNING

Our decision to study the possibility of an involvement of SP in modulating learning and memory was prompted by its uneven distribution pattern in the brain. A number of brain structures that contain high or moderate levels of endogenous SP are also thought to mediate learning and memory processes by virtue of reported amnesic effects resulting from electrical or chemical stimulation. For example, SP is most highly concentrated in the SN (Brownstein et al., 1976; Cuello et al., 1978; Cuello, & Kanazawa, 1978; Hökfelt, Johannson, Kellerth, Ljungdahl, Nilsson, Nygards, & Pernow, 1977; Hökfelt et al., 1975a; Ljungdahl, Hökfelt, & Nilsson, 1978; Powell et al., 1973) and in the medial nucleus of the amygdala (Ben-Ari et al., 1977; Cuello, & Kanazawa, 1978; Ljungdahl, Hökfelt, & Nilsson, 1978); both regions may play a role in modulating memory processes, because posttrial electrical stimulation of these structures has repeatedly been shown to have amnesic effects (Breshnahan & Routtenberg, 1972; Fibiger & Phillips, 1976; Gold, Hankins, Edwards, Chester, & McGaugh; 1975; McDonough & Kesner, 1971; Routtenberg & Holzman, 1973; Stäubli & Huston, 1978). Furthermore, memory impairment is not only caused by electrical stimulation in these areas, but also occurs as a consequence of posttrial drug administration, for example, the injection of cholinergic agonists into the amygdala (Todd & Kesner, 1978), and GABA-ergic antagonists into the SN (Kim & Routtenberg, 1976). The question arose, therefore, as to whether posttrial application of SP in the SN and amygdala would influence learning and memory. Having subsequently found evidence for an amnesic effect of SP applied posttrial in the SN (Huston & Stäubli, 1978) and the amygdala (Huston & Stäubli, 1979), as summarized in the following, we considered it important to examine the effects of exogenously administered SP in other areas known to contain SP and presumed, on the basis of other posttrial manipulations, to be involved in learning and memory. Facilitation of learning induced by posttrial electrical stimulation has been repeatedly demonstrated in the lateral hypothalamus (Destrade & Jaffard, 1978; Huston & Mueller, 1978; Huston, Mueller, & Mondadori, 1977; Major & White, 1978), where a moderate concentration of SP has been found (Brownstein et al., 1976; Cuello & Kanazawa, 1978; Ljungdahl, Hökfelt, & Nilsson, 1978). Another site that contains endogenous SP (Brownstein et al., 1976; Cuello & Kanazawa, 1978;

Ljungdahl, Hökfelt, & Nilsson, 1978), and is known for its memory-facilitating influence when stimulated electrically (Landfield, 1977; Wetzel, Ott, & Matthies, 1977) is the septum.

Thus, the choice of brain sites to test the influence of SP on learning was determined, on the one hand, by their endogenous SP content, and on the other hand, by their susceptibility to either amnesic or facilitating influences on learning as a consequence of posttrial electrical brain stimulation or drug application. The following injection sites in the brain were chosen for posttrial injection of SP: (1) substantia nigra, (2) medial amygdaloid nucleus, (3) medial septal nucleus and (4) the lateral and ventromedial hypothalamus, each at an anterior and a posterior level. The ventromedial hypothalamus was selected as a control injection site to the lateral hypothalamus because, traditionally, these two brain areas have been considered to function antagonistically (Hoebel, 1969).

The following methodology was employed to examine the effects of intracranially injected SP on learning. Assuming that hypothetical short-term memory traces are most susceptible to exogenous influences immediately following a learning trial, SP was injected 30 sec posttrial into one of several brain regions described in the previous section. This strategy precludes possible effects of the treatment upon performance during the training trial itself and permits the reasonable assumption that learning is influenced as a result of an action on memory processing, for instance, on consolidation or on retrieval process (McGaugh, 1966; McGaugh & Herz, 1972).

Throughout all the studies, the paradigm used to test for the effects of SP on learning was a form of passive avoidance training. Passive avoidance paradigms generally employ a defined environment arranged in such a way as to reliably elicit a high probability response, which the animal learns to suppress as a consequence of punishment. The three passive avoidance tasks that were used were (1) the step-down avoidance; (2) the alcove avoidance; and (3) the uphill avoidance task as will be described.

A. Experimental Apparatus

The *step-down avoidance* apparatus consisted of a round platform fixed in the middle of the electrifiable grid floor of a rectangular box. The animal learns, against its natural tendency, to avoid stepping down onto the grid floor as a consequence of receiving a punishing footshock (1mA, 1 sec, scrambled).

The *alcove avoidance* apparatus consisted of a large open chamber connected to a small enclosed chamber. The animal's preference for the dark small alcove is abolished by administration of a punishing (1mA, 3 sec, scrambled) footshock in the alcove (Bureš, Burešova, & Huston, 1976; Kurtz & Pearl, 1960).

The *uphill avoidance* apparatus (Stäubli & Huston, 1979b) consisted of an inclined plane. An animal, which is placed head downwards on the plane, usually, within 5 sec, turns around and locomotes up the incline. This response is suppressed by the punishing stimulation which, in this task, was tailshock (2mA, 1 sec).

Male adult Sprague-Dawley rats were chronically implanted unilaterally with a cannula (0.41 mm i.d.: 0.71 mm o.d.). The coordinates were, according to Hurt, Hanaway, and Netsky (1971) for the substantia nigra: AP: + 3.0 mm/ L:2.0 mm/ IAL: + 1.9 mm, and, according to Pellegrino and Cushman (1967) for the medial nucleus of the amygdala: AP: + 5.0 mm/ L:3.5 mm/DV: − 3.1 mm; for the anterior lateral hypothalamus + 6.2/1.8/ − 2.0; for the posterior lateral hypothalamus + 4.6/1.8/ − 3.0; for the anterior ventromedial hypothalamus + 6.0/0.9/ − 2.9; for the posterior ventromedial hypothalamus + 5.4/0.7/ − 3.5; and, finally for the medial septum: + 8.2/1.6/6.2 at an angle of 15° to the medial plane.

Three doses of SP were administered: 50.0 ng (37.0 pmol), 500.0 ng (370.0 pmol) and 6.7 μg (5.0 nmol). The peptide was dissolved in physiological saline solution and injected intracranially in a volume of 0.5 μl for all doses except for the highest dose (6.7 μg) which was administered in a volume of 1.0 μl. Control animals were administered 0.5 μl of physiological saline. Substance P was injected at a rate of 1.0 μl/min with the injection needle kept in place for an additional 30 sec. The injections were made into hand-held animals, which had been handled daily for 1 wk prior to the start of the experimental sessions. The substances were injected either 30 sec posttrial, or at least 3 hr posttrial in order to control for possible proactive effects on performance.

In all passive avoidance experiments, a few baseline trials were run to familiarize the animals with the apparatus. The training trial itself was identical with the preceding baseline trials, except that after the behavioral response, a footshock or tailshock was administered followed by intracranial injection. The test for retention was always performed 24 hr after training.

The design of all our experiments, including group, dose, and task assignments, is summarized in Fig. 3, which also illustrates the results expressed in terms of percentage of impairment or improvement of learning as compared to saline control groups.

All statistical comparisons were made on the latency scores using the two-tailed randomization test for two independent samples (Siegel, 1956).

B. Amnesia

Compared to saline controls, injection of 500.0 ng SP into the substantia nigra and 50.0 ng into the medial amygdaloid nucleus produced retrograde

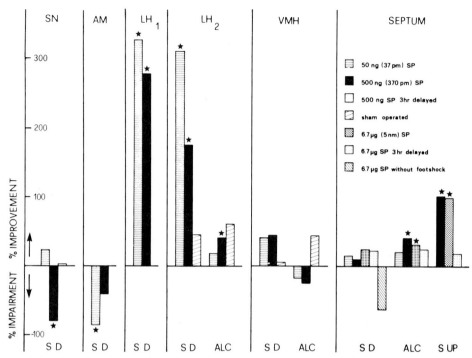

Fig. 3. Percentage impairment or improvement of passive avoidance learning in comparison with posttrial saline control groups. The areas of the brain in which injections were performed, as indicated on the top, include the substantia nigra (SN), the medial nucleus of the amygdala (AM), the lateral hypothalamus (LH₁ and LH₂, referring to two experiments), the ventromedial hypothalamus (VMH), and the medial septal nucleus (septum). The passive avoidance tasks used, as indicated on the bottom, include the step-down avoidance task (SD), the alcove avoidance task (ALC), and the step-up avoidance task (SUP). A significant difference from the saline group as shown by the randomization test is indicated by ★.

amnesia for step-down avoidance learning (Huston & Stäubli, 1978; Huston & Stäubli, 1979).

C. Facilitation

In two studies employing posttrial application of SP to the lateral hypothal-amus, 50.0 ng and 500.0 ng SP enhanced step-down avoidance learning and 500.0 ng SP facilitated alcove avoidance learning in comparison to saline injected animals (Huston & Stäubli, 1979; Stäubli & Huston, 1979a).

Posttrial application of 500.0 ng and 6.7 µg SP into the vicinity of the medial septal nucleus improved uphill avoidance learning, whereas injection of 50.0

ng, 500.0 ng, and 6.7 μg SP facilitated alcove avoidance learning (Stäubli & Huston, 1980). The lack of effect on step-down avoidance learning can probably be accounted for by a ceiling effect due to the almost optimal performance of the saline controls. Injection of SP into the ventromedial hypothalamus influenced neither step-down nor alcove avoidance learning.

The failure of the 3 hr delayed injection of SP into the substantia nigra and into the septum, respectively, to influence learning, indicates, first, that the amnesic and facilitating effects of the 30 sec delay SP groups were time-dependent, presumably because of an action on labile short-term memory processes, and second, that the facilitation and amnesia produced by SP were most likely retrograde effects rather than proactive ones.

R. Kesner (personal communication, 1979) has confirmed our results with posttrial injection of SP into the amygdala. He obtained amnesia with 100 ng (in 1 μl) SP in a one-trial passive avoidance task involving the suppression of licking behavior.

A comparison of the performance levels of the saline control groups across each specific task reveals rather large differences between some groups of animals implanted at different brain sites. This is particularly evident in the case of the LH and amygdala in the step-down avoidance task in which the amygdala–saline group performed very well, whereas the lateral hypothalamus–saline group performed very poorly (Huston & Stäubli, 1979). Whether this variation among the saline control groups can be accounted for by differences in cannula implantation sites per se remains unclear because of the lack of unoperated controls in each experiment. Some evidence pointing toward an implantation-induced deficit in performance of the saline controls in case of the groups with cannulae in the LH comes from the second step-down study in which a sham-operated group was employed (Stäubli & Huston, 1979a). This seems to be especially true for the alcove avoidance study of the LH experiment, in which the performance of sham-operated animals was significantly superior to that of the saline controls, suggesting that implantation of the cannula impaired learning, and that posttrial injection of SP compensated for this deficit.

It should be pointed out, as this example illustrates, that the performance of the control group, which varies according to the specific task parameters, can potentially bias the direction of the effect shown by the experimental treatment. Thus, the learning exhibited by the amygdala–saline group obviously favored our finding an amnesic effect of posttrial SP, insofar as a possible facilitation of learning by SP in this experiment would hardly have occurred because of the ceiling latencies of the control animals. The opposite was the case in the LH experiments, where the relatively poor learning shown by the saline controls favored our finding of a facilitation of learning by posttrial SP, and any possible amnesic treatment effect would not have been observed.

However, the fact that the performance levels of the experimental groups across the same task compared between the various brain sites were also different indicates that the effects of SP on learning are, at least to some extent, a function of locus of application of SP.

Obviously, our results to date with posttrial Substance P must be considered to be preliminary, and further studies with unoperated control groups using varied parameters, such as different footshock levels, are necessary.

III. DISCUSSION

To recapitulate the known effects of posttrial SP on passive avoidance learning, evidence for a facilitation of learning has been demonstrated with injection into the medial septal nucleus (Stäubli & Huston, 1980) and the lateral hypothalamus (Huston & Stäubli, 1979; Stäubli & Huston, 1979a). Retrograde amnesia was shown with injection into the substantia nigra (Huston & Stäubli, 1978) and amygdala (Huston & Stäubli, 1979). No effect was observed following injection into the ventromedial nucleus of the hypothalamus (VMH) (Stäubli & Huston, 1979a) and the prefrontal cortex (unpublished results).

A look at the distribution pattern of SP in the brain shows that the sites in which SP influenced learning contain moderate or high levels of endogenous SP. The VMH, the one site in which SP had no effect on learning, is devoid of SP terminals and its SP seems to be located mostly in cell bodies (Ljungdahl, Hökfelt, & Nilsson, 1978). Conversely, the LH, the SN, the amygdala, and the septum probably contain most of their SP in nerve terminals (Ljungdahl, Hökfelt, & Nilsson, 1978). Thus, it appears that exogenous SP is most effective in its influence on learning in areas where SP is predominantly located in SP terminals.

The neurochemical mechanisms by which SP influences memory are not known. Speculation is difficult because only a few biochemical and anatomical data demonstrating interrelations of SP with other systems exist. However, we can expect rapid progress in this area in the future, and a beginning has been made. There is, for example, some evidence for a possible interaction between catecholamine- and SP-containing neurons. Anatomical studies tracing SP-containing systems indicate that there are many places in the brain where SP terminals contact catecholaminergic dendrites, cell bodies, and terminals (Ljungdahl, Hökfelt, Nilsson, & Goldstein, 1978; Pickel et al., 1979). In addition, some catecholaminergic terminals appear to be localized around SP-containing cell bodies (Ljungdahl, Hökfelt, Nilsson, & Goldstein, 1978). Biochemical evidence of an increased turnover of both DA and NE induced by intraventricularly injected SP (Carlsson, Magnusson, Fisher, Chang, & Folkers,

1977), as well as electrophysiological evidence for an excitatory action of SP on catecholaminergic neurons (Davies, & Dray, 1976; Guyenet & Aghajanian, 1977; Reubi et al., 1978; Walker et al., 1976) is consistent with these anatomical findings.

In the SN, the application of SP presumably excites nigroneostriatal DA neurons, thus inducing a release of striatal DA (Chéramy, Nieoullon, Michelot, & Glowinski, 1977). Taking into account the hypothesis that the retrograde amnesia produced by electrical stimulation of the SN (Fibiger & Phillips, 1976; Routtenberg & Holzman, 1973; Stäubli & Huston, 1978) is caused by stimulation of the nigroneostriatal DA bundle and the resulting release of excessive DA in the caudate nucleus (Fibiger & Phillips, 1976; Routtenberg & Holzman, 1973), it follows that the SP-induced amnesia in the SN could be a result of the activation of the dopaminergic nigrostriatal bundle.

Interactions between SP and catecholamines may also account for the effects of intrahypothalamic administration of SP on learning. In recent studies (Ljungdahl, Hökfelt, & Nilsson, 1978; Ljungdahl, Hökfelt, Nilsson, & Goldstein, 1978), few catecholaminergic and SP terminals could be localized in the VMH, whereas in the LH, high concentrations of SP as well as catecholaminergic fibers and terminals were seen. Similarly, in the amygdala, interactions between catecholamines and SP cannot be excluded, because dense SP and catecholaminergic terminals have been described there (Ljungdahl, Hökfelt, Nilsson, & Goldstein, 1978).

Considering the large number of putative neurotransmitters and neuromodulators in the brain, it is very likely that interactions between SP and some types of neurons, other than catecholaminergic, will be discovered that may account for some effects of SP on learning. Acetylcholine (ACH), for example, is widely considered to be important for memory processing. A link, in fact, does exist between ACH and SP, as 5.0 nmol SP (which corresponds to the 6.7 μg SP injected intraseptally in our study) (Stäubli & Huston, 1980) injected into the medial septal nucleus decreased the rate of hippocampal turnover of ACH (Malthe-Sorenssen et al., 1978). Similar results were reported in another recent study (Wood et al., 1979). In addition, there is evidence that hippocampal theta rhythm, found to correlate positively with learning capacity (Bloch, 1970; Landfield, 1976a,b) is a cholinergic response that can be blocked by lesion of the medial septal nucleus (Teitelbaum, Lee, & Johannessen, 1975). These data can be considered as a basis for further investigations aimed at determining whether the memory facilitating effect of intraseptal SP is mediated by ACH.

The concept of "arousal" could conceivably be invoked to account for the influence of intraseptal SP on learning, as other manipulations of the septum that led to facilitation of memory have been interpreted in terms of influences on hippocampal arousal (Landfield, 1977; Wetzel et al., 1977). The hippocampal theta rhythm, generally considered to be an index of arousal, can be driven through septo–hippocampal pathways by electrical stimulation of the medial septal nucleus, the septum being a possible pacemaker for hippocampal theta activity (von Euler & Green, 1960; Gray, 1972; Stumpf, 1965). It was found that posttrial memory facilitating drugs increase cortical theta activity (which provides an index of activation of hippocampal theta)(Landfield, 1976b).

Intraseptal SP could, therefore, facilitate learning by influencing "arousal" or theta activity. Although no direct evidence for this suggestion exists, an investigation of the effects of intraseptal SP on hippocampal EEG is needed.

We wish to emphasize that, in its effects on learning, posttrial SP mimics the action of posttrial electrical stimulation of the various brain sites that have been tested. Notably, a facilitation of learning has been obtained with both posttrial injection of SP and posttrial electrical stimulation of the medial septal area (Landfield, 1977; Stäubli & Huston, 1980; Wetzel et al., 1977) as well as the lateral hypothalamus (Destrade & Cazala, 1979; Destrade & Jaffard, 1978; Huston & Mueller, 1978; Huston et al., 1977; Huston & Stäubli, 1979; Major & White, 1978; Stäubli & Huston, 1979a). Conversely, both posttrial SP injection and posttrial electrical stimulation of the substantia nigra (Fibiger & Phillips, 1976; Huston & Stäubli, 1978; Routtenberg & Holzman, 1973; Stäubli & Huston, 1978) and amygdala (Berman & Kesner, 1976; Breshnahan & Routtenberg, 1972; Gold et al., 1975) have predominantly amnesic effects (although both facilitation and amnesia have been obtained with amygdaloid stimulation, depending on footshock level) (Gold et al., 1975). Perhaps most intriguing is the fact that electrical stimulation of the medial septal area as well as the lateral hypothalamus is well known to be rewarding. Therefore, because posttrial *reinforcing* stimulation of the lateral hypothalamus has repeatedly been shown to facilitate avoidance learning (Huston & Mueller, 1978; Huston et al., 1977), and because administration of SP parallels the effects of posttrial stimulation of these structures, it may be that SP, in fact, mimics the *rewarding* action of electrical stimulation of these areas. That is, in the septum and LH, the facilitating effects of the electrical stimulation may be due to release of Substance P, or alternatively, the effects of Substance P may be due to its reinforcing action. An interaction between SP and reward has, in fact, been reported in the LH (Goldstein & Malick, 1977). Although we consider it unlikely, the reinforcing effects of SP may also be relevant to its amnesic effects in the SN and amygdala. In one study, posttrial amnesic electrical stimulation of the SN was demonstrated to have rewarding effects (Stäubli & Huston, 1978), and self-stimulation can also be obtained with electrical stimulation of the amygdaloid nuclei, although no direct evidence for a linkage between amnesic amygdaloid stimulation and reward exists so far. Such a general relationship between the amnesic action of Substance P and reward would require that the direction of the effect of posttrial reward on memory is not invariably one of facilitation, but depends on the brain site involved. That this is indeed the case is evidenced by the finding that posttrial reinforcing electrical stimulation facilitates avoidance learning when applied in the lateral hypothalamus (Huston & Mueller, 1978; Huston et al., 1977) and hippocampus (Caudarella, Campbell, & Milgram, 1978), but inhibits such learning when administered to the SN (Stäubli & Huston, 1978) and periaqueductal grey (Kesner, in press).

However, as an alternative to the view of a unitary reinforcing action of SP in the telencephalon, we favor the hypothesis that SP can have reinforcing as well as punishing and neutral effects depending on the site of injection (and of course the dosage) in the brain. Thus, posttrial facilitation of learning is obtained from sites, such as the LH and septum, where SP, like electrical stimulation, is predominantly reinforcing. Amnesia is obtained from sites where SP has principally aversive, or mixed properties, such as in the amygdala, SN, and periaqueductal grey, where electrical stimulation can have reinforcing, aversive, or mixed effects. This hypothesis is in concert with the theory put forward earlier (Huston et al., 1977) that posttrial reinforcers lead to consolidation and facilitation of learning, whereas posttrial nonreinforcers, including punishing stimulation, are likely to disrupt consolidation and, therefore, inhibit learning.

Implicit in the proposal linking the amnesic effects of SP with reward and punishment is that, like electrical stimulation of the brain sites under consideration, SP would have reinforcing or punishing effects when directly injected into these areas. Stein and Belluzzi (1978, 1979) reported that intraventricularly administered SP decreased the operant level of lever pressing. These authors, therefore, suggested that SP may have a role in the mediation of punishment. One problem with results based on intraventricular injection of SP (and possibly other neuroactive substances) is that, as we have shown, SP can have opposite effects on behavior, depending on the site in the brain to which it is applied. If neurotransmitters and neuromodulators can have site-dependent effects on behavior, the application of such substances into the brain's ventricles for purposes of assessing their behavioral action is obviously of limited value and may not be defensible. Therefore, evaluation of our hypothesis regarding the effects of posttrial SP on learning, awaits the results of a direct test of possible site-dependent reinforcing and punishing properties of SP injected into the forebrain areas under consideration. In fact, we have recently completed such a study with the following results (unpublished): As predicted, injections of Substance P into the lateral hypothalamus (100.0 ng) or medial septal nucleus (500.0 ng) served as a positive reinforcer for T-maze learning. Injections into the amygdala (50.0 ng) or substantia nigra (100.0 ng) did not have such reinforcing properties.

Finally, although it seems reasonably well-established that intracranial posttrial SP can influence passive avoidance learning, facilitating or inhibiting performance depending on the site of injection, there is as yet no evidence available to suggest that endogenous SP is actually involved in memory processing, although this possibility surely warrants further investigation.

ACKNOWLEDGMENTS

This work was supported in part by Swiss National Science Foundation Grant No. 3.227-0.77.

REFERENCES

Ben-Ari, Y., Le Gal La Salle, G., & Kanazawa, I. (1977). Regional distribution of substance P within the amygdaloid complex and bed nucleus of the stria terminalis. *Neuroscience Letters*, **4**, 299–302.

Ben-Ari, Y., Pradelles, P., Gros, C., & Dray, F. (1979). Identification of authentic substance P in striatonigral and amygdaloid nuclei using combined high performance liquid chromatography and radioimmunoassay. *Brain Research*, **173**, 360–363.

Berman, R. F., & Kesner, R. P. (1976). Posttrial hippocampal, amygdaloid, and lateral hypothalamic electrical stimulation: Effects on short- and long-term memory of an appetitive experience. *Journal of Comparative and Physiological Psychology*, **90**, 260–267.

Bloch, V. (1970). Facts and hypotheses concerning memory consolidation. *Brain Research*, **24**, 561–575.

Breshnahan, E., & Routtenberg, A. (1972). Memory disruption by unilateral low-level, sub-seizure stimulation of the medial amygdaloid nucleus. *Physiology and Behavior*, **9**, 513–525.

Brownstein, M. J., Mroz, E. A., Kizer, J. S., Palkovits, M., & Leeman, S. E. (1976). Regional distribution of substance P in the brain of the rat. *Brain Research*, **116**, 299–305.

Brownstein, M. J., Mroz, E. A., Tappaz, M. L., & Leeman, S. E. (1977). On the origin of substance P and glutamic acid decarboxylase (GAD) in the substantia nigra. *Brain Research*, **135**, 315–323.

Bureš, J., Burešova, O., & Huston, J. P. (Eds.). (1976). *Techniques and Basic Experiments for the Study of Brain and Behavior*, pp. 111–114. Amsterdam: Elsevier.

Carlsson, A., Magnusson, T., Fisher, G. H., Chang, D., & Folkers, K. (1977). Effect of synthetic substance P on central monoaminergic mechanisms. *In* U.S. von Euler & B. Pernow (Eds.), *Substance P*, pp. 201–205. New York: Raven.

Caudarella, M., Campbell, K. A., & Milgram, N. W. (1978). Effect of hippocampal stimulation on learning in the rat. *Proceedings of the American Psychological Association Meetings*, Toronto, Canada, August 1978.

Chan-Palay, V., Jonsson, G., & Palay, S. L. (1978). Serotonin and substance P coexist in neurons of the rat's central nervous system. *Proceedings of the National Academy of Sciences, U.S.A.*, **75**, 1582–1586.

Chang, M. M., & Leeman, S. E. (1970). Isolation of a sialogogic peptide from bovine hypothalamic tissue and its characterization as substance P. *Journal of Biological Chemistry*, **245**, 4784–4790.

Chang, M. M., Leeman, S. E., & Niall, H. D. (1971). Amino acid sequence of substance P. *Nature: New Biology*, **232**, 86–87.

Chéramy, A., Nieoullon, A., Michelot, R., & Glowinski, J. (1977). Effects of intranigral application of dopamine and substance P on the in vivo release of newly synthesized (^{3}H) dopamine in the ipsilateral caudate nucleus of the cat. *Neuroscience Letters*, **4**, 105–109.

Cuello, A. C., Emson, P. C., Del Fiacco, M., Gale, J., Iversen, L. L., Jessel, T. M., Kanazawa,

I., Paxinos, G., & Quik, M. (1978). Distribution and release of substance P in the central nervous system. In J. Hughes (Ed.), Centrally Acting Peptides, pp. 135–155. London: MacMillan.

Cuello, A. C., Emson, P. C., Paxinos, G., & Jessel, T. M. (1978). Substance P containing and cholinergic projections from the habenula. Brain Research, 149, 413–429.

Cuello, A. C., & Kanazawa, I. (1978). The distribution of substance P immunoreactive fibers in the rat central nervous system. Journal of Comparative Neurology, 178, 129–156.

Davies, J., & Dray, A. (1976). Substance P in the substantia nigra. Brain Research, 107, 623–627.

DeCaro, G., Massi, M., & Micossi, L. G. (1978). Antidipsogenic effect of intracranial injections of substance P in rats. Journal of Physiology, 279, 133–140.

Destrade, C., & Cazala, P. (1979). Aversive and appetitive properties of lateral hypothalamic stimulation in mice. Behavioral and Neural Biology, 27, 398–412.

Destrade, C., & Jaffard, R. (1978). Post-trial hippocampal and lateral hypothalamic electrical stimulation. Behavioral Biology, 22, 354–374.

Diamond, B. I., Comaty, J. E., Sudakoff, G. S., Havdala, H. S., Walter, R., & Borison, R. L. (1979). Role of substance P as a "transducer for dopamine in model choreas." In T. N. Chase, N. S. Wexler, & A. Barbeau (Eds.), Huntington's Chorea—Advances in Neurology (Vol. 23), pp. 505–515. New York: Raven.

Duffy, M. J., & Powell, D. (1975). Stimulation of brain adenylate cyclase activity by the undecapeptide substance P and its modulation by the calcium ion. Biochimica et Biophysica Acta, 385, 275–280.

Emson, P. C., Jessel, T. M., Paxinos, G. & Cuello, A. C. (1978). Substance P in the amygdaloid complex, bed nucleus, and stria terminalis of the rat brain. Brain Research, 149, 97–105.

Evered, M. D., Fitzsimons, J. T., & DeCaro, G. (1977). Drinking behaviour induced by intracranial injections of eledoisin and substance P in the pigeon. Nature, 268, 332–333.

Fibiger, H. C., & Phillips, A. G. (1976). Retrograde amnesia after electrical stimulation of the substantia nigra: Mediation by the dopaminergic nigroneostriatal bundle. Brain Research, 116, 23–33.

Gale, J. S., Bird, E. D., Spokes, E. G., Iversen, L. L., & Jessel, T. M. (1978). Human brain substance P: Distribution in controls and Huntington's chorea. Journal of Neurochemistry, 30, 633–634.

Gold, P. E., Hankins, L., Edwards, R. M., Chester, J., & McGaugh, J. L. (1975). Memory interference and facilitation with posttrial amygdala stimulation: Effect on memory varies with footshock level. Brain Research, 86, 509–513.

Goldstein, J. M., & Malick, J. B. (1977). Effect of substance P on medial forebrain bundle self-stimulation in rats following intracerebral administration. Pharmacology, Biochemistry and Behavior, 7, 475–478.

Goldstein, J. M., & Malick, J. B. (1979). Lack of analgesic activity of substance P following intraperitoneal administration. Life Sciences, 25, 431–436.

Graeff, F. G., & Arisawa, E. A. (1978). Effect of intracerebroventricular bradykinin, angiotensin II, and substance P on multiple fixed-interval fixed-ratio responding in rabbits. Psychopharmacology, 57, 89–95.

Gray, J. A. (1972). Effects of septal driving of the hippocampal theta rhythm on resistance to extinction. Physiology and Behavior, 8, 481–490.

Guyenet, P. G., & Aghajanian, G. K. (1977). Excitation of neurons in the nucleus locus coeruleus by substance P and related peptides. Brain Research, 136, 178–184.

Hayes, A. G., & Tyers, M. B. (1979). Effects of intrathecal and intracerebroventricular injections of substance P on nociception in the rat and mouse. British Journal of Pharmacology, 66, 488.

Hecht, K., Oehme, P., Poppei, M., & Hecht, T. (1979). Conditioned-reflex learning of normal juvenile and adult rats exposed to action of substance P and of a SP analogue. Pharmazie, 34, 419–423.

Henry, J. L.. (1976). Effects of substance P on functionally identified units in cat spinal cord. *Brain Research*, **114**, 439–451.

Henry, J. L. (1980). Substance P and pain: An updating. *Trends in NeuroSciences*, **3**, 95–97.

Henry, J. L., Sessle, B. J., Lucier, G. E., & Hu, J. W. (1980). Effects of substance P on nociceptive and non-nociceptive trigeminal brain stem neurons. *Pain*, **8**, 33–45.

Hoebel, B. G. (1969). Feeding and self-stimulation. *Annals of the New York Academy of Sciences*, **157**, 758–778.

Hökfelt, T., Johansson, O., Kellerth, J.-O., Ljungdahl, A., Nilsson, G., Nygards, A., & Pernow, B. (1977). Immunohistochemical distribution of substance P. *In* U.S. von Euler & B. Pernow (Eds.), *Substance P*, pp. 117–145. New York: Raven.

Hökfelt, T., Kellerth, J.-O., Nilsson, G., & Pernow, B. (1975a). Substance P: Localization in the central nervous system and in some primary sensory neurons. *Science*, **190**, 889–890.

Hökfelt, T., Kellerth, J.-O., Nilsson, G., & Pernow, B. (1975b). Experimental immunohistochemical studies on the localization and distribution of substance P in cat primary sensory neurons. *Brain Research*, **100**, 235–252.

Hökfelt, T., Ljungdahl, A., Steinbusch, H., Verhofstad, A., Nilsson, G., Brodin, E., Pernow, B., & Goldstein, M. (1978). Immunohistochemical evidence of substance P-like immunoreactivity in some 5-hydroxytryptamine-containing neurons in the rat central nervous system. *Neuroscience*, **3**, 517–538.

Hong, J. S., Yang, H.-Y. T., Racagni, G., & Costa, E. (1977). Projections of substance P containing neurons from neostriatum to substantia nigra. *Brain Research*, **122**, 541–544.

Hurt, C. A., Hanaway, J., & Netsky, M. G. (1971). Stereotaxic atlas of the mesencephalon in the albino rat. *Confina Neurologica*, **33**, 93–115.

Huston, J. P., & Mueller, C. C. (1978). Enhanced passive avoidance learning and appetitive T-maze learning with post-trial rewarding hypothalamic stimulation. *Brain Research Bulletin*, **3**, 265–270.

Huston, J. P., Mueller, C. C., & Mondadori, C. (1977). Memory facilitation by posttrial hypothalamic stimulation and other reinforcers: A central theory of reinforcement. *Biobehavioral Reviews*, **1**, 143–150.

Huston, J. P., and Stäubli, U. (1978). Retrograde amnesia produced by post-trial injection of substance P into substantia nigra. *Brain Research*, **159**, 468–472.

Huston, J. P., & Stäubli, U. (1979). Post-trial injection of substance P into lateral hypothalamus and amygdala, respectively, facilitates and impairs learning. *Behavioral and Neural Biology*, **27**, 244–248.

James, T. A., & Starr, M. S. (1977). Behavioural and biochemical effects of substance P injected into the substantia nigra of the rat. *Journal of Pharmacy and Pharmacology*, **29**, 181–182.

James, T. A., & Starr, M. S. (1979). Effects of substance P injected into the substantia nigra. *British Journal of Pharmacology*, **65**, 423–429.

Jessel, T. M. (1978). Substance P release from the rat substantia nigra. *Brain Research*, **151**, 469–478.

Jessel, T. M., Emson, P. C., Paxinos, G., & Cuello, A. C. (1978). Topographic projections of substance P and GABA pathways in the striato- and pallido-nigral system: A biochemical and immunohistochemical study. *Brain Research*, **152**, 487–498.

Jessel, T. M., & Iversen, L. L. (1977). Opiate analgesics inhibit substance P release from rat trigeminal nucleus. *Nature*, **268**, 549–551.

Kanazawa, I., Bird, E., O'Connell, R., & Powell, D. (1977). Evidence for a decrease in substance P content of substantia nigra in Huntington's chorea. *Brain Research*, **120**, 387–392.

Kanazawa, I., Bird, E. D., Gale, J. S., Iversen, L. L., Jessel, T. M., Muramoto, O., Spokes, E. G., & Sutoo, D. (1979). Substance P: Decrease in substantia nigra and globus pallidus in Huntington's disease. *In* T. N. Chase, N. S. Wexler & A. Barbeau (Eds.), *Huntington's Disease—Advances in Neurology*, (Vol. 23), pp. 495–504. New York: Raven.

Kanazawa, I., Emson, P. C., & Cuello, A. C. (1977). Evidence for the existence of substance P-containing fibers in striato-nigral and pallido-nigral pathways in rat brain. *Brain Research*, 119, 447–453.

Katz, R. J. (1979). Central injection of substance P elicits grooming behavior and motor inhibition in mice. *Neuroscience Letters*, 12, 133–136.

Kesner, R. P. (in press). Brain stimulation: Effects on memory. *In* J. L. McGaugh & R. F. Thompson (Eds.), *Neurobiology of Learning and Memory*. New York, Plenum.

Kelley, A. E., & Iversen, S. D. (1978). Behavioural response to bilateral injections of substance P into the substantia nigra of the rat. *Brain Research*, 158, 474–478.

Kim, H.-J., & Routtenberg, A. (1976). Retention disruption following post-trial picrotoxin injection into the substantia nigra. *Brain Research*, 113, 620–625.

Kurtz, K. H., & Pearl, J. (1960). The effect of prior fear experience on acquired-drive learning. *Journal of Comparative and Physiological Psychology*, 53, 201–206.

Landfield, P. W. (1976a). Synchronous EEG rhythms: Their nature and their possible functions in memory, information transmission and behavior. *In* W. H. Gispen (Ed.), *Molecular and Functional Neurobiology*, pp. 389–424. Amsterdam: Elsevier.

Landfield, P. W. (1976b). Computer-determined EEG patterns associated with memory-facilitating drugs and with ECS. *Brain Research Bulletin*, 1, 9–17.

Landfield, P. W. (1977). Different effects of posttrial driving or blocking of the theta rhythm on avoidance learning in rats, *Physiology and Behavior*, 18, 439–445.

Lembeck, F. (1953). Zur Frage der zentralen Uebertragung afferenter Impulse III. Das Vorkommen und die Bedeutung der Substanz P in den dorsalen Wurzeln des Rückenmarks. *Naunyn-Schmiedeberg's Archiv für Experimentelle Pathologie und Pharmakologie*, 219, 197–213.

Ljungdahl, A., Hökfelt, T., & Nilsson, G. (1978). Distribution of substance P-like immunoreactivity in the central nervous system of the rat. I. Cell bodies and terminals. *Neuroscience*, 3, 861–943.

Ljungdahl, A., Hökfelt, T., Nilsson, G., & Goldstein, M. (1978). Distribution of substance P-like immunoreactivity in the central nervous system of the rat. II. Light microscopic localization in relation to catecholamine-containing neurons. *Neuroscience*, 3, 945–976.

Major, R., & White, N. (1978). Memory facilitation by self-stimulation reinforcement mediated by the nigro-neostriatal bundle. *Physiology and Behavior*, 20, 723–733.

Malick, J. B., & Goldstein, J. M. (1978). Analgesic activity of substance P following intracerebral administration in rats. *Life Sciences*, 23, 835–844.

Malthe-Sorenssen, D., Cheney, D. L., & Costa, E. (1978). Modulation of acetylcholine metabolism in the hippocampal cholinergic pathway by intraseptally injected substance P. *Journal of Pharmacology and Experimental Therapeutics*, 206, 21–28.

McDonough, J. H., Jr., & Kesner, R. P. (1971). Amnesia produced by brief electrical stimulation of amygdala or dorsal hippocampus in rats. *Journal of Comparative and Physiological Psychology*, 77, 171–178.

McGaugh, J. L. (1966). Time-dependent processes in memory storage. *Science*, 153, 1351–1358.

McGaugh, J. L., & Herz, M. J. (1972). *Memory Consolidation*. San Francisco: Albion.

Mroz, E. A., Brownstein, M. J., & Leeman, S. E. (1977). Evidence for substance P in the striato-nigral tract. *Brain Research*, 125, 305–311.

Neckers, L. M. (1979). Substance P afferents from the habenula innervate the dorsal raphé nucleus. *Experimental Brain Research*, 37, 619–623.

Olpe, H.-R., & Koella, W. P. (1977). Rotatory behavior in rats by intranigral application of substance P and an eledoisin fragment. *Brain Research*, 126, 576–579.

Otsuka, M., & Konishi, S. (1977). Electrophysiological and neurochemical evidence for substance P as a transmitter of primary sensory neurons. *In* U.S. von Euler & B. Pernow (Eds.), *Substance P*, pp. 207–214. New York: Raven.

Otsuka, M., Takahashi, T. (1977). Putative peptide neurotransmitters. *Annual Review of Pharmacology and Toxicology*, 17, 425–439.

Paxinos, G., Emson, P. C., & Cuello, A. C. (1978a). The substance P projections to the frontal cortex and the substantia nigra. *Neuroscience Letters*, 7, 127–131.

Paxinos, G., Emson, P. C., & Cuello, A. C. (1978b). Substance P projections to the entopeduncular nucleus, the medial preoptic area and the lateral septum. *Neuroscience Letters*, 7, 133–136.

Pellegrino, L. J., & Cushman, A. J. (1967). *A Stereotaxic Atlas of the Rat Brain*, New York: Appleton-Century-Crofts.

Pickel, V. M., Tong, H. J., Reis, D. J., Leeman, S. E., & Miller, R. J. (1979). Electron microscopic localization of substance P and enkephalin in axon terminals related to dendrites of catecholaminergic neurons. *Brain Research*, 160, 387–400.

Powell, D., Leeman, S. E., Tregear, G. W., Niall, H. D., & Potts, J. T., Jr. (1973). Radioimmunoassay for substance P. *Nature: New Biology*, 241, 252–254.

Reubi, J. C., Emson, P. C., Jessel, T. M., & Iverson, L. L. (1978). Effects of GABA, dopamine, and substance P on the release of newly synthesized ^{3}H-5-hydroxytryptamine from rat substantia nigra in vitro. *Naunyn-Schmiedeberg's Archives of Pharmacology*, 304, 271–275.

Rondeau, D. B., Jolicoeur, F. B., Belanger, F., & Barbeau, A. (1978). Motor activity induced by substance P in rats. *Pharmacology, Biochemistry and Behavior*, 9, 769–775.

Routtenberg, A., & Holzmann, N. (1973). Memory disruption by electrical stimulation of the substantia nigra, pars compacta. *Science*, 181, 83–86.

Siegel, S. (1956). *Nonparametric Statistics for the Behavioral Sciences*, pp. 152–156. New York: McGraw-Hill.

Starr, M. S., James, T. A., & Gaytten, D. (1978). Behavioural depressant and antinociceptive properties of substance P in the mouse: Possible implication of brain monoamines. *European Journal of Pharmacology*, 48, 203–212.

Stäubli, U., & Huston, J. P. (1978). Effects of post-trial reinforcing vs. subreinforcing stimulation of the substantia nigra on passive avoidance learning. *Brain Research Bulletin*, 3, 519–524.

Stäubli, U., & Huston, J. P. (1979a). Differential effects on learning by ventromedial vs. lateral hypothalamic posttrial injection of substance P. *Pharmacology, Biochemistry and Behavior*, 10, 783–786.

Stäubli, U., & Huston, J. P. (1979b). Up-hill avoidance: A new passive-avoidance task. *Physiology and Behavior*, 21, 775–776.

Stäubli, U., & Huston, J. P. (1980). Facilitation of learning by post-trial injection of substance P into the medial septal nucleus. *Behavioural Brain Research*, 1, 245–255.

Stein, L., & Belluzzi, J. D. (1978). Brain endorphins and the sense of well-being: A psychobiological hypothesis. In E. Costa and M. Trabucchi (Eds.), *The Endorphins: Advances in Biochemical Psychopharmacology* (Vol. 18), pp. 299–311. New York: Raven.

Stein, L., & Belluzzi, J. D. (1979). Brain endorphins: Possible role in reward and memory formation. *Federation Proceedings*, 38, 2468–2472.

Stewart, J. M., Getto, C. J., Neldner, K., Reeve, E. B., Krivoy, W. A., & Zimmermann, E. (1976). Substance P and analgesia. *Nature (London)*, 262, 784–785.

Stinus, L., Kelley, A. E., & Iversen, S. D. (1978). Increased spontaneous activity following substance P infusion into A 10 dopaminergic area. *Nature (London)*, 276, 616–618.

Stumpf, C. (1965). Drug action on the electrical activity of the hippocampus. *International Review of Neurobiology*, 8, 77–137.

Takahashi, T., & Otsuka, M. (1975). Regional distribution of substance P in the spinal cord and nerve roots of the cat and the effect of dorsal root section. *Brain Research*, 87, 1–11.

Teitelbaum, H., Lee, J. F., & Johannessen, J. N. (1975). Behaviorally evoked hippocampal theta waves: A cholinergic response. *Science*, 188, 1114–1116.

Todd, J. W., & Kesner, R. P. (1978). Effects of posttraining injection of cholinergic agonists and antagonists into the amygdala on retention of passive avoidance training in rats. *Journal of Comparative and Physiological Psychology*, **92**, 958–968.

Tregear, G. W., Niall, H. D., Potts, J. T. Jr., Leeman, S. E., & Chang, M. M. (1971). Synthesis of substance P. *Nature: New Biology*, **232**, 87–89.

von Euler, C., & Green, J. D. (1960). Excitation, inhibition and rhythmical activity in hippocampal pyramidal cells in rabbit. *Acta physiologica Scandinavica*, **48**, 110–125.

von Euler, U. S. (1977). Historical notes. *In* U.S. von Euler & B. Pernow (Eds.), *Substance P*, pp. 1–3. New York: Raven.

von Euler, U. S. & Gaddum, J. H. (1931). An unidentified depressor substance in certain tissue extracts. *Journal of Physiology*, **72**, 74–87.

Walker, R. J., Kemp, J. A., Yajima, M., Kitagawa, K., & Woodruff, N. G. (1976). The action of substance P on mesencephalic reticular and substantia nigra neurons of the rat. *Experientia*, **32**, 214–215.

Wetzel, W., Ott, T., & Matthies, H. (1977). Post-training hippocampal rhythmic slow activity ("theta") elicited by septal stimulation improves memory consolidation in rats. *Behavioral Biology*, **21**, 32–40.

Wood, P. L., Cheney, D. L., & Costa, E. (1979). An investigation of whether septal γ-aminobutyrate-containing interneurons are involved in the reduction in the turnover rate of acetylcholine elicited by substance P and β-endorphin in the hippocampus. *Neuroscience*, **4**, 1479–1484.

Zetler, G. (1970). Distribution of peptidergic neurons in mammalian brain. *In* W. Bargmann & B. Scharrer (Eds.), *Aspects of Neuroendocrinology*, pp. 287–295. Berlin: Springer.

Zetler, G., & Schlosser, L. (1955). Ueber die Verteilung von Substanz P und Cholinacetylase im Gehirn. *Naunyn-Schmiedeberg's Archiv für Experimentelle Pathologie und Pharmakologie*, **224**, 159–175.

CHAPTER 25

Specific Anatomical and Synaptic Sites of Neuropeptide Action in Memory Formation

Jacqueline Sagen and Aryeh Routtenberg

Studies of the effects of Substance P, ACTH fragments, opiates, and angiotensin II on memory suggest roles for these compounds in the memory consolidation processes of the medial amygdala and nigroneostriatal system. This role has been elucidated by local injection of these peptides into these brain regions following passive avoidance learning. Retention disruption by these injections may be a result of specific disruption of normal physiological processes occurring during memory formation. The mechanism of action of these peptides remains elusive, yet coordinated evidence suggests interaction at the receptor level, producing biochemical alterations (phosphorylation, fucosylation). An essential feature of these studies is that memory involves the participation of specific anatomical substrates at specific times during the mnemonic process. Any attempt to clarify the role of peptides in learning and memory should therefore focus on these specific neuronal substrates.

541

ENDOGENOUS PEPTIDES
AND LEARNING AND MEMORY PROCESSES

I. BACKGROUND: REWARD SYSTEMS AS NEURONAL
SUBSTRATES FOR MEMORY CONSOLIDATION

A. Electrically Induced Disruption by Overactivation of the Physiological Processes of Memory Formation in the Amygdala and in Substantia Nigra

The traditionally held view that memory mechanisms are diffuse and re-dundant (Lashley, 1950) is now being challenged. For example, Bresnahan and Routtenberg (1972) showed that nonepileptogenic stimulation of the medial nucleus of the rat amygdala during or immediately following learning of a passive avoidance task disrupted retention when the animals were tested 24 hr later. This stimulation, although given during learning, had no effect on original learning. Retention disruption was localized to the medial amygdaloid nucleus, as stimulation of other nuclei within the amygdala was without effect. This brain region was shown by Wurtz and Olds (1963) to support brain stimulation reward.

Another region involved in brain stimulation reward, the substantia nigra, has also been implicated in consolidation processes. Again, electrical stimulation of a particular nucleus, the pars compacta, but not surrounding regions (e.g., pars reticulata), during or immediately after learning, produced disruptive effects on retention measured 24 hr later (Routtenberg & Holtzman, 1973). In a related study, brief posttrial stimulation of the neostriatum caused retrograde amnesia (Wyers, Peeke, Willeston, & Hertz, 1968). These studies suggest an important role for the nigroneostriatal system in memory processes. Support for the hypothesis of nigroneostrial involvement in memory comes from neu-rophysiological studies demonstrating short latency alterations in unit activity in the pars compacta of the substantia nigra, during performance of a learning task (Olds, Disterhoft, Segal, Kornblith, & Hirsh 1972). Similar alterations in the monkey caudate nucleus occur during performance of a delayed response task (Soltiysik, Hull, Buchwald, & Fekete, 1975). Thus, alterations in the activity of nigroneostriatal system neurons occur during and after learning. Presumably, electrical stimulation in these regions at crucial times in the learning process alters this normal pattern of activity, in effect jamming circuits and leading to an impairment in memory (for review, see Routtenberg & Kim, 1978).

Thus, time-dependent memory processes may be localized to specific brain regions. It is our view that attempts to elucidate the roles of neurotransmitters and peptides in learning and memory should exploit this precise localization by using local injections into specific brain nuclear groupings. This approach can be conveniently coordinated with histochemical information concerning the localization of specific transmitters and peptides.

B. Relevant Neurotransmitter Systems

With this in mind, the roles of putative neurotransmitters intrinsic to the nigroneostriatal system—dopamine, acetylcholine, and gamma-aminobutyric acid (GABA)—were studied in this laboratory.

1. GABA

There is a descending inhibitory GABA-ergic input to the substantia nigra (Precht & Yoshida, 1971), originating, in part, from the neostriatum (Jessell, Emson, Paxinos, & Cuello, 1978). Picrotoxin, a GABA-ergic antagonist, should block this inhibitory effect, leading to overactivation of the nigroneostriatal system and disruption of memory. Indeed, Kim and Routtenberg (1976a) found that *posttrial injection* of picrotoxin in the substantia nigra produced a retention impairment. This effect is time-dependent, as it is observed when injection is given immediately after learning, but not with injection 24 hr after learning, immediately before the retention trial.

2. Dopamine

Because GABA-ergic input to the substantia nigra terminates in relation to dopamine-containing dendrites in the substantia nigra (Kim & Routtenberg, 1978), dopamine projections to the neostriatum may play a role in memory consolidation. In fact, posttrial injection of dopamine into the neostriatum disrupted retention performance (Kim & Routtenberg, 1976b). Furthermore, 6-hydroxydopamine (6-OHDA) injected into the substantia nigra immediately after learning produced dramatic disruptive effects on retention (Kim & Routtenberg, 1980). There is some evidence (Fuxe, personal communication 1976) that intranigral 6-OHDA injection initially causes a profuse release of dopamine in the neostriatum.

3. Acetylcholine

Acetylcholine may also participate in the consolidation process, because it is found in high levels in the neostriatum, and because nigroneostriatal dopamine fibers may terminate on cholinergic interneurons in this region (Butcher & Bilezikjian, 1975; Butcher, Talbot, & Bilezikjian, 1975). Studies by Haycock, Deadwyler, Sideroff, and McGaugh (1973) indicated that scopolamine, a muscarinic receptor antagonist, injected into the neostriatum disrupts passive avoidance retention. Also, Deadwyler, Montgomery, and Wyers (1972) showed that injection of carbachol, a cholinergic agent, also produces disruption of passive avoidance retention. It is unlikely that retention impairment produced by these agents is caused by excess dopamine release. Thus, disruption of the normal physiological process—a reversible lesion—is likely.

These studies illustrate the important point that memory is a time-dependent process involving specific neuronal substrates containing identifiable neuro-transmitters. In particular, the nigroneostriatum and medial amygdala appear to be involved in the consolidation of memory, as electrical or chemical manipulations of these regions are effective only during or immediately after learning. Other brain regions may be involved in initial learning and extinction (see Clavier & Routtenberg, 1980).

II. PEPTIDES AND MEMORY

It has been suggested that peptides play an important role in learning and memory. For example, the systemic injection of adrenocorticotropin (ACTH) or enkephalin alters consolidation processes. Furthermore, Substance P, an-giotensin, and enkephalin are intrinsic to brain regions implicated in mnemonic processes. Substance P is found in high concentrations in the medial amygdala and within the pars compacta of the substantia nigra (Ben-Ari, Pradelles, Gros, & Dray, 1979). Enkephalin and angiotensin are both present in neostriatum (see the following).

We have studied the effects of posttrial injection of these peptides in either amygdala or neostriatum in an attempt to discern their potential roles in con-solidation processes. In our view, disruptive effects might be expected, if ex-cessive concentrations of peptides prevent normally occurring processes.

A. Behavioral and Intracranial Injection Procedures Used

All of the experiments to be described used the same procedure, which will be briefly summarized. Male albino rats (200–300 g) were unilaterally implanted with a 24-gauge guide cannula aimed at the head of the caudate nucleus, or a 26-gauge guide cannula aimed at the medial amygdala. We used chemical injection procedures described previously (Kim & Routtenberg, 1976a). After allowing 4 days for recovery, the animals were placed on the platform of a step-down passive avoidance learning apparatus (Bresnahan & Routtenberg, 1972). The animals received a 0.5 mA footshock through the grid floor whenever they stepped off the platform. The acquisition session lasted until the rats remained on the platform for 2 min without stepping down.

Within 5 min after learning, the animals received a single 1.0 μl intracranial (i.c.) injection of a test solution delivered over 4 min (0.5 μl delivered over 2 min in the case of the medial amygdala). They were then returned to their home cages.

Retention testing 24 hr later employed the same experimental conditions, except that the grid bars did not deliver footshock when the rat stepped off the platform. If the animal remained on the platform for 3 min, he was placed in his home cage and tested the next day. Testing continued until the rat stepped off the platform on 2 consecutive days during the 3 min retention trial. Two performance measures served as an indication of retention: (1) *percentage disruption*—the proportion of rats in each group that stepped down on the first retention test day; and (2) *latency to first descent*—the average time the rats remained on the platform during retention testing.

The shorter the time spent on the platform, the greater was the disruption of retention. Longer times for the experimental animals than for the controls may indicate retention facilitation. Following testing, animals were sacrificed for histological confirmation of cannula placements.

B. Substance P

1. Medial Amygdala

The medial amygdaloid nucleus has at least two unique features that differentiate it from the other nuclei of the amygdala. First, brain stimulation in this region significantly impairs retention performance when compared to the effects of stimulation in other amygdaloid nuclei. Second, Substance P exists in highest concentrations in this region (Ben-Ari et al., 1979). Emanuel & Routtenberg (1978) attempted to determine what role Substance P in the medial amygdala might play in modulating retention performance.

Five minutes after training, 1.0 μg Substance P (Beckman) was injected over a 2-min period. Retention testing was carried out for a 3 min period 24 hr after learning. Animals were tested on 4 succeeding days or until rats descended from the platform on 2 successive days. Four groups of subjects were studied: Group 1: control, unimplanted; Group 2: control, implanted in medial amygdala, saline injections; Group 3: Substance P injections into amygdala, but not medial nucleus; Group 4: Substance P injections into the medial amygdaloid nucleus.

We found that posttrial injections of Substance P into the medial nucleus caused a significant disruption of retention performance. Considering the mean latency to first descent: unimplanted control Group 1 ($n = 4$) 720.0 sec; saline injected Group 2 ($n = 5$) 608.8 sec; nonmedial amygdala Group 3, 494.8 sec; medial amygdala Group 4, 182.8 sec. An overall analysis of variance indicated that there was, indeed, a significant effect of treatment [F (3,21) = 5.69, $p < .01$]. Mann-Whitney U-tests indicated that Group 4 was significantly different from each of the other groups. No other comparisons were statistically significant. In a related study, Huston & Stäubli (1979) demonstrated a disruptive influence of Substance P injected into the amygdala, although no evaluation of nuclear specificity within this structure was made. Emanuel and Routtenberg (1978) showed that posttrial injections of Substance P into the medial amygdala caused retention impairment, whereas injections of this peptide in other amygdaloid nuclei did not.

The evidence from studies employing either electrical stimulation or chemical injection suggest an important role of the medial amygdaloid nucleus in modulating retention of an aversively motivated task. However, its general function in mnemonic processes remains to be determined.

2. Substantia Nigra

This brain region has been shown to be involved in memory (Routtenberg & Holtzman, 1973) and is also rich in Substance P (Ben-Ari et al., 1979). The perikarya containing this peptide may be in the neostriatum (Jessell et al., 1978). Sustance P may interact directly with nigral dopaminergic cells that project to the neostriatum. Application of Substance P to substantia nigra increases dopamine metabolism in the striatum (Waldmeier, Kam, & Stocklin, 1978), and immunoneutralization of nigral Substance P decreases dopamine release from terminals in the caudate nucleus (Cheramy, Michelot, Leviel, Nieoullon, Glowinski, & Kerdelhue, 1978). Furthermore, direct iontophoretic application of Substance P to substantia nigra cells (Dray & Straughton, 1976) increases neuronal firing rate. In view of these studies, and based on our localization of memory-related processes to this region (Kim & Routtenberg, 1976a; Routtenberg & Holzman, 1973), one would expect local injection of Substance P in the substantia nigra to lead to overactivation of the nigroneostriatal system, and thus to disruption of memory. Indeed, Huston and Stäubli (1978) have demonstrated retrograde amnesia produced by posttrial injection of Substance P into substantia nigra.

C. ACTH and Its Role in Neostriatal Memory Functions

1. Injection of ACTH into Neostriatum

There is a large body of evidence supporting the profound role for ACTH fragments in memory processes. However, a discrepancy exists in the time course of effects. Briefly, de Wied's group found that systemic administration of $ACTH_{1-10}$ and $ACTH_{4-10}$ delays extinction of a variety of active avoidance behaviors (de Wied, 1974). These analogs also facilitate passive avoidance behavior when they are adminsterd 1 hr prior to the retention trial, but not if injected immediately after the learning trial. This time course is further supported by van Wimersma Greidanus, van Dijk, de Rotte, Goedemans, Croiset, and Thody (1978), who found that intracerebroventricular (i.c.v.) administration of ACTH$_{1-24}$ antisera disrupted passive avoidance behavior when it was administered 1 hr before the retention trial, but not if given immediately after the learning trial. In contrast, Flood, Jarvik, Bennett, and Orme (1976) did observe a posttrial effect and found that ACTH$_{4-10}$ had to be administered within 60 min of passive avoidance training to improve retention effectively. Martinez, Vasquez, Jensen, Soumireu-Mourat, and McGaugh (1979) found that administration of an ACTH$_{4-9}$ analog 1 hr prior to training markedly facilitated acquisition of an inhibitory avoidance task while having no effect on retention when administered immediately after training or prior to the retention test.

Much of this confusion may be due to the simultaneous activation of different and, perhaps, conflicting components of the learning and memory process, as a result of systemic or i.c.v. administration. It may be noted here that surprisingly few studies exist on direct application of ACTH to selected brain regions. Because we have found that the nigroneostriatal system participates in consolidation of passive avoidance, Sagen and Routtenberg (1976) studied the effects of ACTH$_{1-10}$ and ACTH$_{4-10}$ injected directly into the neostriatum.

The experimental design was similar to that described for the Substance P studies. Approximately 5 min after training, 100.0 ng ACTH$_{1-10}$ (7.8 × 10^{-11} mol) or 100.0 ng ACTH$_{4-10}$ (1.1 × 10^{-10} mol) were injected into the caudate nucleus in a 1.0 μl volume of vehicle. The vehicle in most cases was saline, but because we found slight difficulties dissolving ACTH$_{1-10}$, we also tried the phosphoric acid method of Rigter, van Riezen, and de Wied (1974). Control groups received an equal volume of appropriate vehicle. Neither vehicle nor ACTH$_{1-10}$ plus vehicle had any effect on motor behavior.

Results are summarized in Table 1. A marked disruption of retention performance was caused by 100.0 ng ACTH$_{4-10}$ injected directly into the neostriatum. The mean latency to descend was significantly shorter ($p < .001$) than that of the saline controls. In the control group, only three animals (27%) descended from the platform (retention disruption) within 3 days of retention testing, whereas all of the animals in the ACTH$_{4-10}$ group descended by this time.

These data may at first seem to be in conflict with data from certain other laboratories (although Flood, et al., 1976, have, in fact, shown a posttrial disruptive effect) in that disruption, rather than facilitation, resulted from ACTH treatment. It must be kept in mind, however, that these intracranial injection procedures probably produce disruption via flooding with an excess of extracellular peptide, thus preventing the normal pattern of neuronal activity. Because most of the learning and memory studies have used systemic injection, it is unlikely that a lower dose reaches the sensitive regions and does so over

TABLE 1

Effect of Posttrial Intraneostriatal Injection of ACTH on Step-Down Retention Performance

	Mean latency to first descent (±SD)	Retention test day when first descent was made[a]				
		1	2–3	4–8	9–10	>10
Control (n = 12)	1043 ± 182	2	2	4	3	1
ACTH$_{4-10}$ (n = 9)	208 ± 74**	5	4	0	0	0
ACTH$_{1-10}$ (n = 23)	938 ± 203	8	4	3	2	6

[a] Number of animals descending.

** Significantly different from controls [t (18) = 3.93; $p < .001$].

a longer time so that a similar degree of flooding does not occur. In these latter cases, the peptides may have been present at a low enough concentration to stimulate, but not to disrupt, the normal pattern of events, thus facilitating memory processes.

The $ACTH_{1-10}$ data is more difficult to interpret. At first glance, the results obtained with this drug do not appear to be significantly different from control data, either with saline or acid buffer vehicles. Data are combined in Table 1, because there was no difference between vehicles. However, further analysis of the data revealed that there is a bimodal distribution in latency to the first descent (Table 1). In the control group, only 9% descended on the first retention trial, and only 9% still showed retention after 10 days of testing. However, of the animals receiving $ACTH_{1-10}$, 35% descended on the first retention trial, and 26% showed retention after 10 or more days. Therefore, at least 60% of the animals tested showed differences from control animals in retention behavior at the extremes.

Because descent on the first day is evidence for memory disruption, and retention for greater than 10 days appears to suggest facilitation, it seems that, in the majority of animals tested, either disruption or facilitation of retention occurred following injection of $ACTH_{1-10}$ after learning.

Analysis of the histological material revealed that there was no apparent difference in cannula placements between the $ACTH_{4-10}$ and $ACTH_{1-10}$ groups. Thus, the facilitation observed in some cases and disruption in others is unlikely to be related to differences in the locus of injection within the neostriatum (unlike angiotensin, see the following). Because disruption by $ACTH_{1-10}$ is consistent with our data on $ACTH_{4-10}$, and facilitation is consistent with systemic data from other laboratories, it is possible that the $ACTH_{1-10}$ dose used here produced an effect near the midpoint between a facilitatory and a disruptive dose. In all cases $ACTH_{4-10}$ was disruptive, perhaps because of the higher molar concentration used or of less stereochemical impedence of the shorter peptide. Therefore, a low dose of an ACTH analog may produce facilitation of the consolidation process, whereas a high dose causes disruption of the normal pattern of activity. A careful dose–response study would clearly be of value.

This hypothesis is consistent with the work of Gold and McGaugh (1978) who found that systemic injection of ACTH immediately after training in an inhibitory (passive) avoidance task either *enhanced* or *impaired* later retention, depending on the dose. At low doses (0.03 or 0.30 IU/animal) retrograde amnesia was observed. Since they used whole ACTH, this effect may have been mediated by increased corticosteroid levels.

Of interest in this regard is the fact that subcortical and neostriatal tissue adenylate cyclase and cyclic 3′, 5′ adenosine monophosphate (cAMP) exhibit sensitivity to $ACTH_{1-24}$ in a similar dose-related manner (Wiegant, Dunn, Schotman, & Gispen, 1979). Low doses of $ACTH_{1-24}$ (25.0 μM) stimulated, while 0.1 mM or higher inhibited adenylate cyclase activity. This may explain ACTH facilitation of retention at low doses, and disruption at higher doses.

Van Wimersma Greidanus and de Wied (1971) found that implantation of 10.0 μg $ACTH_{1-10}$ in the posterior thalamus or the cerebrospinal fluid (CSF) inhibited exinction of avoidance. We also looked at effects of $ACTH_{4-10}$ in this region using our behavioral paradigm and a 100.0 ng dose. The results suggested a trend toward disruption of retention. These data are difficult to compare with those from the study of van Wimersma Greidanus and de Wied (1971) because time course, behavioral test conditions, and dosages were different.

It may be the case that 100.0 ng facilitated, whereas 10.0 μg disrupted the function of that site. However, because disruption of function leads to facilitation of retention, it must be assumed that the posterior thalamus is part of a system important for forgetting (e.g., facilitates extinction). Disruption of normal forgetting mechanisms in this region would, therefore, enhance memory. A similar argument has been made with regard to the sulcal frontal cortex (Clavier & Routtenberg, 1980).

2. Interaction of ACTH with Neurotransmitter Systems

The data considered to this point indicate that different neuronal systems play roles in modulating certain components of learning and memory processes and that peptides within these systems participate at different times during such processes. Again, the importance of anatomical specificity in these studies is underscored.

Our data suggest a role for ACTH in the consolidation processes of the neostriatal system. If these analogs produce disruption by altering the physiology of certain components, some effects of ACTH on neurotransmitter metabolism may be observable. Current reviews of this area have been provided by Dunn (in press) and Versteeg (1980). Van Loon, Sole, Kamble, Kim, and Green (1977) have noted that hypophysectomy results in decreased dopamine turnover in the striatum with no effect on tyrosine hydroxylase activity. Iuvone, Morasco, Delaney, and Dunn (1978) found that $ACTH_{4-10}$ and other analogs stimulate the conversion of tyrosine to dopomine, without altering cerebral dopamine content. No detectable effect on norepinephrine turnover was noted. Furthermore, certain behavioral effects produced by either $ACTH_{1-24}$ or morphine (see the following) can be suppressed by intrastriatally applied haloperidol, a dopamine receptor blocker (Cools, Wiegant, & Gispen, 1978). Thus, it is conceivable that the effects of ACTH observed in our study may have been mediated through an action on dopamine metabolism in neostriatum.

Finally, a decrease in cerebral GABA is produced by $ACTH_{4-10}$ (Leonard, 1974). Recalling that picrotoxin, a GABA blocker, disrupted retention when injected into the substantia nigra (Kim & Routtenberg, 1976a), we suggest that ACTH-induced disruption of the function of GABA-ergic pathways may be one mechanism by which impairments in retention performance are produced.

3. Regional Distribution of ACTH

Immunohistochemical localization of ACTH in cell bodies and axons suggest that ACTH may have a transmitter-like function (Larsson, 1978; Orwoll, Kendall, Lamorena, & McGilvra, 1979; Pelletier & Leclerc, 1979). Furthermore, the distrubution of immunoreactivity was not altered by hypophysectomy. ACTH-containing cell bodies are located in the supraoptic and arcuate nuclei. Fibers containing ACTH are distributed throughout regions of the hypothalamus, thalamus, and mesencephalon, but only a low level of immunoreactivity is found in the neostriatum. One possible problem in the interpretation of such results is that antibodies used in these studies react with the C-terminal end of large ACTH fragments, but not with smaller behaviorally active fragments such as $ACTH_{4-10}$. The distribution of ACTH and smaller ACTH fragments is not necessarily identical. A parallel case may be found in the differences in distribution of endorphin and enkephalin.

The central nervous system (CNS) origin of ACTH fragments is not a necessary prerequisite to an explanation of the effects of these molecules on behavior, as either systemic or i.c.v. administration is sufficient to produce effects on memory. Gispen, Reith, Schotman, Wiegant, Zwiers, and de Wied (1977) suggested that these peptides may reach brain effector cells through the circulation or through a pituitary stalk–CSF channel. Mezey, Palkovits, de Kloet, Verhoef, and de Wied (1978) further suggested that a portion of the ACTH fragments found in the brain originate in the pituitary, as intrapituitary injection of ^{3}H-$ACTH_{4-9}$ results in a distribution of radioactivity in the brain with clear regional differences. This evidence lends support to the view that the CSF distributes these peptides to various brain regions. Finally, Allen, Kendall, McGilvra, and Vancura (1974) demonstrated the presence of immunoreactive ACTH in human CSF at concentrations nearly equal to or exceeding those in plasma, again suggesting a direct pituitary–CSF route. Thus, ACTH fragments may reach the neostriatum via the CSF to produce behavioral effects. The fact that the medial wall of the neostriatum is apposed to the lateral ventricular surface may be of functional relevance.

4. Role of Brain Protein Phosphorylation in the Effect of ACTH on Memory

Gispen and co-workers (Zwiers, Tonnaer, Wiegant, Schotman, & Gispen, 1979; Zwiers, Wiegant, Schotman, & Gispen, 1978) suggested that the effects of ACTH on behavior may be mediated by an effect on synaptosomal protein phosphorylation. *In vitro* phosphorylation of a brain protein of 47k molecular weight (MW) was reduced by the presence of $ACTH_{1-24}$. The protein kinase responsible for phosphorylation of this protein band was inhibited by $ACTH_{1-24}$. Cyclic AMP did not stimulate phosphorylation of this band.

These results are intriguing because our laboratory has shown that *in vitro* phosphorylation of a protein of 47k MW which we term *band* F-1, is probably altered following training (for reviews, see Routtenberg & Benson, 1980; Routtenberg, 1979a,b), and during early handling (Cain & Routtenberg, 1980). Phosphorylation of another protein band of 41k MW, termed *band* F-2, is also altered. Because of the difficulty of comparing with certainty our bands with those of Zwiers et al. (1978, 1979), it is not certain whether our F_1 band corresponds to their ACTH-sensitive band. It is notable that Zwiers et al. (1978), found that their ACTH-sensitive band, like our band F-1, is not stimulated to incorporate more phosphate in the presence of cyclic AMP.

It is interesting to suppose, based on our recent findings, that training alters phosphorylation of certain brain proteins. The relative concentration of ACTH either in the intracellular or extracellular compartment may act to amplify the change of phosphorylation state engendered by the environmental context.

In summary, the effects of ACTH on memory consolidation may be mediated by dopamine, GABA, and phosphorylation in specific brain regions. Unfortunately, no CNS receptor for radiolabeled ACTH has been found thus far. Terenius (1976) showed, however, that ACTH fragments exhibit affinity for opiate binding sites with the characteristics of a partial agonist. Thus, although ACTH has several different physiological actions, it may exert its CNS effects through opiate receptor interactions.

Support for this ACTH–opiate receptor hypothesis comes from electrophysiological, hormonal, and behavioral studies. In peripheral tissue, Plomp and van Ree (1978) found that several NH_2-terminal ACTH peptides depress electrically induced contractions of the mouse vas deferens. This response is dose-dependent and parallel to that found for morphine. Naloxone completely abolishes this peptide-induced inhibition. Yasukawa, Monder, Michael, and Christian (1978) demonstrated that naloxone can block ACTH-induced delay of puberty and also reverse decreases in weights of reproductive organs produced by ACTH. Morphine-induced decrease in sexual function is also reversed by naloxone, suggesting a possible peripheral interaction of ACTH, morphine, and naloxone at similar receptor sites.

An increasing number of investigations suggest that these compounds interact at central locations as well. Electrophysiological studies show that ACTH analogs antagonize morphine-induced inhibition of spinal reflex activity (Zimmerman & Krivoy, 1973). Stereotyped behavioral activation (running, Straub tail, unresponsiveness to environment) by systemically injected morphine is antagonized in a dose-related fashion by i.c.v. administered $ACTH_{4-10}$ (Katz, 1976). Moreover, excessive grooming behavior elicited by i.c.v. ACTH can be completely suppressed by 1.0 mg/kg subcutaneous naloxone (Gispen et al., 1977). Intraventricular morphine also elicits excessive grooming that is reversible by naloxone.

Pain perception studies show that systemic pretreatment with ACTH analogs reduced morphine-induced analgesia by 50% (Gispen, Buitelaar, Wiegant, Terenius, & de Wied, 1976). The potency of the antagonism correlated with their binding affinities for the opiate receptor. Bertolini, Poggioli, and Ferrari (1979) demonstrated that $ACTH_{1-24}$ given i.c.u. markedly reduced hot plate and tail-flick latency (hyperalgesia). This effect was antagonized by morphine and potentiated by naloxone. Thus, these studies provide evidence that ACTH and opioid peptides may interact at the same receptor or at closely related, but different, sites of action.

D. Disruption of Retention by Posttrial Injection of Morphine into Neostriatum

The possibility that ACTH may act via opiate receptors is especially intriguing to us, as the neostriatum has a high concentraion of opiate binding sites. Therefore, it seemed worthwhile to examine the effects of the opiate, morphine, on memory functions. It has previously been shown that systemically administered Met- Leu-enkephalin attenuate CO_2 induced amnesia (Rigter, 1978), and systemic morphine facilitates posttraining memory processing (Castellano, 1975; Mondadori & Waser, 1979). At lower doses, morphine impairs performance (Izquierdo, 1979). In addition, both α- and β-endorphin delay extinction of pole-jumping avoidance and facilitate or impair passive avoidance behavior in much lower doses than those needed to induce analgesia (see de Wied et al., 1978, and Martinez & Rigter, 1980, for recent reports).

Morgan & Routtenberg (1978) investigated the effects of opiate injection into the neostriatum on retention performance. Five minutes after passive avoidance training to a 2 min learning criterion, each rat received 1.0 μl of either saline ($n = 15$) or 100.0 ng morphine sulfate ($n = 15$). Posttrial injection of morphine sulfate significantly disrupted retention performance measured 24 hr later, both with respect to percentage of animals descending on the first retention day (6% versus 20%, $p < .05$) and latency to first descent measured over a 5-day retention period (141.5 sec versus 509.1 sec, $p < .01$). In a hot plate test conducted 2 wk later, no analgesic effect was observed, using the same dose of morphine injected into neostriatum. These data demonstrate that morphine injected into the neostriatum disrupts posttrial memory consolidation processes. It is conceivable, then, that posttrial manipulation of opiate receptors in neostriatum leads to an alteration in retention performance of the step-down task.

In the amygdala, another brain region important in memory processes, injection of levorphanol, an opiate receptor agonist, following passive avoidance conditioning also impaired retention, whereas naloxone facilitated performance. Similar posttrial injection of opiates into tissue dorsal to the amygdala had no effect on retention (Gallagher & Kapp, 1978).

These effects of opiates on retention may have been produced by disruption of normally occurring processes. In support of this view, opiate receptor agonists are known to increase the turnover rate of GABA in the substantia nigra and decrease turnover in the caudate nucleus (Moroni, Cheney, Peralta, & Costa, 1978). Concurrently, there is an increase in turnover of dopamine in the neostriatum, perhaps via presynaptic receptors on dopamine terminals. Furthermore, long-term narcotic exposure decreases phosphorylation of protein bands in the 40–47k MW range (Davis & Ehrlich, 1979). We have shown that these phosphoproteins were altered following learning (for review, see Routtenberg, 1979b), suggesting a possible mechanism for the effect of morphine on learning.

Anatomical substrates for self-stimulation are postulated to be involved in memory processes (Routtenberg, 1979a). Since both ACTH fragments and enkephalin may play roles in consolidation, it is of great interest that self-stimulation sites are rich in enkephalin, and furthermore, that self-administration of both enkephalin and ACTH analogs is rewarding (Jouhaneau-Bowers & LeMagnen, 1979; Stein, 1980). Finally, of possible importance is the parallel regional distribution of endogenous opiates and ACTH in the brain (Larsson, 1978; Nilaver, Zimmerman, Defendini, Motta, Krieger, & Brownstein, 1979). Recent evidence indicates that ACTH and β-lipotropin may have the same 31,000 dalton precursor molecule (Liotta, Gildersleeve, Brownstein, & Krieger, 1979). It is thus likely that certain of these smaller peptides are contained within the same neurons (Nilaver et al., 1979) and perhaps even the same vesicles (Pelletier, 1979).

E. Posttrial Injection of Angiotensin II into Neostriatum

Angiotensin II regulates fluid uptake and blood pressure by action on neurons in the subfornical organ (Felix & Schlegel, 1977; Simpson, Epstein & Camardo, 1978; Simpson & Routtenberg, 1973). These actions are probably mediated through angiotensin binding sites, found in high concentrations in the subfornical organ (Sirett, McClean, Bray, & Hubbard, 1977).

Because a renin–angiotensin system endogenous to the brain apparently exists (Ganten, Boucher, & Genest, 1971) and brain angiotensin-converting enzyme is in the highest concentration in the neostriatum (Poth, Heath, Heath, & Ward, 1975), our laboratory investigated the effects of angiotensin II on passive avoidance retention (Morgan & Routtenberg, 1977). Retention performance measured 24 hr later was disrupted by 100.0 ng injected into the dorsal, but not ventral, neostriatum 5 min after learning. If the same dosage of angiotensin was given 2 hr before retention testing, no effect on this behavior was observed. Thus, as in other treatments of the nigroneostriatal system, only manipulations that are given close in time to the original training appear to have effects on retention performance.

The effect of angiotensin could not be ascribed to a nonspecific action of the peptides, as thyrotropic-releasing hormone (TRH) and lysine vasopressin administered at an equimolar concentration as angiotensin II had no effects on retention performance.

One might expect that if the angiotensin system in the neostriatum participates in memory formation, the metabolism of this system should be altered by training. Morgan and Routtenberg (1979a,b) have shown that the metabolism of a glycoprotein of 120k MW, similar to angiotensin converting enzyme, is altered following training. In our study, ^{3}H-fucose was injected *in vivo* into neostriatum. This precursor selectively labels the terminal side chain of glycoproteins in the cell body. The influence of training on the labeling of electrophoretically separated protein moieties was studied from 2 hr to 5 days after training. Of the 10 glycoproteins studied at the five different time intervals, this particular training-dependent change was one of only three that were observed.

The possibility exists that, following training, alterations occur in the synthesis or fucosylation of angiotensin-converting enzyme in the neostriatum. These biochemical alterations would provide a net change in the level of angiotensin II, thus altering the functional activity of neurons in the neostriatum. The nature of this alteration may be of interest in relation to mnemonic processes.

A recent study by Koller, Krause, Hoffmeister, and Ganten (1979) lends support to this view. These authors elevated endogenous angiotensin by injecting renin, which stimulates the biosynthesis of angiotensin in brain. This procedure produced a disruption of passive avoidance behavior. Furthermore, blockade of angiotensin-converting enzyme with the inhibitor SQ 14225 abolished the effects of renin. Renin injected 22 hr after learning was without effect. The authors suggest that their results indicate "a role for the endogenous peptide in mnemonic processes [Koller et al., 1979, p. 72]."

F. Some Principles of the Peptide–Memory Association

1. Anatomical Specificity

We have shown that certain brain regions may be implicated in memory processing using local electrical brain stimulation techniques. It may be relevant that the injection of peptides existing in these same brain locations also disrupts learning. It is possible that such localized effects are mediated by a quantal transmitter-like release mechanism (Snyder & Innis, 1979). Even if peptide action were not via such a punctate mechanism, there is no reason to believe a priori that peptide action would be anatomically diffuse, as restricted sites of action could be conferred by postsynaptic receptor localization.

As we have pointed out, biphasic responses to peptides may be a function of the multiple sites of action at specific brain regions with different functional roles in the mnemonic process. We strongly advocate the analysis of *peptide dose–memory retention response* at specific brain loci.

2. Peptide Manipulations and Endogenous Biochemical Response

Although the evidence is still suggestive, the paradigm presented here will be crucial for understanding peptide participation in mnemonic processes. That is, effects of exogenous administration of peptides on memory do not prove that peptides participate in the process. To demonstrate peptide participation, it will be necessary to show that a component of that enzyme system is metabolically altered following the same event, which is altered by peptidergic manipulation. The effects of ACTH on both learning and brain protein phosphorylation, as well as the effects of angiotensin on both learning and brain protein fucosylation stand as paradigms, not proofs, for the approach we advocate. Clearly, then, the proof of the paradigm is in the protein. The next section deals with this issue in more detail by providing a set of coordinated criteria for approaching proof state.

3. Criteria for Establishing the Role of Peptides in Memory

Based on the considerations already discussed, it seems worthwhile to propose a tentative list of criteria that may be useful in deciding whether a particular peptide participates in the memory formation process.

1. Electrical stimulation in anatomically defined loci alters retention performance in a learning situation.

2. A particular peptide is demonstrated to exist in this anatomically defined brain region with histo- or immunochemical methods.

3. Injection of this peptide into this anatomically defined region, but not into adjacent regions where the peptide is not found, alters retention performance.

4. A related (e.g., molecular weight) peptide that does *not* exist in this brain region, has no influence on retention performance.

5. Control points in the biochemical pathways involved in the metabolism of this peptide should be selectively altered by behavioral training in the same learning situation.

ACKNOWLEDGMENTS

For the generous supply of ACTH$_{4-10}$ and ACTH$_{1-10}$, we wish to thank Dr. H. van Riezen of Organon Laboratories, Dr. David de Wied of the University of Utrecht, and Dr. Rittel and Dr. R. Maier of CIBA-Geigy in Basel. We are most grateful for their kind assistance.

Gratitude is expressed to Betty Wells for assistance in preparation of the manuscript. Supported, in part, by MH 25281 to A. R. Helpful comments on the manuscript were provided by Dr. David de Wied and Dr. Adrian Dunn.

REFERENCES

Allen, J. P., Kendall, J. W., McGilvra, R., & Vancura, C. (1974). Immunoreactive ACTH in cerebrospinal fluid. *Journal Medical Endocrinology*, **38**, 586–593.

Ben-Ari, Y., Pradelles, P., Gros, C., & Dray, F. (1979). Identification of authentic substance P in striatonigral and amygdaloid nuclei using combined high performance liquid chromatography and radioimmunoassay. *Brain Research*, **173**, 360–363.

Bertolini, A., Poggioli, R., & Ferrari, W. (1979). ACTH-induced hyperalgesia in rats. *Experentia*, **35**, 1216–1217.

Bresnahan, E., & Routtenberg, A. (1972). Memory disruption by unilateral low-level subseizure stimulation of the medial amygdaloid nucleus. *Physiology and Behavior*, **9**, 513–525.

Butcher, L., & Bilezikjian L. (1975). Acetylcholinesterase containing neurons in the neostriatum and substantia nigra revealed after punctate intracerebral injection of di-isopropyl-fluorophosphate. *European Journal of Pharmacology*, **34**, 115–125.

Butcher, L. L., Talbot, K., & Bilezikjian, L. (1975). Localization of acetylcholinesterase within dopamine containing neurons of the zona compacta of the substantia nigra. *Proceedings Western Pharmacological Society*, **18**, 256–259.

Cain, S. T., & Routtenberg, A. (1981). Endogenous phosphorylation *in vitro* effects of neonatal handling. *Federation Proceedings*, **40**, 250.

Castellano, C. (1975). Effects of morphine and heroin on discrimination learning and consolidation in mice. *Psychopharmacologia*, **42**, 235–242.

Cheramy, A., Michelot, R., Leviel, V., Nieoullon, A., Glowinski, J., & Kerdelhue, B. (1978). Effect of immunoneutralization of substance P in the cat substantia nigra on the release of dopamine from dendrites and terminals of dopaminergic neurons. *Brain Research*, **155**, 404–408.

Clavier, R. M., & Routtenberg, A. (1980). In search of reinforcement pathways: An anatomical odyssey. *In* A. Routtenberg (Ed.), *Biology of Reinforcement: Facets of Brain Stimulation Reward*, pp. 81–107. New York: Academic Press.

Cools, A. R., Wiegand, V. M., & Gispen, W. H. (1978). Distinct dopaminergic systems in ACTH induced grooming. *European Journal of Pharmacology*, **50**, 265–268.

Davis, L. G., & Ehrlich, Y. H. (1979). Opioid peptides and protein phosphorylation. *Advances in Experimental and Medical Biology*, **110**, 233–244.

Deadwyler, S. A., Montgomery, D., & Wyers, D. J. (1972). Passive avoidance and carbachol excitation of the caudate nucleus. *Physiology and Behavior*, **8**, 631–635.

de Wied, D. (1974) Pituitary-adrenal system hormones and behavior. *In* F. O. Schmitt and F. G. Worden (Eds.), *The Neurosciences, Third Study Program*. pp. 653–666. Cambridge, Massachusetts: MIT Press.

de Wied, D., Bohus, B., Van Ree, J. M., & Urban, I. (1978). Behavioral and electrophysiological effects of peptides related to lipotropin. *Journal of Pharmacology and Experimental Therapeutics*, **204**, 570–580.

Dray, A., & Straughton, D. W. (1976). Synaptic mechanisms in the substantia nigra. *Journal of Pharmacy and Pharmacology,* **28,** 400–405.

Dunn, A. J. (1981). Central nervous system effects of adrenocorticotropin (ACTH), β-lipotropin (β-LPH) and related peptides. *In* A. J. Dunn & C. B. Nemeroff (Eds.), *Molecular and Behavioral Neuroendocrinology.* Jamaica, New York: Spectrum Publications, in press.

Emanuel, D., & Routtenberg, A. (1977). Post-trial substance P injected into medial amygdala but not other amygdaloid nuclei disrupts retention performance: A selective cytoarchitectonic-peptidergic effect. Northwestern University.

Felix, D., & Schlegel, W. (1977). Angiotensin receptive neurones in the subfornical organ: Structure-activity relations. *Brain Research,* **149,** 107–116.

Flood, J. F., Jarvik, M. E., Bennett, E. L., & Orme, A. E. (1976). Effects of ACTH peptide fragments on memory formation. *Pharmacology, Biochemistry and Behavior, (Supplement 1),* **5,** 41–51.

Gallagher, M., & Kapp, B. S. (1978). Manipulation of opiate activity in the amygdala alters memory processes. *Life Sciences,* **23,** 1973–1978.

Ganten, D., Boucher, R., & Genest, J. (1971). Renin activity in brain tissue of puppies and adult dogs. *Brain Research,* **33,** 557–559.

Gispen, W. H., Buitelaar, J., Wiegant, V. M., Terenius, L., & de Wied, D. (1976). Interaction between ACTH fragments, brain opiate receptors and morphine-induced analgesia. *European Journal of Pharmacology,* **39,** 383–387.

Gispen, W. H., Reith M.E.A., Schotman, P., Wiegant, V. M., Zwiers, H., & de Wied, D. (1977). CNS and ACTH-like peptides: Neurochemical response and interaction with opiates. *In* L. H. Miller, C. A. Sandman, & A. J. Kastin (Eds.), *Neuropeptide Influences on the Brain and Behavior,* pp. 61–80. New York: Raven.

Gold, P. E., & McGaugh, J. L. (1978). Endogenous modulators of memory storage processes. *In* L. Corenza, P. Pancheri, & L. Zichella (Eds.), *Clinical Psychoneuroendocrinology in Reproduction,* pp. 25–46. New York: Academic Press.

Haycock, J. W., Deadwyler, S. A., Sideroff, S. J., & McGaugh, J. L. (1973). Retrograde amnesia and cholinergic systems in the caudate-putaman complex and dorsal hippocampus of the rat. *Experimental Neurology,* **41,** 201–213.

Huston, J. P., & Stäubli, U. (1978). Retrograde amnesia produced by post-trial injection of substance P into substantia nigra. *Brain Research,* **159,** 468–472.

Huston, J. P., & Staubli, U. (1979) . Behavioural and biochemical effects of substance P injected into the substantia nigra. *Behavioral and Neural Biology,* **27,** 244–248.

Iuvone, P. M., Morasco, J., Delanoy, R. L., & Dunn, A. J. (1978). Peptides and the conversion of ^{3}H-tyrosine to catecholamines: Effects of ACTH-analogs, melanocyte stimulating hormones, and lysine vasopressin. *Brain Research,* **139,** 131–139.

Izquierdo, I. (1979). Effect of naloxone and morphine on various forms of memory in the rat: Possible role of endogenous opiate mechanisms in memory consolidation. *Psychopharmacology,* **66,** 199–203.

Jessell, T. M., Emson, P. C., Paxinos, G., & Cuello, A. C. (1978). Topographic projections of substance P and GABA pathways in the striato- and pallido-nigral system: A biochemical and immunohistochemical study. *Brain Research,* **152,** 487–498.

Jouhaneau-Bowers, M., & Le Magnen, J. (1979). ACTH self-administration in rats. *Pharmacology, Biochemistry and Behavior,* **10,** 325–334.

Katz, R. J. (1976) ACTH$_{4-10}$ antagonism of morphine-induced behavioral activation in the mouse. *European Journal of Pharmacology,* **53,** 393–397.

Kim, H.-J., & Routtenberg, A. (1976a). Retention disruption following posttrial picrotoxin injection into the substantia nigra. *Brain Research,* **152,** 620–625.

Kin, H.-J., & Routtenberg, A. (1976b). Retention deficit following post-trial dopamine injection into the rat neostriatum. *Society for Neuroscience Abstracts,* **3,** 445.

Kim, H.-J., & Routtenberg, A. (1978). Fluorescence microscopic mapping of substantia nigra dopamine somata and their dendrites: Relation to dopamine and non-dopamine thionin-stained cells in identical vibratome sections. *Society for Neuroscience Abstracts*, **5**, 275.

Kim, H.-J., & Routtenberg, A. (1980). Dual role of substantia nigra dopamine neurons in mnemonic and motor functions. *Society for Neuroscience Abstracts*, **7**, 723.

Koller, M., Krause, H. P., Hoffmeister, F., & Ganten, D. (1979). Endogenous brain angiotensin II disrupts passive avoidance behavior in rats. *Neuroscience Letters*, **14**, 71–75.

Larsson, L. I. (1978). Distribution of ACTH-like immunoreactivity in rat brain and gastrointestinal tract. *Histochemistry*, **55**, 225–233.

Lashley, K. S. (1950). In search of the engram. *Symposia for the Society of Experimental Biology*, **4**, 454–482.

Leonard, B. E. (1974). The effect of two synthetic ACTH analogs on the metabolism of biogenic amines in the rat brain. *Archives Internationales de Pharmacodynamie et de Therapie*, **207**, 243.

Liotta, A. S., Gildersleeve, D., Brownstein, M. J., & Krieger, D. T. (1978). Biosynthesis *in vitro* of immunoreactive 31,000 dalton corticotropin β-endorphin like material by bovine hypothalamus. *Proceedings of the National Academy of Sciences, U.S.A.*, **76**, (3), 1448–1452.

Martinez, Jr., J. L., Vasquez, B. J., Jensen, R. A., Soumireu-Mourat, B., & McGaugh, J. L. (1975). ACTH$_{4-9}$ analog (ORG 2766) facilitates acquisition of an inhibitory avoidance response in rats. *Pharmacology, Biochemistry and Behavior*, **10**, 145–147.

Martinez, Jr., J. L., & Rigter, H. (1980). Endorphins alter acquisition and consolidation of an inhibitory avoidance response in rats. *Neuroscience Letters*, **19**, 197–201.

Mazey, E., Palkovits, M., de Kloet, E. R., Verhoef, J., & de Wied, D. (1978). Evidence for pituitary-brain transport of a behaviorally potent ACTH analog. *Life Sciences*, **22**, 831–838.

Mondadori, C., & Waser, P. G. (1979). Facilitation of memory processing by post-trial morphine: Possible involvement of reinforcement mechanisms? *Psychopharmacology*, **63**, 297–300.

Morgan, J. M., & Routtenberg, A. (1977). Angiotensin injected into the neostriatum after learning disrupts retention performance. *Science*, **196**, 87–89.

Morgan, J., & Routtenberg, A. (1978). Morphine sulfate injected into neostriatum disrupts retention without producing analgesia. Northwestern University.

Morgan, D. G., & Routtenberg, A. (1979a). The incorporation of intrastriatally injected ^{3}H-fucose into electrophoretically separated synaptosomal glycoproteins. I. Time course of incorporation and molecular weight estimations. *Brain Research*, **179**, 329–341.

Morgan, D. G., & Routtenberg, A. (1979b). The incorporation of intrastriatally injected ^{3}H-fucose into electrophoretically separated glycoproteins of the neostriatal P$_2$ fraction. II. The influence of passive avoidance training. *Brain Research*, **179**, 343–354.

Moroni, F., Cheney, D. L., Peralta, E., & Costa, E. (1978). Opiate receptor agonists as modulators of γ-aminobutyric acid turnover in the nucleus caudatus, globus pallidus and substantia nigra of the rat. *Journal of Pharmacology and Experimental Therapeutics*, **207**, 870–877.

Nilaver, G., Zimmerman, E. A., Defendini, R., Motta, A. S., Krieger, D. T., & Brownstein, M. J. (1979). Adrenocorticotropin and β-lipotropin in the hypothalamus. *Journal of Cell Biology*, **81**, 50–58.

Olds, J., Disterhoft, J. F., Segal, M., Kornblith, C. L., & Hirsch, R. (1972). Learning centers of rat brain mapped by measuring latencies of conditioned unit responses. *Journal of Neurophysiology*, **35**, 202–219.

Orwoll, E., Kendall, J. W., Lamorena, L., & McGilvra, R. (1979). Adrenocorticotropin and melanocyte stimulating hormone in the brain. *Endocrinology*, **104**, 1845–1852.

Pelletier, G. (1979). Ultrastructural immunohistochemical localization of adrenocorticotropin and beta lipotropin in the rat brain. *Journal of Histochemistry and Cytochemistry*, **27**, 1046–1048.

Pelletier, G., & Leclerc, R. (1979). Immunohistochemical localization of adrenocorticotropin in the rat brain. *Endocrinology*, **104**, 1426–1433.

Plomp, G. J. J., & van Ree, J. M. (1978). Adrenocorticotropic hormone fragments mimic the effect of morphine *in vitro*. *British Journal of Pharmacology*, **64**, 223–227.

Poth, M. M., Heath, R., Heath, G., & Ward, M. (1978). Angiotensin-converting enzyme in human brain. *Journal of Neurochemistry*, **25**, 83–85.

Precht, W., & Yoshida, M. (1971). Blockage of caudate evoked inhibition of neurons in the substantia nigra by picrotoxin. *Brain Research*, **32**, 229–233.

Rigter, H. (1978). Attenuation of amnesia in rats by systemically administered enkephalins. *Science*, **200**, 83–85.

Rigter, H., van Riezen, H., & de Wied, D. (1974). The effects of ACTH and vasopressin analogues on CO_2-induced retrograde amnesia in rats. *Physiology and Behavior*, **13**, 381–388.

Routtenberg, A. (1979a). Participation of brain stimulation reward substrates in memory: Anatomical and biochemical evidence. *Federation Proceedings*, **38**, 2446–2453.

Routtenberg, A. (1979b). Anatomical localization of phosphoprotein and glycoprotein substrates of memory. *Progress in Neurobiology*, **12**, 85–113.

Routtenberg, A., & Benson, G. (1980). *In vitro* phosphorylation of a 41,000-MW protein band is selectively increased 24 hr after footshock or learning. *Behavioral and Neural Biology*, **29**, 168–175.

Routtenberg, A., & Holtzman, N. (1973). Memory disruption by electrical stimulation of substantia nigra, pars compacta. *Science*, **181**, 83–86.

Routtenberg, A., & Kim, H.-J. (1978). The substantia nigra and neostriatum: Substrates for memory consolidation. In L. L. Butcher (Ed.), *Cholinergic-Monoaminergic Interactions in the Brain*, pp. 305–311. New York: Academic Press.

Sagen, J., & Routtenberg, A. (1976). Effect of ACTH fragments injected posttrial into neostriatum on passive avoidance learning. Northwestern University.

Simpson, J. B., & Routtenberg, A. (1973). Subfornical organ: Site of drinking elicitation by angiotensin-II. *Science*, **181**, 1172–1174.

Simpson, J. B., Epstein, A. N., & Camardo, J. S., Jr. (1978). Localization of receptors for the dipsogenic action of angiotensin II in the subfornical organ of rat. *Journal of Comparative and Physiological Psychology*, **92**, 581–608.

Sirett, N. E., McLean, A. S., Bray, J. J., & Hubbard, J. I. (1977). Distribution of angiotensin II receptors in rat brain. *Brain Research*, **149**, 107–116.

Snyder, S. H., & Innis, R. B. (1979). Peptide neurotransmitters. *Annual Review of Biochemistry*, **48**, 755–782.

Soltysik, S., Hull, C. D., Buchwald, N. A., & Fekete, T. (1975). Single unit activity in basal ganglia of monkeys during performance of a delayed response task. *Electroencephalography and Clinical Neurophysiology*, **39**, 65–78.

Stein, L. (1980). The chemistry of reward. In A. Routtenberg (Ed.), *Biology of Reinforcement: Facets of Brain Stimulation Reward*, pp. 109–130. New York: Academic Press.

Terenius, L. (1976). Somatostatin and ACTH are peptides with partial antagonist-like selectivity for opiate receptors. *European Journal of Pharmacology*, **38**, 211–213.

van Loon, G. R., Sole, M. J., Kamble, A., Kim, C., & Green, S. (1977). Differential responsiveness of central noradrenergic and dopaminergic neuron tyrosine hydroxylase to hypophysectomy, ACTH, and glucocorticoid administration. *Annals of the New York Academy of Sciences*, **297**, 284–294.

van Wimersma Greidanus, Tj. B., & de Wied, D. (1971). Effects of systemic and intracerebral administration of two opposite acting ACTH-related peptides on extinction of conditioned avoidance behavior. *Neuroendocrinology*, **7**, 291–301.

van Wimersma Greidanus, Tj. B., van Dijk, A. M. A., de Rotte, A. A., Goedemans, J. H. J., Croiset, G., & Thody, A. J. (1978). Involvement of ACTH and MSH in active and passive avoidance behavior. *Brain Research Bulletin*, **3**, 227–230.

Versteeg, D. H. G. (1980). Interactions of peptides related to ACTH, MSH, and β-LPH with neurotransmitters in the brain. *Pharmacology and Therapeutics* **11**, 535–558.

Waldemeier, R. C., Kam, R., & Stocklin, K. (1978). Increased dopamine metabolism in rat striatum after infusions of substance P into the substantia nigra. *Brain Research*, **159**, 223–227.

Wiegant, V. M., Dunn, A. J., Schotman, P., & Gispen, W. H. (1979). ACTH-like neurotropic peptides: Possible regulation of rat brain cyclic AMP. *Brain Research*, **168**, 565–584.

Wurtz, R. H., & Olds, J. (1963). Amygdaloid stimulation and operant reinforcement in the rat. *Journal Comparative and Physiological Psychology*, **56**, 941–949.

Wyers, E. J., Peeke, H. V. S., Willeston, J. S., & Hertz, M. J. (1968). Retroactive impairment of passive avoidance learning by stimulation of the caudate nucleus. *Experimental Neurology*, **22**, 350–366.

Yasukawa, N., Monder, H., Michael, S. D., & Christian, J. J. (1978). Opiate antagonist counteracts reproductive inhibition by porcine ACTH extract. *Life Sciences*, **22**, 1381–1391.

Zimmerman, E., & Krivoy, W. A. (1973). Antagonism between morphine and the polypeptides ACTH, ACTH$_{1-24}$, and β-MSH in the nervous system. *In:* E. Zimmermann, W. H. Gispen, B. H. Marks, & D. de Wied (Eds.), *Drug Effects on Neuroendocrine Regulation: Progress in Brain Research*, (Vol. 39) pp. 383–394. Amsterdam: Elsevier.

Zwiers, H., Tonnaer, J., Wiegant, V. M., Schotman, P., & Gispen, W. H. (1979). ACTH-sensitive protein kinase from rat brain membranes. *Journal of Neurochemistry*, **33**, 247–256.

Zwiers, H., Wiegant, V. M., Schotman, P., & Gispen, W. H. (1978). ACTH-induced inhibition of endogenous rat brain protein phosphorylation *in vitro*: Structure activity. *Neurochemical Research*, **3**, 455–463.

CHAPTER 26

Multiple Independent Actions
of Neuropeptides on Behavior

Abba J. Kastin, Richard D. Olson, Curt A. Sandman,
and David H. Coy

This chapter discusses and illustrates four different but related aspects of our general theory of multiple independent actions of neuropeptides on behavior. The first shows a dissociation of the pituitary and central nervous system (CNS) effects of hypothalamic peptides. The second shows the independent pigmentary and extrapigmentary effects of melanocyte-stimulating hormone (MSH). The third shows a dissociation of the behavioral and narcotic effects of opiate peptides. The fourth shows a situational dissociation of peptides including different aspects of the same process. Each dissociation illustrates the unifying concept of multiple independent actions of peptides.

I. INTRODUCTION

The use of a naturally occurring peptide for more than one purpose would seem to be very efficient. This principle has guided our work for many years in four different areas of research (Kastin, Olson, Schally, & Coy, 1979b). In each area, described in detail in the following paragraphs, we found effects of the peptides that differed from the actions by which they were originally described. In one area, we introduced the concept that hypothalamic peptides exert direct effects on the central nervous system (CNS), effects that are in-

563

ENDOGENOUS PEPTIDES
AND LEARNING AND MEMORY PROCESSES

dependent of their effects on the release of hormones from the pituitary gland. In a second area, we showed that melanocyte-stimulating hormone (MSH) affects behavior, particularly in the human being, independent of its pigmentary actions. In the third area, we found that the brain opiates affect behavior independently of their narcotic actions. In the fourth area, we showed that similar peptide sequences can have different effects in different tests. This means that the increased potency of an analog in one test does not necessarily imply increased activity in other tests, or, of course, other species. Most of these areas were novel but unpopular when introduced into the scientific literature. All four areas were initially poorly understood, clouded in skepticism, and only reluctantly accepted. Even though the study of peptides has engrossed scientific interest in the past few years, an integrative interpretation is lacking. This chapter will discuss each of these four related areas and further develop the principle of multiple, independent actions of naturally occurring peptides.

II. PITUITARY–CNS DISSOCIATION OF HYPOTHALAMIC PEPTIDES

The tripeptide Pro-Leu-Gly-NH$_2$ (MSH-release-inhibiting factor 1, MIF-1) was found in hypothalamic tissue on the basis of its capacity to inhibit the release of MSH in rats (Celis, Taleisnik, & Walter, 1971; Nair, Kastin, & Schally, 1971). More likely to be of future importance are its actions on the CNS. It was shown in 1971 that the pituitary gland is not required for MIF-1 to increase the behavioral excitability induced by dopa (Plotnikoff, Kastin, Anderson, & Schally, 1971). The same effect could be exerted in the absence of other organs as well, indicating their lack of mediation of the CNS effects (Plotnikoff, Minard, & Kastin, 1974). A large number of centrally active drugs are effective at doses about 100 times larger than the effective dose of MIF-1 in the dopa-potentiation test (Plotnikoff & Kastin, 1977). These include tricyclic antidepressants and antiparkinsonian agents.

Another animal system in which antiparkinsonian drugs are active is the oxotremorine test. We showed in 1972 that the pituitary gland is not required for MIF-1 to reverse the tremors induced by oxotremorine and that the reversal could not be accounted for by an anticholinergic action of MIF-1 (Plotnikoff, Kastin, Anderson, & Schally, 1972). Several other reports have confirmed the efficacy of MIF-1 in these two tests (Bjorkman, Castensson, & Sievertsson, 1979; Bjorkman & Sievertsson, 1977; Castensson, Sievertsson, Lindeke, & Sum, 1974; Huidobro-Toro, Scotti De Carolis, & Longo, 1974; Huidobro-Toro, Scotti De Carolis & Longo, 1975; Voith, 1977). These and other studies of the CNS effects of MIF-1 in animals have been reviewed elsewhere (Kastin, Olson, Schally, & Coy, 1979b; Kastin, Plotnikoff, Schally, & Sandman, 1976b).

Because it now appears that inhibition of MSH release by MIF-1 is evident only in some assay systems, the Roman numeral "I" or arabic numeral "1" has been affixed to the abbreviation to designate that it probably is not the physiological inhibitor of MSH release. This is shown by studies of the actions of MIF-1 in human beings. We failed to find any effect of MIF-1 on the release of MSH in man as determined by bioassay (Kastin, Gual, & Schally, 1972), and this finding has been confirmed by radioimmunoassay (Ashton, Millman, Telford, Thompson, Davies, Hall, Shuster, Thody, Coy, & Kastin, 1977; Donnadieu, Laurent, Luton, Bricaire, Girard, & Binoux, 1976; Faglia, Paracchi, Ferrari, Beck-Peccoz, Ambrosi, Travagline, Spada, & Oliver, 1976; Thody, Shuster, Plummer, Bogie, Leigh, Goolamali, & Smith, 1974). In view of the controversy concerning the nature of MSH activity in human blood, there are some advantages in the use of a bioassay that measures total MSH-like activity. Yet, doses of MIF-1 comparable to those ineffective in reducing MSH levels were found to reduce symptoms of Parkinson's disease; the degree of the effect varied according to the particular study. Small doses also appeared to be beneficial in mental depression. The effects of MIF-1 in these clinical conditions thus illustrate the principle of pituitary–CNS dissociation of hypothalamic peptides. These results have been recently summarized in greater detail (Kastin, et al., 1979b; Kastin, Sandman, Schally, & Ehrensing, 1978a).

Other hypothalamic hormones also exert CNS effects in hypophysectomized animals. Thyrotropin-releasing hormone (TRH), luteinizing-hormone-releasing hormone (LH-RH), and somatostatin potentiate the motor activity induced by threshold amounts of dopa in mice lacking a pituitary gland (Plotnikoff & Kastin, 1977). Increased sexual behavior was observed in rats injected with LH-RH under conditions in which hormonal release was stable (Moss & McCann, 1973), or in which the pituitary and ovaries had been removed (Pfaff, 1973). Somatostatin and TRH applied intracerebroventricularly (i.c.v.) to normal as well as to hypophysectomized rats has a considerable effect on motor activity and results in profound alterations in the sleep–waking pattern (Havlicek, Rezek, & Friesen, 1976). The actions of TRH in causing tremors (Schenkel-Hullinger, Koella, Hartman, & Maitre, 1974) or antagonizing ethanol-induced sleep (Cott, Breese, Cooper, Barlow, & Prange, 1976) were also unaltered in hypophysectomized animals. Thus, many central effects of hypothalamic peptides have been found that are not mediated by the pituitary. The effects be might called "extrapituitary" or even "extraendocrine."

III. PIGMENTARY–EXTRAPIGMENTARY DISSOCIATION OF MSH

Applezweig and Baudry (1955) were the first to demonstrate behavioral effects of ACTH in hypophysectomized rats tested for conditioned avoidance responding (CAR). In the same year, Murphy and Miller (1955) showed delayed extinction of the CAR in a larger group of rats receiving ACTH. Similar effects were found by Miller & Ogawa (1962) in adrenalectomized rats. Several years later, de Wied and co-workers extended these results with ACTH by showing that similar effects could be observed with the MSH portion of the ACTH molecule (Greven & de Wied, 1973).

The only study of the effects of MSH in human beings without a functional pituitary gland was published in 1971 (Kastin, Miller, Gonzalez-Barcena, Hawley, Dyster-Aas, Schally, Velasco-Parra, & Velasco, 1971). The pituitaries of two patients in this study had been removed surgically because of acromegaly and one patient had spontaneous hypopituitarism due to Sheehan's syndrome. Although each patient served as his own control, an additional two hypopituitary patients (one woman hypophysectomized for acromegaly and another for Sheehan's syndrome) received diluent instead of α-MSH. The amplitude of a component of the averaged somatosensory cortical evoked responses in each of the three hypopituitary patients receiving MSH increasedly about 2.5, 5, and 5 times. These patients also improved their performances on the Benton Visual Retention test by 50–200% after the administration of α-MSH. The two hypopituitary patients receiving control solution showed no improvement (Kastin, et al., 1971). The findings demonstrate an extrapituitary effect of the pituitary hormone MSH. This principle is similar in many ways to that of the hypothalamic hormones discussed in the previous section of this chapter.

Because MSH did not cause visible pigmentation in any of the reported behavioral experiments, each of the studies can be said to show that the pigmentary and extrapigmentary actions of MSH are independent. The same consideration applies to several early experiments of the extrapigmentary effects of MSH that predate the efforts of the groups in Utrecht and in New Orleans. In 1961, two other groups reported results with MSH apparently involving the nervous system. In one study, the active core of MSH (and ACTH) caused stretching activity in dogs after intracisternal injection (Ferrari, Gessa, & Vargiu, 1961). The other study that year was the first of several by Krivoy and co-workers that demonstrated an effect of purified β-MSH preparations on spinal cord activity (Krivoy & Guillemin, 1961) as well as on the spontaneous electric discharge of the transparent knife fish (Krivoy, Lane, Childers, & Guillemin, 1962). In addition, as one of an elegant series of early studies of the effects of MSH on the ocular aqueous flare phenomenon, Dyster-Aas and Krakau (1965) found a number of other extrapigmentary effects.

During the decade in which we were performing studies in human beings with MSH and its active core, we were also performing concomitant studies in rats. These early animal investigations showed a number of previously unknown actions, perhaps the most important of which was the demonstration that improved attention (Sandman, Miller, Kastin, & Schally, 1972) rather than facilitated memory processes explained most of the effects of MSH on behavior. Instead of using a single task, such as the CAR (de Wied, 1971) as a bioassay, we used a variety of test systems in an attempt to answer a number of questions (Kastin, Miller, Nockton, Sandman, Schally, & Stratton, 1973b; Kastin, Sandman, Stratton, Schally, & Miller, 1975).

These papers, discussed in Kastin, et al. (1973b, 1975), provided the first demonstrations that MSH exerts effects on appetitive behavior (Sandman, Kastin, & Schally, 1969), passive avoidance behavior (Dempsey, Kastin, & Schally, 1972; Sandman, Kastin, & Schally, 1971b), electroencephalographic patterns (Sandman, Denman, Miller, Knott, Schally, & Kastin, 1971a), attention (Kastin, et al., 1971; Sandman, et al., 1972), operant appetitive behavior (Kastin, Dempsey, LeBlanc, Dyster-Aas, & Schally, 1974), overtraining (Sandman, Beckwith, Giddis, & Kastin, 1974) acquisition (Stratton & Kastin, 1975), tonic immobility (Panksepp, Reilly, Bishop, Meeker, Vilberg, & Kastin, 1976; Stratton & Kastin, 1976), in newborn rats (Beckwith, O'Quin, Petro, Kastin, & Sandman, 1977a; Beckwith, Sandman, Hothersall, & Kastin, 1977b), and on a variety of neuropharmacological tasks (Plotnikoff & Kastin, 1976) including effects not explainable by actions on general motor activity (Kastin, Miller, Ferrell, & Schally, 1973a; Nockton, Kastin, Elder, & Schally, 1972). Effects of MSH were also shown to interact with light–dark conditions (Sandman, et al., 1972; Stratton, Kastin, & Coleman, 1973), animal strain (Sandman, Alexander, & Kastin, 1973a; Stratton, et al., 1973), sex (Beckwith, et al., 1977a, 1977b), different levels of electrical shock (Stratton & Kastin, 1974), and with d-amphetamine (Sandman & Kastin, 1978). In addition, MSH levels were found to be elevated after physical and psychological stress (Sandman, Kastin, Schally, Kendall, & Miller, 1973b).

In other animal studies, we examined the effects of MSH on regional blood perfusion of the brain (Goldman, Sandman, Kastin, & Murphy, 1975), 3′,5′adenosine monophosphate (cylic AMP) levels (Christensen, Harston, Kastin, Kostrzewa, & Spirtes, 1976), cyclic guanosine 5′-phosphoric acid (GMP) levels (Spirtes, Christensen, Harston, & Kastin, 1978), local evoked potentials (Kastin, et al., 1976b), sleep (Panksepp, et al., 1976), and mesenteric vasodilatation (Kadowitz, Chapnick, & Kastin, 1976). Observations after administration of MSH were also made on the aqueous flare response (Dyster-Aas, Kastin, Vidacovich, & Schally, 1970), cardiac actions (Aldinger, Hawley, Schally & Kastin, 1973), biogenic amines in the brain (Kastin, Schally, & Kostrzewa, 1980; Kostrzewa, Hardin, Snell, Kastin, Coy, & Bymaster, 1979; Kostrzewa, Fukushima, Harston, Perry, Fuller, & Kastin, 1979; Kostrzewa, Kastin, & Spirtes, 1975; Kostrzewa, Joh, Kastin, Christensen, & Spirtes, 1976; Spirtes, Kostrzewa, & Kastin, 1975; Spirtes, Plotnikoff, Kostrzewa, Harston, Kastin, & Christensen, 1976), interactions with apomorphine (Cox, Kastin, & Schnieden, 1976; Davis, Kastin, Beilstein, & Vento, in press; Kostrzewa,

Kastin, & Sobrian, 1978), distribution in the brain and rest of the body (Dupont, Kastin, Labrie, Pelletier, Puviani, & Schally, 1975; Kastin, Nissen, Nikolics, Medzihradszky, Coy, Teplan, & Schally, 1976a), half-time disappearance in blood (Kastin, et al., 1976a; Redding, Kastin, Nikolics, Schally, & Coy, 1978), and biodegradation (Marks, Stern, & Kastin, 1976). These were pursued in an attempt to discover the still largely elusive mechanism by which the extrapigmentary effects of MSH are exerted.

IV. BEHAVIORAL–NARCOTIC DISSOCIATION OF OPIATE PEPTIDES

Using the paradigm of a complex maze, with which an effect of MSH on behavioral acquisition was first shown (Stratton & Kastin, 1975), we demonstrated behavioral effects of enkephalin administered intraperitoneally (i.p.) at 80.0 μg/kg body weight (Kastin, Scollan, King, Schally, & Coy, 1976c). Met-enkephalin has no analgesic effects by this route of administration and doses many times larger are required for analgesia even when it is injected into the cerebral ventricles. Substitution of the stereoisomer of phenylalanine in the fourth position resulted in the [D-Phe^4]-Met-enkephalin analog, which is essentially inactive in opiate receptor binding or the *vas deferens* assay (Coy, Kastin, Schally, Morin, Caron, Labrie, Walker, Fertel, Berntson, & Sandman, 1976); yet, this analog appeared equipotent to the parent Met-enkephalin in inducing the hungry rats to run the maze faster and with fewer errors than the controls injected with diluent (Kastin, et al., 1976c). On the other end of the narcotic spectrum, the [D-Ala^2]-Met-enkephalin analog has much stronger analgesic properties than Met-enkephalin after central administration (Walker, Berntson, Sandman, Coy, Schally, & Kastin, 1977). When this potent enkephalin analog was given i.p. to rats running the maze, the effect appeared to be similar to that seen with Met-enkephalin and the analgesically weak [D-Phe^4]-Met-enkephalin. The behavioral actions of Met-enkephalin did not appear to be due to effects of appetite, general motor activity, arousal, or olfaction. At the same doses, morphine tended to result in slower running of the maze. Together, these findings strongly suggest that classical narcotic activity had nothing to do with the behavioral results obtained. At that time, we reiterated the suggestion that peptides are involved in neurotransmission and suggested a reevaluation of opiate actions (Kastin, et al., 1976c).

Effects of enkephalin in the maze test were observed beginning 15 min after i.p. injection. In the dopa-potentiation test, the observational period began 1 hr after i.p. injection of the opiate peptide and continued for 1 hr. Met-enkephalin was extremely active in this test (Plotnikoff, Kastin, Coy, Christensen, Schally, & Spirtes, 1976). It also seemed to reduce footshock-induced fighting in mice, but the relationship to dose was bell-shaped (Plotnikoff, et al., 1976). We have also seen this bell or inverted U-shaped response with other brain peptides and have commented upon it in many papers. It indicates the importance of the appropriate dose for evaluation of CNS and other effects.

The two studies just mentioned were the only ones published in 1976 involving behavior after peripheral injection and were generally ignored by capable investigators whose many years of experience was in fields not involving peptides. These researchers tended to be misled by the lack of analgesia after peripheral administration of the opiate peptides and by the rapid degradation of the material; it had been recognized several years earlier that other brain peptides are rapidly metabolized but still can exert CNS effects when injected systemically (Kastin, et al., 1979b).

The following year saw only two additional studies in which opiate peptides were injected peripherally. The first showed attenuation of amnesia for an avoidance response (Rigter, Greven, & van Riezen, 1977). The second was a controversial clinical study, lacking a control group, which reported beneficial effects of β-endorphin in treating schizophrenia (Kline, Li, Lehmann, Lajtha, Laskie, & Cooper, 1977).

In 1978, other groups such as that of de Wied (de Wied, Kovacs, Bohus, van Ree, & Greven, 1978) and Li (Gorelick, Catlin, George, & Li, 1978) found similar dissociative effects between behavioral and narcotic actions of the brain opiates by utilizing the peripheral route of administration. The results we obtained in 1976 with a maze were extended to an immobility test in which analogs superactive in terms of causing analgesia were ineffective even though the parent compound and its weak analogs were effective (Kastin, Scollan, Ehrensing, Schally, & Coy, 1978b). Peripheral effects of the opiate peptides were also reported after injection into fish (Olson, Kastin, Montalbano-Smith, Olson, Coy, & Michel, 1978), chickens (Panksepp, Vilberg, Bean, Coy, & Kastin, 1978), and monkeys (Olson, Olson, Kastin, Castellanos, Kneale, Coy, & Wolf, 1978).

By 1979, a number of other papers began to appear confirming our suggestion that peripheral administration of the opiate peptides can result in behavioral changes dissociated from classical narcotic effects. This included work in which investigators reversed a previous position so as to lend additional credence to the concept. One type of experiment published during 1979 should have great impact on future research in this area. It involves the long-term effects of β-endorphin on adult rats injected during the first week of life (Sandman, McGivern, Berka, Walker, Coy, & Kastin, 1979). After early treatment with β-endorphin, both male and female adult rats were found to be insensitive to thermal pain. Long-term effects were also found in adult rats whose mothers received β-endorphin during pregnancy (McGivern, Sandman, Kastin, & Coy, 1979). The studies of newborns resemble those involving neonatal injections of α-MSH (Beckwith, et al., 1977a, 1977b) or TRH (Stratton, Gibson, Kolar, & Kastin, 1976) and have far-ranging implications.

Some of the results just mentioned could represent a partial effect of Met-enkephalin on classical opiate receptors as well as on different receptors, activation of which may be critical to changes in behavior. It may be naive to consider that one peptide acts on only one type of receptor. Unless the concept of opiate receptors is broadened to include newer types or states not sensitive to naloxone, it is reasonable to assume that the main, although not the exclusive, receptors mediating most of the behaviors seen after peripheral administration of analgesically weak opiate peptides are not classical opiate receptors. The use of analgesically potent analogs of enkephalin and endorphin tend to obscure this distinction. Moreover, peptides such as MIF-1 may act at the same opiate receptors as does naloxone in some, but not all, situations (Kastin, Olson, Ehrensing, Berzas, Schally, & Coy, 1979a). The existence of an endogenous naloxone-like material, therefore, is possible. We do not expect the behavioral and analgesic activity of opiate peptides and their analogs necessarily to parallel each other.

It follows that related peptides could exert different effects in certain test situations. We showed this in the open field where α-, β-, and γ-endorphins exerted different effects (Veith, Sandman, Walker, Coy, & Kastin, 1978). De Wied, et al., (1978) found a similar dissociation in a different paradigm, in which γ- and Des-Tyr-γ-endorphin exerted effects opposite to those of α- and β-endorphin, but similar to those of the neuroleptic haloperidol. However, the concept that Des-Tyr-γ-endorphin is an endogenous neuroleptic related to haloperidol was subsequently disproved by studies performed by two groups (Le Moal, et al., 1979; Weinberger, Arnsten, & Segal, 1979). The study by Le Moal, Koob, & Bloom (1979) also supports our view that the effects of peptides may depend upon the situation in which they are tested.

V. SITUATIONAL DISSOCIATION OF PEPTIDES

The use of several different types of behavioral tests for studying the CNS effects of peptides permits observations of different profiles of responses to different substances. The point is illustrated by the pattern or "fingerprinting" of the neuropharmacological effects we have been finding since 1971 for the hypothalamic peptides (Kastin, et al., 1976b, 1978b, 1979b). The tests usually consisted of behavioral observations of dopa-potentiation, serotonin potentiation, oxotremorine antagonism, footshock-induced fighting, audiogenic seizure reduction, and swimming immobility reduction. MIF-1, TRH, LH-RH, somatostatin, α-MSH, and enkephalin are all active to varying degrees in the dopa-potentiation test. Only TRH is highly active in the serotonin potentiation test, whereas LH-RH and enkephalin are weakly active. Only MIF-1 reduces the tremors induced by oxotremorine, whereas only α-MSH and enkephalin are strongly active in reversing the swimming immobility. Thus, any one peptide was only active in certain of the tests. Even though the activities in any test were shared by several peptides, the general pattern for a given peptide appeared to be different. Additional tests with other peptides are needed to confirm the concept of a unique profile.

More recently, this concept has been further refined using a group of compounds related to MSH and ACTH. These all shared the common MSH–ACTH$_{4-10}$ core of 7 amino acids and additionally include α-MSH (13 amino acids), β$_p$-MSH (18 amino acids), β$_h$-MSH (LPH$_{37-58}$; 22 amino acids), and the endocrinologically active portion of ACTH (ACTH$_{1-24}$; 24 amino acids). As the peptide chain is enlarged, it was generally assumed that the additional sequences added little additional information relevant to behavior and that the peptides therefore produce the same behavioral effects. However, it now seems clear that, rather than being redundant messengers for the same function, the different peptides may vary in relative potency depending upon the task. This point is illustrated with different elements of a visual discrimination learning problem divided into three phases: original learning, reversal learning, and extinction. Whereas MSH–ACTH$_{4-10}$ improved the initial phase of the learning process the most and ACTH$_{1-24}$ the least, α-MSH and β-MSH were more effective in reversal learning, which was sensitive to the attentional state (Sandman, Beckwith, & Kastin, 1980). A third and different relationship was found for extinction. These results again indicate that the functional relationships among peptides are not always satisfactorily ascertained by a single task. The specificity of different sequences of any peptide may be manifest only by examination of different tasks.

Thus, a variety of situational dissociations of peptides emerges. Different peptides (e.g., MIF-1 and TRH) containing completely different amino acids can produce the same effect in some situations and different effects in other situations. A similar condition exists with peptides (e.g., α-MSH and ACTH$_{4-10}$) containing an identical amino acid sequence. The additional amino acid sequences in peptides with shared cores may or may not modify the response, depending upon the situation. With none of these peptides can it be necessarily assumed that exogenous administration will exert the same effects as those exerted by the same peptide endogenously. Metabolism, degradation, dose, and accessibility of target sites must be taken into consideration.

ACKNOWLEDGMENTS

Many of the findings reported in this chapter were supported in part by the Medical Research Service of the Veterans Administration Medical Center and NIH (NS07664). The authors thank Dr. James Zadina and Ms. Debra Bayhi for help with the manuscript.

REFERENCES

Aldinger, E. E., Hawley, W. D., Schally, A. V., & Kastin, A. J. (1973). Cardiovascular actions of melanocyte-stimulating hormone (MSH) in the dog. *Journal of Endocrinology*, 56, 613–614.

Applezweig, M. H., & Baudry, F. D. (1955). The pituitary–adrenocortical system in avoidance learning. *Psychological Reports*, 1, 417–420.

Ashton, H., Millman, J. E., Telford, R., Thompson, J. W., Davies, T. F., Hall, R., Shuster, S., Thody, A. J., Coy, D. H., & Kastin, A. J. (1977). Psychopharmacological and endocrinological effects of melanocyte-stimulating hormones in normal man. *Psychopharmacology*, 55, 165–172.

Beckwith, B. E., O'Quin, R. K., Petro, M. S., Kastin, A. J., & Sandman, C. A. (1977a). The effects of neonatal injections of alpha-MSH on the open field behavior of juvenile and adult rats. *Physiological Psychology*, 5, 295–299.

Beckwith, B. E., Sandman, C. A., Hothersall, D., & Kastin, A. J. (1977b). Influence of neonatal injections of alpha-MSH on learning, memory and attention in rats. *Physiology and Behavior*, 18, 63–71.

Bjorkman, S., Castensson, S., & Sievertsson, H. (1979). Tripeptide analogues of melanocyte-stimulating hormone release inhibiting hormone (Pro-Leu-Gly-NH$_2$) as inhibitors of oxotremorine-induced tremor. *Journal of Medicinal Chemistry*, 22, 931–935.

Bjorkman, S., & Sievertsson, H. (1977). On the optimal dosage of Pro-Leu-Gly-NH$_2$ (MIF) in neuropharmacological tests and clincial use. *Naunyn-Schmiedeberg's Archives of Pharmacology*, 298, 79–81.

Castensson, S., Sievertsson, H., Lindeke, B., & Sum, C. Y. (1974). Studies on the inhibition of oxotremorine induced tremor by a melanocyte-stimulating hormone release-inhibiting factor, thyrotropin-releasing hormone and related peptides. *Federation for Experimental Biology Society Letters*, 44, 101–105.

Celis, M. E., Taleisnik, S., & Walter, R. (1971). Regulation of formation and proposed structure of the factor inhibiting the release of melanocyte stimulating hormone. *Proceedings of the National Academy of Sciences, USA*, 68, 1428–1433.

Christensen, C. W., Harston, C. T., Kastin, A. J., Kostrzewa, R. M., & Spirtes, M. A. (1976). Investigations on α-MSH and MIF-I effects on cyclic AMP levels in rat brain. *Pharmacology, Biochemistry and Behavior (Supplement 1)*, 5, 117–120.

Cott, J. M., Breese, G. R., Cooper, B. R., Barlow, T. S., & Prange, A. J., Jr. (1976). Investigations into the mechanism of reduction of ethanol sleep by thyrotropin-releasing hormone (TRH). *Journal of Pharmacology and Experimental Therapeutics*, 196, 594–604.

Cox, B., Kastin, A. J., & Schnieden, H. (1976). A comparison between a melanocyte-stimulating hormone inhibitory factor (MIF-I) and substances known to activate central dopamine receptors. *European Journal of Pharmacology*, 36, 141–147.

Coy, D. H., Kastin, A. J., Schally, A. V., Morin, O., Caron, N. G., Labrie, F., Walker, J. M., Fertel, R., Berntson, G. G., & Sandman, C. A. (1976). Synthesis and opioid activities of stereoisomers and other D-amino acid analogs of methionine-enkephalin. *Biochemical and Biophysical Research Communications*, 73, 632–638.

Davis, K. L., Kastin, A. J., Beilstein, B. A., & Vento, A. L. (in press). MSH and MIF-I in animal models of tardive dyskinesia. *Brain Research Bulletin*.

Dempsey, G. L., Kastin, A. J., & Schally, A. V. (1972). The effects of MSH on a restricted passive avoidance response. *Hormones & Behavior*, 3, 333–337.

de Wied, D. (1971). Long term effect of vasopressin on the maintenance of a conditioned avoidance response in rats. *Nature (London)*, 232, 58–60.

de Wied, D., Kovacs, G. L., Bohus, B., van Ree, J. M., & Greven, H. M. (1978). Neuroleptic activity of the neuropeptide β-LPH$_{62-77}$ (Des-Tyr^1-γ-endorphin; DTγE). *European Journal of Pharmacology*, 49, 427–436.

Donnadieu, M., Laurent, M. F., Luton, J. P., Bricaire, H., Girard, F., & Binoux, M. (1976). Synthetic MIF has no effect on β-MSH and ACTH hypersecretion in Nelson's syndrome. *Journal of Clinical Endocrinology and Metabolism*, **42**, 1145–1148.

Dupont, A., Kastin, A. J., Labrie, F., Pelletier, G., Puviani, R., & Schally, A. V. (1975). Distribution of radioactivity in the organs of the rat and mouse after injection of ^{125}I α-melanocyte-stimulating hormone. *Journal of Endocrinology*, **64**, 237–241.

Dyster-Aas, H. K., & Krakau, C. E. T. (1965). General effects of α-melanocyte-stimulating hormone in the rabbit. *Acta Endocrinology*, **48**, 609–618.

Dyster-Aas, H. D., Kastin, A. J., Vidacovich, R. P., & Schally, A. V. (1970). Melanocyte-stimulating activity in serum and the aqueous flare response in rabbits. *Journal of Endocrinology*, **46**, 285–286.

Faglia, G., Paracchi, A., Ferrari, C., Beck-Peccoz, P., Ambrosi, B., Travagline, P., Spada, A., & Oliver, C. (1976). Influence of L-Prolyl-L-Leucyl-Glycine amide on growth hormone secretion in normal and acromegalic subjects. *Journal of Clinical Endocrinology and Metabolism*, **42**, 991–994.

Ferrari, W., Gessa, G. L., & Vargiu, L. (1961). Stretching activity in dogs intracisternally injected with a synthetic melanocyte-stimulating hexapeptide. *Experientia*, **17**, 90.

Goldman, H., Sandman, C. A., Kastin, A. J., & Murphy, S. (1975). MSH affects regional perfusion of the brain. *Pharmacology, Biochemistry and Behavior*, **3**, 661–664.

Gorelick, D. A., Catlin, D. H., George, R., & Li, C. H. (1978). Beta-endorphin is behaviorally active in rats after chronic intravenous administration. *Pharmacology, Biochemistry and Behavior*, **9**, 385–386.

Greven, H. M., & de Wied, D. (1973). The influence of peptides derived from corticotrophin (ACTH) on performance structure–activity studies. *In* E. Zimmermann, W. H. Gispen, B. H. Marks, & D. de Wied (Eds.), *Drug Effects on Neuroendocrine Regulation: Progress in Brain Research* (Vol. 39), pp. 429–442. Amsterdam: Elsevier.

Havlicek, V., Rezek, M., & Friesen, H. (1976). Somatostatin and thyrotropin-releasing hormone: Central effect on sleep and motor system. *Pharmacology, Biochemistry and Behavior*, **4**, 455–459.

Huidobro-Toro, J. P., Scotti de Carolis, A., & Longo, B. G. (1974). Action of two hypothalamic factors (TRH, MIF) and angiotensin II on the behavioral effects of L-DOPA and 5-hydroxytryptophan in mice. *Pharmacology, Biochemistry and Behavior*, **2**, 105–109.

Huidobro-Toro, J. P., Scotti de Carolis, A., & Longo, B. G. (1975). Intensification of central catecholaminergic and serotonergic processes by the hypothalamic factors MIF and TRH and by angiotensin II. *Pharmacology, Biochemistry and Behavior*, **3**, 235–242.

Kadowitz, P. J., Chapnick, B. M., & Kastin, A. J. (1976). Comparison of alpha-MSH and several vasoactive substances on vascular resistance in the feline mesenteric vascular bed. *Pharmacology, Biochemistry and Behavior*, **5**, 219–221.

Kastin, A. J., Dempsey, G. L., LeBlanc, B., Dyster-Aas, K., & Schally, A. V. (1974). Extinction of an appetitive operant response after administration of MSH. *Hormones & Behavior*, **5**, 135–139.

Kastin, A. J., Gual, C., & Schally, A. V. (1972). Clinical experience with hypothalamic releasing hormones II: Luteinizing hormone releasing hormone and other hypophysiotropic hormones. *Recent Progress in Hormone Research*, **28**, 201–227.

Kastin, A. J., Miller, M. C., Ferrell, L., & Schally, A. V. (1973a). General activity in intact and hypophysectomized rats after administration of melanocyte-stimulating hormone (MSH), melatonin, and Pro-Leu-Gly-NH$_2$. *Physiology and Behavior*, **10**, 339–401.

Kastin, A. J., Miller, L. H., Gonzalez-Barcena, D., Hawley, W. D., Dyster-Aas, K., Schally, A. V., Velasco-Parra, M. L., & Velasco, M. (1971). Psycho–physiologic correlates of MSH activity in man. *Physiology and Behavior*, **7**, 893–896.

Kastin, A. J., Miller, L. H., Nockton, R., Sandman, C. A., Schally, A. V., & Stratton, L. O. (1973b). Behavioral aspects of melanocyte-stimulating hormone (MSH). *In* E. Zimmermann,

W. H. Gispen, B. H. Marks, & D. de Wied (Eds.), *Drug Effects on Neuroendocrine Regulation: Progress in Brain Research* (Vol. 39), pp. 461–470. Amsterdam: Elsevier.

Kastin, A. J., Nissen, C., Nikolics, K., Medzihradszky, K., Coy, D. H., Teplan, I., & Schally, A. V. (1976a). Distribution of ³H-alpha-MSH in rat brain. *Brain Research Bulletin*, 1, 19–26.

Kastin, A. J., Olson, R. D., Ehrensing, R. H., Berzas, M. C., Schally, A. V., & Coy, D. H. (1979a). MIF-I's differential actions as an opiate antagonist. *Pharmacology, Biochemistry and Behavior*, 11, 721–723.

Kastin, A. J., Olson, R. D., Schally, A. V., & Coy, D. H. (1979b). CNS effects of peripherally administered brain peptides. *Life Sciences*, 25, 401–414.

Kastin, A. J., Plotnikoff, N. P., Schally, A. V., & Sandman, C. A. (1976b). Endocrine and CNS effects of hypothalamic peptides and MSH. *In* S. Ehrenpreis & I. J. Kopin (Eds.), *Reviews of Neuroscience*, pp. 111–148. New York: Raven.

Kastin, A. J., Sandman, C. A., Schally, A. V., & Ehrensing, R. H. (1978a). Clinical effects of hypothalamic pituitary peptides upon the central nervous system. *In* H. L. Klawans (Ed.), *Clinical Neuropharmacology*, pp. 133–152. New York: Raven.

Kastin, A. J., Sandman, C. A., Stratton, L. O., Schally, A. V., & Miller, L. H. (1975). Behavioral and electrographic changes in rat and man after MSH. *In* W. H. Gispen, Tj. B. van Wimersma Griedanus, B. Bohus, & D. de Wied (Eds.), *Hormones and Homeostasis in the Brain: Progress in Brain Research*, pp. 143–150. Amsterdam: Elsevier.

Kastin, A. J., Schally, A. V., & Kostrzewa, R. M. (1980). Possible aminergic mediation of MSH release and of the CNS effects of MSH and MIF-I. *Federation Proceedings*, 39, 2931–2936.

Kastin, A. J., Scollan, E. L., King, M. G., Schally, A. V., & Coy, D. H. (1976c). Enkephalin and a potent analog facilitate maze performance after intraperitoneal administration in rats. *Pharmacology, Biochemistry and Behavior*, 5, 691–695.

Kastin, A. J., Scollan, E. L., Ehrensing, R. H., Schally, A. V., & Coy, D. H. (1978b). Enkephalin and other peptides reduce passiveness. *Pharmacology, Biochemistry and Behavior*, 9, 515–520.

Kline, N. S., Li, C. H., Lehmann, H. E., Lajtha, A., Laskie, E., & Cooper, T. (1977). Beta-endorphin induced changes in schizophrenic and depressed patients. *Archives of Genetic Psychology*, 34, 1111–1113.

Kostrzewa, R. M., Fukushima, H., Harston, C. T., Perry, K. W., Fuller, R. W., & Kastin, A. J. (1979). Striatal dopamine turnover and MIF-I. *Brain Research Bulletin*, 4, 799–802.

Kostrzewa, R. M., Hardin, J. C., Snell, R. L., Kastin, A. J., Coy, D. H., & Bymaster, F. (1979). MIF-I and postsynaptic receptor sites for dopamine. *Brain Research Bulletin*, 4, 657–662.

Kostrzewa, R. M., Joh, T. H., Kastin, A. J., Christensen, C., & Spirtes, M. A. (1976). Effects of L-prolyl-L-leucyl-glycine amide (MIF-I) on dopaminergic neurons. *Pharmacology, Biochemistry and Behavior*, 5, 125–127.

Kostrzewa, R. M., Kastin, A. J., & Sobrian, S. (1978). Potentiation of apomorphine action in rats by L-prolyl-L-leucyl-glycine amide. *Pharmacology, Biochemistry and Behavior*, 9, 375–378.

Kostrzewa, R. M., Kastin, A. J., & Spirtes, M. A. (1975). Alpha-MSH and MIF-I effects on catecholamine levels and synthesis in various rat brain areas. *Pharmacology, Biochemistry and Behavior*, 3, 1017–1023.

Krivoy, W. A., & Guillemin, R. (1961). On a possible role of beta-melanocyte-stimulating hormone (beta-MSH) in the central nervous system of the mammalia: An effect of beta-MSH in the spinal cord of the cat. *Endocrinology*, 69, 170–174.

Krivoy, W. A., Lane, M., Childers, H. E., & Guillemin, R. (1962). On the action of beta-melanocyte stimulating hormone (beta-MSH) on spontaneous electric discharge of the transparent knife fish, *E. eigenmannia. Experientia*, 18, 1–4.

Le Moal, M. L., Koob, G. F., & Bloom, F. E. (1979). Endorphins and extinction: Differential actions of appetitive and aversive tasks. *Life Sciences*, 24, 1631–1636.

Marks, N., Stern, F., & Kastin, A. J. (1976). Biodegradation of alpha-MSH and derived peptides by rat brain extracts, and by rat and human serum. *Brain Research Bulletin*, 1, 591–593.

McGivern, R. F., Sandman, C. A., Kastin, A. J., & Coy, D. H. (1979). Behavioral effects of prenatal administration of beta-endorphin. Tenth International Congress of the International Society of Psychoneuroendocrinology, Park City, Utah. Abstracts, p. 41.

Miller, R. E., & Ogawa, N. (1962). The effect of adrenocorticotrophic hormone (ACTH) on avoidance conditioning in the adrenalectomized rat. *Journal of Comparative and Physiological Psychology*, **55**, 211–213.

Moss, R. L., & McCann, S. M. (1973). Induction of mating behavior in rats by luteinizing hormone-releasing factor. *Science*, **181**, 177–179.

Murphy, J. V., & Miller, R. E. (1955). The effect of adrenocorticotrophic hormone (ACTH) on avoidance conditioning in the rat. *Journal of Comparative and Physiological Psychology*, **48**, 47–49.

Nair, R. M. G., Kastin, A. J., & Schally, A. V. (1971). Isolation and structure of hypothalamic MSH release inhibiting hormone. *Biochemical and Biophysical Research Communications*, **43**, 1376–1381.

Nockton, R., Kastin, A. J., Elder, S. T., & Schally, A. V. (1972). Passive and active avoidance responses at two levels of shock after administration of melanocyte-stimulating hormone. *Hormones and Behavior*, **3**, 339–344.

Olson, R. D., Kastin, A. J., Montalbano-Smith, D., Olson, G. A., Coy, D. H., & Michel, G. F. (1978). Neuropeptides and the blood–brain barrier in goldfish. *Pharmacology, Biochemistry and Behavior*, **9**, 521–524.

Olson, G. A., Olson, R. D., Kastin, A. J., Castellanos, F. X., Kneale, M. T., Coy, D. H., & Wolf, R. H. (1978). Behavioral effects of D-ala^2-beta-endorphin in squirrel monkeys. *Pharmacology, Biochemistry and Behavior*, **9**, 687–692.

Panksepp, J., Reilly, P., Bishop, P., Meeker, R. B., Vilberg, T. R., & Kastin, A. J. (1976). Effects of alpha-MSH on motivation, vigilance and brain respiration. *Pharmacology, Biochemistry and Behavior, (Supplement 1)*, **5**, 59–64.

Panksepp, J., Vilbert, T., Bean, N. J., Coy, D. H., & Kastin, A. J. (1978). Reduction of distress vocalization in chicks by opiate-like peptides. *Brain Research Bulletin*, **3**, 663–667.

Pfaff, D. W. (1973). Luteinizing hormone-releasing factor potentiates lordosis behavior in hypophysectomized ovariectomized female rats. *Science*, **182**, 1148–1149.

Plotnikoff, N. P., & Kastin, A. J. (1976). Neuropharmacological tests with alpha-melanocyte-stimulating hormones. *Life Sciences*, **18**, 1217–1222.

Plotnikoff, N. P., & Kastin, A. J. (1977). Neuropharmacological review of hypothalamic releasing factors. *In* L. H. Miller, C. A. Sandman, & A. J. Kastin (Eds.), *Neuropeptide Influences on Brain and Behavior*, pp. 81–107. New York: Raven.

Plotnikoff, N. P., Kastin, A. J., Anderson, M. D., & Schally, A. V. (1971). Dopa potentiation by a hypothalamic factor, MSH release-inhibiting hormone (MIF). *Life Sciences*, **10**, 1279–1283.

Plotnikoff, N. P., Kastin, A. J., Anderson, M. D., & Schally, A. V. (1972). Oxotremorine antagonism by a hypothalamic, melanocyte-stimulating hormone release-inhibiting factor, MIF. *Proceedings of the Society for Experimental Biology and Medicine*, **140**, 811–814.

Plotnikoff, N. P., Kastin, A. J., Coy, D. H., Christensen, C. W., Schally, A. V., & Spirtes, M. A. (1976). Neuropharmacological actions of enkephalin after systemic administration. *Life Sciences*, **19**, 1283–1288.

Plotnikoff, N. P., Minard, F. N., & Kastin, A. J. (1974). Dopa potentiation in ablated animals and brain levels of biogenic amines in intact animals after prolyl-leucylglycinamide. *Neuroendocrinology*, **14**, 271–279.

Redding, T. W., Kastin, A. J., Nikolics, K., Schally, A. V., & Coy, D. H. (1978). The disappearance and excretion of labeled alpha-MSH in man. *Pharmacology, Biochemistry and Behavior*, **9**, 207–212.

Rigter, H., Greven, H., & van Riezen, H. (1977). Failure of naloxone to prevent reduction of amnesia by enkephalins. *Neuropharmacology*, **16**, 545–547.

Sandman, C. A., Alexander, W. D., & Kastin, A. J. (1973a). Neuroendocrine influences on visual discrimination and reversal learning in the albino and hooded rat. *Physiology and Behavior*, 11, 613–617.

Sandman, C. A., Beckwith, W., Giddis, M. M., & Kastin, A. J. (1974). Melanocyte-stimulating hormone (MSH) and overtraining effects on extradimensional shift (EDS) learning. *Physiology and Behavior*, 13, 163–166.

Sandman, C. A., Beckwith, B. E., & Kastin, A. J. (1980). Are learning and attention related to the sequence of amino acids in ACTH/MSH peptides? *Peptides*, 1, 277–280.

Sandman, C. A., Denman, P. M., Miller, L. H., Knott, J. R., Schally, A. V., & Kastin, A. J. (1971a). Electroencephalographic measures of melanocyte-stimulating hormone activity. *Journal of Comparative Physiology*, 76, 103–109.

Sandman, C. A., & Kastin, A. J. (1978). Interaction of alpha-MSH and MIF-I with *d*-amphetamine on open field behavior of rats. *Pharmacology, Biochemistry and Behavior*, 6, 759–762.

Sandman, C. A., Kastin, A. J., & Schally, A. V. (1971b). Behavioral inhibition as modified by melanocyte-stimulating hormone (MSH) and light–dark condition. *Physiology and Behavior*, 6, 45–48.

Sandman, C. A., Kastin, A. J., & Schally, A. V. (1969). Melanocyte-stimulating hormone and learned appetitive behavior. *Experientia*, 25, 1001–1002.

Sandman, C. A., Kastin, A. J., Schally, A. V., Kendall, J. W., & Miller, L. H. (1973b). Neuroendocrine responses to physical and psychological stress. *Journal of Comparative and Physiological Psychology*, 84, 386–390.

Sandman, C. A., McGivern, R. F., Berka, C., Walker, M., Coy, D. H., & Kastin, A. J. (1979). Neonatal administration of beta-endorphin produced "chronic" insensitivity to thermal stimuli. *Life Sciences*, 25, 1755–1760.

Sandman, C. A., Miller, L. H., Kastin, A. J., & Schally, A. V. (1972). A neuroendocrine influence on attention and memory. *Journal of Comparative and Physiological Psychology*, 80, 54–58.

Schenkel-Hulliger, L., Koella, W. P., Hartman, A., & Maitre, L. (1974). Tremorogenic effect of thyrotropin releasing hormone in rats. *Experientia*, 30, 1168–1170.

Spirtes, M. A., Christensen, C. W., Harston, C. T., & Kastin, A. J. (1978). Alpha-MSH and MIF-I effects on cGMP levels in various rat brain regions. *Brain Research*, 144, 189–193.

Spirtes, M. A., Kostrzewa, R. M., & Kastin, A. J. (1975). MSH and MIF-I effects on serotonin levels and synthesis in various rat brain areas. *Pharmacology, Biochemistry and Behavior*, 3, 1011–1015.

Spirtes, M. A., Plotnikoff, N. P., Kostrzewa, R. M., Harston, C. T., Kastin, A. J., & Christensen, C. W. (1976). Possible association of increased rat behavioral effects and increased striatal dopamine and norepinephrine levels during the DOPA-potentiation test. *Pharmacology, Biochemistry and Behavior, (Supplement 1)*, 5, 121–124.

Stratton, L. O., Gibson, C. A., Kolar, K. G., & Kastin, A. J. (1976). Neonatal treatment with TRH affects development, learning, and emotionality in the rat. *Pharmacology, Biochemistry and Behavior, (Supplement 1)*, 5, 65–67.

Stratton, L. O., & Kastin, A. J. (1974). Avoidance learning at two levels of motivation in rats receiving MSH. *Hormones and Behavior*, 5, 149–155.

Stratton, L. O., & Kastin, A. J. (1975). Increased acquisition of a complex appetitive task after MSH and MIF. *Pharmacology, Biochemistry and Behavior*, 3, 901–904.

Stratton, L. O., & Kastin, A. J. (1976). Melanocyte stimulating hormone and MSH/ACTH$_{4-10}$ reduce tonic immobility in the lizard. *Physiology and Behavior*, 16, 771–774.

Stratton, L. O., Kastin, A. J., & Coleman, W. P. (1973). Activity and dark-preference responses of albino and hooded rats receiving MSH. *Physiology and Behavior*, 11, 907–909.

Thody, A. J., Shuster, S., Plummer, N. A., Bogie, W., Leigh, R. J., Goolamali, S. K., & Smith,

A. G. (1974). The lack of effect of MSH release inhibiting factor (MIF) in the secretion of beta-MSH in normal men. *Journal of Clinical Endocrinology and Metabolism*, **38**, 491–494.

Veith, J. L., Sandman, C. A., Walker, J. M., Coy, D. H., & Kastin, A. J. (1978). Systemic administration of endorphins selectively alters open field behavior of rats. *Physiology and Behavior*, **20**, 539–542.

Voith, K. (1977). Synthetic MIF analogues II: Dopa potentiation and fluphenazine antagonism. *Arzneimittel Forschung*, **27**, 2290–2293.

Walker, M. J., Berntson, G. G., Sandman, C. A., Coy, D. H., Schally, A. V., & Kastin, A. J. (1977). An analogue of enkephalin having a prolonged opiate-like effect *in vivo*. *Science*, **196**, 85–87.

Weinberger, S. B., Arnsten, A., & Segal, D. S. (1979). Des-tyrosine[1]-gamma-endorphin and haloperidol: Behavioral and biochemical differentiation. *Life Sciences*, **24**, 1637–1644.

Index